PHYSICAL AND GEOTECHNICAL PROPERTIES OF SOILS

PHYSICAL AND GEOTECHNICAL PROPERTIES OF SOILS

Second Edition

Joseph E. Bowles

Consulting Engineer/Software Consultant
Engineering Computer Software

McGraw-Hill Book Company

New York   St. Louis   San Francisco   Auckland   Bogotá   Hamburg
Johannesburg   London   Madrid   Mexico   Montreal   New Delhi
Panama   Paris   São Paulo   Singapore   Sydney   Tokyo   Toronto

DISCLAIMER

This book is intended as a textbook and guide based on the author's interpretation of currently accepted practice. Throughout the user is cautioned that soil is highly unpredictable, and recommendation for use is made only after all pertinent factors are considered. Examples are used to illustrate methodology, and while extensive effort has been made to eliminate computational errors through checking and proofreading, no warranty is claimed that some errors do not exist. Similarly, correlations are given which represent general value ranges; however, the reader is cautioned that no correlations exist to date which do not contain large scatters. For these several factors, any design recommendations based in whole or part on textbook methodology and examples, or on design correlations displayed are the user's sole responsibility for correctness, and neither the author nor publisher shall be held liable.

This book was set in Times Roman.
The editors were Kiran Verma and Susan Hazlett;
the production supervisor was Charles Hess.
New drawings were done by J & R Services, Inc.
The cover was designed by Nicholas Krenitsky.
Halliday Lithograph Corporation was printer and binder.

PHYSICAL AND GEOTECHNICAL PROPERTIES OF SOILS

1234567890 HALHAL 8987654

ISBN 0-07-006772-4

Library of Congress Cataloging in Publication Data

Bowles, Joseph E.
    Physical and geotechnical properties of soils.

    Bibliography: p.
    Includes indexes.
    1. Soil mechanics. 2. Soil physics. I. Title.
TA710.B684   1984   624.1′5136   83-17575
ISBN 0-07-006772-4

CONTENTS

Preface   xi

Chapter 1 Introduction to Geotechnics, Statistics,
               and SI   1

1-1   General Introduction   1
1-2   The Study of Soil Mechanics   2
1-3   Text Objectives   4
1-4   Some Typical Soil Problems   4
1-5   Historical Development of Soil Mechanics   5
1-6   Soil Failures   9
1-7   Sources of Information for Geotechnical Engineers   13
1-8   Elements of Statistics and Probability   16
1-9   SI Units   23

Chapter 2 Soil Properties—Physical and Index   26

2-1   Introduction   26
2-2   Soil Composition and Terms   26
2-3   Basic Definitions and Mass-Volume Relationships   27
2-4   Noncohesive and Cohesive Soils   39
2-5   Atterberg (or Consistency) Limits   40
2-6   Soil Moisture   41
2-7   Indices of Soil Consistency   42
2-8   Specific Surface   43
2-9   Specific Gravity   44
2-10   Soil Texture   46
2-11   Soil Phases   46
2-12   Grain Size   48
2-13   Unit Weight of Soils (Dry, Wet, and Submerged)   50
2-14   Effective or Intergranular Pressures   52
2-15   Intergranular Pressures in Partially Saturated Soils   59
2-16   Summary   60

Chapter 3 Geologic Properties, Formations of
Natural Soil Deposits, and Groundwater   64

  3-1   Introduction   64
  3-2   The Earth   65
  3-3   Physical Properties of Minerals   66
  3-4   The Rock and Soil Cycle   67
  3-5   Igneous Rocks   70
  3-6   Sedimentary Rocks   74
  3-7   Metamorphic Rocks   79
  3-8   Crustal Movements   80
  3-9   Rock Weathering and Soil Formation   85
  3-10 General Considerations in Rock Weathering   92
  3-11 Soil Formations Produced by Weathering   92
  3-12 Running Water and Alluvial Deposits   96
  3-13 Glacial Deposits   100
  3-14 Wind Deposits   106
  3-15 Gravity Deposits   109
  3-16 Subsurface Water   109
  3-17 Sources of Geologic Information   116
  3-18 Summary   117

Chapter 4 Soil Classification Tests and
Common Systems   119

  4-1   General   119
  4-2   Soil Tests for Classification   121
  4-3   The Unified Soil Classification System   128
  4-4   The AASHTO Soil Classification System   135
  4-5   General Description of AASHTO Soil
Classification Subgroups   137
  4-6   The AASHTO Group Index   138
  4-7   Field Identification Tests   140
  4-8   Summary   141

Chapter 5 Soil Structure and Clay Minerals   145

  5-1   Soils and Soil Formation   145
  5-2   Soil Structure and Fabric   146
  5-3   Soil Cohesion and Friction   146
  5-4   Granular Soil Structure   147
  5-5   Other Considerations of Granular Soil Structure;
Relative Density   149
  5-6   Structure of Cohesive Soils   154
  5-7   Clay and Clay Minerals   158
  5-8   General Clay Mineral Properties   163
  5-9   Summary   167

Chapter 6 Soil Exploration and Sampling   170

  6-1   Introduction   170
  6-2   Site Investigation   171
  6-3   Site Geology   172

6-4   Determining Subsurface Conditions   174
6-5   Location, Spacing, and Depth of Borings   177
6-6   Sampling   179
6-7   Water Table   184
6-8   Sample Quality and Engineering Practice   184
6-9   The Standard Penetration Test (SPT)   185
6-10 SPT Correlations   186
6-11 Boring Logs   188
6-12 Cone Penetration Test (CPT)   189
6-13 Rock Drilling   192
6-14 Seismic Exploration   194
6-15 Resistivity Exploration   199
6-16 The Total Stress Cell and Dilatometer   200
6-17 Summary   202

Chapter 7   Compaction and Soil Stabilization   204

7-1   General Concept of Soil Stabilization   204
7-2   Soil as a Construction Material   205
7-3   Soil Stabilization   205
7-4   Soil Compaction   207
7-5   Theory of Compaction   208
7-6   Compaction of Cohesionless Soils   212
7-7   Structure and Properties of Compacted
      Cohesive Soils   213
7-8   Stabilization of Expansive Clays   215
7-9   Other Comments on Soil Compaction   216
7-10 Excavation and Compaction Equipment   217
7-11 Compaction Specifications   223
7-12 Field Control of Compaction   225
7-13 Statistical Field Unit Weight Control   230
7-14 Deep Compaction of In Situ Soils   232
7-15 Static Stabilization   234
7-16 Fabric Stabilization   234
7-17 Soil-Cement and Lime–Fly-Ash Soil Stabilization   236
7-18 Special Problems in Soil Compaction   238
7-19 Summary   239

**Chapter 8   Soil Hydraulics, Permeability,
              Capillarity, and Shrinkage**   241

8-1   Water in Soil   241
8-2   Permeability   242
8-3   Soil Water Flow and the Bernoulli Energy Equation   245
8-4   Determination of the Coefficient of Permeability   249
8-5   Limitations and Other Considerations in Determining k 252
8-6   Effective Coefficient of Permeability of
      Stratified Soils   254
8-7   Capillarity and Capillary Effects in Soil   257
8-8   Seepage Forces and Quick Conditions   267
8-9   Summary   274

Chapter 9   Seepage and Flow Net Theory   277

9-1   Introduction   277
9-2   Seepage Flow through Soil—The LaPlace Equation   278
9-3   Flow Nets   280
9-4   Flow Nets for Sheet-Pile Cut-Off Walls   282
9-5   Flow Nets for Earth Dams   286
9-6   Methods for Obtaining the Phreatic Line for Earth Dams   289
9-7   Flow Net Construction   293
9-8   Direct Computation of Seepage Quantity   296
9-9   The Flow Net When $k_x \neq k_y$ 297
9-10   Control of Seepage through Dams   299
9-11   Sudden Drawdown and Seepage Forces   301
9-12   Piping and Control of Piping   302
9-13   Other Methods for Seepage Quantities   304
9-14   Elements of Well Hydraulics   305
9-15   Summary   307

Chapter 10   Stresses, Strains, and Rheological Concepts   310

10-1   General Considerations   310
10-2   General Stresses and Strains at a Point   312
10-3   Theory of Elasticity Concepts Used in Soil Mechanics Problems   313
10-4   Octahedral Stresses   314
10-5   Stress-Strain Modulus and Hooke's Law   320
10-6   Anisotropic Soil   325
10-7   Two-Dimensional Stresses at a Point   326
10-8   Mohr's Stress Circle   328
10-9   Mohr's Circle for Inclined Principal Axes   332
10-10   Boussinesq Stresses in an Elastic Half Space   333
10-11   Boussinesq Solution for Round Plates   335
10-12   Numerical Integration of the Boussinesq Equations   336
10-13   Other Solutions for the Boussinesq Equations   337
10-14   Pressure Bulbs and Stress Concentration Factors   341
10-15   The Vertical Pressure Profile   342
10-16   The Average Pressure   344
10-17   The Newmark Influence Chart   345
10-18   Westergaard's Stress   348
10-19   Soil Deformation or Settlement   349
10-20   Rheological Models   351
10-21   Summary   353

Chapter 11   Consolidation and Consolidation Settlements   356

11-1   Soil Consolidation and Settlement Problems   356
11-2   Soil Consolidation   358
11-3   Interpreting the Consolidation Test   360
11-4   Correction of the Compression Curve for Disturbance   365

11-5	Preconsolidation and Estimation of the Preconsolidation Pressure	365
11-6	The Compression Parameters	367
11-7	Compression Parameters for Arithmetic Plots	369
11-8	Empirical Approximations for C_c and C_c'	370
11-9	Soil Structure and Consolidation	372
11-10	Settlement Computations	373
11-11	Secondary Compression (Creep) Settlements	374
11-12	Illustrative Examples	376
11-13	Controlling Consolidation Settlements	382
11-14	Reliability of Consolidation Settlement Computations	382
11-15	Summary	383

Chapter 12   **Rate of Consolidation**   387

12-1	The Coefficient of Consolidation	387
12-2	Percent Consolidation	391
12-3	Methods of Obtaining the Time of Interest for Computing c_v	393
12-4	Rate of Consolidation Based on Strain	396
12-5	Illustrative Examples	402
12-6	Consolidation Rates for Layered Media	410
12-7	Three-Dimensional Consolidation	411
12-8	Summary	411

Chapter 13   **Shear Strength of Soils**   414

13-1	Introduction	414
13-2	Shear Strength Parameters c and ϕ	416
13-3	Soil Failure, Critical Void Ratio, and Residual Strength	418
13-4	Soil Tests to Determine Shear Strength Parameters	420
13-5	Stress Paths	435
13-6	Shear Strength of Cohesionless Soil	440
13-7	Shear Strength of Cohesive Soils	442
13-8	Pore Pressure Effects in Consolidated-Drained Tests	455
13-9	Sensitivity of Cohesive Soils	455
13-10	Empirical Methods for Shear Strength	456
13-11	In-Situ Direct Measurement of Shear Strength	459
13-12	Factors Affecting Shear Strength	462
13-13	Normalized Soil Parameters	463
13-14	The s_u/p_o Ratio	464
13-15	Pore Pressure Parameters	466
13-16	Summary	471

Chapter 14   **Static and Dynamic Stress-Strain Characteristics**   476

14-1	Stress-Strain Data	476
14-2	The Stress-Strain Modulus	476
14-3	Poisson's Ratio	479

14-4   Factors Affecting the Stress-Strain Modulus
and Approximations   482

14-5   Resilient Modulus   482

14-6   Dynamic Soil Stress-Strain Modulus   483

14-7   Cyclic Modulus of Deformation and Liquefaction   486

14-8   Summary   496

Chapter 15   Lateral Pressures, Bearing Capacity,
and Settlement   498

15-1   Introduction   498

15-2   Soil Stresses at a Point-K_o Conditions   498

15-3   Active and Passive Earth Pressures   501

15-4   Pressures against Walls   506

15-5   Inclined Cohesionless Ground   507

15-6   Lateral Earth Pressure for Cohesive Soils   510

15-7   The Trial Wedge Solution   513

15-8   Logarithmic Spiral and ϕ-Circle Methods for
Passive Pressure in Cohesionless Soil   514

15-9   Bearing Capacity—Theoretical   517

15-10   Bearing Capacity by Empirical Methods   522

15-11   Other Bearing Capacity Problems   524

15-12   Deep Foundations   525

15-13   Immediate Settlement Computations   529

15-14   Summary   533

Chapter 16   Stability of Slopes   536

16-1   General Considerations in Stability of Slopes   536

16-2   Infinite Slopes   537

16-3   Stability of Infinite Cohesive Slopes   539

16-4   Circular Arc Analysis   540

16-5   The ϕ-Circle Method   542

16-6   Slope Stability Charts   544

16-7   Slope Analysis by Method of Slices   551

16-8   Wedge Block Analysis   555

16-9   Validity of Slope Stability Analyses   558

16-10   Summary   559

Bibliography   563

Indexes   571
     Name Index
     Subject Index

PREFACE

This text is an up-to-date assemblage of material needed for a basic understanding of geotechnical engineering. The in-depth coverage of physical properties of soils, soil origins, and geotechnical properties needed for flow strength and stability analyses has been retained from the first edition. Based on user feedback, I have expanded the coverage of soil exploration, mechanics—including Mohr's circle, soil stresses, bearing capacity, and settlement analyses. This makes the text suitable for those schools with only a general first course in geotechnical engineering and also applicable for more in-depth coverage of fundamentals when one or more additional courses are required. This expanded treatment also makes the text very useful as a reference work for the professional practitioner.

I have substantially increased the number of worked examples so that the text can be used either as a traditional textbook or in a self-study environment. More than half of the end-of-chapter problems are new or revised, and, as in the first edition, I have provided answers, or partial answers, to a large number to build user confidence. To reinforce the self-study aspect and to give authority to the material, I have greatly expanded the bibliography. This is particularly important in an area where the subject matter is often controversial, and it is not unusual for researchers to present conflicting work.

I have reordered the chapter on geology to immediately follow Chapters 1 and 2. This ordering of fundamental soil definitions allows laboratory work to progress nearly in parallel with the classroom and allows Chapter 3 (on Geology) to be omitted where the students have already taken a required course. In this latter case, the chapter should be assigned as required reading for the refresher value and for the engineering orientation.

I have included a short section on statistics and probability in Chapter 1 to put the remaining text material into proper perspective. This section, expanded somewhat from the first edition, should be assigned reading if it is not covered in class since assessments of data reliability are regularly made in subsequent chapters. This material is used to emphasize that geotechnical engineering is based heavily on parameter estimates using limited numbers of samples so that the risk factor often cannot be quantified by use of formal statistics and probability methods.

The text organization is such that there is a logical progression of subject matter. Extensive cross-referencing and reuse of topics in later chapters makes the text most useful through reinforcement learning. I suggest the following sequence where a course in foundation engineering is also required: Chapters 1, 2, and 3 (required or to read); Chapters 4, 5, and 6 (Sections 6-1 to 6-12) 7 to 14, 15 (Sections 15-1 to 15-8) (read remainder); and Chapter 16 be covered in depth, with Chapters 13 and 15 being somewhat superficially covered. Where the text is used as a general first course, all the chapters (with the possible exception of Chapter 3) should receive coverage with the depth dependent on the instructor's interest. In a construction program I would suggest Chapters 1 and 2 (Sections 2.1 to 2.5 and 2.12 to 2.14; reading assignment in Chapters 3 to 6, Chapter 7 (Sections 7-3 to 7-8 and 7-10 to 7-14), and Chapter 8 (Sections 8-3 to 8-5 and 8-8); read Chapters 9 and 10 (Sections 10-1 to 10-3 and 10-6 to 10-14); and read Chapters 11 to 13 (Sections 13-1 to 13-3 and 13-10 to 13-12), Chapter 15 (Sections 15-1 to 15-6 and 15-9 to 15-13), and Chapter 16. The classroom discussion should briefly touch on the "reading" material as well as the material covered in more detail.

The bibliography is not intended as a literature survey. There is a very real problem of scanning the voluminous amount of material currently published to include that which is deemed pertinent to reinforce statements or methodology introduced in the text. In the process, some valuable work will inevitably be overlooked, and very probably some work of lesser caliber will be included. If I have overlooked a significant contribution or failed to properly credit any original work, I sincerely apologize. In many cases I included only the most recent work since its bibliography would reference earlier applicable works. It is simply not possible in a textbook to include everything and credit everyone who has ever made a contribution. I have generally limited the bibliographical entries to sources which are fairly easy to obtain for two reasons: (1) most material of value is published in the cited sources, and (2) it is most frustrating to be referred to a reference which is almost impossible to obtain.

I should like to express appreciation to Richard J. Fragaszy, San Diego State University and William Gotolski, Pennsylvania State University who provided overall review and suggestions for the revisions.

I wish to express thanks to my wife Faye for helping with the typing and manuscript assembly.

Joseph E. Bowles

PHYSICAL AND GEOTECHNICAL PROPERTIES OF SOILS

INTRODUCTION TO GEOTECHNICS, STATISTICS, AND SI

1-1 GENERAL INTRODUCTION

Geotechnical engineering is concerned with the multidisciplined coordination of:

1. Mechanics—the response of masses to forces (statics, mechanics of materials concepts)
2. Material properties—physical, such as particle size and structure composition; index, used for classification or sorting; and engineering, including strength, angle of internal friction, cohesion, stress-strain modulus, Poisson's ratio, etc. and used for stability analyses and water flow
3. Fluid flow—where the fluid is usually water and principles of fluid mechanics are used
4. Environmental effects—climate, rainfall, gravitational, and chemical
5. Both soil and rock—with little practical difference between "soil" and "rock" despite some early efforts to distinguish between them

Items 1, 2, and 3 have been loosely grouped under the heading of "soil mechanics," but at present the more descriptive term is "geotechnical engineering." A principal factor for the name expansion was that as the science of soil mechanics developed, it was found that natural soil and rock formations and their associated physical, index, and engineering properties are very often dependent on the geologic processes of formation as well as time and environmental effects. Thus soil mechanics is a subset of geotechnical engineering.

Environmental effects include both an assessment of geologic history on mass response as well as future effects caused by the intended construction. Typical problems concern lowering of groundwater levels, avoidance of groundwater pollution, utilization of the soil to produce safe waste disposal systems, and estimation of earthquake effects on soil and rock masses, as well as more traditional problems of settlement and other stability problems. All these problems may influence the environment or may be influenced by the environment.

Rock mechanics evolved from mining works and was initially concerned with the stability of rock masses in these and tunneling operations. Stability and water flow were the principal concerns since the rock was often jointed. Cracks both introduce great uncertainty in stability computations and allow large quantities of water to flow compared to negligible amounts through sound rock. Items 1, 2, and 3 above were used in the solutions to these problems; the essential difference was substitution of the rock block for the aggregation of soil particles.

In general, the reader may substitute "rock" for "soil" in the remainder of the text as necessary to encompass the study of rock mechanics.

The geotechnical engineer thus has far more responsibility and is involved in substantially more complex multidisciplinary decisions than the "soils engineer" of 15 or 20 years ago.

The geotechnical engineer is called on to predict the behavior and performance of soil as a construction material or as a support for engineered works. Predictions of load-deformation characteristics of the natural as well as compacted fill beneath structures or of a soil structure such as a dam are required. These predictions usually require obtaining representative soil samples for testing to obtain physical and index properties. Additional samples or field tests may be required to obtain engineering properties. Since this sampling and testing is of necessity on small sample populations, some kind of statistical method is required to estimate the reliability of the predictions. Until very recently this reliability has been estimated more by a "feel" for the data than by any type of quantifiable analysis. Basic elements of statistics are introduced later in this chapter for use in making quantifiable test reliability estimates.

Geologically, soil mechanics is concerned with the unconsolidated mantle of weathered rock material overlying solid rock. A distinction is generally made between soil and rock in that soil is a particulate mass forming a skeleton structure, whereas rock is a dense structure with the constituent particles so firmly bonded together that great effort is required to separate them. Size is also a consideration with small rocks and boulders being considered "soil."

1-2 THE STUDY OF SOIL MECHANICS

Virtually all civil, environmental, transportation, structural, and geotechnical engineers are intimately concerned with soil mechanics principles. This is because almost all the construction endeavors of these individuals are concerned with soil behavior; either the soil is used as a construction material or the structure is placed on it. Foundation engineering is the design specialty that is specifically concerned with soil behavior and performance at the interface of the super-structure (above-ground element) and substructure (the foundation) with the ground.

The study of soil mechanics is of considerable economic importance since soil is the most readily available construction material at any site. All above-ground structures are supported by either soil or rock, and much of the public water supply moves through soil (to wells) or is retained by it in reservoirs using earthen dams. The most common method of waste disposal is in the ground (landfills), and only recently have long-term adverse environmental effects been associated with early landfills which were often "state of the art" at the time they were started. The acceptable risk level of some of these early designs was not set sufficiently high.

A study of soil mechanics, and of the various standardized testing procedures available for determining the several soil properties, enables the engineer to rapidly gain experience and obtain a "feel" for soil behavior. Of course, no matter how standardized the test or how carefully it is performed, if the soil sample used is not *representative* of the mass, the results will not be of much value.

Geotechnical engineering is considerably more "state of the art-" or judgment-dependent than the traditional scientifically oriented engineering disci-plines. This results from the heterogeneity of natural soil deposits; the large quantities of materials involved; and little control over environmental factors to which the mass is exposed, such as rainfall, temperature ranges, and gravitational effects which can affect the soil properties greatly. In these cases experience—both of the individual and of others through literature searches—is a very important factor.

A problem encountered by many students just starting the study of geotech-nical engineering is the introduction of a substantial terminology and exposure to a series of seemingly unrelated and diverse topics. Generally some relationship between the topic and engineering practice is made as the subjects are presented, but the varied nature of geotechnical work is such that this must of necessity be brief in a textbook. A student who has mastered the material in this text should be able (with some practice) to integrate the various seemingly unrelated topics into a unified body of knowledge which can be used as a tool for any future geotechnical study or work.

Because of the very complex nature of geotechnical engineering work and the fact that may universities have only a few credits for a course in "soil mechanics," most geotechnical firms now require documented additional course work and/or experience or an M.S. degree in candidates for a position. Thus any reader seriously considering geotechnical engineering should elect additional courses or plan on an advanced degree.

1-3 TEXT OBJECTIVES

This text introduces the reader to:

1. Soil mechanics terminology
2. The physical, index, and engineering properties of soils and some of the methods of measurement
3. Classification of soils in the several widely used systems and in terms of geologic formation
4. Methods of determining suitability of soils for various types of engineered construction
5. Evaluation of soil response to changes in loading and soil moisture
6. Effects of water on soil properties and movement of water through soil
7. Basic elements of statistics and probability as a tool to evaluate a soil testing program or data reliability
8. Methods of solution of certain soil mechanics problems

This list indicates that there is considerable diversity in the subject matter. Students should continually strive to integrate the material; the real problem often requires meshing the effects of several factors to develop an overall solution.

1-4 SOME TYPICAL SOIL PROBLEMS

To put the discussion into perspective, some typical soil problems with which a geotechnical engineer might be involved include:

1. In a soil exploration program to investigate site conditions, how many borings are necessary, and how deep? How many samples are required? What soil tests will need to be performed?
2. What is the stress in the soil at a given depth from the imposed super-structure or fill load? Can the soil carry this stress without a shear failure?
3. How much settlement can be expected for a structure as a result of the increase in soil stress? How long will it take for this settlement to occur?
4. Is this soil suitable for a highway or railroad fill? For use as a dam where water will be retained? For an embankment to retain industrial waste without leaking environmental pollutants?
5. Can this soil be used directly in fills, or will it require admixtures to modify certain undesirable index properties prior to use? What additives can be used? Can we use additives which alone are environmental pollutants, such as fly ash (power plant byproduct of coal burning), paper mill wastes, or mine wastes?
6. What happens to the soil structure if the ground water table fluctuates? Will pumping to dewater an excavation cause environmental problems?
7. What is the effect of frost or ice formation? Can the effects be avoided or reduced?

8. What is the effect of soil moisture change on the volume of the soil mass? How can volume change be controlled for pavements? For other structures, including residential construction?
9. What is the rate of water movement through a soil mass, i.e., can it be easily drained? Will a well provide an adequate supply? Will a dam built over this soil hold water?
10. What kind of excavation slope can be cut in the soil—1 : 2 (1 vertical on 2 horizontal), 1 : $\frac{1}{2}$, 1 : 1, or what? This can cause serious economic problems for highway and railroad work in particular because of the additional cut volume and additional right of way required.
11. Can a site be used as a landfill or for impoundment of industrial wastes without pollution of the groundwater?
12. How can an environmentally safe disposal be made for mine wastes, industrial waste solids, sewage sludge, etc.?
13. Is a site safe for a radiation-producing plant? Can settlements be controlled so that no leakage occurs? Will an earthquake produce a disaster?

Working solutions for some of these problems are shown in Fig. 1-1. It should be obvious that if some of these questions are inadequately answered or if the risk factor is too high, a failure may occur. Failure may take the form of:

1. Structural damage to buildings from excessive settlements or differential settlements
2. Bumpy roads resulting from differential settlements within a fill or at the junction of a cut and fill
3. Embankment failures, which may be slope (landslide) failures or excessive settlements or the underlying foundation or within the fill itself
4. Dam failures of various types, including "embankment" as in item 3 as well as excessive leakage through the embankment or through the underlying soil.

Not all the problems cited here will be considered in this text, since neither time nor space is available. The soil properties, both physical and engineering, which will be needed to solve these problems will, however, be considered in some detail. Topics beyond the scope of this textbook can be found in texts on foundation engineering (e.g., Bowles, 1982).

1-5 HISTORICAL DEVELOPMENT OF SOIL MECHANICS

Most authorities date the beginning of soil mechanics as an engineering science to the publication of *Erdbaumechanik auf bodenphysikalischer Grundlage* by Karl Terzaghi (the first textbook in soil mechanics) in Germany in 1925. Because of this publication, Terzaghi's early work in Europe, Asia, and the United States, and his over 250 technical papers, Terzaghi is often called the "father of soil mechanics." The Terzaghi Lecture is given at the annual meeting of the American Society of Civil Engineers in honor of Terzaghi. A historical perspective of Ter-

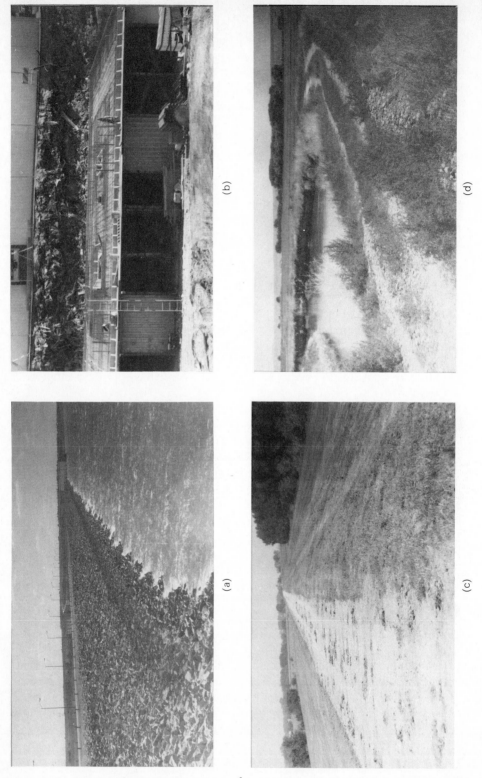

(a)

(b)

(c)

(d)

6

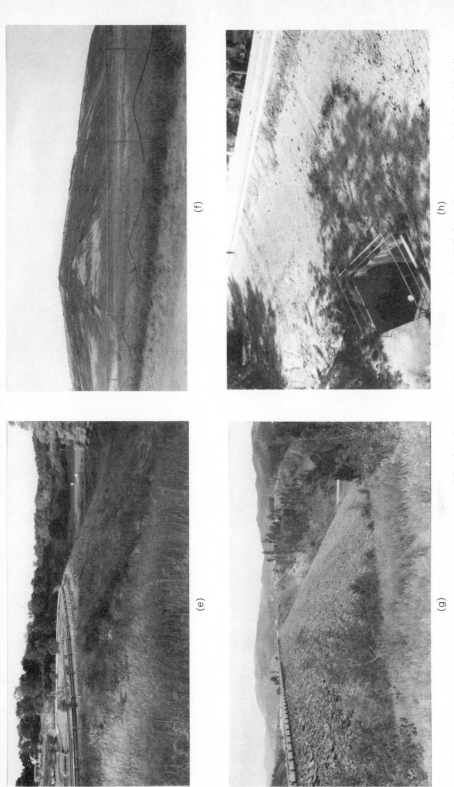

(e)

(f)

(g)

(h)

Figure 1-1 Some soil engineering projects. (*a*) Upstream face of Garrison Dam a rockfill dam near Pick City, North Dakota. (*b*) Basement of a high-rise building with excavation through a rubbish fill (in background) which supported the old building removed from the site and the existing street behind security shack. (*c*) Portion of Mississippi levee in Missouri. River is several hundred meters behind trees on right. Levee is built from soil obtained between levee location and river rather than from farm land on land side. (*d*) Irrigation canal in Montana. (*e*) Small dam in a residential land development project near Peoria, Illinois. Concrete overflow weir can be seen at far shore line. (*f*) Large above ground sanitary landfill in Michigan. Fill is more than 30 m high, with an anticipated future height of about 50 m. (*g*) Rockfill dam being used for a roadway. Dam is a reservoir for Rapid City, South Dakota. (*h*) Deep highway fill over large twin concrete culverts.

7

zaghi as a soils engineer and a bibliography of most of his publications, including those considered most noteworthy, are given in *From Theory to Practice in Soil Mechanics*, by Bjerrum et al. (1960).

Soil construction and the associated problems have been with people ever since they began digging caves and building mud huts for shelter. The Bible makes reference to a preference for building on a rock foundation as opposed to sand (Matt. 7:24–27). The Egyptians were aware of soil problems and even used caissons in pre-Biblical times to sink shafts through very soft Nile River sediments. Later the Romans were involved with building roads, parts of which are presently in use as subgrades, and foundations for aqueducts, some of which carried water for many kilometers. The Romans also constructed many large buildings, some of which are still in existence, such as the Forum and the Colosseum. These and later medieval constructions, including the notable St. Peter's Church, were built by utilizing principles of solid, well-drained foundations which are as valid today as they were in historical times. In Asia the Chinese made considerable use of soil from early times. A notable construction is the "Great Wall of China," first built during the Ch'in Dynasty (221–207 B.C.).

During medieval times many huge religious structures such as churches and bell towers were constructed in Europe. One which became famous because of an unsolved soil settlement problem is the Leaning Tower of Pisa in the city of that name in central Italy. This tower was begun in 1174 and discontinued after uneven settlement began to occur. Construction was later resumed, and the tower was finished in 1350. It is now some 5 m out of plumb in a 60-m height and is expected to turn over in about 200 years as a result of the continuing differential settlement. This settlement has now been continuing for more than 600 years. Recently the city of Pisa offered a prize (money) to anyone who could design a feasible method of halting the settlement without damage to the tower. As of this writing no satisfactory scheme seems to have been proposed. A brief engineering analysis of this project is given by Mitchell et al. (1977).

Substantial construction on, and utilizing, soil during early times was not limited to Europe and Asia. Many large earthen mounds, pyramids, and temples were constructed during the period from about 2000 B.C. to 1500 A.D. by the American Indians in the United States and by the Mayan and Incan peoples in Mexico and throughout Central America (Sowers, 1981). Earth dams and other substantial waterworks were also constructed by both these latter peoples from about 1000 B.C. (Schnitter, 1982). These works are particularly notable in the absence of any organized writing or numbering system.

1-6 SOIL FAILURES

To indicate both the scope and seriousness of the problem, a number of failures that have occurred fairly recently will be cited. This list is by no means exhaustive but was selected to indicate both the broad range of problems and the substantial damage which can occur.

General Failures

Mexico City settlements. These settlements range from 1 to 4 m and are primarily due to massive pumping of underground water from an extremely porous subsoil aquifer (void ratio e up to 14 and natural water contents up to 650 percent) as reported by Hiriart and Marsal (1969). The Palace of Fine Arts in Mexico City, begun in 1904 and completed in 1944, has settled some $3\frac{1}{2}$ m, according to Leggett (1962).

Houston, Texas, settlements. These are areal (the whole area) settlements underlying Houston and adjacent towns. Parts of the area have settled as much as 3 m, and the current settlement rate is some 150 mm/yr. Recently in a bayou location a bridge redesign was necessary because of the 2.4-m settlement at the site between the time the bridge was designed in 1970 and the beginning of construction in 1977 (ENR† 6/9/77, p. 11). These settlements are attributed to pumping of underground water for local water supplies. Some also believe that pumping of oil from deeper strata is contributing to the areal subsidence.

Transcona elevator. In 1914 a 1-million-bushel grain elevator at Winnipeg, Canada, consisting of 5 rows of 13 bins each and 30 m high suddenly (in approximately 12 hours) tilted to about 30° from the vertical after filling. The structure was later tilted back to the vertical. The tilt was attributed to a shear failure caused by overloading the foundation soil (Peck and Bryant, 1952; White, 1952).

Settlements Due to Lateral Flow of Soil from Beneath Foundation

Vertical movement of soil resulting from loss of lateral support, such as that caused by an adjacent excavation, is still a common problem. For example:

1. Construction of a depressed section of Interstate Highway in California caused a lateral flow of soil into the cut and subsidence of adjacent ground and buildings along the cut (ENR, 10/10/68, p. 22).
2. During construction of a 32-story office building in Los Angeles, California, the bulkheads retaining the excavation slipped laterally on the order of 75 mm because of excessive lateral pressure buildup from soil and rainfall conditions not properly accounted for in design. This much slip can cause large settle-

---

† ENR, 6/9/77, p. 11 = *Engineering News-Record*, June 9, 1977, page 11.

ments called *ground loss* around the perimeter, which, in an urban area, means ruined pavements and cracks in the closer buildings (ENR, 9/26/68).
3. During construction of a nine-story office building in Osaka, Japan, a loss of soil beneath footings on one side resulted in the building tilting about 5° out of plumb. The cause was believed to be excavation for a foundation on the adjoining lot (ENR, 10/17/68, p. 30).

Differential Settlements

These are unequal settlements beneath different areas of a building or embankment; they result in cracking when they are sufficiently large. The notable exception to differential settlements being unwanted is the Leaning Tower of Pisa, which has become a tourist attraction—but even here the owner wants the settlement to stop! Examples of unwanted differential settlements include:

1. The Charity Hospital in New Orleans, Louisiana, where the addition of a new wing to the original hospital resulted in differential settlements between the two parts of some 380 mm (a ramp was finally used to bridge the discontinuity).
2. Library-office building on the Cleveland State University campus in Cleveland, Ohio, had differential settlements of some 25 mm between the interior high-rise portion and the attached perimeter low-rise part. While this amount of settlement does not seem large, it was sufficient to cause detrimental cracking in the masonry walls and required remedial measures (ENR, 2/18/71, p. 12).
3. A building collapse in Akron, Ohio, was due to an isolated sinkhole formation missed during boring operations. The failure was initiated by the buildup of a roof water load which caused settlement in the sinkhole area. This caused the roof water to "pond" at that point and collapse the roof. The building was about 6 years old at the time of the roof collapse (ENR, 12/11/69, p. 23).

Slope Failures

Slope failures are extremely numerous—especially small ones involving 5 to 50 m^3 of earth—along the sides of road cuts. Other failures, including dam embankment failures, are far less numerous but often cause considerable property damage and loss of life when they do occur. A few major slope failures to indicate the scope of the problem include:

1. Panama Canal slides. These slides are still occurring but were particularly troublesome during construction of the canal. The original earthwork estimate increased from 79 to 177 million m^3 of earth because of the slides. Fortunately, increased efficiency in earth moving kept the final cost within the original appropriation of $375 million according to Mills (1913).

2. Fort Peck Dam. This earth fill dam being built on the Missouri river in 1938 failed during construction. The failure involved some 5 million m^3 of fill material. It was apparently caused by an excess pore pressure buildup from filling too rapidly (Casagrande, 1965; Middlebrooks, 1942). The Middlebrooks paper (including the discussions) should be read to obtain a perspective on how "experts" may disagree as to the cause of a problem given the same set of data.

3. A massive landslide on the slopes of Mt. Huascaran in the Peru Andes mountains killed 4000 to 5000 people in 1962. Eight years later (1970) a repeat slide, believed to have been initiated by an earthquake, killed over 18 000 people (Cluff, 1971).

4. A slide into the Vaiont Reservoir in northeastern Italy involving about 250 million m^3 of soil destroyed five downstream villages and killed between 2000 and 3000 people. The slide filled the reservoir which had a volume of only about 150 million m^3. The resulting downstream rush of displaced water caused the damage (Kiersch, 1964).

5. Slides (and mudflows) are particularly numerous in California with several million dollars of annual damage.

6. Landslides occurring on Interstate I-40 in Tennessee resulted in some $10 million in cost overruns (ENR, 5/13/73, p. 15).

Through 1975 the Federal Highway Administration estimated that approximately $50 million was spent annually to repair landslide damage on just the federally funded portion of the national highway system (Chassie and Goughnour, 1976). The reader can often observe landslides along highway cuts and on the sides of hills when traveling along highways in more rugged terrain in many areas of the world. Although the more common occurrences of landslides are in these areas, they are by no means limited to rugged terrain.

Undoubtedly many roadway, and other, slides can be prevented at the current state of the art. In those cases where slides occur 5 to 10 years after construction, long-term nonquantifiable effects are involved so that success depends on design as well as intangible factors such as climate, time, even luck.

Dam Failures

Earth dam failures may occur because of slope failures, but more often the failure is caused by overtopping, which erodes a channel from the top down. The water velocity increases with channel depth, and the subsequent rapid increase in erosion produces a washout. Overtopping can be initiated from flooding and an inadequate spillway size or from internal seepage forming an initial channel (called piping) through the dam at, or near, the base—perhaps aided by a rodent den—which "caved," resulting in the crest channel being formed. It is extremely difficult to pinpoint the exact cause of dam failures—even more so where the washout is extensive so that little of the original structure remains. Several recent dam failures include:

1. Kelly Barnes Lake Dam near Toccoa, Georgia. This was an earth dam about 150 m long by 8 m high washed out during severe flooding, killing approximately 37 persons (ENR, 10/11/77, p. 13).
2. Mine Tailing Pile. This was a "dam" produced by an accumulation of mine wastes across a waterway near Man, West Virginia, washed out during flooding and killing 66 people. The "dam" was 14 m high, 91 m wide at the base, and 76 m wide at the top (ENR, 3/2/72, p. 10).
3. Dam washouts in Connecticut. Torrential rains which dumped more than 254 mm of water over 2 days washed out or partly washed out 19 small earth dams. The tallest dam was 9 m high, and this failure carried out two smaller downstream dams (ENR, 6/17/82). No loss of life was reported with these failures.
4. Teton Dam. This failure involved a 93-m-high by 930-m-long dam with approximately 7.2 million m^3 of earth fill located in eastern Idaho. The failure was observed in progress as a piping failure believed to have been caused by inadequate grouting (leaving flow channels) of the badly fractured rocks near the abutments (ENR, 6/15/76). This dam only partly washed out (Fig. 1-2d).

Many of the dam failures result when flooding occurs. This indicates that the level of acceptable risk was not sufficient for the failure event. Many of these dams (and others worldwide) have been in existence for very long periods of time. Many were privately constructed and often not given any risk assessment—particularly if there was little (or no) downstream development at the time.

Other Failures

Other types of foundation failure include excessive settlement of pile foundations [see those cited by Blessey (1970) and Miller (1938)]. Another problem of widespread occurrence is caused by expansive soils. Soils which expand on wetting and shrink on drying are "expansive." The amount of volume change is very difficult to evaluate; however, methods to be presented later will alert one to the problem. This problem involves residential construction as well as larger structures and pavements. Gromko (1974) gives a review of this problem which should be read since the problem is so widespread.

Many soil failures are of a much lesser order of magnitude than those just cited; several of these are shown in Fig. 1-2. Small failures may be just as damaging to a small client as a major failure is for a governmental agency. Also, small failures such as those caused by settled fills, the maintenance of highway slopes for slide removal, and repair of cracked masonry in buildings are expensive.

This list of failures and photographs showing "failures" is not to emphasize failures but rather to show that:

1. Failures do occur in spite of the considerable advances in soil mechanics technology over the past 20 years.
2. Soil is an uncertain material with which to work or to take any excessive risk. While some risk is nearly always inherent in geotechnical engineering work, the risk factor should be assessed, and unnecessarily high risks must be avoided.

Actually, for the quantity of geotechnical work which is, or has been, done, the percentage of failures is quite small. Note that failures are "newsy" and tend to be reported, whereas the successes are generally hidden from view and almost never reported. When failures do occur, they are almost always due (in spite of disclaimers) to carelessness or to taking an excessive risk.

1-7 SOURCES OF INFORMATION FOR GEOTECHNICAL ENGINEERS

Geotechnical engineering relies heavily on the reported experiences of others as found in technical literature. This is both good and bad—good in that considerable confidence is obtained from using an established solution. This is somewhat bad from the viewpoint that many persons will simply repeat the earlier solution without attempting to introduce a novel and possibly better solution.

The following list of publications is by no means exhaustive, but represents the major sources of useful information in English. Many foreign publications are omitted—particularly if local or not in English—since much of that work is retitled and ends up published again in the cited journals where reviewers consider it worthwhile. Most of the following publications are not overly difficult to obtain and most university libraries carry most—if not all—listed.

1. *Journal of the Geotechnical Engineering Division, ASCE.* The Geotechnical Engineering Division sponsors periodic specialty conferences and publishes the conference proceedings. The eleventh specialty conference was held at Pasadena, California, in June 1978.
2. *Canadian Geotechnical Journal (Canada).* A publication similar to the *ASCE Geotechnical Journal;* it began publication in 1963.
3. *Geotechnique (United Kingdom).* Journal which in 1948 originated the idea of a separate publication in geotechnical engineering. Papers are primarily British (although authors from other countries are published) and often highly theoretical.
4. *Soils and Foundations (Japan).* A publication similar to *Geotechnique* with the authors primarily Japanese, although authors from other countries are published. This journal began publication in 1960.

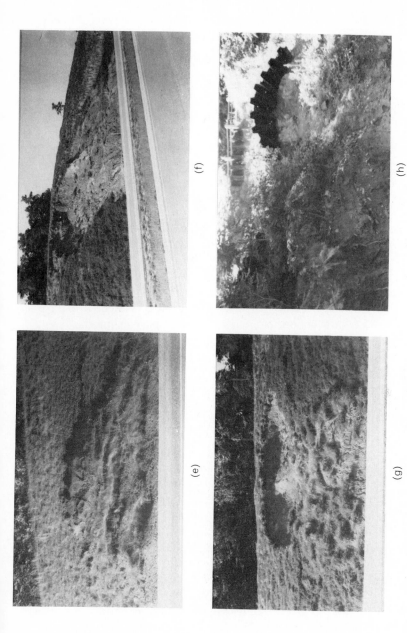

(e)

(f)

(g)

(h)

Figure 1-2 Some soil failures. (*a*) Large landslide near Vicksburg, Mississippi, which completely closed both lanes of U.S. 61 south of Vicksburg. (*b*) Large piping failure which temporarily closed Route 9 near Pekin, Illinois. The piping was caused during the flood stage of the Illinois River, which is about 300 m to the right, by an inadequately plugged drill hole. Piling is around initial piping; it was driven in a futile attempt to halt the piping. Piping enlarged the hole and carried material from beneath Route 9, causing the pavement to settle about 0.8 m. (*c*) Large pavement settlement in Kentucky. (*d*) Teton dam. Light material is dam fill. Failure caused erosion of a channel through the dam at far end. Close inspection can reveal the poor quality of the rock in the far abutment. (*e–h*) Typical local roadway slope failures: These vary in size as shown and represent considerable maintenance expense. A large number are shown to give an indication of the magnitude of the problem. (*e*) Illinois; (*f*) Kentucky; (*g*) West Virginia; and (*h*) roadway in background is moving laterally downhill toward viewer. Earlier remedial measures using H piles (upper) and sheet piles (lower) have not halted movement. Road overlies approximately 3 m of fill and 12 to 15 m of soft clay (which is flowing downhill).

5. Proceedings of the International Conference of Soil Mechanics and Foundation Engineering held by the International Society of Soil Mechanics and Foundation Engineering (ISSMFE). The international conferences have been held every 4 years since 1948 and the first in 1936. The 10th conference was held in Stockholm, Sweden in 1981. Papers are published in the conference proceedings from every country with membership. Content varies from trivial to very good.
6. Proceedings of Regional Soil Conferences. Some of these include (a) European—eighth held in Helsinki, Finland, 1983, (b) African—eighth held in Salisbury, Zimbabwe, 1983, (c) Asian—seventh held in Haifa, Israel, 1983, (d) Asian, Southeast—seventh held in Hong Kong, 1982, (e) Pan-American—seventh held in Vancouver, B.C., 1983, and (f) Australia–New Zealand—seventh held in Brisbane, Australia in 1975. This is now the Australia–New Zealand Geomechanics Conference, with the third held in 1980.
7. University Engineering Experiment Station Bulletins. These publications provide soils data for the local state.
8. American Society for Testing and Materials (ASTM). Periodic conferences sponsored by Committee D-18 (Soil and Rock) with proceedings published in Special Technical Publications (STPs).
9. Transportation Research Board (formerly Highway Research Board) publications. Areas coded by TRB which pertain to soils are Nos. 61 through 64.
10. Civil Engineering. This is a monthly magazine published by ASCE. It often contains notes on other publications of interest to geotechnical engineers.
11. Proceedings, Institution of Civil Engineers of London, Australia, and India. Publications correspond to item 1 of this list, and the Institution of Civil Engineers corresponds to ASCE.

1-8 ELEMENTS OF STATISTICS AND PROBABILITY

Statistics is the collection, tabulation, and analysis of data so that intelligent decisions may be made. Because of the inevitable uncertainties in every undertaking, probability is concerned with the probable accuracy of the data analysis (or conclusions drawn). Both statistics and probability are dependent on the sample size—either quantity, number of data sets, or tests.

For example, if 100 ball-point pens are manufactured for sale and each is tested and 10 require repair prior to shipment, the sample (all 100) is 100 percent reliable in that each pen was capable of writing when it left the plant. Looking further, however, we can also say that since 10 out of 100 were defective, there is a probability of 1 in 10 pens being defective, both in this run and in any subsequent production runs. Note that the number of defectives in future runs is speculative at this point since we are basing the estimate on statistical data from the first run only.

Suppose that we had tested only the first 10 pens and found 2 defective (the 8 remaining bad ones would be found when the remaining 90 were tested). On this basis we might conclude the probability that 1 in 5 pens will be bad in this run. It is also possible that no bad pens would be found in the first 10 or (remotely) that all 10 are bad. Further, if the first 10 tested were all bad due to a production malfunction (say, the ink supply had not been filled), this would not be a statistics-probability problem.

When the bad devices (pens here) are a random (not predictable) occurrence, we have a statistics-probability problem. For example, if there are unavoidable defects in the material extruded to make the pen bodies so that a void occasionally occurs, we have a statistics problem. In this case, in theory, we should be able to test 10 pens selected randomly and obtain the same ratio of 1 in 10 bad units. Practically, however, we might also find 2, 3, or even all 10 bad units—this is a chance we must take unless we wish to test every pen. Here for 100 pens it would be quite practical to test all of them; if 10 million are ultimately produced, it is obviously impractical, and some means must be found to test a sufficient number to determine the acceptance level. If the acceptance level (risk) set by management is too high (perhaps 1 bad out of each 1000), the quality control effort may make the pens too costly. If the level is set too low (perhaps 5 bad out of 100), the company may also go out of business as users go to another pen brand.

Two problems are thus posed: (1) how to establish an acceptable level of reliability (or risk) at an affordable cost and (2) how to reliably predict the risk level—whatever it may be.

The following short discussion will attempt to address these problems as they might pertain to a soil exploration and testing program. The interaction of this probability with estimated (or otherwise determined) probabilities from other sources such as settlement, loads, climate, etc., produces a probability model which is beyond the scope of this textbook. As a matter of fact, entire textbooks are devoted to these types of problems.

Since soils generally obey random laws of distribution in terms of:

1. Grain size, type of soil, and horizontal and vertical variations.
2. Physical and engineering properties—that is, the cohesion, angle of internal friction, or modulus of elasticity should vary from sample to sample in a random manner. The properties listed here will be taken up in later chapters, since the reader is presumed not to know what they are at this point.

If soil is naturally altered in any manner, such as drying, wetting, or increasing or decreasing in density, the process should be random. Note carefully, however, that:

1. Testing techniques can introduce nonrandom errors (termed *bias*).
2. Random distribution of effects pertains to a particular soil. That is, where a soil mass consists of several layers of different soils, as clay layers, sand layers, silty-sand layers, etc., we must consider each layer separately in applying statistical concepts of the mean, standard deviation, and numbers of tests as

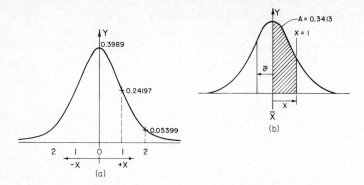

Figure 1-3 Normal distribution curves. (*a*) Normal distribution curve; (*b*) normal distribution curve mean and standard deviation. Shaded area is given in Table 1-1.

introduced in the following paragraphs. If all the layers are combined, erroneous statistical conclusions will be obtained—from incorrectly applying the statistical methods.

With random laws of distribution assumed to apply, the results produce a normal distribution curve which is symmetrical about the centroid, or mean, as in Fig. 1-3*a*.

In statistics we will be concerned with the following terms (some of which are identified in Fig. 1-3):

N = number of tests or the size of the sample population
V = test value of interest (such as unit weight, water content, or any other numerical quantity)
$\bar{X}$ = sample mean computed as

$$\bar{X} = \frac{\sum V}{N}$$

Note that this differs from:

Median = midvalue so chosen that half of the tests are above and half are below it
  Mode = maximum value from the peak of the distribution curve (is not maximum value of all numbers in the test series of N values)
    $\bar{\sigma}^2$ = sample variance computed as

$$\bar{\sigma}^2 = \frac{1}{N} \left(\sum_{1}^{N} V^2 - N\bar{X}^2 \right)$$

In the above equation, $N - 1$ is often used in the denominator–particularly when N is large. However, for the sample sizes commonly available in a soil testing program, N rather than $N - 1$ is recommended and used in this textbook.

$\bar{\sigma}$ = standard deviation = $\sqrt{\bar{\sigma}^2}$. [The mean deviation is computed simply as $(1/N) \sum (V - \bar{X})$ and is generally less than the standard deviation $\bar{\sigma}$.]

$\bar{C}$ = coefficient of variance computed as

$$\bar{C} = \frac{\bar{\sigma}}{\bar{X}}$$

This value is the percent (as a decimal) change per unit of $\bar{X}$.

Y = ordinate of the normal distribution curve (may use symmetrical with the origin to simplify the mathematics for drawing the curve)

$$Y = \frac{1}{\bar{\sigma}\sqrt{2\pi}} \exp\left[-\frac{1}{2}\left(\frac{X - \bar{X}}{\bar{\sigma}}\right)^2 \right]$$

where X = any abscissa value such as 0, ± 1, ± 2, ± 3, etc.

$Y_0 = 0.39894$ (see Fig. 1-3a)

$Y_1 = 0.24197$

$Y_2 = 0.05399$ when $\bar{\sigma} = 1.0$ and $\bar{X} = 0.00$ (origin)

This equation may be used to plot a normal distribution curve (symmetrical) as shown in Fig. 1-3a. The total area of this curve is 1.00, and the area under the curve bounded by any two ordinates represents the probability that X is between these two points. Table 1-1 gives selected values of curve areas between $X = 0$ and $X = 3.9$ which can be doubled to give the total area on both sides. A table such as this is most readily obtained by integrating the equation above for Y and programming on the computer. Note that it is a simple matter to transform the curve of Fig. 1-3b to put the origin to the left so that the curve peak is at a normal distribution for the actual numerical values of V, such as 10.0, 100.0, etc., depending on what the curve represents, such as water contents, densities, specific gravities, or whatever.

From the table, we find that at $X = 1.0$ the area is 0.3413; thus, doubling for symmetry, we have 0.6826, or the percent of the curve included between ± 1.0 is

$$\text{Percent} = \frac{0.6826}{1}(100) = 68.26 \text{ percent of total curve area}$$

and the area excluded is $1.0000 - 0.6826 = 0.3174$ or 31.74 percent. For a standard deviation $\bar{\sigma} = 1.0$, the probability of any test N_i being between the values of

$$\bar{X} - \bar{\sigma} \leq V \leq \bar{X} + \bar{\sigma}$$

is 68.26 percent.

t = Student's t-distribution values. This is a widely used series of numbers for normal random distributions. The equation can be found in most textbooks on statistics and requires a computer to obtain tables such as Table 1-2.

Table 1-1 Areas under the standard distribution curve for X as indicated. Double table values for $\pm X$ to get total area

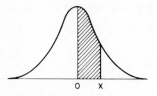

X	0.00	0.01	0.02	0.03	0.04	0.05	0.06	0.07	0.08	0.09
0.00	0.0000	0.0040	0.0080	0.0120	0.0160	0.0199	0.0239	0.0279	0.0319	0.0359
0.10	0.0398	0.0438	0.0478	0.0517	0.0557	0.0596	0.0636	0.0675	0.0714	0.0753
0.20	0.0793	0.0832	0.0871	0.0910	0.0948	0.0987	0.1026	0.1064	0.1103	0.1141
0.30	0.1179	0.1217	0.1255	0.1293	0.1331	0.1368	0.1406	0.1443	0.1480	0.1517
0.40	0.1554	0.1591	0.1628	0.1664	0.1700	0.1736	0.1772	0.1808	0.1844	0.1879
0.50	0.1915	0.1950	0.1985	0.2019	0.2054	0.2088	0.2123	0.2157	0.2190	0.2224
0.60	0.2257	0.2291	0.2324	0.2357	0.2389	0.2422	0.2454	0.2486	0.2517	0.2549
0.70	0.2580	0.2611	0.2642	0.2673	0.2703	0.2734	0.2764	0.2793	0.2823	0.2852
0.80	0.2881	0.2910	0.2939	0.2967	0.2995	0.3023	0.3051	0.3078	0.3106	0.3133
0.90	0.3159	0.3186	0.3212	0.3238	0.3264	0.3289	0.3315	0.3340	0.3365	0.3389
1.00	0.3413	0.3437	0.3461	0.3485	0.3508	0.3531	0.3554	0.3577	0.3599	0.3621
1.10	0.3643	0.3665	0.3686	0.3708	0.3729	0.3749	0.3770	0.3790	0.3810	0.3830
1.20	0.3849	0.3869	0.3888	0.3906	0.3925	0.3943	0.3962	0.3980	0.3997	0.4015
1.30	0.4032	0.4049	0.4066	0.4082	0.4099	0.4115	0.4131	0.4147	0.4162	0.4177
1.40	0.4192	0.4207	0.4222	0.4236	0.4251	0.4265	0.4279	0.4292	0.4306	0.4319
1.50	0.4332	0.4345	0.4357	0.4370	0.4382	0.4394	0.4406	0.4418	0.4429	0.4441
1.60	0.4452	0.4463	0.4474	0.4484	0.4495	0.4505	0.4515	0.4525	0.4535	0.4545
1.70	0.4554	0.4564	0.4573	0.4582	0.4591	0.4599	0.4608	0.4616	0.4625	0.4633
1.80	0.4641	0.4648	0.4656	0.4664	0.4671	0.4678	0.4686	0.4693	0.4699	0.4706
1.90	0.4713	0.4719	0.4726	0.4732	0.4738	0.4744	0.4750	0.4756	0.4761	0.4767
2.00	0.4772	0.4778	0.4783	0.4788	0.4793	0.4798	0.4803	0.4808	0.4812	0.4817
2.10	0.4821	0.4826	0.4830	0.4834	0.4838	0.4842	0.4846	0.4850	0.4854	0.4857
2.20	0.4861	0.4864	0.4868	0.4871	0.4874	0.4878	0.4881	0.4884	0.4887	0.4890
2.30	0.4893	0.4896	0.4898	0.4901	0.4904	0.4906	0.4909	0.4911	0.4913	0.4916
2.40	0.4918	0.4920	0.4922	0.4924	0.4927	0.4929	0.4930	0.4932	0.4934	0.4936
2.50	0.4938	0.4940	0.4941	0.4943	0.4945	0.4946	0.4948	0.4949	0.4951	0.4952
2.60	0.4953	0.4955	0.4956	0.4957	0.4958	0.4960	0.4961	0.4962	0.4963	0.4964
2.70	0.4965	0.4966	0.4967	0.4968	0.4969	0.4970	0.4971	0.4972	0.4973	0.4974
2.80	0.4974	0.4975	0.4976	0.4977	0.4977	0.4978	0.4979	0.4979	0.4980	0.4981
2.90	0.4981	0.4982	0.4982	0.4983	0.4984	0.4984	0.4985	0.4985	0.4986	0.4986
3.00	0.4986	0.4987	0.4987	0.4988	0.4988	0.4988	0.4989	0.4989	0.4990	0.4990
3.10	0.4990	0.4991	0.4991	0.4991	0.4991	0.4992	0.4992	0.4992	0.4993	0.4993
3.20	0.4993	0.4993	0.4994	0.4994	0.4994	0.4994	0.4994	0.4995	0.4995	0.4995
3.30	0.4995	0.4995	0.4995	0.4996	0.4996	0.4996	0.4996	0.4996	0.4996	0.4996

From these tables, how many tests N would be required to obtain a reliability of 68.26 percent? Solution: Inspecting the tables at $N = 1$, we find that a single test will produce a condition of 75 percent reliability that the value V will not be different from $\bar{X}$ by more than a maximum of $\bar{\sigma} = 1$, or

$$\bar{X} - 1 \leq V \leq \bar{X} + 1$$

Table 1-2 Values of Student's t distribution for several percentages (reliability) R indicated and for numbers of test values N as shown

R \ N	0.999	0.995	0.975	0.950	0.900	0.850	0.800	0.750	0.700	0.600
1	318.309	63.657	12.706	6.314	3.078	1.963	1.376	1.000	0.727	0.325
2	22.327	9.925	4.303	2.920	1.886	1.386	1.061	0.816	0.617	0.289
3	10.215	5.841	3.182	2.353	1.638	1.250	0.978	0.765	0.584	0.277
4	7.173	4.604	2.776	2.132	1.533	1.190	0.941	0.741	0.569	0.271
5	5.893	4.032	2.571	2.015	1.476	1.156	0.920	0.727	0.559	0.267
6	5.208	3.707	2.447	1.943	1.440	1.134	0.906	0.718	0.553	0.265
7	4.785	3.499	2.365	1.895	1.415	1.119	0.896	0.711	0.549	0.263
8	4.501	3.355	2.306	1.860	1.397	1.108	0.889	0.706	0.546	0.262
9	4.297	3.250	2.262	1.833	1.383	1.100	0.883	0.703	0.543	0.261
10	4.144	3.169	2.228	1.812	1.372	1.093	0.879	0.700	0.542	0.260
12	3.930	3.055	2.179	1.782	1.356	1.083	0.873	0.695	0.539	0.259
14	3.787	2.977	2.145	1.761	1.345	1.076	0.868	0.692	0.537	0.258
16	3.686	2.921	2.120	1.746	1.337	1.071	0.865	0.690	0.535	0.258
18	3.610	2.878	2.101	1.734	1.330	1.067	0.862	0.688	0.534	0.257
20	3.552	2.845	2.086	1.725	1.325	1.064	0.860	0.687	0.533	0.257
22	3.505	2.819	2.074	1.717	1.321	1.061	0.858	0.686	0.532	0.256
24	3.467	2.797	2.064	1.711	1.318	1.059	0.857	0.685	0.531	0.256
26	3.435	2.779	2.056	1.706	1.315	1.058	0.856	0.684	0.531	0.256
28	3.408	2.763	2.048	1.701	1.313	1.056	0.855	0.683	0.530	0.256
30	3.385	2.750	2.042	1.697	1.310	1.055	0.854	0.683	0.530	0.256
35	3.340	2.724	2.030	1.690	1.306	1.052	0.852	0.682	0.529	0.255
40	3.307	2.704	2.021	1.684	1.303	1.050	0.851	0.681	0.529	0.255
45	3.281	2.690	2.014	1.679	1.301	1.049	0.850	0.680	0.528	0.255
50	3.261	2.678	2.009	1.676	1.299	1.047	0.849	0.679	0.528	0.255
55	3.245	2.668	2.004	1.673	1.297	1.046	0.848	0.679	0.527	0.255
60	3.232	2.660	2.000	1.671	1.296	1.045	0.848	0.679	0.527	0.254
65	3.220	2.654	1.997	1.669	1.295	1.045	0.847	0.678	0.527	0.254
70	3.211	2.648	1.994	1.667	1.294	1.044	0.847	0.678	0.527	0.254
75	3.202	2.643	1.992	1.665	1.293	1.044	0.846	0.678	0.527	0.254
80	3.195	2.639	1.990	1.664	1.292	1.043	0.846	0.678	0.526	0.254
85	3.189	2.635	1.988	1.663	1.292	1.043	0.846	0.677	0.526	0.254
90	3.183	2.632	1.987	1.662	1.291	1.042	0.846	0.677	0.526	0.254
95	3.178	2.629	1.985	1.661	1.291	1.042	0.845	0.677	0.526	0.254
100	3.174	2.626	1.984	1.660	1.290	1.042	0.845	0.677	0.526	0.254
105	3.170	2.623	1.983	1.659	1.290	1.042	0.845	0.677	0.526	0.254

Example 1-1 Given are the following numbers from a series of plastic limit tests (see Chap. 3 for definition):

$$V = 18.2 \quad 16.0 \quad 19.1 \quad 17.3 \quad 20.1 \quad 17.9 \quad 20.3 \quad 21.4$$
$$20.7 \quad 20.5 \quad 21.5 \quad 20.6 \quad 24.6$$

REQUIRED

(a) Compute the mean, median, and mode values.
(b) Compute the standard deviation and coefficient of variance.
(c) What is the percent probability that any test would be within the standard deviation?

SOLUTION

(a) Compute the mean, median, and mode values:

The mean (after summing the 13 values of V) is

$$\bar{X} = \frac{\sum V}{N} = \frac{258.2}{13} = 19.86 \text{ percent}$$

The median is obtained by ranking the 13 tests in order to obtain

16.0   17.3   17.9   18.2   19.1   20.1   20.3

20.5   20.6   20.7   21.4   21.5   24.6

to obtain 20.3 percent by inspection.
The mode value will be approximately

$$(20.1 + 20.3 + 20.5 + 20.6 + 20.7)/5 = 20.44$$

without plotting a distribution curve.
(b) Compute the standard deviation and coefficient of variance:
Using the just computed mean $\bar{X}$ and the given equation, obtain the following:

$$\bar{\sigma}^2 = \frac{1}{N} \left(\sum_1^N V^2 - N\bar{X}^2 \right) = \frac{1}{13} \left[\sum_1^{13} V^2 - 13(19.86)^2 \right]$$

$$= \frac{1}{13} (5186.32 - 5127.45) = 4.528$$

and the standard deviation $\bar{\sigma}$:

$$\bar{\sigma} = \sqrt{\bar{\sigma}^2} = \sqrt{4.528} = 2.128$$

The coefficient of variance:

$$\bar{C} = \frac{2.128}{19.86} = 0.01072$$

(c) What is the percent probability that any test is within 1 standard deviation, i.e., $19.86 \pm 1(2.13)$?
Referring to Table 1-2 and using $N = 12$ to avoid four-way interpolation, locate 1 between 0.8 and 0.85 and obtain

At 85 percent = 0.85   = 1.083

80 percent = 0.80   = 0.873

Differences = $\overline{0.05}$ = $\overline{0.210}$

$$\text{Percent} = 0.80 + \frac{0.05(1.000 - 0.873)}{0.210}$$

$$= 0.830 \rightarrow 83 \text{ percent}$$

This says that there is an 83 percent chance that any one of the 13 tests would be some value between $19.86 - 2.13 = 17.73$ and 21.99. Inspecting the data, we find 16.0, 17.3, and 24.6 out of this range or

$$\text{Percent} = (13 - 3) \times \frac{100}{13} = 77 \text{ percent} < 83$$

This is not too bad a correlation between theory and reality. Now in theory one should be able to state *for this soil* that the plastic limit if correctly done has an 83 percent chance of being between 17.73 and 21.99. We may also state that the expected error is the standard deviation (or 2.13). Thus for this soil, the plastic limit can be determined within only about 2.13 units of the "true" value.

////

In most soil work a range of ± 1 standard deviation is usually used. Between Tables 1-1 and 1-2 we find a single test for whatever purpose has between a 68 and 75 percent (say, 70 percent) chance of being within $\pm 1\bar{\sigma}$ of the mean value of $\bar{X}$. Still to be determined are both the mean $\bar{X}$ and $\bar{\sigma}$. These values require substantially more than one test to obtain "reliable" values. As the number of tests increases $\bar{X}$ can change; usually $\bar{\sigma}$ decreases toward some minimum value. Note that these values may be altered by rejection of selected tests. It is common to reject extreme values as being "obviously wrong."

1-9 SI UNITS

This textbook uses SI units and/or preferred usage metric units throughout. Most soil laboratory equipment will last for many years if properly maintained; thus, mass units of grams and kilograms will continue to be used to obtain the mass (formerly—and likely to continue to be commonly—called "weight") of a sample. Grams and kilograms are valid SI mass units, so no problems will arise in making mass measurements. A problem will arise with the unit weight term due to the common practice of using the terms "weight" and "mass" interchangeably. In this text the term "unit weight" will be used to define a body force unit (force per unit volume) of kilonewtons per cubic meter (kN/m^3). The kilonewton will be used exclusively since the newton unit is too small. Density will be taken as mass per unit volume of kilograms per cubic meter (kg/m^3) or grams per cubic centimeter (g/cm^3); this is the definition presently used by the scientific community. Unfortunately, in soil mechanics, not only have mass and weight been used interchangeably, but density and unit weight have been also. It is the author's opinion that if the reader carefully adheres to using density for mass/volume and unit weight for force/volume and recognizes that laboratory "weights" are in mass units, no problems will arise. The use of unit weight as a force unit is necessary in order to be able to compute lateral and vertical pressures in a soil mass due to gravitational body forces. Pressure is defined to be a force per unit area (kilonewtons per square meter = kilopascals = kPa in this text).

Most soil samples will be on the order of 50 to 75 mm in diameter × 100 to 200 mm in length. An area using 50 to 75-mm diameter is not practical in the standard SI units of millimeters (it results in too large a number) or meters (too small); thus square centimeters (cm^2) is the only practical unit and is often used by the author in this text.

Weighing measurements will be most convenient in grams (g) or kilograms (kg), and for density the best laboratory measuring unit is grams per cubic centimeter. Note that grams is not strictly SI, but most laboratory scales measure in either grams or kilograms and are not likely to be changed in the near future. Volumetric devices not calibrated in cubic centimeters can be converted (soft conversion) and marked, once and for all. The use of grams per cubic centimeter is particularly advantageous since:

In SI: $1 \text{ g/cm}^3 \times 9.807 = 9.807 \text{ kilonewtons/m}^3 \text{ (kN/m}^3)$

Also $1 \text{ g/cm}^3 = 1 \text{ Mg/m}^3 = 1000 \text{ kg/m}^3 = 1 \text{ tonne/m}^3$

In fps: $1 \text{ g/cm}^3 \times 62.4 = 62.4 \text{ pounds/ft}^3$

both of which are the *unit weight* of water (Mg = megagrams). This conversion factor(s) and the length units of:

$$3.2808 \text{ ft} = 1 \text{ m}$$

$$1 \text{ in} = 25.4 \text{ mm} = 2.54 \text{ cm}$$

and force units of

$$1 \text{ g} = 980.7 \text{ dynes} \quad \text{and} \quad 1 \text{ newton (N)} = 1 \times 10^5 \text{ dynes}$$

and the gravitational constant

$$g = 980.7 \text{ cm/s}^2$$

will enable the reader to make any conversions needed for derived SI units. Note that these conversion factors have been located at the end of this chapter for rapid reference (see also inside back cover).

PROBLEMS

1-1 A material has a density of 1.83 g/cm^3. What is the unit weight in fps and SI?

1-2 A soil sample has a diameter of 62.3 mm. What is the area in fps and SI?

1-3 Given the following soil compaction unit weights (kN/m^3) from a student laboratory section:

1 17.9	5 18.3	9 18.4
2 18.0	6 17.3	10 17.1
3 17.9	7 17.7	11 17.2
4 17.0	8 16.9	12 18.1

Compute the mean, standard deviation, and coefficient of variation for these 12 tests. Use Table 1-2 to estimate the percent chance that any single test will be within 1 standard deviation of the mean, and compare this estimate to the actual results given.

1-4 Given the following plastic limit water content values (percents) from a student laboratory section:

| 24.9 | 19.3 | 24.1 | 24.2 | 26.2 | 18.1 | 24.3 | 23.8 | 25.1 |

Compute the mean, standard deviation, and coefficient of variance. Estimate the percent chance that a single test will be within 1 standard deviation of the mean, and compare this estimate to the actual results given.

1-5 Estimate the minimum number of tests N needed to give an 85 percent reliability that the true value is within 1 standard deviation.

1-6 What is the best (minimum) possible value of k to given $\bar{X} \pm k\bar{\sigma}$ to obtain a 96 percent estimated reliability and not perform more than 10 tests?

TWO

SOIL PROPERTIES—PHYSICAL AND INDEX

2-1 INTRODUCTION

This chapter will introduce the reader to a number of terms and definitions routinely used by geotechnical engineers. These terms, and equations, are primarily for description of the physical and index properties of soils. Several of the equations will be used so frequently throughout the text that it will be most worthwhile to memorize them. They will be most easily remembered if, in subsequent work, they are used in deriving the needed relationships (equations) rather than spending time looking either herein or elsewhere for an equation which might be applicable.

Since index properties are easier to determine, much effort has been expended to attempt to "correlate" engineering properties to them. Often correlations are little better than estimates (dignified guesses) but are useful in preliminary work where the additional expense and effort to obtain somewhat better values for the needed engineering property are not justified.

2-2 SOIL COMPOSITION AND TERMS

Soil is a particulate mixture comprised of any or all of the following:

Boulders—large pieces of rock generally taken as larger than 250 to 300 mm. In the size range of 150 to 250 mm the rock fragments may be *cobbles* or *pebbles*.

Gravel—rock particles from 150 down to about 5 mm.

Sand—rock particles from about 5 down to about 0.074 mm. Ranges from *coarse* (5 to 3 mm) to *fine* (<1 mm).

Silt—rock particles from about 0.074 down to 0.002 mm. Large quantities of silt (and clay) are found in deposits sedimented into lakes and near shorelines at river exits (along Gulf Coast and Atlantic and Pacific oceans). A *loess* deposit results when winds transport silt particles to a location. Wind transport limits the particle size so that the resulting deposit is nearly one grain size.

Clay—mineral particles of less than 0.002 mm. These particles are the primary source of cohesion in a "cohesive" soil.

Colloids—inert mineral particles smaller than 0.001 mm.

Many soil deposits contain varying percentages of all of the above-named particles. Where a particular particle size predominates, the deposit may be given the name, sand, gravel, sandy gravel, clay, etc. An exception occurs with clay and silts where a predominantly silt deposit with 10 to 25 percent clay may be called "clay."

Deposits of sand, gravel, or sand-gravel mixtures may be loose, medium, or dense, based on particle packing as ascertained visually or by tests. These materials and colloids are generally inert (cohesionless) but may be naturally cemented by geologic processes such as submergence in a marine environment and later emergence, rainfall and decomposition of organic surface matter, and numerous other factors. Clay minerals are the primary cementing agent in some deposits and in all "clays."

Cohesive soil deposits may be soft, stiff, hard, etc., depending on the instant water content and geologic history producing any particle packing.

Supplemental descriptive terms will be added as necessary, but the principal "soils" have been named. Sometimes local terminology may be used to describe a deposit of one or more of these soils with some distinctive feature (color, stickiness, hardness, stratification). Terms with which the reader may not be familiar will be subsequently introduced in this and later chapters.

A more systematic method of soil classification will be presented in Chap. 4. The clay minerals whose presence causes the most trouble for the geotechnical engineer will be considered in detail in Chap. 5.

2-3 BASIC DEFINITIONS AND MASS-VOLUME RELATIONSHIPS

In soil mechanics, as in any other area of engineering, certain relationships are so fundamental that they become definitions even though the relationships are expressed as mathematical formulas. In other cases, the definitions are used in the usual context. Several of the more fundamental soil mechanics definitions will be introduced or developed in this and following sections.

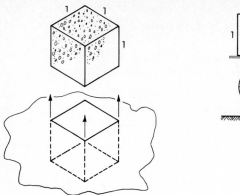

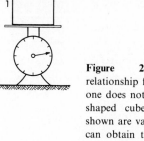

Figure 2-1 Weight-volume relationship for soil. Note that one does not require a regular shaped cube; the conditions shown are valid as long as one can obtain the volume of the hole and the weight of soil removed from the hole.

If one were to go into the field, clear off the ground surface, and (if it were physically possible) extract a cube of soil of unit dimensions (say, $1 \times 1 \times 1$ cm), as shown in Fig. 2-1, a visual inspection would reveal the block to be made up of

1. *Pores* or *voids*, which are the open spaces between soil grains, of various sizes.
2. *Soil grains*, which may be macroscopic† or microscopic in size. Obviously microscopic sizes cannot be observed with the unaided eye, but it is not unreasonable to expect that they would be present in varying amounts.
3. *Soil moisture*, which may cause the soil to appear wet to damp to somewhat dry. The water in the pores or voids, or pore water, may be of sufficient quantity to completely fill the voids (a saturated soil) or may only surround the soil grains to some extent.

The soil pores which do not contain water are, of course, full of air and/or water vapor. Referring again to Fig. 2-1, if we place the cube (of unit volume) of soil on a scale before any water drains from the block, the resulting weight may be taken as the wet unit weight γ_m of the soil. A special case exists when *all* the voids are filled with water; the resulting weight is the *saturated unit weight* γ_{sat} of the soil. If the cube of soil is placed in an oven and dried to a constant weight and reweighed, the *dry unit weight* γ_d of the soil is obtained. These, and other symbols used in this text, are consistent with those recommended by the Geotechnical Committee on Nomenclature for Soil Mechanics. It might be observed that even if the cube crumbles in the oven as it dries, the weights obtained are still unit weights. For the case shown in Fig. 2-1, the unit weight will be X grams per cubic centimeter (g/cm^3). If the foot-pound-second (fps) system of units had been used, the unit weight would have been given in pounds per cubic foot (lb/ft^3)

---

† "Macroscopic" refers to particles which are visible to the eye, whereas "microscopic" refers to particles which can be seen only with the aid of a microscope or other magnifying device.

and in SI as kilonewtons per cubic meter. From the preceding discussion, it follows that the unit weight of soil (or any material) is

$$\gamma = \frac{\text{weight of material}}{\text{volume of material}} \tag{2-1}$$

By analogy the density is

$$\rho = \frac{\text{mass of material}}{\text{volume of material}} \tag{2-1a}$$

Since water at 4°C has a density of 1 g/cm^3 (or approximately 62.38 lb/ft^3), the unit weight of water at 4°C is

$$\gamma_w = \frac{62.4}{1} = 62.4 \text{ lb/ft}^3$$

or in SI units

$$\gamma_w = 1 \frac{\text{g}}{\text{cm}^3} \times 980.7 \frac{\text{dynes}}{\text{g}} \times \left(1 \times 10^{-5} \frac{\text{N}}{\text{dyne}}\right)\left(1 \times 10^6 \frac{\text{cm}^3}{\text{m}^3}\right) \times \frac{1 \text{ kN}}{1000 \text{ N}}$$

$$= 1 \times 9.807 = 9.807 \text{ kN/m}^3$$

The density (and corresponding unit weight) of water varies slightly depending on the temperature, but for all practical engineering purposes, γ_w may be taken as 9.807 kN/m^3 (or a density of 1 g/cm^3) for a temperature range of 0 to about 20°C unless the problem context indicates otherwise.

It should be observed at this point that the 1-cm^3 block being used for illustration in Fig. 2-1 represents only a minuscule portion of the total volume of soil at the location from which it was extracted. Whether the unit weight obtained is valid for the entire mass will depend on how well the cube represents the soil mass.

If the water is collected and the soil grains are melted down so that they fuse into a nonporous solid, one can depict the cube of Fig. 2-1 in the form of a block diagram (Fig. 2-2c) which shows the same unit of bulk soil volume.

From Fig. 2-2 we will define several useful terms.

1. Void ratio e is defined as

$$e = \frac{V_v}{V_s} \tag{2-2}$$

where in this and following equations the terms are defined on Fig. 2-2c. The void ratio e is normally expressed, and used, as a decimal. This definition is widely used in geotechnical engineering. Inspection of Fig. 2-2 and Eq. (2-2) indicates that the possible range of e is

$$0 < e \ll \infty$$

Typical values of void ratios for natural sands may range from 0.5 to 0.8, typical values for cohesive (sticky when wet) soils from 0.7 to 1.1.

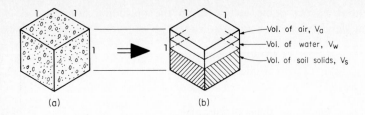

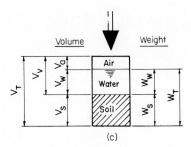

Figure 2-2 Volumetric and weight relationships for a soil mass. (*a*) Volume of soil removed from the ground; it may be any size or shape but is shown here as a unit cube. (*b*) If constituent parts of the soil sample are separated and full recovery of air and water in soil pores is possible, the soil sample may be shown as above. (*c*) It is usual practice (and more convenient) to use a line diagram with the relationship as shown rather than a three-dimensional block as in part *b*; it is understood that the line diagram is one unit in the third dimension. The left side of the line diagram displays volume relationships; the right side of the line diagram displays weight relationships. Note that the volume of air is significant but contributes no weight.

2. Porosity *n* is defined as

$$n = \frac{V_v}{V_T} \times 100 \qquad (2\text{-}3)$$

Porosity is often used by agricultural and to a lesser extent by geotechnical engineers. It is expressed most often as a percentage, although it is used in engineering computations as a decimal. Inspection of Fig. 2-2 and Eq. (2-3) indicates that the range of *n* is

$$0 \le n \le 1$$

A combination of Eqs. (2-2) and (2-3) and noting $V_T = V_v + V_s$ results in

$$e = \frac{n}{1 - n} \qquad (2\text{-}4)$$

3. Water content *w* is defined as

$$w = \frac{W_w}{W_s} \times 100 \qquad (2\text{-}5)$$

This equation gives the water content as an independent variable, since W_s is a constant for a steady state soil condition. A few authorities have used a water content definition

$$w' = \frac{W_w}{W_T} = \frac{W_w}{W_s + W_w} \times 100$$

but this is a dependent equation with the weight of water present in both the numerator and denominator. This equation is not generally used by geotechnical engineers for this reason.

The range of water content is

$$0 \leq w, \text{ percent} \ll \infty$$

It is not unusual for marine and organic lake soils to have water content values up to 300–400 percent, but the natural water content for most soils is under 60 percent. Soils which appear dry often have 2 to 3 percent water content.

4. Degree of saturation S is defined as

$$S = \frac{V_w}{V_v} \times 100 \qquad (2\text{-}6)$$

This equation expresses the ratio of water present in the soil pores to the total amount which could be present if all the pores were full of water. The degree of saturation is the percentage of total void volume that contains water. Inspection of Eq. (2-6) indicates that if the soil is dry (no water), $S = 0$ percent, and if the pores are full of water, the soil is *saturated* and $S = 100$ percent; therefore, the range of S is

$$0 \leq S, \text{ percent} \leq 100$$

5. Specific gravity G_i. This uses some of the terms from Fig. 2-2. There are two definitions of specific gravity which may be used. The basic definition can be found in any physics text and computed according to the following equation:

$$G = \frac{\text{weight of unit volume of any material}}{\text{weight of unit volume of water at } 4°C} \qquad (2\text{-}7)$$

Generally, geotechnical engineers need the specific gravity of the soil grains (or solids) G_s, and this should be assumed as the value under consideration when no qualification is used. The specific gravity of the soil solids G_s is computed as

$$G_s = \frac{\gamma_s}{\gamma_w} = \frac{W_s}{V_s \gamma_w} \qquad (2\text{-}8)$$

where γ_s = unit weight of the soil solids (no pores). Typical values of G_s for soil solids are 2.65 to 2.72. The specific gravity of mercury is 13.6; of gold, 19.3, or 19.3 times heavier than water.

The bulk specific gravity is rarely used, but it can be computed as

$$G_m = \frac{\gamma_m}{\gamma_w}$$

Obtaining the specific gravity of the soil grains G_s requires finding the weight and volume of a representative portion of the soil grains and using Eq. (2-8). The exact method is in any soil mechanics laboratory manual.†

From the preceding basic definitions, many other useful relationships and/or quantities can be computed, as the following examples illustrate.

Example 2-1 Given: 1870 g of wet soil compacted into a mold with a volume of 1000 cm³. The soil is put into the oven and dried to a constant weight of 1677 g. The specific gravity G_s is assumed to be 2.66.

REQUIRED Compute the following quantities:

(a) Water content
(b) Dry unit weight γ_d
(c) Porosity n
(d) Degree of saturation S
(e) Saturated unit weight γ_{sat}

SOLUTION It is suggested that the student draw a block diagram when working problems such as this as an aid in noting what is given and what is required. Initially the block diagram is as shown in Fig. E2-1a.

Step 1 Compute the water content w:

When the soil is dried in the oven, the loss in weight is only the loss of the water which evaporates. The oven temperature is not high enough to evaporate any soil; therefore,

$$W_w = 1870 - 1677 = 193 \text{ g}$$

$$w = \frac{W_w}{W_s} = \frac{193}{1677} \times 100 = 11.5 \text{ percent}$$

† See *Engineering Properties of Soils and Their Measurement*, 2d ed., by J. E. Bowles, McGraw-Hill Book Co., New York, 1978.

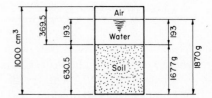

Figure E2-1a   **Figure E2-1b**

Step 2 Compute the dry unit weight:

$$\gamma_d = \frac{W_s}{V_T} = \frac{1677}{1000} \times 9.807 = 16.45 \text{ kN/m}^3$$

Note that generally the unit weight in SI should be reported to the nearest 0.01 kN/m³, which is equivalent to 0.1 lb/ft³ used in fps, and water content to no more than the nearest 0.1 percent

Step 3 Compute the porosity n:
Rearrange Eq. (2-8) to obtain

$$V_s = \frac{W_s}{G_s \gamma_w} = \frac{1677}{2.66(1)} = 630.5 \text{ cm}^3$$

The total volume was 1000; therefore, the volume of voids is

$$V_v = 1000 - 630.5 = 369.5 \text{ cm}^3$$

$$n = \frac{V_v}{V_T} = \frac{369.5}{1000} \times 100 = 36.95 \text{ percent}$$

A similar computation gives the void ratio as 0.586 (the reader should verify this value). At this point redraw the block diagram with new entries (Fig. E2-1*b*).

Step 4 Compute the degree of saturation S:
The weight of water from Step 1 was 193 g; using Eq. (2-8) rearranged,

$$V_w = \frac{W_w}{G_w \gamma_w} = \frac{193}{1(1)} = 193 \text{ cm}^3$$

From this, the reader should note that when using grams and centimeters, $V_w = W_w$.

$$S = \frac{V_w}{V_v} = \frac{193}{369.5} \times 100 = 52.2 \text{ percent}$$

Step 5 Compute the saturated unit weight γ_{sat}:
The saturated unit weight is obtained when the voids are completely full of water; thus,

$$\gamma_{sat} = \gamma_d + \text{weight of water in } V_v$$

$$\gamma_{sat} = 16.45 + \frac{(369.5)(1)(9.807)}{1000} = 20.07 \text{ kN/m}^3$$

////

Example 2-2 Redo Example 2-1 using fps units.

SOLUTION We will work the problem slightly differently to illustrate alternative methods of accomplishing the same result.

Step 1 Find the water content:

This can be worked only one way; therefore, the answer of $w = 11.5$ percent is independent of units but would most likely be computed as in Example 2-1 since most laboratories use scales calibrated in grams.

Step 2 Compute the dry unit weight:

We will use the water content and wet unit weight as an alternative computation procedure.

$$\gamma_{\text{wet}} = \gamma_d + \text{wt of water}$$

$$\gamma_{\text{wet}} = \frac{W_T}{V_T} = \frac{1870}{1000} = 1.87 \text{ g/cm}^3$$

$$\gamma_d + w\gamma_d = 1.87$$

Solving for γ_d, obtain

$$\gamma_d = \frac{1.87}{1 + 0.115}(62.4) = 104.7 \text{ lb/ft}^3$$

Step 3 The porosity would be computed exactly as in Example 2-1.

Step 4 The degree of saturation would be computed as in Example 2-1.

Step 5 Compute the saturated unit weight.

$$\gamma_{\text{sat}} = \gamma_d + \text{weight of water in filled voids}$$

$$\gamma_{\text{sat}} = 104.7 + 0.3695(62.4) = 127.7 \text{ lb/ft}^3$$

////

Example 2-3 Express the porosity in terms of the void ratio $[n = f(e)]$.

SOLUTION Referring to the block diagram of Fig. E2-3, and since V_s is not defined, let us arbitrarily assume $V_s = 1$ (as shown on the figure).

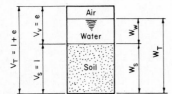

Figure E2-3

With $V_s = 1$, the void ratio is

$$e = \frac{V_v}{V_s} = \frac{V_v}{1} = V_v$$

With $V_s = 1$ and $e = V_v$, the total volume is

$$V_T = V_s + V_v = 1 + e$$

From Eq. (2-3) we have by substitution

$$n = \frac{V_v}{V_T} = \frac{e}{1 + e} \tag{2-9}$$

Note that this equation could have been obtained from Eq. (2-4) showing that there is often more than one way to obtain the desired relationship.

////

Example 2-4 Given are the following quantities:

w, water content *when S = 100 percent* (i.e., saturated condition)

G_s, specific gravity of soil solids

γ_w, unit weight of water

REQUIRED Express the dry unit weight γ_d in terms of w, G_s, and γ_w, or

$$\gamma_d = f(w, G_s, \gamma_w)$$

SOLUTION Construct a block diagram as Fig. E2-4 for $S = 100$ percent and note that

$$\gamma_d = \frac{W_s}{V_T}$$

The following steps are used to derive the required formula:

$$V_w = V_v \qquad \text{when } S = 100 \text{ percent}$$

also,

$$V_w = \frac{W_w}{G_w \gamma_w} = W_w \tag{a}$$

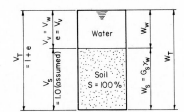

Figure E2-4

Assume $V_s = 1.0$, from which

$$e = \frac{V_v}{V_s} = \frac{V_v}{1} = V_v \qquad (b)$$

From the definition of water content and Eq. (a)

$$V_w = W_w = wW_s \qquad (c)$$

From the definition of specific gravity,

$$G_s = \frac{W_s}{V_s \gamma_w}$$

The weight of solids, with $V_s = 1.0$, is

$$W_s = G_s \gamma_w \qquad (d)$$

Now, substituting for W_s and V_T, obtain

$$\gamma_d = \frac{W_s}{V_T} = \frac{G_s \gamma_w}{V_T} = \frac{G_s \gamma_w}{1 + V_w} \qquad (e)$$

The volume of water from Eqs. (a) and (c) is

$$V_w = \frac{W_w}{G_w \gamma_w} = \frac{wG_s \gamma_w}{1(1)} = wG_s \qquad (f)$$

Now, substituting Eq. (f) into Eq. (e), we obtain the desired equation as

$$\gamma_d = \frac{G_s \gamma_w}{1 + wG_s} \qquad (2\text{-}10)$$

Note from the definition of void ratio in Eq. (b) that

$$\boxed{e = wG_s} \qquad \text{when } S = 100 \text{ percent}$$

and substitution into Eq. (2-10) gives the dry unit weight as

$$\gamma_d = \frac{G_s \gamma_w}{1 + e} \qquad (2\text{-}11)$$

$////$

Example 2-5 Given: w for any S, G_s, γ_w.

REQUIRED

(a) $e = f(w, S, G_s)$
(b) $\gamma_d = f(w, S, G_s, \gamma_w)$

SOLUTION Draw the block diagram of Fig. E2-5. The following steps are used in the derivations required.

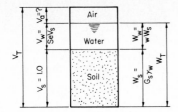

Figure E2-5

(a) For void ratio e the dry unit weight is defined as

$$\gamma_d = \frac{W_s}{V_T} \qquad (a)$$

From the definition of degree of saturation, obtain

$$V_w = SV_v \qquad (b)$$

From the definition of void ratio, obtain

$$V_v = eV_s \qquad (c)$$

Substituting (c) into (b), obtain

$$V_w = SeV_s \qquad (d)$$

From the definition of water content, obtain

$$W_w = wW_s \qquad (e)$$

Again assume $V_s = 1$ as on the block diagram and

$$W_s = 1(G_s)\gamma_w$$

Also,

$$V_w = \frac{W_w}{\gamma_w} = \frac{wW_s}{\gamma_w} = \frac{wG_s\gamma_w}{\gamma_w} \qquad (f)$$

Now equate (d) and (f) and since $V_s = 1$, obtain

$$Se = wG_s$$

and rearrange to obtain the desired equation for void ratio e as

$$e = \frac{wG_s}{S} \qquad (2\text{-}12)$$

Note that w and S are used as decimals in this equation.

(b) Find the relationship for dry unit weight γ_d. Simply substitute Eq. (2-12) into Eq. (2-11) to obtain

$$\gamma_d = \frac{G_s\gamma_w}{1 + (w/S)G_s} \qquad (2\text{-}13)$$

////

Example 2-6 Given: a *saturated* sample of soil with a specific gravity $G_s =$ 2.67 and water content data as follows:

$$\text{Weight of can + wet soil} = 150.63 \text{ g}$$

$$\text{Weight of can + dry soil} = 131.58 \text{ g}$$

$$\text{Weight of can} = 26.48 \text{ g}$$

REQUIRED

(a) Water content of soil
(b) Dry and saturated unit weights of soil (both SI and fps)

SOLUTION Construct a block diagram (Fig. E2-6) and place selected data and results on the diagram as computed.

Step 1 Compute water content from given water content data:

$$\text{Weight of dry soil} = 131.58 - \text{weight of can}$$

$$= 131.58 - 26.48 = 105.10 \text{ g}$$

$$\text{Weight of water} = 150.63 - 131.58 = 19.05 \text{ g}$$

from which the water content is computed as

$$w = \frac{19.05}{105.10} \times 100 = 18.1 \text{ percent}$$

Step 2 Find dry and saturated unit weights:
Assume $V_T = 1.0$. It may be easier to assume $V_s = 1.0$ for given data, but this is to illustrate that either volume may be assumed. With $V_T = 1.0$, the porosity n is

$$n = \frac{V_v}{V_T} = V_v$$

and the volume of solids is

$$V_s = V_T - V_v = 1 - n \qquad \text{values shown on Fig. E2-6}$$

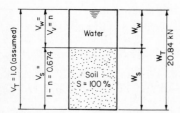

Figure E2-6

From Fig. E2-6, the definition of w, and taking $\gamma_w = 1.0$,

$$V_v = wW_s = wV_sG_s = w(1 - n)G_s$$

$$= 0.181(1 - n)2.67 = 0.483 - 0.483n$$

Since $V_T = 1 = V_v + V_s$, we have on substitution of values

$$0.483 - 0.483n + 1 - n = 1.0$$

$$n = 0.326 \text{ (units)}$$

and the volume of solids is

$$V_s = 1 - n = 1 - 0.326 = 0.674$$

from which the dry unit weight is readily computed as

$$\gamma_d = \frac{W_s}{V_T} = 0.674(2.67)(1)(9.807) = 17.65 \text{ kN/m}^3$$

$$= 0.674(2.67)(1)(62.4) = 112.3 \text{ lb/ft}^3$$

The saturated unit weight is computed as (using the definition of water content)

$$\gamma_{sat} = \gamma_d + W_w$$

$$= 17.65 + 0.181(17.65) = 20.84 \text{ kN/m}^3$$

and in fps units the saturated unit weight is

$$\gamma_{sat} = \frac{20.84}{9.807} (62.4) = 132.6 \text{ lb/ft}^3$$

////

These six examples cover several types and methods of computing volumetric-gravimetric relationships. Combinations of these methods should enable the reader to compute any other needed relationship.

2-4 NONCOHESIVE AND COHESIVE SOILS

If an inherent physical characteristic of the mass of soil grains is that on wetting and/or any subsequent drying, the soil grains stick together so that some force is required to separate them in the dry state, the soil is *cohesive*. If the soil grains fall apart after drying and stick together only when wet because of surface tension forces in the water, the soil is *cohesionless*, or noncohesive. A cohesive soil may be nonplastic, plastic, or a viscous fluid, depending on the instantaneous water content value. A cohesionless soil exhibits no demarcation line between the plastic and nonplastic states, as this type of soil is nonplastic for all ranges of water content. Under certain conditions, however, a *cohesionless* soil with a sufficiently high water content may act as a viscous fluid.

Surface tension gives cohesionless soils an *apparent cohesion,* so called because the cohesion disappears when the soil is either completely dry or saturated. From a practical standpoint, a cohesionless soil with apparent cohesion (damp to wet but not saturated) may be excavated on a vertical cut for shallow depths, or bore holes where the hole will stand open may be made. When the soil dries or becomes saturated, the apparent cohesion disappears and the sides of the cut or bore holes will collapse.

2-5 ATTERBERG (OR CONSISTENCY) LIMITS

A Swedish soil scientist, A. Atterberg, engaged in ceramics and agriculture work proposed (ca. 1911) five states of soil consistency. These limits of soil consistency are based on water content and are:

1. *Liquid limit* w_L. The water content above which the soil behaves as a viscous liquid (a soil-water mixture with no measurable shear strength). In soils engineering the liquid limit is rather arbitrarily defined as the water content at which 25 blows of the liquid limit machine closes a standard groove cut in the soil pat for a distance of 12.7 cm. Casagrande (1958) and others have modified the test as initially proposed by Atterberg so that it is less operator-subjective and more reproducible. With standard equipment various operators can reproduce liquid limit values to within 2 to 3 percent (i.e., say $w_L = 39 \pm 2$ percent and not 39×0.02). This test is considered in additional detail in Sec. 4-2.

2. *Plastic limit* w_P. The water content below which the soil no longer behaves as a plastic material. It is in the range of water contents between w_L and the plastic limit w_P that the soil behaves as a plastic material. This range is termed the *plasticity* (or *plastic*) *index* and is computed as

$$I_P = w_L - w_P \tag{2-14}$$

By definition of I_P it is impossible to obtain a negative value.

   The plastic limit is arbitrarily defined as the water content at which a thread of soil, when rolled down to a diameter of 3 mm, will just crumble. This test is more operator-subjective than the liquid limit test.

3. *Shrinkage limit* w_S. That water content, defined at degree of saturation = 100 percent, below which no further soil volume change occurs with further drying. This limit is of considerable importance in arid areas and for certain soil types which undergo considerable volume change with changes in water content. One should note that the smaller the shrinkage limit, the more susceptible a soil is to volume change—that is, the smaller the w_S, the less water is required to start the soil to change in volume. If the shrinkage limit is 5 percent, then when the in situ water content exceeds this value, the soil will begin to expand. The relative locations of w_L, w_P, and w_S on a water content scale are shown in Fig. 2-3.

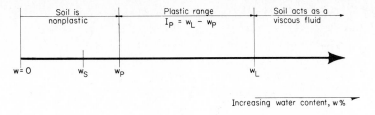

Figure 2-3 Relative locations of the plastic and liquid limits of a soil.

4. *Sticky limit.* That water content at which a soil loses its adhesive property and ceases to stick to other objects such as the hands or the smooth metal surface of a spatula blade. This limit is of some significance in agriculture and to earthwork contractors, as the drag on the plow moldboard is increased if the soil is wet enough to stick to it.
5. *Cohesion limit.* That water content at which the soil grains just cease to stick together, e.g., at which cultivation of the soil does not result in clods or lumps forming. This limit also has more significance for the agriculturist than for the soil engineer.

The liquid, plastic, and shrinkage limits are well known worldwide. The sticky limit has been used in Europe, but by and large, the sticky and cohesion limits have not been used by geotechnical engineers.

This discussion, together with the definitions for cohesive and cohesionless soils of the preceding section, indicates that the Atterberg limits are applicable only for cohesive soils.

The index properties of liquid limit and plasticity index are required for the several systems of soil classification considered in Chap. 4.

2-6 SOIL MOISTURE

Moisture or water content of a soil has previously been defined as the ratio of the weight of water in the pores of a soil to the weight of soil solids (grains). A distinction has been made between the water content determination made in the laboratory on laboratory specimens and the water content which indicates the instantaneous field value. This latter value is termed the *natural moisture* or water content of the soil and given the symbol w_N. The value of natural field moisture w_N will vary depending on the location of the soil sample, i.e., at or near the ground surface, deep in the ground, beneath a lake; recency of rainfall; etc. It should be evident that samples obtained from beneath a permanent ground water level will probably not change in water content from day to day or year to year. On the other hand, samples of soil near the ground surface or above some permenent water level will have a varying natural moisture content due to climatic factors such as temperature; amount, duration, and recency of rainfall; and length of dry periods. The maximum water content depends on the void ratio.

2-7 INDICES OF SOIL CONSISTENCY

The state or potential state of consistency of a natural soil can be established through a relationship termed the *liquidity index* I_L,

$$I_L = \frac{w_N - w_P}{I_P} \tag{2-15}$$

where w_N is the natural moisture or water content of the soil at the field site or in situ. The relationship between water content and I_L is illustrated in Fig. 2-4.

From this expression it can be seen that if

$$0 < I_L < 1$$

the soil is in the plastic range. If

$$I_L \geq 1.0$$

the soil is in a liquid state or potentially liquid. While it may be currently stable, a sudden shock or remolding may transform the mass to a liquid. Such a soil would be known as a sensitive clay. Such clay deposits exist in southwest Canada and in Scandinavia.

Another relationship occasionally used is the *index of consistency*, defined as

$$I_C = \frac{w_L - w_N}{I_P} \tag{2-16}$$

with all terms as previously defined. Both Eqs. (2-15) and (2-16) give an index value between 0 and 1 when the field moisture is between w_P and w_L. The essential differences are in the numerical values when the field soil has a natural moisture content greater than the liquid or less than the plastic limit.

The most important consistency index is the plastic index (or plasticity index) I_P, previously defined by Eq. (2-14). In general, the larger the plasticity index, the greater will be the engineering problems associated with using the soil as an engineering material, such as foundation support for residential building, road subgrades, etc.

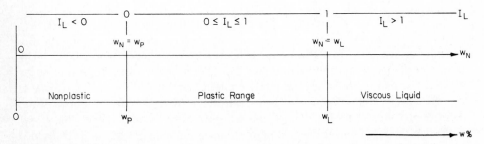

Figure 2-4 Relationship between w_P, w_L, and w_N in computing I_L.

There are many engineering correlations of soil properties and behavior associated with the plasticity index. These include soil strength parameters, horizontal in situ (or in place) earth pressure, and shrinkage-swell potential. Most of these correlations should be used cautiously; some are reasonably reliable, but many are little better than guesses.

2-8 SPECIFIC SURFACE

Specific surface relates the surface area of a material to either weight or volume of the material, with volume being generally preferred. Using this latter definition, specific surface is:

$$\text{Specific surface} = \frac{\text{surface area}}{\text{volume}} \tag{2-17}$$

Physically the significance of specific surface can be demonstrated using a $1 \times 1 \times 1$ cm cube as follows:

$$\text{Specific surface} = \frac{\text{surface area}}{\text{volume}} = \frac{6}{1} = 6$$

Now let us subdivide the cube so that each side is 0.5 cm.

$$\text{Number of cubes} = 2 \times 2 \times 2 = 8$$

$$\text{Surface area} = (0.5)^2(6)(8) = 12 \text{ cm}^2$$

and

$$\text{Specific surface} = \frac{12}{1} = 12$$

Now divide the sides by 10:

$$\text{Number of cubes} = 10 \times 10 \times 10 = 1,000$$

$$\text{Surface area} = (0.1)^2(6)(1,000) = 60 \text{ cm}^2$$

and

$$\text{Specific surface} = \frac{60}{1} = 60$$

This illustrates that large particles, whether cubes or soil particles, have smaller surface areas per unit of volume and thus smaller specific surfaces than small soil grains.

Now if sufficient water were present to just dampen the surface area in the preceding example, it would take 10 times as much water to wet the surface of all the grains when the cubes were $0.1 \times 0.1 \times 0.1$ cm as when the same volume occupied a single cube of 1 cm^3. Note also that if one were trying to remove

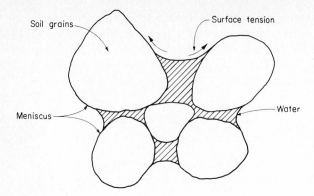

Figure 2-5 Surface tension effects in pulling soil grains together.

water from the surface-wet soil, there would be 10 times as much water to remove from the smaller grains.

This discussion illustrates that specific surface is inversely related to the grain size of a soil. We generally do not compute the value for practical cases, since naturally occurring soil grains are too irregular in shape, but one can extrapolate from the preceding paragraphs that a soil mass made up of many small particles will have a larger specific surface than the same mass made up of large particles.

From the concept of specific surface, one could expect larger moisture contents for small-grained soils than for coarse-grained soils, other things being equal; however, one must also consider the effect of grain size on the void ratios of the soils and exterior factors such as climate and location of the sample.

Specific surface is a primary factor in concrete and asphalt mix design, as in both types of design it is necessary to provide sufficient cement paste or asphalt to coat the particle surfaces.

Specific surface is a primary factor in soils subject to volume change and to the surface tension effects of water at particle interfaces, as shown in Fig. 2-5. With large numbers of particles and small sizes, the cumulative effect of surface tension of the water film in holding or pulling the particles together is very large (even though the actual amount of surface tension per unit of area is very small).

2-9 SPECIFIC GRAVITY

Specific gravity was defined by Eq. (2-8) as the ratio of a unit weight of material to the unit weight of water.

Table 2-1 lists the specific gravity of a number of minerals which are common in soils. Most soils (individual grains in aggregate) contain large amounts of quartz and feldspars and to a lesser extent mica and iron-based minerals.

Results of many specific gravity determinations on large numbers of soils indicate that values of 2.55 to 2.80 will bracket nearly all soils, with values for

Table 2-1 Typical values of specific gravity for soil minerals

Mineral	Specific gravity	Mineral	Specific gravity
Bentonite	2.13–2.18	Muscovite (mica)	2.80–2.90
Gypsum	2.30	Dolomite	2.87
Gibbsite	2.30–2.40	Aragonite	2.94
Montmorillonite	2.40	Anhydrite	3.00
Orthoclase feldspar	2.56	Biotite (mica)	3.00–3.1
Illite	2.60	Hornblende	3.00–3.47
Quartz	2.60	Augite	3.20–3.40
Kaolinite	2.60–2.63	Olivine	3.27–3.37
Chlorite	2.6 –3.0	Limonite	3.8
Plagioclase feldspar	2.62–2.76	Siderite	3.83–3.88
Talc	2.70–2.80	Hematite	4.90–5.30
Calcite	2.80–2.90	Magnetite	5.17–5.18

most soils being between 2.60 and 2.75. As a matter of fact, the specific gravity test is not often performed, and values are arbitrarily taken as follows:

Sands, gravels, coarse-grained materials $\quad G_s = 2.65–2.67$
Cohesive soils, as mixtures of clay, silt
    sand, etc. $\qquad\qquad\qquad\qquad\qquad\quad G_s = 2.68–2.72$
Clay $\qquad\qquad\qquad\qquad\qquad\qquad\qquad$ Use values in Table 2-1
$\qquad\qquad\qquad\qquad\qquad\qquad\qquad\qquad\quad$ for specific type

An indication of the computation error in the void ratio if a G_s of 2.65 is used when a true value of 2.60 would have been obtained from a laboratory test is computed as follows:

Given a soil of $\gamma_d = 1.80$ g/cm^3 (should be representative, but implies error of ± 0.005 g/cm^3).

From the definition of specific gravity,

$$G_s = \frac{W_s}{V_s \gamma_w}$$

Therefore, the volume of solids is

$$V_s = \frac{1.80}{2.60(1)} = 0.692 \text{ cm}^3 \text{ (true)}$$

$$V'_s = \frac{1.80}{2.65(1)} = 0.679 \text{ cm}^3 \text{ (assumed)}$$

The volume of voids V_v for the two cases is

$$V_v = 1 - 0.0692 = 0.308 \rightarrow e = \frac{0.308}{0.692} = 0.445$$

and

$$V'_v = 1 - 0.679 = 0.321 \rightarrow e' = \frac{0.321}{0.679} = 0.473$$

The percent increase in void ratio due to the use of the erroneous value of G_s is

$$\text{Percent} = \frac{e'}{e} \times 100 = \frac{0.473}{0.445} \times 100 = 106 \text{ percent}$$

The void ratio is 6 percent too large due to using the erroneous value 2.65. Since this solution also depends on $\gamma_d = 1.80$ being statistically correct, there may be some compensating error. If similar calculations are performed for $\gamma_d = 2.0$ and 1.5 g/cm^3, the percent increase is found to be 8.2 and 4.5 percent, respectively. Also the "true" value of 2.60 will be true only if the small quantity of soil (generally about 150 g) used to determine G_s is truly representative of the soil mass.

2-10 SOIL TEXTURE

Soil texture may be defined as the visual appearance of a soil based on a qualitative composition of soil grain sizes in a given soil mass. Large soil particles with some small particles will give a coarse-appearing or *coarse-textured* soil. A conglomeration of smaller particles will give a *medium-textured* material, and a conglomeration of fine-grained particles yields a *fine-textured* soil. It can be observed, however, that lumps of fine-grained materials will give a *coarse* texture, so we must also relate texture to the state of elemental soil particles. Figure 2-6 indicates the textural classification of several soils.

Texture based on visual appearance is often used in soil classification of cohesionless materials such as coarse sand, medium coarse sand and gravel, fine sand, etc. Texture is not used for cohesive soils, since the soil state is a factor in the texture (i.e., lumps can be pulverized), as illustrated with soils 8 and 9 of Fig. 2-6.

2-11 SOIL PHASES

The definition of phase as used by the chemist is "any homogeneous part of the material system separated from other parts by physical boundaries, as water in ice, with water vapor above." In this context it is evident that a soil mass may be a:

1. Two-phase system consisting of soil and air ($S = 0$ percent), soil and water ($S = 100$ percent), or soil and ice ($S = 100$ percent)

Figure 2-6 Soil texture of several soils.

2. Three-phase system (refer to Fig. 2-1) consisting of soil, water, and air ($0 < S < 100$ percent); soil, ice, and air ($0 < S < 100$ percent); or soil, water, and ice ($S = 100$ percent)
3. Four-phase system consisting of soil, water, ice, and air

Of course, the "air" may contain appreciable amounts of water vapor.

2-12 GRAIN SIZE

Grain size of a soil refers to the diameters of the soil particles making up the soil mass. Since a macroscopic examination of a mass of soil grains indicates that few, if any, of the particles are round and thus possess a diameter, one can conclude that this is a rather loose description of the soil.

Grain size is determined by sieving a quantity of soil through a stack of sieves of progressively smaller mesh openings from top to bottom of the stack. The quantity of soil retained on a given sieve in the stack is termed one of the grain sizes of the soil sample. Actually this operation only brackets a portion of the soil as being between two sizes, as shown in Fig. 2-7—the size indicated by the particular sieve under consideration and the size of the one immediately above it in the stack.

Figure 2-8 illustrates both the shape and size of particle ranges bracketed. All the soil particles except soil No. 2 were sieved through the No. 10 and retained on the No. 20 sieve, and they clearly show a range of particle sizes. Soil 2 is a fine beach sand from Atlantic City, New Jersey, and was retained on the No. 60 sieve.

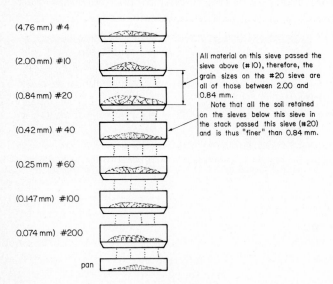

Figure 2-7 Sieve analysis of a soil. Note that the sieve arrangement is such that sieve openings decrease in size from top to bottom of the stack.

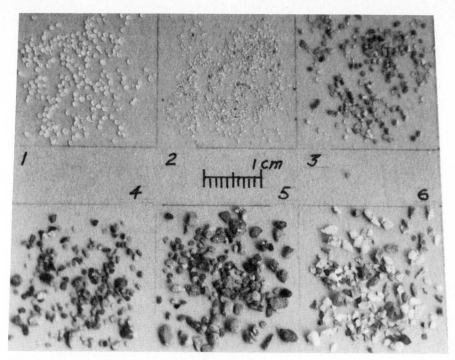

Figure 2-8 Several soils to illustrate grain size and range.

Soils 1 through 3 are transported soils and are well rounded, soil 1 being Ottawa sand and soil 3 being a sand from a river near Peoria, Illinois. Soils 4 and 5 are what is called residual (Sec. 3-11) soil and are typified by angular particles. Soil 4 is from Kentucky and soil 5 from Georgia. Soil 6 is crushed rock and is shown for comparison.

Sieve sizes range from openings of 101.6 mm (4 in) down to 0.037 mm (No. 400). All the openings are square; thus, what constitutes the diameter of a soil particle is somewhat academic since the likelihood of a particle passing a given sieve opening depends on both its size and its orientation with respect to the sieve opening.

The sieve sizes are referred to mesh openings from 101.6 mm down to 6.35 mm; then the sieves are designated by numbers (as shown in Table 4-2). Using sieves of numbers larger than No. 200 (0.074 mm)† is largely impractical, as soil can be sieved through this size mesh only with some difficulty. This mesh is fine enough to just begin to provide resistance to water flow, and soil provides considerably more resistance to passage through the mesh than does water.

To determine approximately particle diameters smaller than 0.07 mm, an analysis based on the velocity of fall of spheres through a viscous fluid (Stokes'

---

† Or closest corresponding sieve opening for other standards, as shown in Table 4-2.

law) is used. One method of using Stokes' law uses a hydrometer to measure the specific gravity of a soil-water suspension and is called hydrometer analysis.

The grain-size analysis is useful, as it helps identify such soil properties as

1. Whether a given soil can be easily drained
2. Whether the soil is suitable for use in construction projects such as dams, levees, and roads
3. Potential frost heave
4. Estimated height of capillary rise (Chap. 8)
5. Whether the soil can be used in asphalt or concrete mixes (It should be understood that the definition of "soil" includes the sand and gravel used in the manufacture of concrete.)
6. Filter design, to prevent fine-grained material from being washed out of a soil mass and lost (see Sec. 9-12)

The grain-size analysis is an important part of most of the engineering classifications of soil to be considered in Chap. 4.

2-13 UNIT WEIGHT OF SOILS (DRY, WET, AND SUBMERGED)

The basic definition of soil unit weight was given in Sec. 2-2 as weight per unit volume of material. From an inspection of Fig. 2-1, one arrives at the conclusion that, in general, unit weight is

$$\gamma = f(G_s, e, w)$$

That is, the unit weight of a soil can only depend on the weight of the individual soil grains [$f(G_s)$], the total number of soil particles present [$f(e)$], and the amount of water present in the voids [$f(w)$]. It should be kept in mind that unit weight can be altered only by changing the void ratio and/or the water content of the soil mass (since G_s = constant for a given soil mass). Strictly, the unit weight is a state vector and should include the void ratio and water content in the description; however, except for such qualifying terms as "wet" or "dry" unit weight, this is seldom done, it being commonly understood that the unit weight is for the correct soil state; i.e., the state (condition) is assumed to be a constant.

A special unit weight of considerable interest to the soils engineer is the *buoyant* (or *submerged*) *unit weight*, given the symbol γ'. Referring to Fig. 2-9, take a cube of saturated soil $1 \times 1 \times 1$ cm and weigh it. This weight is, of course, γ_{sat}, since it is the weight of a unit volume of saturated material. Now suspend the cube under water, as shown in Fig. 2-9b. The question is, What is the weight showing on the scales?

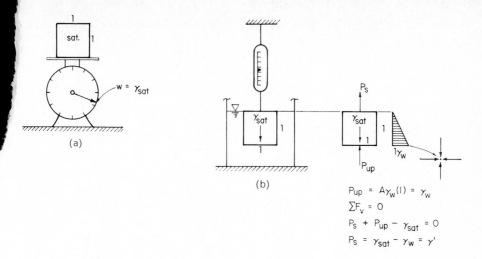

$$P_{up} = A\gamma_w(1) = \gamma_w$$
$$\Sigma F_v = 0$$
$$P_s + P_{up} - \gamma_{sat} = 0$$
$$P_s = \gamma_{sat} - \gamma_w = \gamma'$$

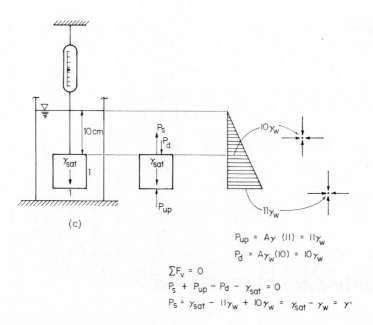

$$P_{up} = A\gamma(11) = 11\gamma_w$$
$$P_d = A\gamma_w(10) = 10\gamma_w$$

$$\Sigma F_v = 0$$
$$P_s + P_{up} - P_d - \gamma_{sat} = 0$$
$$P_s = \gamma_{sat} - 11\gamma_w + 10\gamma_w = \gamma_{sat} - \gamma_w = \gamma'$$

Figure 2-9 Development of the concept of buoyant weight of a soil. From parts *b* and *c* it can be seen that depth of submergence does not affect the submerged (buoyant) unit weight of a soil. (*a*) Unit cube of saturated soil placed on scales. (*b*) Unit cube of soil submerged from scales. Note that the soil will saturate if it is not already saturated. (*c*) Unit cube of soil submerged 10 cm to illustrate that the effect of buoyancy is independent of depth of submergence.

From a free-body analysis (note even if soil is not initially saturated it will be on submergence),

$$\sum F_v = 0$$

$$P_s + P_{up} - \gamma_{sat} = 0$$

but from basic fluid statics we obtain

$$P_{up} = \sigma A = \gamma_w(h)(A) = \gamma_w(1)(1) = \gamma_w$$

Therefore,

$$P_s + \gamma_w - \gamma_{sat} = 0$$

or

$$P_s = \gamma_{sat} - \gamma_w$$

but with a unit of volume, P_s is the unit weight of the submerged soil; therefore,

$$P_s = \gamma'$$

and the *submerged* (also called *buoyant*) *unit weight* is

$$\gamma' = \gamma_{sat} - \gamma_w \qquad (2\text{-}18)$$

The computations at the bottom of Fig. 2-9 illustrate the soil cube submerged 10 cm. This shows that the submergence effect is independent of water surface location above the soil element but assumes the submerged soil is saturated.

The submerged unit weight of a soil has particular importance in soil mechanics stability computations as shown in the next section and in later chapters.

2-14 EFFECTIVE OR INTERGRANULAR PRESSURES

Figure 2-10a illustrates a soil mass in situ. The soil skeleton is supported both vertically (and laterally) by the intergrain contacts. The pore fluid–air for dry soils—may affect the contact stresses in varying amounts. If the pore fluid is air at atmospheric pressure, the effect is negligible since the shear resistance is so small. A liquid pore fluid may exert substantial effects, which will be examined in the following paragraphs. Commonly the pore liquid is water but the following principles apply for any fluid.

The grain-to-grain (or intergranular) contact pressure which balances the vertical load P_t above "plane" A-A in Fig. 2-10b is termed the *effective* pressure. It is this pressure which develops a friction resistance F_f to particle movements such as rolling, slipping, sliding, etc., as in any friction analysis.

$$F_f = vN$$

where v = coefficient of friction between materials (particles)

N = normal contact force (or stress if F_f is stress)

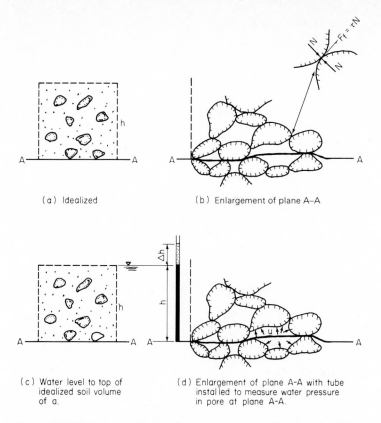

(a) Idealized

(b) Enlargement of plane A–A

(c) Water level to top of
idealized soil volume
of a.

(d) Enlargement of plane A-A with tube
installed to measure water pressure
in pore at plane A-A.

Figure 2-10 Idealization of soil volume to develop concept of effective stress and effect of pore pressure u on effective stress.

Friction resistance is usually the significant factor in mass stability and the ability of a soil to carry foundation loads.

In the case of a flat steel block of weight P and contact area A on top of a second block the normal effective stress is distributed evenly and is directly computed as

$$\sigma' = \frac{P}{A}$$

Figure 2-10b is an enlargement of the horizontal plane A-A in Fig. 2-10a. This plane is taken such that it intersects the nearest grain contacts. The effective stress σ' is computed *nominally* from Fig. 2-10a as

$$\sigma' = \frac{P_t}{A} = \gamma_{\text{soil}} \cdot \frac{h}{A}$$

Strictly, the intergranular stress is computed as

$$\sigma'_a = \frac{P_t}{A_c}$$

where A_c is the sum of all of the grain contact areas across "plane" A-A.

It is evident that if the grains were spheres the contact areas are points so that A_c (as a sum) would be a very small value and the σ'_a stresses would be higher than the nominal σ' stresses. If P_t is sufficiently large, some of the point contacts are undoubtedly crushed. Fortunately, it is not necessary to evaluate σ'_a since the shear stress can be evaluated using the nominal value and an adjusted value for the friction coefficient (which includes any grain crushing). As long as the product of friction coefficient and nominal stress-force describes the measured shear resistance, or predicts it adequately on the basis of past practice, a practical solution is obtained.

Now what is the effective stress if the unit of soil is placed under water as shown with the dashed line in Fig. 2-10c? It is evident that when the water level stabilizes as shown, all the interconnected soil voids will be full of water. This is a saturated soil state and now

$$P_t = \gamma_{\text{sat}} \cdot h$$

This value of P_t is resisted by a combination of pore fluid pressure u and effective stress σ' as in Fig. 2-10d; thus

$$\sigma'A + uA_w = P_t$$

where u = pore fluid pressure = $\gamma_f h$ (for water, use $9.807h$)

A_w = projected area of pores and neglecting the small variations in elevation along plane A-A

The effective pressure is obtained as

$$\sigma' = \frac{P_t}{A} - \frac{uA_w}{A}$$

Define P_t/A as the "total" pressure acting on plane A-A = σ_t; also note that the area of water A_w is related to the cross-sectional area A as

$$A_w + A_c = A$$

Solving for A_w and substituting in the equation for σ', we obtain

$$\sigma' = \sigma_t - u\left(1 - \frac{A_c}{A}\right) \tag{2-19}$$

If we assume that A_c is very small, for practical purposes we have

$$\sigma' = \sigma_t - u \tag{2-20}$$

which is one of the most critical concepts in geotechnical engineering. Referring to Fig. 2-10*d*, we see that raising the water level by Δh produces a new pore pressure u_i of

$$u_i = h\gamma_w + \Delta h\gamma_w = u + \Delta u$$

Thus at any instant the effective stress is

$$\sigma' = \sigma_t - (u + \Delta u) \qquad (2\text{-}20a)$$

If the friction resistance is $\tau = F_f/A$ (nominal stress), we have

$$\tau = v\sigma' = v[\sigma_t - (u + \Delta u)]$$

Any number of combinations of σ_t and pore pressure change Δu can reduce σ' to very nearly zero. A value of $\sigma' < 0$ is meaningless since the contact stress is defined as compressive. A value of $\sigma' < 0$ would be a tension state at the grain interface, and while some soils may be able to develop very small tension stresses, most cannot, and it is usual to assume no tension; hence $\sigma' \geq 0$.

Referring again to Fig. 2-10*c* and with $A = 1$ unit^2, the effective stress is

$$\sigma' = \frac{\gamma_{\text{sat}} hA}{A} - \frac{\gamma_w hA}{A} = h(\gamma_{\text{sat}} - \gamma_w)$$

In geotechnical literature the symbol γ' of Eq. (2-18) is used for the unit weight difference shown above in parentheses or

$$\gamma' = \gamma_{\text{sat}} - \gamma_w$$

This value is termed the "effective" or buoyant unit weight of the soil. The effective unit weight can also be derived from fluid statics as shown in Fig. 2-11*a*. The hypothetical volume of soil as shown (note that the use of a height of h

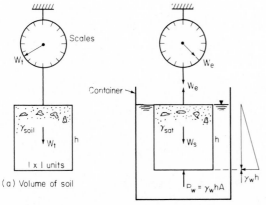

(a) Volume of soil

(b) Volume of soil submerged as shown

Figure 2-11 Concepts used to develop expression for submerged unit weight. It would not be easy to produce a volume of soil in the geometric shape shown in any practical case.

allows any number of unit cubes to be stacked (i.e., $h = 0.5, 1, 2, 10$, etc.). In air the scales show a weight of

$$W_t = \gamma_{\text{soil}} hA \qquad [\text{lb, kilopounds (k), or kN}]$$

and in water the effective weight W_e can be obtained from the free-body diagram of Fig. 2-11b as

$$W_e - W_s + P_w = 0$$

and substituting values (from the figure) and noting again that the soil saturates on submergence obtain

$$W_e = \gamma_{\text{sat}} hA - \gamma_w hA$$

and collecting terms

$$W_e = (\gamma_{\text{sat}} - \gamma_w)hA = \gamma' hA$$

and the effective pressure on releasing the soil from the scales and letting it rest on the bottom of the container is

$$\sigma' = \frac{W_e}{A} = \frac{\gamma' hA}{A} = \gamma' h$$

This illustrates that unit weight multiplied by depth always produces stress. Observe that h and γ must be properly interpreted and in most situations the stress will be a summation of several h and γ values (as in Example 2-7).

These several concepts will be illustrated by the following three examples.

Example 2-7 Given the soil profile shown in Fig. E2-7.

REQUIRED What are the total and effective pressures at point X?
Display results on a pressure profile.

SOLUTION

Step 1 Find γ_{dry} and γ_{sat} of the sand. Let $V_T = 1.0$, from which

$$n = V_v$$

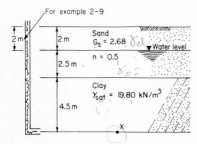

Figure E2-7a

The volume of solids is

$$V_s = 1 - V_v = 1 - n$$

The weight of solids is

$$W_s = G_s V_s \gamma_w$$

Substituting $1 - n$ for V_s, obtain

$$W_s = 2.68(1 - 0.5)(9.807) = 13.14 \text{ kN/m}^3$$

But $\gamma_{dry} = W_s$ since $V_T = 1$. The saturated unit weight is

$$\gamma_{sat} = \gamma_{dry} + W_w$$

$$\gamma_{sat} = 13.14 + 0.5(9.807) = 18.04 \text{ kN/m}^3$$

The total pressure σ is (also shown in Fig. E2-7b)

$$2(13.14) = 26.28$$
$$+ 2.5(18.04) = 45.10$$
$$+ 4.5(19.80) = 89.10$$
$$\overline{\sigma_t = 160.48 \text{ kPa}}$$

The effective pressure σ' is

$$\sigma' = 160.48 - 9.807(7) = 91.83 \text{ kPa}$$

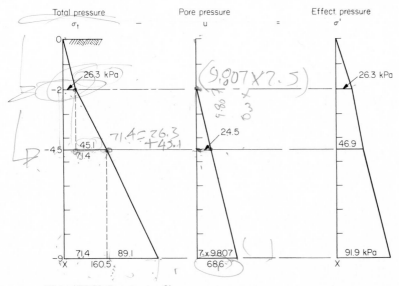

Figure E2-7b Pressure profiles.

The effective pressure can also be computed as follows:

$$2(13.14) = 26.28$$

$$+2.5(18.04 - 9.807) = 20.58$$

$$+4.5(19.80 - 9.807) = 44.97$$

$$\sigma' = \overline{91.83} \text{ kPa} \quad \text{(checks)}$$

Note that values are rounded on Fig. 2-7b.

////

Example 2-8 Redo Example 2-7 if the water level rises to the ground surface.

SOLUTION When the water surface rises soil A saturates to the unit weight of soil B, and the total pressure at point X is

$$\sigma_t = 4.5 \times 18.04 = 81.18 \text{ kPa}$$

$$+4.5 \times 19.80 = 89.10$$

$$\text{Total} = 170.3 \text{ kPa}$$

The pore pressure u from a "head" of 9 m is

$$u = 9 \times 9.807 = 88.3 \text{ kPa}$$

$$\sigma' = 170.3 - 88.3 = 82.0 \text{ kPa}$$

////

Example 2-9 Redo Example 2-7 if water in the pressure tube (piezometer) rises 2 m while the ground water table remains as shown. Assume that the excess pore pressure varies from zero at the top of the clay layer to 2 m at the tube tip according to

$$\Delta u = 2\gamma_w \sin \frac{\pi y}{2 \times 4.5}$$

(*Note:* This is a state which can occur in the field from applying a fill over point X at a fairly rapid rate; pore drainage from pressure gradient occurs more rapidly near the sand interface than at a greater depth to produce the sine wave pressure profile.)

SOLUTION This situation fits Eq. (2-20a) (refer to Fig. E2-7a):

$$\sigma' = \sigma_t - (u + \Delta u)$$

This problem can be solved for the effective pressure at point A directly to obtain $\sigma' = 160.5 - (68.6 + 2 \times 9.807) = 72.3$ kPa using data shown at point

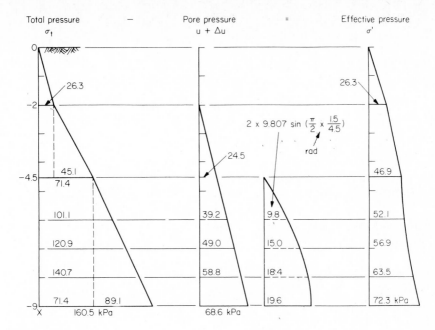

Figure E2-9

X from Fig. 2-7b. If it is desired to obtain other values of effective pressure within the clay layer, it will be useful to reproduce the total and pore pressure profiles in Fig. 2-7b and add the pressure profile for the excess pore pressure Δu to produce the profiles shown in Fig. E2-9.

////

2-15 INTERGRANULAR PRESSURES IN PARTIALLY SATURATED SOILS

The preceding section has shown that intergranular pressures for saturated soils can be computed by using Eq. (2-20). Considerable engineering experience to date indicates that this equation is sufficiently accurate for most geotechnical work. When the soil is partially saturated, laboratory tests indicate that Eq. (2-20) can be considerably in error (Skempton, 1961). A more general expression can be written as

$$\sigma' = \sigma - [u_a - \psi(u_a - u_w)] \tag{2-21}$$

where u_a = pore air pressure

ψ = parameter relating to degree of saturation; it is 1 when $S = 100$ percent and must be experimentally determined for $S < 100$ percent

Other terms have been previously defined.

Equation (2-21) is little better than an estimate at the present level of technology. This is because at $S < 100$ percent, the actual pore water distribution and the resulting pore pressures are indeterminate. In small laboratory samples the distribution may be less complex than in situ. For this reason the only reliable means of obtaining in situ pore pressures is by use of a piezometer system.

2-16 SUMMARY

This chapter has introduced several fundamental definitions of soil mechanics which the student should memorize: *void ratio, porosity, water content,* and *degree of saturation.* The student should already be familiar with the concept of *unit weight* and *specific gravity* relationships from physics courses. All the weight-volume relationships needed in soil mechanics can be derived from appropriate combinations of these six quantities. The assumption of a volume = 1.0 for either the volume of solids V_s or total volume V_T of a soil mass, as used in Examples 2-3 and 2-7, is a convenience, and the user should verify that this is not necessary.

The reader should form the habit of deriving any needed weight-volume relationships using the fundamental definitions rather than searching the literature (or this text) for a "formula" which fits the given data.

The physical significance of the liquid and plastic limits, as well as the plastic index relationship, should be studied to obtain a careful understanding. Note particularly that there are both a definition of the liquid and plastic limits and the arbitrary means of obtaining them. Be especially sure you understand the difference.

The factor 9.807 should be memorized since it is so commonly used in converting mass density to unit weight. Note the number of significant digits used in reporting unit weight (0.01 in SI and 0.1 in fps) and water content (0.1).

The concept of effective pressure is one of the most important factors in stability analysis in geotechnical work. The role of pore water, or excess pore water, pressures in developing effective pressure should be thoroughly understood. A very large number of soil failures have been, and still are, caused by buildup of excess pore pressures. This chapter has considered excess pore pressure in its simplest mode—a column of water in a piezometer above the point of interest. Later chapters will examine the mechanisms which might produce this column of water other than physically pouring water down the piezometer tube.

PROBLEMS

2-1 Define the following terms:
   (a) Void ratio
   (b) Porosity
   (c) Water content
   (d) Specific gravity of a soil
   (e) Unit weight of a soil
   (f) Wet unit weight of a soil
   (g) Dry unit weight of a soil

2-2 What is soil *texture* and what is its significance?

2-3 Redo Examples 2-3 and 2-4 without assuming $V_s = 1$.

2-4 A 110.0-cm³ volume of clay weighs 145.0 g in its undisturbed state. When the clay specimen is dried it weighs 110.5 g. What is the natural water content of the clay, and what is the degree of saturation? Assume $G_s = 2.70$.

    Partial Ans: S = 49.9 percent.

2-5 The dry density of a soil is 1.78 g/cm³. It is determined that the void ratio is 0.55. What is the wet unit weight if $S = 50$ percent? If $S = 100$ percent? What would the unit weight be if the voids were filled with oil of $G = 0.9$?

    Partial Ans: $\gamma_{wet} = 19.20$ kN/m³ at $S = 50$ percent.

2-6 The moisture content of a *saturated* clay is 160.0 percent. The specific gravity of the soil solids G_s is 2.40. What are the wet and dry unit weights of the saturated clay?

2-7 Derive an expression for $\gamma_{dry} = f(G_s, n, \gamma_w)$.

    Ans: $\gamma_{dry} = G_s \gamma_w (1 - n)$

2-8 Derive an expression for $\gamma_{sat} = f(G_s, e, \gamma_w)$.

2-9 Derive an expression for $e = f(\gamma_{sat}, \gamma_w, w)$.

2-10 Plot a curve of $n = f(e)$. Comment on the curve and establish the valid range of the curve.

2-11 Prove that $e = wG_s$ when and only when $S = 100$ percent.

2-12 The following data were obtained from a liquid limit (w_L) test:

No. of blows	17	22	28	34
Water content w, percent	63.8	63.1	60.6	60.5

Two-plastic limit (w_p) tests gave values of 28.3 percent and 29.2 percent, respectively. The natural water content of the soil is 78 percent. Required are:

    (*a*) Liquid and plastic limits

    (*b*) Liquidity index I_L and consistency index I_c

    (*c*) Appropriate comments on the soil in the natural state

2-13 A saturated sample of inorganic clay has a volume of 22.4 cm³ and weighs 36.7 g. After drying at 105°C to constant weight, the volume is found to be 14.0 cm³. The weight of dry soil is 23.2 g. For the soil in its *natural* state, find:

    (*a*) Water content w, percent

    (*b*) Specific gravity, G_s

    (*c*) Void ratio e

    (*d*) Saturated unit weight γ_{sat}

    (*e*) Dry unit weight γ_d

    (*f*) Shrinkage limit w_s

2-14 A soil specimen with a volume of 60.0 cm³ weighs 105 g. Its dry weight is 80.2 g, and G_s is 2.65. Compute:

    (*a*) Water content w, percent

    (*b*) Void ratio e and porosity n

    (*c*) *Mass* specific gravity G_m

    (*d*) Degree of saturation S, percent

2-15 A soil (medium sand) has a damp density of 1.81 g/cm³. The moisture content was found to be 16.1 percent. Find:

    (*a*) Void ratio e

    (*b*) Dry density γ_d

    (*c*) Degree of saturation S

2-16 Given the soil profile shown in Fig. P2-16, find:

    (*a*) Intergranular pressure at point A

    (*b*) Effective and total pressure at point A if the water table drops to A and the water content averages 10 percent

    (*c*) Values found in (*b*) in fps units

    Ans: 2.75 ksf

$$W = \frac{W_w}{W_s}$$

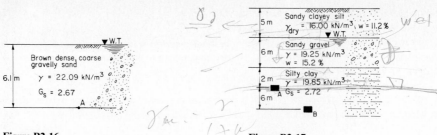

Figure P2-16

Figure P2-17

2-17 Given the soil profile of Fig. P2-17, find:

    (a) Total and effective soil pressures at point A for the water table shown

    (b) Total and effective soil pressures at point B if the water table drops to point A with an average water content of 15 percent for the silty clay above the water table

2-18 Given the soil profile of Fig. P2-18, compute the total and effective soil pressures at the clay-rock interface.

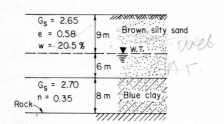

Figure P2-18

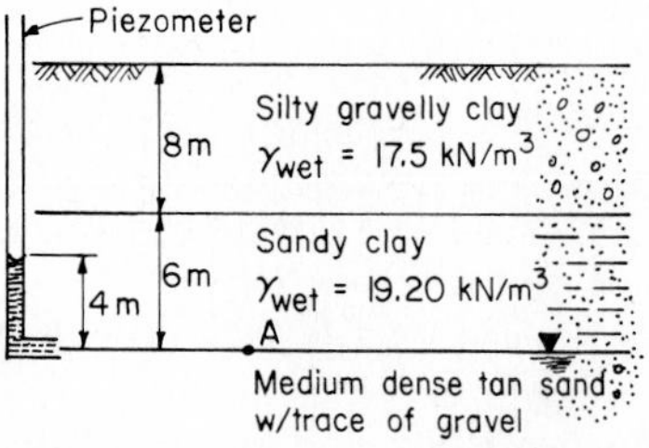

Figure P2-19

2-19 Given the soil profile of Fig. P2-19. Required:

    (a) Effective pressure at A

    Ans: 215.98 kPa

    (b) Effective pressure at A if the piezometric head drops 0.5 m

    (c) Piezometric head needed to make the effective pressure at $A = 0.0$

    Ans: (c) 26.02 m

2-20 A saturated soil has $w = 45$ percent and $e = 1.21$. Find:

    (a) Specific gravity G_s

    (b) Saturated unit weight γ

    Ans: $G_s = 2.69$; $\gamma = 17.31$ kN/m³

2-21 A sand has a porosity of 45 percent and $G_s = 2.65$. Find:

    (a) e

    (b) γ_d and γ_{sat}

    (c) γ_{wet} if $S = 30$ percent

    Ans: (a) $e = 0.818$; (c) $\gamma_{30} = 15.62$ kN/m³

2-22 A 1-m³ volume of dry soil has a mass of 1.7 tons (metric). The water content is 20 percent, and $G_s = 2.70$. Find:

    (a) e and n

    (b) S

    (c) γ in both kN/m³ and lb/ft³

    Ans: $e = 0.587$; $n = 0.37$

2-23 A 580-cm^3 volume of soil weighs 1100 g. Its dry weight is 1010 g. $G_s = 2.67$. Find:

(a) e, n, w

(b) γ both wet and dry (in both SI and fps)

Ans: (a) 0.533, 34.8 percent, 8.9 percent; (b) $\gamma_{dry} = 17.07$ kN/m^3 or 108.6 lb/ft^3

2-24 Given $G, w, S = 100$ percent, derive:

(a) $n = f(w, G)$

(b) $e = f(w, G)$

(c) $\gamma_{sat} = f(G, w, \gamma_w)$

THREE

GEOLOGIC PROPERTIES, FORMATIONS OF NATURAL SOIL DEPOSITS, AND GROUNDWATER

3-1 INTRODUCTION

Soil may be defined for engineering purposes as "the unconsolidated material above solid rock." Within this, we may particularly distinguish *topsoil*, the top 0.01 to 0.5 m of unconsolidated material, which contains organic materials and plant nutrients and which supports plant life. Topsoil is of particular interest to the agricultural engineer.

Soil is the deposited byproducts from weathering of the rock crust and/or rocks suspended, or exposed, in the soil matrix. Since the unconsolidated soil material constitutes such a large part of the earth's surface, both on the continents and beneath the oceans, lakes, and other water-covered areas, few engineering projects except perhaps rock tunneling operations can be conducted without encountering some type of soil. As it is generally impractical to carry building foundations to rock below the soil mantle, the founding of foundation structures in or on soil is one of the most important aspects of geotechnical engineering.

This chapter will be concerned with the geologic aspects of soil formation in terms of landforms, soil formation, and groundwater. The reader is encouraged to supplement the following material by obtaining a textbook of geology, preferably one oriented to engineering geology, and carefully studying it. In any case the reader should make a habit of observing exposed soil and rock formations when traveling highways, when hiking, when on surveying projects, or whenever there are other visual opportunities.

This chapter will present a brief introduction to the geology of rocks, including some additional detail of the several rock groups. This will be followed by a discussion of soil formation via rock weathering. The geologic factors in the formation of the several types of soil deposits will be considered, and finally, a discussion of groundwater will be presented.

3-2 THE EARTH

According to generally accepted theories, the earth was formed about 4.5 billion years ago from a huge molten ball of cosmic gases and debris. The cooling of this mass formed the *atmosphere*, *hydrosphere*, and *lithosphere*. The atmosphere is the gaseous envelope surrounding the hydrosphere, or zone of water (as in the ocean basins and lakes), and the lithosphere, or earth's crust and interior mass.

The earth's crust consists of both rock and weathered rock (as soil) and is generally believed to extend downward 10 to 15 kilometers (km) or more, as shown in Fig. 3-1. Figure 3-1 also illustrates several other facts generally accepted by geologists.

The principal elements which make up the earth's *outer crust* are approximately as follows:

Element	Symbol	Percent by weight	Percent by volume
Oxygen	O	46.6	93.8
Silicon	Si	27.7	0.9
Aluminum	Al	8.1	0.5
Iron	Fe	5.0	0.4
Magnesium	Mg	2.1	0.3
Calcium	Ca	3.6	1.0
Sodium	Na	2.8	1.3
Potassium	K	2.6	1.8

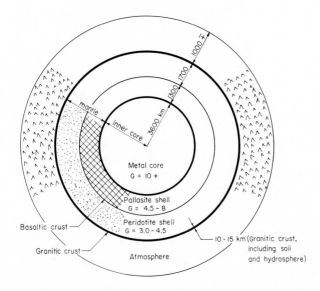

Figure 3-1 The earth, including atmosphere, with approximate dimensions.

These elements seldom exist alone but rather occur in combination, as *minerals*. The principal minerals which tend to be exposed to weathering to produce soil at or near the earth's surface are:

Mineral	Approximate percentage
Feldspar 　　Orthoclase $[K(Al)Si_3O_8]$—pink, white, and gray-to-green 　　Plagioclase $[Na(Al)Si_3O_8]$—white, gray, green, and red; and 　　　may contain Ca instead of Na	30
Quartz (SiO_2, or silicon dioxide)	28
Clay minerals (see also Chap. 6) and micas 　　Muscovite $[K(Al_2)Si_3Al(O_{10})(OH)_2]$—light-colored mineral 　　Biotite $[K_2(Mg, Fe)_6(SiAl)_8O_{20}(OH)_4]$—black, brown, or green color	18
Calcite (as $CaCO_3$) or dolomite [as $CaMg(CO_3)_2$]	9
Iron oxides 　　Hematite (Fe_2O_3)—red shades 　　Limonite ($2Fe_2O_3 \cdot 3H_2O$)—varying shades of yellow	4
Pyroxene and amphibole 　　Pyroxene—Calcium, magnesium, iron, and aluminum silicate 　　Amphibole (hornblende)—sodium, calcium, magnesium, iron, and 　　　aluminum silicate	1
Others including 　　Kaolinite (clay)—hydrous aluminum silicate $[Al_2Si_2O_5(OH)_4]$ as a 　　　principal weathering byproduct of feldspar 　　Olivine (greenish color)—magnesium, iron silicate $[(MgFe)_2SiO_4]$	10

3-3 PHYSICAL PROPERTIES OF MINERALS

The physical properties especially useful in mineral identification are:

Hardness—what materials a mineral will scratch and what materials will in turn scratch it

Color—green, white, colorless, gray, etc.

Streak—the color of the line of mineral powder formed when the surface is scratched with a hard object

Luster—the appearance of a freshly broken surface as seen in reflected light (bright, greasy, shiny, metallic, dull, etc.)

Specific gravity—related to the weight of a quantity of mineral

Cleavage—breaking along defined planes

Fracture—breaking along irregular fracture lines

The Mohs hardness scale is used as a basis for evaluating the hardness of minerals as follows, in order of increasing hardness:

1. Talc (softest)
2. Gypsum
3. Calcite
4. Fluorite
5. Apatite

6. Feldspar
7. Quartz
8. Topaz
9. Corundum
10. Diamond (hardest)

Any mineral in the hardness scale will scratch the minerals below it, i.e., diamond will scratch all nine minerals below it. Hardness kits are available containing small specimens of the ten minerals in the Mohs hardness scale. In lieu of the kit the following may be used:

	Hardness
Fingernail	$2\frac{1}{2}$ (will scratch talc and gypsum)
Copper penny	3
Glass	5–$5\frac{1}{2}$ (scratch apatite to talc)
Knife blade	$5\frac{1}{2}$–6 (may scratch feldspar)
Steel file	$6\frac{1}{2}$–7 (scratch feldspar to talc)

Figure 3-2 illustrates three of the more common minerals.

3-4 THE ROCK AND SOIL CYCLE

Geologists classify all rocks into three basic groups: *igneous*, *sedimentary*, and *metamorphic*. Rocks are mixtures of several minerals or compounds, and vary greatly in composition. Limestone, for example, is primarily calcite, whereas granite contains feldspars, quartz, and varying amounts of ferromagnesians.

The approximately 1 billion years of documented geologic history (Table 3-1) indicates that the earth is continually changing. Weathering processes aided by crustal deformities (hills, valleys, etc.) reduce the solid rock(s) to fragments, creating *residual* soils, or in-place products of rock weathering. Initially the weathering process was applied to igneous rocks and/or deposits of mineral precipitates formed during the cooling of the molten rock. Gravity through sliding and creep, moving water as surface runoff, or wind and ice action may transport these weathered rock byproducts to new locations, producing sediments, or *transported* soil deposits.

These sedimented deposits through geologic time became indurated by consolidation due to the weight of the overlying sediments and/or cementation into *sedimentary* rocks. Much of the sedimentation took place in marine environments, so that considerable calcium, sodium, and magnesium salts (as carbonates, sulfates, chlorides, etc.) were present both as solution precipitates and from shell life which provided both sediments and cementing agents. Uplifts and other crustal movements either allowed additional sedimentation and induration pressures or exposed the sediments, along with underlying igneous and sedimentary rocks, to new weathering.

Figure 3-2 Several of the more commonly occurring minerals. Paper-clip-size scale is 2.3 cm. (*a*) Orthoclase (pink) feldspar; (*b*) plagioclase (white) feldspar; (*c*) quartz—the white piece on the left is from a granite intrusion with the width (vertical) shown; the piece on the right is pink, found in a stream bed.

Where crustal movements caused increased overburden pressures and heat via energy dissipation and via cracks in the crust which allowed molten magma to flow close, some of the sedimentary (and some igneous) rocks metamorphosed into *metamorphic* rocks. Later crustal movements have exposed some of these rocks to renewed weathering and in some cases, at sufficient depth and geologic conditions, have returned them to molten magma to start the cycle anew. Figure 3-3 illustrates the rock-soil cycle just described.

The earth's crust consists of approximately 95 percent igneous rock and only about 5 percent sedimentary and metamorphic rocks. However, of the rocks exposed to weathering at the surface, 75 percent are sedimentary rocks, and of these some 22 percent consist of limestones and dolomites. In order of area covered, the most important rocks (ranked in order of potential geotechnical

Table 3-1 Geological time scale

Relative geologic time					
Era	Period		Epoch	Time B.P.*	Type of life
Cenozoic	Quaternary Neogene		Holocene		Man
			Pleistocene		
				2–3	
			Pliocene		
				12	
	Tertiary		Miocene		
				26	
		Paleogene	Oligocene		
				37–38	
			Eocene		
				53–54	
			Paleocene		
				65	
Mesozoic	Cretaceous		Late		Mastodons
			Early		
				136	
	Jurassic		Late		Mammals
			Middle		
			Early		
				190–195	
	Triassic		Late		Dinosaurs
			Middle		
			Early		
				225	
Palezoic	Permian		Late		
			Early		
				280	
	Carbon-iferous	Pennsylvanian	Late		Coal-forming Swamps
			Middle		
			Early		
		Mississippian	Late		
			Early		
				345	
	Devonian		Late		
			Middle		
			Early		
				395	
	Silurian		Late		
			Middle		
			Early		
				430–440	
	Ordovician		Late		Fish
			Middle		
			Early		
				500	
	Cambrian		Late		
			Middle		
			Early		
				570	
Precambrian				3600+	

* Estimated time *before* the *present* (B.P.), millions of years.

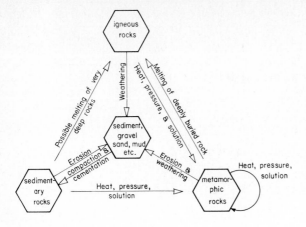

Figure 3-3 The rock-soil cycle.

problems) and their percentages are as follows:

1. Shales   52 percent   4. Granites   15 percent
2. Limestones and   5. Basalt   3 percent
   dolomites   7 percent
3. Sandstones   15 percent   6. All other rocks 8 percent

Available geologic evidence indicates that the sedimentary record is on the order of 5000 to 6000 m in depth. That is, sufficient weathering has taken place to place a depth of sediments of this thickness over much of the earth's surface. Had uplift and other crustal movements not taken place, this depth would have reduced the earth's surface to such an extent that a sheet of water would cover the entire surface. Much of the early sediments has long ago reformed into sedimentary rocks, so that the unconsolidated material is of much lesser thickness, generally well under 600 m.

3-5 IGNEOUS ROCKS

Igneous rocks are those rocks formed by the cooling of molten magma. Most of the magma now existing is at a considerable depth below the earth's crust, as qualitatively illustrated in Fig. 3-1, except in the active volcanic areas such as Yellowstone National Park, Wyoming; Hawaii; and Japan. As periodic stress adjustments produce cracks and faults in the rock crust, magma may find a path either part way (producing hot springs and geysers under certain conditions) or in some cases all the way to the surface (producing volcanoes). Part-way flows into the crust form intrusive or plutonic rocks as illustrated in Fig. 3-4; this figure also gives the geological terms associated with these types of rock intrusions. Figure 3-5 illustrates two cases of later erosion exposing these rock intrusions.

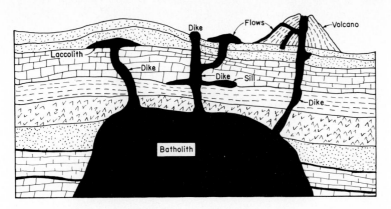

Figure 3-4 Igneous rock intrusions and extrusions.

Igneous rocks are classified according to texture, composition, color, and source. Several igneous rocks are:

Coarse-grained	Fine-grained	Lava rocks
Granite—light-colored	Rhyolite—light-colored	Obsidian—black and glassy
Diorite—intermediate color	Basalt—dark-colored	Pumice—light, frothy, and glassy
Gabbro—dark-colored		Scoria—reddish to black with large voids

Figure 3-6 illustrates several of the more commonly occurring igneous rocks. Granite tends to be the predominant igneous rock; it is best known for its use as a building and monument stone. It ranges from grayish white to medium red, is rich in quartz, and tends to be intermixed with darker grains of mica and hornblende. A large range of grain sizes exists from relatively fine to very coarse-grained visible mineral crystals in the rock matrix. If the rock is very coarse-grained, it may also be called a porphyry, and if the crystals are abnormally large, the rock is termed a *pegmatite*. Rhyolite is essentially a granite with a fine-grained structure. The rate of magma cooling determines the size of the crystal structure—coarseness is due to slow cooling and fineness to rapid cooling.

Extrusive igneous rocks are formed when the molten rock hardens after reaching the surface. The most common extrusives are formed during volcanic eruptions, which, in addition to liquid lava, eject solid particles of volcanic ash and rock fragments termed *bombs*. The crystal structure of extrusive rocks tends to be fine-grained as a result of the rapid cooling. Some volcanic rocks may be quite porous (pumice and scoria) because they solidified while steam and other gases were still bubbling out. Figure 3-6 also illustrates samples of several extrusive igneous rocks.

(a)

(b)

Figure 3-5 Igneous rock intrusions which have been exposed by erosion: (*a*) dike on the order of 15 m wide by about 10 km in length; (*b*) Stone Mountain, Georgia; large granite laccolith (light-colored); weathering has made surface dark gray.

(a)

(b)

(c)

(d)

(e)

Figure 3-6 Several of the more common igneous rocks. Paper clip scale is 2.3 cm: (*a*) light gray granite from Stone Mountain, Georgia; (*b*) black fine-grained basalt; (*c*) granites (porphyrys)—upper is red, left is pink, lower right is mottled; (*d*) rock with seams of obsidian; (*e*) lava rock. Note coarse-grained, porous structure.

Basalt, a fine-grained, dark-colored mineral aggregation often called *trap rock*, is one of the most abundant extrusive rocks. Basalts are rich in ferromagnesium minerals and are typically dark gray, dark green, brown, or black. It is a very hard, fine-grained rock, and if crushed is excellent for road construction.

Obsidian is a lustrous, glassy black to reddish brown, extremely fine-grained (there is actually no visible grain structure) rock formed by rapid cooling of molten lava. Pumice is a porous, light-colored rock with such a low mass that it may float on water. It is primarily glass and is formed as frothy lava is thrown into the air during a volcanic eruption.

3-6 SEDIMENTARY ROCKS

Rocks exposed through the earth's soil mantle are especially vulnerable to the agents of weathering. Weathering reduces the rock mass to fragmented particles which can be more easily transported by wind, water, and ice. When dropped by the agents of transportation, they are termed *sediments*. Sediments are typically deposited in layers or beds termed *strata*, and when compacted and cemented together (a process called *lithification*), they form sedimentary rocks. These rocks, of which the most common are shale, sandstone, and limestone, make up about 75 percent of the rocks exposed at the earth's surface.

Sedimentary rocks are generally classified as *clastic* or *chemical*. Clastic rocks are formed from rock grains of varying size. Typical clastic rocks include:

Shale. This is the most abundant of the sedimentary rocks. It is formed from silts and clays which have hardened into rock, with the principal induration agent being pressure. Shale may be *arenaceous*, with large amounts of sand; *argillaceous*, with large amounts of clay; *carbonaceous*, with large amounts of organic matter; or *calcareous*, with large amounts of lime as from shell life. Calcareous shale is used in the manufacture of portland cement, and carbonaceous shale may yield petroleum or coal. Shale may also be called claystone or siltstone based on the primary constituents.

Sandstone. This rock is composed essentially of pressure-cemented grains of sand (quartz). Sandstone may also contain grains of calcite, gypsum, feldspar, or iron compounds. Sandstone is used as an abrasive, as a building stone, and, when composed mostly of quartz, for glass making. A widely used sandstone is the St. Peter sandstone found from Minnesota through Wisconsin and underlying most of the state of Illinois. This sandstone varies from a few meters to more than 200 m thick but is commonly 30 to 60 m thick. It is widely found at subsurface depths of 50 to 100 m. It is believed to be a marine sand deposited during the Ordovician period some 400 million years B.P., when much of the central U.S. was, or was being, covered with a sea. Outcrops occur at Ottawa, Illinois, along the Illinois River, and at several other locations in the state. This sandstone is a major groundwater aquifer. It

Figure 3-7 Outcrop of St. Peter (Ottawa sand) sandstone near Ottawa, Illinois. Note the extreme weathering that has taken place. The near white color is due to the sand being nearly pure quartz (silica).

has considerable commercial value, being almost pure quartz, and the out-crops are extensively mined. It is very porous and loosely cemented, and often can be crushed by hand. It is quarried by blasting and using hydraulic washing to break the grains apart and to remove the few impurities coating the grains, such as iron oxide. It is the widely used "Ottawa sand" standard for civil engineering testing laboratories. It is also widely used for glass making and as molds for metal castings. Figure 3-7 shows an outcrop of St. Peter sandstone from near Ottawa, Illinois. Figure 2-8 shows grains of Ottawa sand compared with several other sands, and it can be seen that the sand is particularly well rounded; the light color indicates nearly pure quartz.

Conglomerate. This is a rock composed of cemented pebbles intermixed with sand. If the grains are angular, the rock is termed *breccia*; if it is formed from glacial deposits, it may be called *tillite*.

Chemical sedimentary rocks include:

Limestone. This is a chemical sediment consisting primarily of calcite (calcium carbonate, $CaCO_3$). There are several varieties of limestone depending on the makeup and physical appearance, as containing shells, fossils, sand, etc. Limestone quickly reacts with dilute (say, 0.1 N) hydrochloric acid. The acid reaction may be used as a principal identification test for limestone. Limestone may contain silica precipitates or nodules of flint (dark color) or chert (light color).

Dolomite. This is limestone in which some of the calcite has been replaced with magnesium [$CaMg(CO_3)_2$]. Dolomite is very similar to limestone, and due to this similarity the only reliable determination, for any but the more experienced geologist, is the acid reaction test, since for dolomite the reaction is slow to nonexistent with dilute hydrochloric acid. Both limestone and dolomite tend to have the same grain structure and color; colors range from white to very dark gray, including greens, yellows, etc., depending on mineral impurities.

Evaporites. These are sedimentary rocks produced by minerals precipitated from sea water. They include gypsum ($CaSO_4 \cdot 2H_2O$), anhydride ($CaSO_4$), and rock salt ($NaCl$ and $CaCl_2$). Travertine limestone is a porous calcite precipitate from freshwater.

Biochemical or organic sedimentary rocks include:

Coquina. This is limestone containing shells and shell fragments (also called *fossil limestone*).

Reef limestone. This is limestone containing coral fragments.

Chalk. This is limestone consisting of calcareous shells of microorganisms.

Coral. This is marine limestone formed from the skeletons of marine invertebrate animals.

Coal. The carbonized plant remains; various stages include:

1. Peat—decaying and semicompact organic matter
2. Lignite—second stage, more compact, may be called brown coal
3. Bituminous (soft coal)
4. Anthracite (hard coal and final stage)

Sedimentary rocks are generally characterized by stratification as in Fig. 3-8. The typical texture of several rock samples is shown in Fig. 3-9.

Ripple marks, indicating sedimentation under water or from wind deposits, are often found in sedimentary rocks. Mud cracks, which occur when a mud dries and shrinks, may be found also. If the cracks fill with sediments during a sudden rain before the mud adsorbs water and swells the crack closed, the sediments in the cracks will later form identifying inclusions in the rock. Many sedimentary rocks are brilliantly colored, as in the Grand Canyon, the Painted Desert, and Yellowstone Park. Color is produced by the weathering of hematite, limonite, and manganese compounds contained in the rocks. Some sedimentary rocks contain fossils which are used by the geologist in dating the rock, i.e., since certain types of shell life existed at certain times, a rock containing a shell could only have sedimented during the period of existence of that particular shell.

(*a*) Stratified limestone. Note that the upper layers are dark gray and the lower layers light gray. Vertical stains are from wet weather springs leaching iron oxides.

(*b*) Stratified shale, limestone, and coal. Geologist's hammer rests against a coal seam approximately 40 cm thick. Ridges about 1.5 m above the lower coal seam and above the upper seam are limestone layers. Other material is shale.

(*c*) Stratified limestone which is well weathered after less than 10 years.

Figure 3-8 Representative stratification of sedimentary rocks. (Refer also to Figs. 3-15 and 3-16.)

Figure 3-9 Several of the more common sedimentary rocks. Paper-clip-size scale is 2.3 cm: (*a*) banded reddish brown sandstone; (*b*) sample of St. Peter sandstone with iron oxide staining; (*c*) conglomerate from near Peoria, Illinois; (*d*) fossilized limestone (contains fossil shells); (*e*) shales—left is clayey, right is sandy; (*f*) limestone conglomerate; (*g*) limestones; the one on the right is very dark gray, the one in the middle is light gray, and the one on the left is travertine limestone—note the extreme porosity.

It is very important to identify shales and limestones in a construction site. Shales tend to quickly weather back to soil on minimal exposure. Only small amounts of atmospheric moisture (or moisture at the interface of wet concrete) are necessary to soften and greatly alter the strength of shale. Limestone is dissolved by surface water as well as by subsurface water percolating through the soil. Caves and sinkholes are the result of this latter. Figure 3-8*a* and particularly 3-8*c* illustrates these statements.

3-7 METAMORPHIC ROCKS

Metamorphism through high temperatures and pressures acting on either sedimentary or—less commonly—igneous rocks that have been buried deep in the earth produces metamorphic rocks. During the process of metamorphism the original rock undergoes both chemical and physical alterations which change the texture, as well as the mineral and chemical composition.

The rearrangement of minerals during metamorphism results in two basic rock textures: *foliated* and *nonfoliated*. Foliation results in the rock minerals becoming flattened or platy and arranged in parallel bands or layers. Typical foliated rocks include:

Slate. This is metamorphosed shale, characterized by a very fine texture, splitting into thin slabs; typically colors are gray, black, reds, and greens. Slate is widely used for roofing, blackboards, sidewalks, and pool tables.

Schist. Schist is a medium- to coarse-grained rock containing considerable mica. Although they are commonly formed from shale, schists may also be formed from igneous rock. *Mica schists* are rock with mica as the predominating mineral; chlorite schist has the mineral chlorite, etc. Figures 3-10*d* and 3-16*d* illustrate mica schist.

Gneiss. Gneiss is a highly metamorphosed (generally from granite) coarse-grained and banded rock. The rock is characterized by alternating bands of darker minerals, such as chlorite, biotite, mica, and graphite. The bands are typically folded and contorted and may resemble schists, but cleavage is very difficult, whereas with schist, slab separations may sometimes be effected with a knife blade.

Nonfoliated metamorphic rocks include:

Quartzite. This is metamorphosed quartz sandstone. It is one of the most resistant of all rocks. When formed of pure quartz, the rock is white; impurities may give red, yellow, or brown tints.

Marble. Marble is metamorphosed limestone or dolomite. Marble may be white when pure, but impurities cause a wide range of colors and tints. Marble is commonly used for building stone and monuments.

Anthracite. This is metamorphosed bituminous or soft coal.

Figure 3-10 illustrates the typical physical features of several of the more common metamorphic rocks.

Figure 3-10 Several common metamorphic rocks. Paper-clip-size scale is 2.3 cm. (*a*) Pink quartzite with surface worn somewhat smooth; (*b*) yellow quartzite freshly fractured; (*c*) gneiss; (*d*) dark green mica schist; (*e*) slate—note two thin pieces on the edge in the upper left; (*f*) white marble—note the fine-grained texture.

3-8 CRUSTAL MOVEMENTS

The crust of the earth has undergone considerable structural change during past periods of earth history. Geologic evidence indicates that large land areas of all the continents have been covered periodically by shallow seas. This evidence has been obtained from study of the fossils found in both sediments and exposed rocks. Figure 3-11 illustrates the approximate outline of the present land area of the North American continent which was covered by the sea at some time between the Cambrian to about the Pliocene periods (570 to 12 million years B.P.).

NOT
PREVIOUSLY
COVERED
BY INLAND
SEAS

0   1200
KILOMETERS

Figure 3-11 Map of North America showing approximate zones not covered by the sea at some time in the geologic past. Zones once covered by the seas and now uplifted are characterized by sedimentary rocks, particularly limestones, sandstones, and large quantities of shale.

Geologic evidence indicates that the Appalachian Mountains were formed and reformed during the Paleozoic and approximately up to the Cenozoic era (225 to about 63 million years B.P.). Considerable erosion has taken place, and subsequent uplift has formed the Piedmont plateau and moved the sediments overlying the coastal plain outward onto the continental shelf eastward into and under the Atlantic Ocean. The Rocky Mountains are of somewhat later origin (believed to be from the late Mesozoic or early Cenozoic era). Large mountain masses are found on all, the continents. Figure 3-12 is an approximate surface outline of the United States which illustrates the locations of the several crustal movements locating the present mountains.

Crustal movements produce structural deformities termed folds, faults, and joints, as identified in Figs. 3-13 and 3-14. A *syncline* bends the rock layers into a concave shape upward; the *anticline* is convex upward. A *geosyncline* is an areal depression, often adjacent to a mountain, which fills with sediments and volcanic debris and later may be uplifted as a mountain. A *monocline* is a single fold, and it should be noted that both anticlines and synclines may be adjacent depending on the amount of folding. Figure 3-15 illustrates field observations of both small and large folding of strata.

Terms used to orient the geometry of inclined and folded rock beds include *strike*, the angle of the bed axis from compass north, and *dip*, the angle of the bed axis from the horizontal plane measured at right angles to the direction of the strike. Both of these terms are illustrated in Fig. 3-13.

When the stresses within the rock crust exceed the ultimate strength of the rock, fracturing occurs. If very little movement occurs along the fracture zone, the fracture is termed a *joint*. *Normal faults* occur when movements have taken place along the fracture in the vertical direction, as indicated in Fig. 3-13, with the

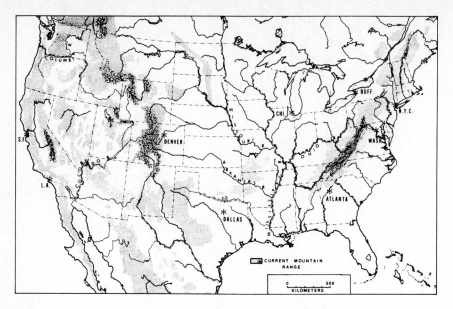

Figure 3-12 Outline of current mountain ranges in the United States. In mountainous areas outcrops of igneous rocks are likely to be found. Also, these areas are characterized by rock faults and fractures.

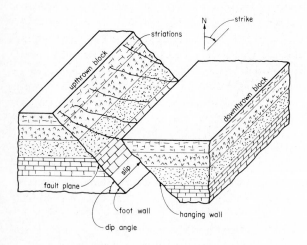

Upthrown block appears to have moved upward.
Downthrown block appears to have moved downward.
If hanging wall is on the upthrown side we have a reverse fault.
If hanging wall is as above we have a normal fault.

Figure 3-13 Fault elements. Only a fresh fault might appear as above. Old faults will have erosion across the zone so that only a gradual difference in elevation might appear (if any).

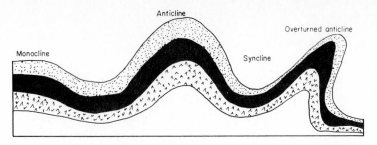

Figure 3-14 Crustal folding and some of the terms used to describe the configurations.

hanging wall over the fracture and on the downthrown block. *Strike-slips* occur when the fault movements are lateral. Many faults are slip faults, where both vertical and lateral movements occur. In east Africa there is a slip fault with a fracture zone some 6000 km in length. The Beartooth Mountains of Montana and Wyoming are partially due to a fault block 64 by 128 km raised vertically some 1000 m. The term "scarp" is used to describe the edge of the "cliff" formed by the abrupt difference in elevation of this type of fault. The San Andreas fault in California, beginning near the Salton Sea near the Mexican border and running along the coast northwesterly some 960 km to Point Arena in northern California, where it appears to enter the Pacific Ocean, is a strike-slip fault since the relative movement is primarily parallel to the fracture zone (and ground surface). The cracked and fractured rock zone extends several kilometers on each side of this fault. Varying opinions place relative displacements of the San Andreas fault as from 10 to 540 km laterally. A maximum relative lateral movement, on the order of 7 m, appears to have occurred during the 1906 earthquake, with some earth movements all along approximately 400 km of the fault line. Faults of this type are worldwide, including the Great Glenn fault in Scotland, the Alpine fault in New Zealand, and the Dead Sea Rift. A fault movement in the Cayman Trough south of Cuba in the Caribbean Sea running roughly east-west on the order of 2000 km to Guatemala is believed to be the cause of the great earthquake in Guatemala, Central America, on February 4, 1976, which killed more than 20 000 persons, injured some 77 000, and left more than 1 million homeless.

Lesser-known faults may be the most trouble for geotechnical engineers, since the great faults are more commonly known and usually there is more surface evidence of their existence. Lesser faulting is extremely widespread; for example, both northern and southern Illinois, southeastern Missouri, Tennessee, Pennsylvania, northern Georgia, and other locations in the United States contain small to medium-sized fault zones. If the relative movement is small and/or took place so far in the past that erosion has removed any possible surface evidence, the only way to detect a fault may be a careful study of borings *taken into the bedrock*. In any case, a fault is an existing structural weakness that increases the probability of crustal movements in these zones as crust stresses build up over a

Figure 3-15 Examples of stratum folding: (*a*) limestone folding near the Ozarks in Missouri; (*b*) limestone and shale folding in Montana; (*c*) folding on a large scale in the Big Horn Mountains.

period of time. Earthquakes are produced when the stresses become too large and relative rock mass movements suddenly occur. Figure 3-16 illustrates two large local faults in the Beartooth mountains and two cases of localized rock jointing.

From a geologic viewpoint fault zones are a mixed blessing, since many valuable ore deposits are found as precipitates along the fault line.

3-9 ROCK WEATHERING AND SOIL FORMATION

Weathering of rocks is one of the most important of all geologic processes. It provides the material from which sedimentary rocks are formed and produces soil, without which both plant and animal life on earth would be impossible. Rock fragments produced by weathering are removed by *erosion*. Weathering may be either mechanical (or physical) or chemical.

Mechanical Weathering

Mechanical weathering takes place when rock is reduced to smaller fragments without any chemical change taking place. Rock weathering is very dependent on the type of rock and on time. It may be caused by any or all of the following factors acting for significant periods of time.

Climate effects (including both temperature and rainfall). These are probably the principal factors involved in rock disintegration. Daily temperature fluctuations may not be too important, but freeze-thaw cycles over a long period of time cause rock fatigue even in milder climates. Severe temperatures producing local freezing of short duration may be significant, since water in rock pores will increase in volume approximately 9 percent at 0°C and will exert tremendous pressures. As the freezing pressure will tend to extrude ice from the pores and reduce the expansion pressures, local effects will be greater when the temperature drops considerably below 0°. Differential temperatures, not necessarily below freezing, coupled with the different thermal coefficients of the constituent rock minerals can have a fatiguing effect and produce rock fragments. In fact, some believe that temperature effects are one of the most significant mechanical agents in the weathering process.

Exfoliation. Exfoliation is the spalling off of the exterior surface of exposed rocks. Rocks underlying thick soil strata are under large compressive forces. Surface stress adjustments accompanying regional uplift, coupled with erosion from surface water runoff reducing the overburden stresses, cause the outer rock shell to separate (or spall) from the main rock. Again the different stress responses of the constituent rock minerals may accelerate the process. Exfoliation, especially of igneous rocks, may also be caused by relatively sudden temperature changes.

(a)

(b)

Figure 3-16 Earth discontinuities. (*a*, *b*) Large faults in Montana and Wyoming. Note particularly in part *a* the pronounced scarp line. (*c*) A joint in a limestone formation. (*d*) Jointing through a micaceous schist (dark band) underlying a granite formation in the Black Hills of South Dakota.

(c)

(d)

Erosion by wind and rain. This is a very important topography-dependent factor and a continuing event. Flowing water carrying tiny particles of rock in suspension can erode or abrade the most solid of rock over geologic time periods. This is especially significant in areas of rugged topography where high velocities may be obtained, as in mountainous areas. This is evidenced by the fact that stones found in stream beds tend to be subangular to highly rounded. Extreme cases of erosion are the Grand Canyon of the Colorado River in Utah, Arizona, Nevada, and New Mexico and the Cheddar Gorge of the River Avon in the south of England. Lesser erosion models include Niagara Falls, where the Niagara River flows over a bed of Niagara limestone which is relatively hard but is underlain by shale and soft Clinton limestone, which has eroded away to form the falls lying between the United States and Canada. Large canyons or gorges are found, even with small streams, in the western United States, Canada, Australia, Africa, and elsewhere which display the eroding effects of water acting over geologic time periods.

Figure 3-17a and b illustrates that erosion is not limited to rocks and that large areas can be involved, and that with sufficient erosion an area can be rendered uninhabitable.

Abrasion. Strictly, abrasion is the wear caused when two hard materials undergo relative movement while in contact. This can be caused by one of the materials being suspended in water, as sand, for example, but in the context of this text the term will be used to describe the pushing of large quantities of soil or ice under pressure across the underlying rock by glaciers, grinding or abrading both materials to smaller sizes.

Organic activity. Cracking forces exerted by growing plants and roots in voids and crevasses of rock can force fragments apart. Animals, such as insects and worms, burrowing into the ground may bring rock fragments to the surface or otherwise expose the fragments to additional weathering.

Chemical Weathering

Chemical weathering involves alteration of the rock minerals into new compounds. It may include the following processes.

Oxidation. A chemical reaction may take place when rocks are in contact with rainwater. It is readily noticeable as the brown to red staining of the weathered surface in rocks containing iron. Oxidation has produced the stains on the rock surfaces shown in Figs. 3-8a and 3-15a and the bright colors (bandings) shown in Fig. 3-17. Reactions may yield hydrated iron oxides, carbonates, and sulfates. If these reactions result in an increased volume, there will be a subsequent disintegration of the rock.

(a)

(b)

Figure 3-17 Sedimentation and erosion. (*a*) Severe erosion of sedimented deposits laid some 25 to 35 million years ago on the "Badlands" of South Dakota. Deposits consist of sands, gravels, fossils, volcanic dust, and loess materials as well as silts and clays. (*b*) Closeup of one of the erosion faces of part *a*, showing typical sedimentations. The darker band near the top is volcanic dust. A thin black band is believed to be carbon contamination from a prairie fire. The deposits have been laid so long that many of the lower deposits have lithified to shale. The thin band is limestone. This deposit extends about 150 m above the floor.

Solution. Certain rocks, notably limestones, are partially to completely dissolved in rainwater, especially if the rainwater contains appreciable carbon dioxide in the form of weak carbonic acid or has a pH < 7. Even a very weak acid solution acting over geologic time periods can decompose many rocks. In the case of limestones, the reader may readily observe that over time periods of only 5 to 10 years there can be considerable weathering, as along highway cuts. Figures 3-8c, 3-15b, and 3-16c illustrate limestone deterioration after periods of less than 10 years.

Caves are widely formed, as are limestone sinkholes (karst formations) in areas with many limestone formations and considerable rainfall. Figure 3-18 illustrates typical karst topography as found in parts of north central Kentucky and south central Indiana. Land sinking and subsequent erosion tend to produce the rolling topography shown in Fig. 3-18c rendering the land unsuitable for anything but grazing.

Leaching. Water reacting with the cementing material of sedimentary rocks may cause the particles to loosen, with the smaller particles and the cementing agents carried away either to deeper strata or as surface runoff. Cementing agents carried to deeper strata by percolating rainwater may be a factor in future formation of new sedimentary rocks. In areas of little rainfall, water vapor may carry the cementing agents such as sulfates, carbonates, etc., to the ground surface, creating a salt crust which may make the soil unfit to support plant life.

Hydrolysis (formation of H^+ ions). Chemical weathering agents may be acting simultaneously. Consider, for example, the formation of clay from the weathering of orthoclase (usually pink in color) feldspar in the presence of ordinary water and carbonic acid formed by water mixing with carbon dioxide:

$$
\begin{array}{ccc}
\text{2 parts} & \text{1 part} & \text{1 part} \\
2(K)AlSi_3O_8 + & H_2CO_3 + & H_2O \rightarrow
\end{array}
$$

$$
\begin{array}{ccc}
\text{1 part} & \text{1 part} & \text{4 parts} \\
Al_2Si_2O_5(OH)_4 + & K_2CO_3 & + 4SiO_2 \\
\text{Clay mineral} & \text{Potassium} & \text{Quartz} \\
& \text{carbonate} &
\end{array}
\tag{3-1}
$$

In this case the H^+ ion from the water forces the K^+ ion out of the feldspar. The H^+ ion then combines with the aluminum silicate to form the clay mineral. A plant root in the soil may attract local soil-water and become surrounded by an excess of H^+ ions, which initiates the hydrolysis process. Any fragment of orthoclase feldspar in close proximity can be broken down to form the clay mineral, according to Eq. (3-1). The potassium carbonate may be further broken down and leached away, it may become plant food, or the clay mineral may attract the potassium ions to form *kaolinite* clay.

(a)

(b)

(c)

Figure 3-18 Limestone sinkholes and karst topography: (*a*) small sinkholes in pasture; (*b*) large sinkhole which has just formed with an ideal round shape (seldom obtained); (*c*) typical sinkhole area where erosion has somewhat smoothed out the initial steep sides of parts *a* and *b*.

3-10 GENERAL CONSIDERATIONS IN ROCK WEATHERING

The rate of weathering depends on the particle size. Small particles weather, in general, at a faster rate than large ones due to their larger surface area. Type of material, climate, moisture, exposure conditions, and plant and animal/insect activity are important factors affecting the rate of weathering.

Most weathering takes place near the ground surface; however, exfoliation due to loss of overburden pressure may be taking place at a depth of many meters. Downward-percolating rainwater or underground water may be producing chemical weathering far below the ground surface without the effects ever being exposed.

Generally both the rate and amount of weathering increase with time due to both the reduction of rock size and more material being exposed to the process. The very important effects of rainfall and temperature are summarized here.

Rainfall

Low rainfall areas. Water only penetrates the soil to a limited depth; weathering takes place, but the byproducts (carbonates, sulfates, etc.) are not removed from the soil profile, and the resulting pH tends to be alkaline. Water tends to be removed by evaporation, which tends to concentrate calcium, sodium, and potassium salts in the surface zone of the soil profile.

High rainfall areas. Water percolates through the soil, and weathering material is removed by leaching. Soluble substances are removed, and clay tends to be dispersed in the lower soil profile. The pH of the soil tends to be acidic.

Temperature Effects

High average temperatures in moist areas increase vegetation and chemical weathering. High average temperatures in arid areas decrease vegetation and chemical weathering, and mechanical weathering tends to dominate.

Low average temperatures in moist areas cause soil freezing and permafrost and slow weathering.

The geotechnical engineer will be particularly concerned with the high rate of weathering of exposed shale, and to a lesser extent sandstone. Limestone weathering takes place with considerable rapidity in the presence of water regardless of whether it is exposed.

3-11 SOIL FORMATIONS PRODUCED BY WEATHERING

Soil may be classified according to the method of formation of the deposit as residual soil or transported soil. A *residual* soil is one which was formed in its present location through weathering of the parent (or bed) rock. These soils are rather widespread in tropical areas, where they may be termed *laterites*, and in other less tropical areas where glaciers have not been present, as in the southeastern and southwestern parts of the United States, most of Australia, India, Africa, and southern Europe. Residual soil deposits vary from a few centimeters to 100 or more meters in depth depending on the geologic age and weathering conditions. These soils are formed by weathering and leaching from the top downward of the water-soluble materials. As leaching action naturally diminishes with depth, the residual soil will be less and less altered until the parent rock is reached.

If a vertical cut is made in a residual soil, a horizontal arrangement of layers can sometimes be seen, especially in a fresh cut. The vertical section is a soil profile, and the individual layers are *soil horizons*. Figure 3-19a illustrates a simplified arrangement of the soil horizons for geotechnical engineering use. In general, the horizons are:

Horizon	Comments
A	Top zone consisting of topsoil and organic matter, and in humid areas, highly leached materials; in arid areas it may be rich in various water-soluble salts remaining as water vapor from the lower depths evaporates; it generally is highly weathered, dark-colored material including various shades of blacks and browns of a few centimeters to 1 or 2 m thick and grading into the *B* horizon
B	Zone underlying the *A* horizon and containing considerable leached materials (water-soluble salts such as carbonates, sulfates, and chlorides) and clay minerals; this zone may be on the order of 0.5 to several meters thick and grades into the *C* horizon
C	Transitional zone of freshly weathered parent material (rock); it may consist of considerable rock fragments or may be absent or of very shallow depth and grades into the *D* horizon
D	Parent (or bed) rock

The soil from the *B* horizon is considered the best for borrow since it contains both granular material and binder. The *A* horizon contains too much organic material and too little binder to be of value as a construction material. The *C* horizon may be too open graded, or deficient in material passing the No. 100 sieve and clay sizes, for use as borrow, although it may be blended with *B* horizon material and made satisfactory.

Residual soils tend to be characterized by

1. Presence of minerals that have weathered from the parent rock
2. Particles tending to be angular to subangular as illustrated in Figs. 3-19b and 3-19c as compared with the rounded particles in the transported deposits in Figs. 3-19d and 3-19e
3. Large angular fragments of rock tending to be found dispersed throughout the mass, as in Fig. 3-19c

An important residual soil found in many mountainous areas is termed a *saprolite*. A saprolite is a condition of chemical rock weathering such that the rock tends to crumble but still retains the original structure and texture. A climatic condition of moderately heavy to heavy rainfall, such as in the southern Appalachian mountains, the Australian Alps and areas of western Australia, India, South America, and Hawaii, produces this type of soil. Saprolitic material has been reported to depths of 100 m (Carroll, 1970).

Transported soils were formed from rock weathering at one site and are now found at another site. The transporting agent may be

1. Water (principal transporting agent)
2. Glaciers
3. Wind
4. Gravity

Water, wind, and glacier deposits are very widespread. Often the deposits are given names indicative of the mode of transportation causing the deposit. Deposits formed by these several modes of transportation will be considered in more detail in the following sections.

A question naturally arises: What is the classification of the deposits in areas covered by marine deposits from several million years ago? While in the strictest sense these are transported deposits, the deposition took place so long ago that some to considerable lithification has since taken place. In present conditions the indurated soil is being weathered anew, producing a material that is more residual than transported. With these soils, however, the horizon concept may have little meaning due to the previous stratification, which may include sands,

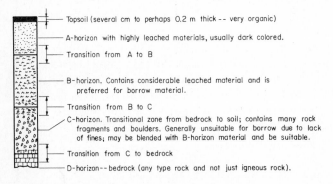

Topsoil (several cm to perhaps 0.2 m thick -- very organic)

A-horizon with highly leached materials, usually dark colored.

Transition from A to B

B-horizon. Contains considerable leached material and is preferred for borrow material.

Transition from B to C

C-horizon. Transitional zone from bedrock to soil; contains many rock fragments and boulders. Generally unsuitable for borrow due to lack of fines; may be blended with B-horizon material and be suitable.

Transition from C to bedrock

D-horizon -- bedrock (any type rock and not just igneous rock).

(a) Hypothetical soil horizon profile with several subdivisions.

Figure 3-19 Soil horizons and several natural soil deposits: (*a*) hypothetical soil horizon profile with several subdivisions; (*b*) residual soil in West Virginia (scale: ball-point pen is 14 cm); (*c*) residual soil in Montana with 30-cm scale shown; (*d*) deposit in dry part of stream bed with 30-cm scale shown; (*e*) moraine deposit near Peoria, Illinois, with 30-cm scale shown.

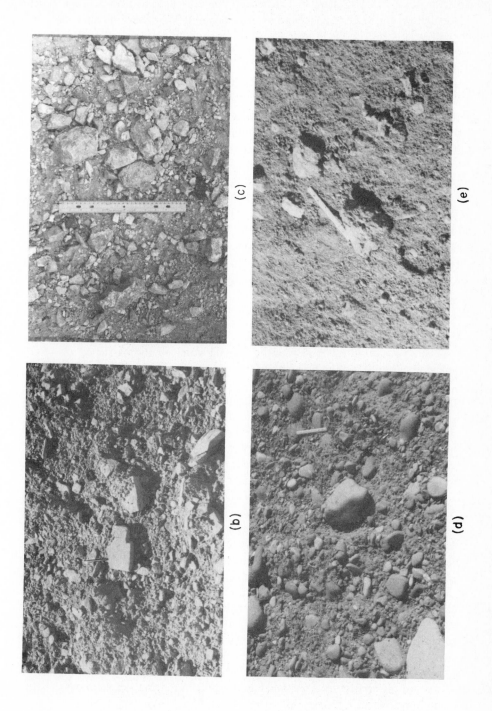

limestones, clay and silt layers or shales, etc. In these soils the value of a layer for construction purposes will depend on the properties of that stratum, and may depend very little on leaching from above or on the properties of the adjacent strata. Figure 3-17 illustrates this particularly well.

3-12 RUNNING WATER AND ALLUVIAL DEPOSITS

Once water has fallen on the land as precipitation, it follows one of the many paths making up the hydrological cycle (see Sec. 3-16). That portion following the path of *runoff* is of interest in this section as an agent causing *erosion* and *transportation*. Erosion and transportation depend on the velocity of the moving water, which in turn depends on the gradient, amount of water passing a point, and nature of the stream banks. In general, the gradient decreases from the headwaters (upper end) to the mouth. The mouth may terminate in another stream, a lake, or the ocean. Terminal velocity in lakes or the ocean will approach zero a short distance from shore based on considerations of continuity and the flow equation

$$Q = Av \qquad (3\text{-}2)$$

where Q = discharge quantity
A = area of flowing water
v = velocity

Erosion is caused by friction of the flowing water, including the effects of any suspended material, on the flow channel. Since the eroded materials contain flaws and have varying degrees of resistance to erosion, no stream will be straight as seen from above, except for very short distances—generally less than 10 times the width of the effective channel. The center of the channel, or principal flow, tends to swing back and forth, or *meander*, from side to side. Over geological time, this usually results in wide valleys cut between rock bluffs or banks, with the valley floor consisting of transported soils or sediments. Sedimentation occurs when the lowered velocity of the water on the inside of a meander will no longer support the suspended material. The outside of the meander, being of higher velocity, erodes into the bank, thus increasing the meander as illustrated in Fig. 3-20*a*. Flood stages may cause formation of *natural levees* parallel to the banks as the stream rises and overflows, and the velocity falls as the channel area suddenly enlarges. The reduced velocity causes transported materials to precipitate along the banks. Trees and shrub growth along the bank may considerably aid the formation of natural levees. The formation of natural levees may raise the river and the levees 3 to 5 m or more above the valley floor until a very large flood overtops the levee, with the resulting erosion cutting a new channel through the levee wall and into the valley floor. River terraces (Fig. 3-21*b*) may be formed when the stream cuts into a previously deposited sediment or as the stream bed is lowered over geologic periods due to normal erosion or to crustal deformation.

Erosion of the neck of a meander may result in cutting the meander, leaving a curved and isolated channel or *oxbow* as in Fig. 3-20*a*. The oxbow may be a

(a)

(b)

Figure 3-20 Stream configurations. (*a*) Well-developed stream meanders, with oxbows shown. Also shown is a location (at the extreme right) where a cutoff will soon develop and a new oxbow form. This stream is about 7 m wide and is essentially a model but shows in a small, easily seen area many of the typical "older" stream features. (*b*) Well-developed stream valley. This stream displays a slough and stream braiding (the several branches of the stream in the center background).

(a)

(b)

Figure 3-21 River bank formations: (a) natural levee along the Wabash River; (b) well-developed stream terrace on Yellowstone River in Montana.

slough if one end remains open to the stream so that the backwater stands as in Fig. 3-20*b*. An *oxbow lake* is formed if the oxbow fills with water.

These several depressions may later fill with fine-grained sediments, muds, and organic material during and between subsequent valley flood stages, and the result is a particularly poor deposit of highly plastic (w_L often 60 to 100 or more) and/or organic silts, silty clays, or peat. Soil exploration for foundation sites should proceed with caution to locate and identify these deposits.

The continuous slow change in channel position results in the entire valley floor consisting of *alluvium* or *sediments*. This continual reworking is gradually moving all the material downstream and reducing the stream gradient. The material downstream is, of course, progressively finer due to several factors, including more abrasion and the lower gradient with its resulting reduced velocity which can only support the finer weathered material. Since the valley floor is nearly flat and near the high water level of the stream, at flood stage the valley is essentially a flood plain and susceptible to widespread shallow flooding. These areas are poor building sites due to the periodic flooding unless the stream channel is confined by supplemental man-made levees. Buildings in a flood plain cannot obtain flood insurance in many areas; thus, sites near streams should always be checked for possible flooding.

Lake deposits are also called *lacustrine* deposits. *Varves* are a particular type of lake deposit formed during glacial periods from seasonal ice melting which temporary increased the runoff velocity so that precipitated sand layers alternate with silt or silt-clay layers of precipitates made at low velocities.

A *marine* deposit is obtained when the sediment precipitates through salt water.

Deltas are sediments (Fig. 3-22) precipitated at the mouths of streams into bays, oceans, or lakes. *Fans* are a similar type of deposit but found in arid areas

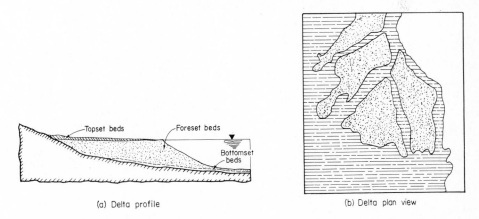

(a) Delta profile   (b) Delta plan view

Figure 3-22 River or lake delta formation. Note that an alluvial fan will be similar in plan to the delta but will be a deposit at the exit from uplands onto a plain, valley, or desert flatland: (*a*) delta profile; (*b*) delta plan view.

where mountain stream runoff flows into wide valleys or onto the plain at the stream mouth.

Lake, marine, and delta deposits tend to be relatively fine-grained, with silt and clay sizes predominating. Most of these deposits will be loose and highly compressible. Organic material is sometimes present, as are seams of fine to medium-coarse sand. Some of these deposits are 75 to 150 m in thickness.

3-13 GLACIAL DEPOSITS

Glacial deposits form a very large group of transported soils. At various times a large part of the North American continent, as illustrated in Fig. 3-23, has been covered by glacial ice, as has northern Europe, including much of Germany, Poland, Northern Russia to the Ural Mountains, all the Scandinavian countries, the British Isles, and Greenland. The approximate outline of the European glaciation is shown in Fig. 3-24. Greenland is still very nearly covered with glacial ice, as are parts of northern Canada and Alaska, most of Antarctica, the higher mountains of Scandinavia, the Swiss Alps, the Himalayas, and some of the Andes of South America; some small glaciers exist on the highest mountains in the United States. A survey by Flint (1970) showed that some 10 percent of the presently existing land surface was covered at one time by glaciers, amounting to some 15×10^6 km^2.

The eroding action of the glacial ice both scraped up soil at the interface of the ice and soil or rock and pulverized, crushed, and abraded the parent bedrock into silt, sand, and gravel-sized material. This could be accomplished due to the

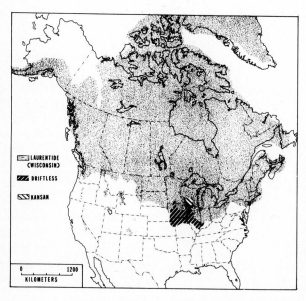

Figure 3-23 Approximate outline of glaciation on the North American continent. The most recent is the Laurentide (or Wisconsin) glaciation, which disappeared some 10 000 to 13 000 years ago. The Kansan glaciation covered about the same area but extended somewhat further south locally into Kansas, Missouri, and Illinois as shown.

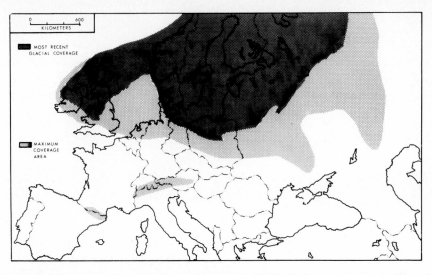

Figure 3-24 General location of European glaciation. The outline is only approximate but gives an indication of the amount of glaciation and about where to expect glacial deposits.

great depths, and resulting enormous pressure, of glacier ice. Thicknesses were on the order of:

Location	Approximate thickness, m
North America and Canada	300–2000
British Isles	350–900
Europe	500–3000

An ice thickness of 1000 m would create an ice pressure of some 8800 kPa on the underlying soil.

Soil deposits pushed into ridges around the periphery of the glacier are called *moraines. Terminal moraines* (Fig. 3-25*b*) are ridges of material scraped or bull-dozed to the front of the glacier; *lateral moraines* develop along the sides. These formations are generally 0.5 to several kilometers wide, may be 25 to 100 or more meters high, and may be 60 to 100 km long. The moraine may not be a single nicely rolled ridge, but rather a highly serrated, above ground level, earth mass. There may be discontinuities in the ridge where glacial melt forms channels carrying outwash, and small lakes may temporarily form in depressions behind the ridge, producing lacustrine sediments. *Ground moraine* (also termed *glacial till* or simply *till*) was the deposit of ice-suspended material through the bottom of the glacier; it ranges from a few centimeters to 150 m or more in thickness. *Eskers* (Fig. 3-26) are ridges formed when water-suspended material flowing in ice tunnels precipitated; they vary from about 10 to 30 m high and are from about

Figure 3-25 Photographs of several glacial features as noted. (*a*) Small mountain glacier; also shown are small valley moraines and glacially formed lakes. (*b*) Larger terminal moraine; note ridges rather than a single ridge. This moraine is about 0.5 km wide by about 5 km long. (*c*) Glacial topography confined to a small area; shown are the terminal moraine, a drumlin, and several erratics.

(a)

(b)

Figure 3-26 Two eskers. (a) This esker is some 6 km in length and in places 27 m high. It is in North Dakota and is one of the largest ones ever reported; (b) Esker (below) is about 2 km long, but higher points are on the order of 30 m.

0.5 to several kilometers in length. *Drumlins* are isolated mounds of glacial debris varying from about 10 to 70 m high and 200 to 800 m long. Most drumlins are on the order of 30 m or less in height and 300 m or less in length. They often occur in drumlin fields (several), as in Fig. 3-27. *Erratics* are large boulders picked up by glaciers, transported to a new location, and dropped, as in Fig. 3-25c.

Eskers, drumlins, and glacial outwash tend to have commercial value as sand or gravel sources, since the material often contains very little (−) No. 200 sieve material. To determine suitability as a sand or gravel source requires some soil exploration, since not all these formations are suitable. Sometimes localized areas of a lateral or terminal moraine may contain suitable borrow, as shown in Fig. 3-28, where a borrow location is established in the Shelbyville Moraine near Peoria, Illinois. Note the characteristic ridge profile and the grading which may be obtained in these deposits.

Figure 3-27 Drumlins. The drumlin field on top is near Sauk Centre, Minnesota; the field directly above is from Wyoming. One can distinguish a drumlin from an eroded rock outcrop because the interior rocks and gravel will be rounded as a result of glacial abrasion.

Melting glacial ice formed streams flowing away from the glacier which carried fine sand, silt, and clay material to lakes to form varves or downstream as fluvial sediments as the ice melting increased or decreased with the seasons and the stream velocity fluctuated. As the glaciers melted, material suspended in the ice precipitated onto the underlying soil or rock to form glacial till. Till deposits are characterized by containing all sizes of particles with no obvious arrangement. One analysis stated that the till around Boston, Massachusetts, consists of 25 percent gravel, 20 percent sand, 40 to 45 percent fine sands, and less than 12 percent clay (Leggett, 1962). The glacial deposit is called *stratified drift* if the profile is sorted according to size.

Glacial till or drift has a highly variable thickness which depends on the location, such as in buried valleys (old stream erosion traces) or in end moraines.

(a)

(b)

(c)

Figure 3-28 A borrow pit in the Shelbyville Moraine near Peoria, Illinois: (*a*) end view of one of the ridges, which is about 20 m high; (*b*) closeup showing material distribution of the left side of part *a*; (*c*) closeup showing material distribution of the right side of part *a*. Darker material is clay which is slightly wet from a rain two days before. The sand-gravel material quickly dried and is light in color.

Some typical till thickness values are as follows:

Location	Estimated thickness, m
Great Lakes region, USA	12
Illinois	0–180 and averaging 35
Central Ohio	29
But in buried valleys	60–230
Ontario, Canada	0–75$^+$
New Hampshire	10
Southeastern Wisconsin	14
Central Quebec	2–3
Denmark	2–40
Sweden	0–200
Finland	2–3

Glacial deposits range from excellent to poor foundation materials. In many locations, even though the deposits are unsorted, the material is dense and contains considerable sand and gravel. Many of these deposits are permanently above the water table. Submerged valleys, lenses of saturated silt and/or clay, and the presence of suspended boulders cause the most problems. The boulders cause difficulty in both soil exploration and pile driving. The presence of small gravel creates problems in obtaining undisturbed soil samples for laboratory testing. *Boulder clay* is a term used to describe deposits containing considerable cohesive material with randomly suspended boulders.

Identification of suspended boulders is particularly important where their presence may produce a misleading bedrock location. It is often necessary to carefully study the area geology to correctly identify suspended boulders above competent bedrock. One or more of a number of borings may miss boulders and locate bedrock but in at least one reported case all four of the site borings encountered boulders which initially set the location of bedrock since no corings were made which would go through the boulders—at least small ones. Later settlement problems developed at this site and a reinvestigation located bedrock at a much greater depth.

Note in passing that while boulders are more common in glacial till, they may also be found in residual soil deposits but here the weathering process is such that bedrock should be in close proximity.

3-14 WIND DEPOSITS

Wind, or aeolin, deposits are primarily *loess* and *dune* sands. Loess covers large areas of the central United States, Russia, Europe, and Asia. These deposits are believed to have been at least partly caused by changes in air density in the vicinity of melting glaciers and flowing outwash streams causing windborne particles to precipitate. Loess deposits are characterized by being of buff color, of low density (often less than 14 kN/m^3), of low wet strength, and with the ability to stand on vertical cuts. Loess deposits range in thickness from a few centimeters to more than 30 m. Commonly, along the Illinois River the loess depth is 5 to 8 m. The thickness tends to be greater near the east side of the streams, and the deposits thin rapidly with distance eastward from the stream, reinforcing the theory of how they were formed.

Loess is a quartzose, somewhat feldspathic, clastic sediment composed of a uniformly sorted mixture of silt, fine sand, and clay particles. Typically the particles range from 0.002 to 1 mm, with the largest percentage between 0.005 and 0.150 mm (No. 100 sieve). It tends to deposit in a loose arrangement which becomes rather stable due to cementation from clay particles, organic activity, and calcium carbonation. This structure is particularly susceptible to collapse on saturation. On vertical cuts, large blocks tend to slough when wet as illustrated along the base of the cut in Fig. 3-29. Water percolating vertically through root and worm holes may cause considerable vertical erosion.

Dune sands are sand deposits formed by wind action rolling the sand, which is too large for air transport, along the ground until an obstruction is met, whereupon a dune (or mound) forms. Later winds may demolish the dune and redeposit it at a new location further downwind. Dune sands tend to be well rounded from abrasion. Dune deposits are found in desert areas such as areas of California, the Sahara Desert in northern Africa, large areas of the Mideast such as Saudi Arabia, and the Gobi Desert in Asia. A few local dunes are found along the southeastern shores of Lake Michigan and in Nebraska.

Figure 3-29 A loess deposit near Vicksburg, Mississippi. Deposits here are considerably more than 30 m thick. This vertical cut, which has been standing more than 10 years, is about 8 m high. Note the typical weathering pattern of vertical chimney formations (grooves) and spalling.

(a)

(b)

Figure 3-30 Gravity deposits as found in regions of rugged topography: (*a*) slope deposit gradually moving to bottom of mountain; (*b*) talus deposit.

3-15 GRAVITY DEPOSITS

Gravity deposits are primarily *talus*, found at the base of cliffs. They may also include landslide deposits if the slide has removed the soil sufficiently from the original site.

Talus is the weathered rock/soil deposit formed at the base of cliffs when rock weathering causes the face of the cliff to loosen and fall away, producing a pile of rock fragments at the cliff base. These fragments are likely to be rather loose and porous and may require removal where a dam is to abut against the cliff. Figure 3-30 illustrates both a talus deposit and a gravity deposit formation on the sides of a mountain.

3-16 SUBSURFACE WATER

Groundwater is one of the most important mineral resources extracted from beneath the earth's surface. Probably 30 percent of the daily water consumption worldwide is obtained from groundwater; the remainder is obtained from surface water in streams or lakes.

The geotechnical engineer is concerned with groundwater when solving problems of water supply, drainage, excavations, foundations, and control of earth movements. Because of the many engineering projects in which goundwater is a significant parameter, the engineer should have a good understanding of the modes of its occurrence and movement.

Subsurface water is derived from several sources, and impurities in it may be indicative of its origin and/or history. Some groundwater is a direct contribution from magmatic or volcanic activity during the process of rock cooling. This water may be termed *juvenile* (just beginning to freely circulate) water. Water trapped in the interstices of sediments which are later covered by more impermeable sediments may be retained until tapped by accident or intent. This water is called *connate* water and is often salty since most of the sediments were deposited beneath seawater.

The most important source of groundwater is that portion of the precipitation which sinks into the ground, called *meteoric* water. Water is drawn into the atmosphere by evaporation and widely distributed by wind currents. Condensation returns this water to the earth's surface as rain, snow, sleet, hail, frost, and dew. That part falling on land surfaces becomes subdivided as follows:

1. Part is reevaporated back to the atmosphere (probably 70 percent).
2. Part runs off into streams and thence to lakes or the ocean.
3. Part is used by plant and animal life.
4. Part sinks into the ground to become groundwater (probably less than 20 percent of the condensation falling on the surface).

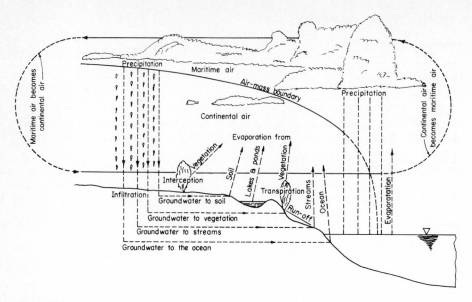

Figure 3-31 The hydrologic cycle. *(After USDA.)*

Figure 3-31 illustrates the hydrologic cycle. The amount of subsurface water obtained depends on:

1. *Surface gradient.* Steep slopes encourage surface runoff in both quantity and rate.
2. *Vegetation.* Thick foliage may intercept large amounts of condensation before it even reaches the ground surface.
3. *Climatic conditions.* Amount of rainfall and daily temperature influence the evaporation rate.
4. *Porosity and permeability of the mantle.* This means the percentage of pore space and the facility with which the water can move through the earth mass.

Water entering the mantle may be partly held by surface tension forces in the upper soil layers (vadose zone), to later evaporate or be used by plant life. Below this zone is the saturation zone, extending to a considerable depth but depending on the stratigraphy, in which the interstices and cracks are completely filled with water. The saturation zone includes (as in Fig. 3-32) a depth in which the water is held by surface tension, or capillary zone, and a lower zone where the water is free to move, or flow, under the influence of gravity. The *phreatic* line delineates between these two zones and defines the *water table*. The water table must be penetrated to provide a dependable well or permanent stream. The water table tends to follow the contours of the ground surface, rising under hills and descending beneath valleys. It tends to be close to the surface in moist climates and at greater depths in arid regions. If the water table is not replenished, usage lowers

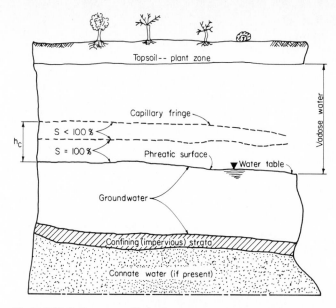

Figure 3-32 Soil-water profile in the upper mantle of the earth.

it. The slope of the water table is the *hydraulic gradient*. Figure 3-33 illustrates groundwater and conditions for streams to replenish (influent) or take from (effluent) the groundwater supply.

Aquifers

A permeable material through which the groundwater actually flows is called an *aquifer*. Sand or sand and gravel strata are particularly excellent aquifer materials due to their large porosity and permeability. Table 3-2 lists typical values of porosity (*n* values) of several types of rocks. Some porous sandstones are important aquifers, such as the St. Peter sandstone of Illinois, Wisconsin, and parts of Indiana with entrance in Wisconsin, and the Dakota sandstone underlying large

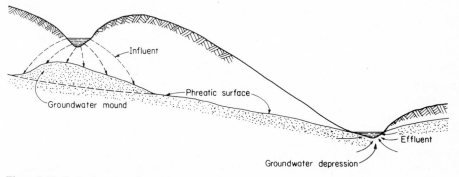

Figure 3-33 Groundwater and streams.

Table 3-2 Typical porosities of some rocks (after Leggett, 1962)

Type	n
Soil and loam	< 60
Chalk	< 50
Sand and gravel	25–35
Sandstone	10–15
Oolitic limestone	10
Limestone and marble	5
Slate and shale	4
Granite	1.5
Crystalline rocks, generally	< 0.5

areas of the Dakotas, Minnesota, Kansas, Nebraska, and parts of Colorado with entrance near the Black Hills of the Dakotas. It should be noted, however, that materials of high porosity may not be good aquifers. Mississippi River sediments often have porosities on the order of 80 to 90 percent, but the permeability is so low that little water would be obtained from a well. This is generally true of all silts, silt-clays, very fine silty and/or clayey sands, and loam soils.

Limestone which has weathered sufficiently to contain large solution cavities may be an excellent source of underground water. Chalk is also an excellent source; it was the source of early artesian water in France and supplies considerable water for domestic use in the southern part of Great Britain. Generally igneous, metamorphic, and other sedimentary rocks are poor aquifers unless they are badly cracked or fissured to provide both a water reservoir and flow channels. Figure 3-34 illustrates conditions for water supply via wells or springs. The

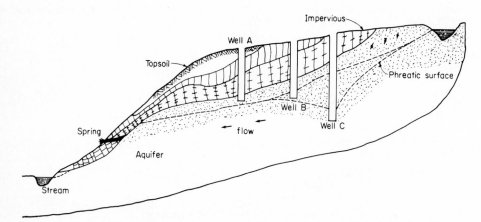

Figure 3-34 Conditions for wells, a spring and stream supplied by groundwater. The spring flows through a crack in the impervious upper layer and may be artesian if some pressure head remains after head loss through the crack. Well A is dry unless the water table rises. Well B becomes nonproductive when well C drawdown lowers the groundwater table as shown.

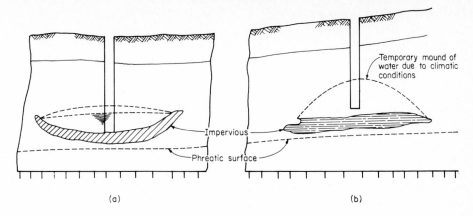

Figure 3-35 Two different conditions producing a perched water table. Note that in part *b* the well may dry up through a combination of production and natural drainage.

"perched" water table of Fig. 3-35 is a common occurrence. In the situation of Fig. 3-35*b*, a well or spring may be intermittent. In the conditions of Fig. 3-35*a*, the supply may be permanent. Note, however, that boring through the impermeable layer containing the perched water table, carelessly or by design, may allow it to drain and be permanently lost.

Artesian Water

Artesian water is obtained from an aquifer which is under a hydrostatic pressure. Conditions necessary to produce artesian water are as follows (see also Fig. 3-36):

1. The water must be contained in a permeable layer so inclined that one end can intake water at the ground surface.

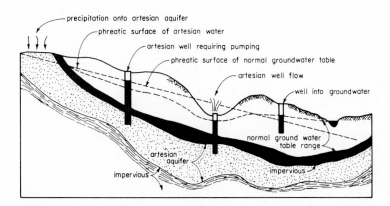

Figure 3-36 Conditions for artesian water.

2. The aquifer is capped by an impermeable layer of clay, shale, or other dense rock.
3. Water cannot escape from the aquifer either laterally or from the lower end.
4. There is sufficient pressure in the confined water to raise the free surface above the aquifer when it is tapped via a well (or any boring).

Thirty to one hundred years ago artesian wells where the water overflowed the well at the ground surface were widely obtained. At present, because of indiscriminate use (or ignorance of the source) and/or allowing continuous flowage from the earlier artesian wells, most of the aquifers are no longer under pressure or the pressure has fallen to such values that pumps must be used. This represents an irrevocable loss of a natural resource, since replenishment is estimated to take on the order of 150 to over 1000 years.

Groundwater Discharge

Large quantities of groundwater are used by artificial wells. Probably even larger quantities are lost through seepage directly into a lake, ocean, or stream, or from springs and through ordinary evaporation. Springs (Fig. 3-37) usually are found on hillsides or at the edges of valleys, but they may issue beneath the sea, lake, or stream. Streams may be mere trickles or torrents. For years the principal source of water for the city of Huntsville, Alabama, was a large spring. Big Spring in Carter County, Missouri, discharges some 11 m^3/s. Silver Springs in Florida, which is an artesian spring, discharges some 23 m^3/s, Thousand Springs along the Snake River in Idaho discharges some 140 m^3/s from the several springs making up the "thousand springs." Several rivers and numerous smaller streams originate as springs.

A *mineral* spring is any spring with considerable mineral content, which gives the water a distinctive taste (all groundwater contains some dissolved minerals). Hot springs are found in volcanic regions or, as in the case of Hot Springs, Arkansas, or Warm Springs, Georgia, are caused by the rising of artesian water from a great depth, where it may be presumed that a deep-seated igneous intrusion has heated the surrounding rock and water. These springs may be fissure springs due to the likelihood of artesian water flowing through a crack in the confining rock.

Geysers are hot springs which erupt intermittently; they are found in areas of dying vulcanism where a substantial heat source is available. Notable geyser regions include Yellowstone National Park in Wyoming, North Island of New Zealand, and Iceland.

Wet-weather or intermittent springs flow during periods of wet weather when the recent rainfall builds the water table to a sufficient height. As the water table recedes, the spring ceases to flow. These springs are nuisances for road construction (see Figs. 3-8a and 3-15a) and are frequently found in subdivisions where landscaping has altered the topography such that springs form in the newly exposed areas and in the streets or at the basement level of houses. The former

Figure 3-37 Several springs. (*a*) Spring from a limestone cave near Decatur, Alabama. It is about 3.5 m wide × 1.5 m deep, and the flow rate is about 0.15 m/s. (*b*) Spring from a limestone hole near Roanoke, Virginia. The flow volume is about 0.25 cm/s. (*c*) Spring through a talus in Montana. (*d*) Hot spring. Steam bubbles can be seen. The spring is surrounded by a travertine limestone cone, precipitated from the mineral-rich water, with outflow in the small channel in the upper background.

condition results in early pavement breakup and visibly wet spots in the street after heavy rains or in the spring due to groundwater recharge from melting snow. The latter springs result in wet basements until the stream flow is intercepted; the condition may be particularly expensive to correct if the basement was constructed during dry weather so that a spring was not evident.

In arid regions groundwater, generally as water vapor, is drawn upward to the surface, where it evaporates, leaving behind a coating of any dissolved salts. These deposits are termed *alkalies* (any bitter-tasting salts), also *caliche* (crusty deposits of calcium carbonate). These deposits will ruin agricultural land and tend to form where land is irrigated and the quantity of irrigation water is not sufficient to leach these materials into the *B* horizon.

Groundwater Erosion

Groundwater is an effective erosion agent because it is charged with carbonic acid (contained in rainwater), which dissolves carbonate rocks such as limestone, dolomite, marble, rock salt, and gypsum. Limestone, being particularly widespread, tends to concentrate the effects of subsurface erosion.

As water percolates into the earth through cracks in the rock, the chemical and physical action of the water enlarges the cracks. This tends to isolate the rock into blocks and form cavities around and in them into which the water drains. Large limestone caves are formed by this process. Where the cavities are more local in extent, the surface may collapse into the underground cavity, forming *sinks* or *sinkholes*. When the outlets are adequately plugged, small lakes may form and later fill with sediment. An area with many sinkholes is termed a *karst* region after a region of Italy and Yugoslavia with this characteristic topography. Karst areas in the United States include the Shenandoah Valley of Virginia, southern Indiana, and north central Kentucky; portions of Missouri in the Ozark Mountain area; and the central portions of Florida.

Karst areas should be treated cautiously, since a foundation site on a potential sinkhole, or on a sinkhole later filled with sediments, may produce a disaster. These sinkholes, often less than 10 000 m² in plan (on the order of 3 to 6 m in initial depth) may be missed in a soil exploration program unless the engineer recognizes that the area contains these filled-in sinkholes.

Groundwater precipitates may fill cracks or fissures to form mineral veins. Calcite and quartz veins are common and may carry concentrations of metallic minerals such as copper, silver, or gold.

Recharge of Groundwater

Most groundwater recharge is obtained from precipitation. For the normal water table in moist areas, the recharge via this mode keeps pace with groundwater removal. In wet years, the water table may even increase, with some loss during dry periods.

Where ground intake is particularly heavy, as in areas of large population or high industrial densities, the water table may suffer irreparable harm unless artificial recharge methods are used, such as putting the used water back into the aquifer via wells or recharge ponds whose bottoms exit onto the aquifer, or pumping river water into the aquifer.

3-17 SOURCES OF GEOLOGIC INFORMATION

From a study of the geologic concepts and photographs illustrating these concepts, it is evident that many landform features are areal rather than local. Identification of large-sized features often requires study of the plan, or at least of an oblique view from above. In many cases a perspective from a hill or mountain top is not possible, and often in the growing season plant growth will obscure the geology.

Because of these several problems it is often necessary to obtain aerial photographs of the area or topographic quadrangle sheets. Aerial photographs may be obtained from the Agricultural Conservation Service of the U.S. Department of Agriculture. Many, if not all, of these photographs may be obtained for stereoscopic viewing (in adjacent pairs). Topographic maps, commonly called *quadrangle sheets*, can be obtained from the U.S. Geological Survey. Annual publications give the areas where quadrangle sheets have been completed.

State geologic survey publications are particularly useful for identifying state geology, mining operations, etc. Often the state geologic publications which have been made for the separate counties are particularly well detailed.

3-18 SUMMARY

This chapter has presented a great deal of geologic data and terms. Terms have been italicized for ease of location, but it is not expected that the reader will be able to learn all these terms nor is it generally necessary.

The reader should have an awareness of:

1. The principal earth-forming minerals
2. The range of specific gravities making up the earth's mass
3. The three rock groups (igneous, sedimentary, and metamorphic) and the mode of formation of each, including recognition of
   a. Granite, gabbro, basalt, and feldspar
   b. Shale, sandstone, some of the limestones, and coal
   c. Marble, quartzite, and slate
4. The processes of rock weathering to produce *residual* and *transported* soil deposits
5. The role of glaciers and typical glacial deposits
6. The role of wind and typical wind-formed deposits
7. The role of water and typical fluvial water deposits
8. Groundwater, groundwater development, the concept of the water table, water well development, sources of springs, and development of artesian water
9. The role of subsurface water in erosion

The reader should supplement this chapter material with a good textbook on geology. The local library usually contains state geologic survey publications which go into considerable detail for that state's geology. In addition to the usual and numerous textbooks on general geology, the reader should be aware that separate geology textbooks are available which describe the geology of the several continents as well as separately considering glaciation and regional or areal geomorphology (landforms).

The reader should make it a habit, whenever the opportunity arises, to observe landforms and rock formations using exposed profiles in highway cuts as well as natural topography and stream problems and to observe rocks exposed at the ground surface and in stream beds and/or attempt to identify them.

PROBLEMS

3-1 Why are some igneous rocks fine-grained whereas others are coarse?

3-2 Discuss the typical profile likely to be obtained in an end or terminal moraine.

3-3 What type of profile may be obtained in an esker? Why?

3-4 Why is the *B* horizon soil preferable as a fill rather than the *C* horizon material?

3-5 Why is the soil-horizon concept not applied to transported soils?

3-6 Sketch the necessary conditions for a boring through a perched water table to drain it.

3-7 Sketch the necessary conditions under which a boring through an excavation would have to be plugged so that the excavation could be made.

3-8 Sketch the conditions showing how in a soil exploration program a filled-in slough or oxbow could be missed.

3-9 Sketch the conditions showing how a soil exploration program may miss a sedimented-in sink-hole. Explain how you might be able to discern that you are in an area of filled-in sinkholes.

3-10 Explain how you could increase the yield from a well into a chalk formation which is 1.5 m in diameter. *Hint:* What is an adit?

3-11 Explain how you might distinguish between a drumlin and a weathered mound of earth.

3-12 Explain how you might distinguish moraine (ridged) topography from ordinary weathering of residual soil.

3-13 How would you proceed to identify a rock fault area?

3-14 What are several problems which may develop in a jointed rock?

3-15 Obtain the state geologic survey bulletin from the library (if available), and write a rough description of the geology of the assigned area of the state.

FOUR

SOIL CLASSIFICATION TESTS AND COMMON SYSTEMS

4-1 GENERAL

Soils may be classified in a general way as cohesionless or cohesive as in Sec. 2-4, or as coarse- or fine-grained as in Sec. 2-10. These terms are too general to provide either a repeatable or reproducible identification of similar soils. Additionally, this classification is not adequate to identify whether the soil is suitable as a construction material.

A number of classification systems have been proposed in the recent past and on occasion some one proposes a new one or modifications to existing ones. Figure 4-1 illustrates several of the classification systems that have been proposed—only two of these are currently used. The most recent modification proposal is that of Mirza (1982) for the USC system following.

Of the several soil classification systems, only the following two will be further considered:

Unified Soil Classification System (USC)—most widely used system (and internationally) for foundation engineering such as dams, buildings, and similar. It is commonly used for airfield design and (outside the United States) for road earthwork specifications.

American Association of State Highway and Transportation Officials (AASHTO, formerly Bureau of Public Roads)—used almost exclusively by the several state Departments of Transportation and the Federal Highway Administration in earthwork specifications for transportation lines.

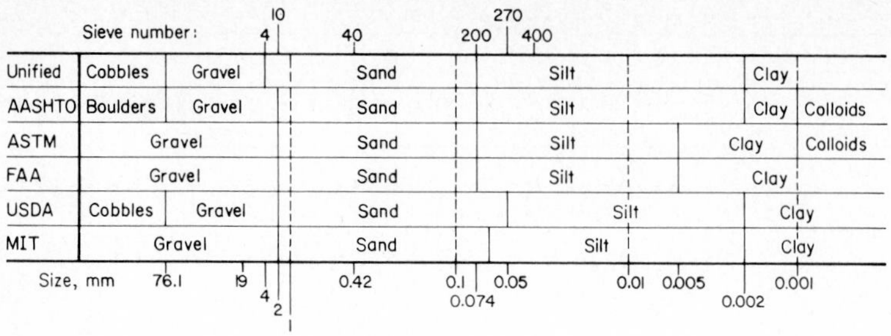

Sieve number:	10 4		40	200	270 400		
Unified	Cobbles	Gravel	Sand		Silt	Clay	
AASHTO	Boulders	Gravel	Sand		Silt	Clay	Colloids
ASTM	Gravel		Sand		Silt	Clay	Colloids
FAA	Gravel		Sand		Silt	Clay	
USDA	Cobbles	Gravel	Sand		Silt	Clay	
MIT	Gravel		Sand		Silt	Clay	
Size, mm	76.1	19 4 2	0.42	0.1 0.074	0.05	0.01 0.005 0.002	0.001

ASTM = American Society for Testing and Materials
USDA = U.S. Department of Agriculture
MIT = Massachusetts Institute of Technology

Figure 4-1 Several soil classification systems and range of grain sizes to identify the several soil particles.

The USC system was originally developed for military airfield construction during World War II and subsequently published with wide acceptance resulting. The AASHTO system was started by the (then) U.S. Bureau of Public Roads during 1927–1929 and revised in 1945 to include group indexing and addition of subgroups within the *A-2* group in Table 4-3. The number of soil tests was reduced from five to the same three used in the USC system.

The Federal Aviation Administration (FAA) had a separate soil classification system used for civil airfield construction where federal funds were used. This system seems to no longer be used, but information can be found in several of the older soil mechanics texts (including the previous edition of this one).

Soil classification systems as used in the context of this chapter are for the purpose of identifying soils in a systematic manner to determine suitability for use in specific applications based on past experience. A classification system is also useful in communicating soils information to other geographic areas and in building a data base from the experience of others. The communications factor is particularly important since all soils have a common origin from rock decomposition as pointed out in Chap. 3.

Those physical properties of use in predicting suitability of a soil as a construction material for fill as in earth dams and levees, for use in building sites as fill, for road fills, and similar are

Percentages of gravel, sand, and fines—requiring a sieve analysis
Shape of the grain size distribution curve—may require plotting the sieve analysis data
Plasticity (w_L, w_P, and I_P)—requiring Atterberg limits

The use of a soil classification system does not eliminate the need for detailed studies of soils or for testing for engineering properties. For example, the unit weight, compaction characteristics, performance under saturated conditions, susceptibility to frost action, strength, and so on, are not directly included in any of the classification systems.

4-2 SOIL TESTS FOR CLASSIFICATION

Both of the classification systems presented here use (1) liquid and plastic limit tests and (2) grain size analysis test.

The method of performing these three tests can be found in ASTM (volume 4.08) and in laboratory manuals (e.g., Bowles, 1978). Several limitations will, however, be discussed here to put "soil classification" into proper perspective.

Atterberg Limits

Terzaghi (1925) is generally credited with recognizing the use of the liquid and plastic limits as consistency index values which could be useful in soil classification. Atterberg's original procedure for determining the liquid limit was modified (Casagrande, 1932) to improve the reproducibility of the test.

The liquid and plastic limits are performed on cohesive soils that are usually air dried, pulverized, and sieved through a No. 40 sieve. Outside the United States similar sieve sizes are used—the No. 40 is 0.422 mm where several alternatives use a 0.400 mm mesh opening.

Oven drying is never done to produce a sievable soil. Air drying is commonly done but may lower the liquid limit from 2 to 6 percent unless the soil to be used is then prewetted for 24 to 48 hours prior to making the tests. Even prewetting may not fully recover the liquid limit for some soils. The plastic limit does not seem to be much affected by air drying.

Some laboratories suggest washing the sample through the No. 40 sieve and using the sediments for the liquid limit test. There are at least two disadvantages to this procedure; one is that the test takes much longer to run. More important, however, is that the sedimentation process segregates the clay particles so that very careful mixing on the part of the technician is necessary to produce the same distribution as in the raw soil. A procedure which is satisfactory in the author's opinion for soils which predominate in (−) No. 40 material but with occasional pieces of gravel is to visually remove the gravel by hand and then make the tests on the wet (or damp) soil.

The liquid and plastic limit tests are reasonably reproducible and repeatable even with inexperienced operators as shown by the two laboratory sections of student values shown in Table 4-1. While the plastic limit test has unacceptable scatter for commercial purposes, the liquid limit values show little scatter. Actually there is often little to base correctness of either the w_L or w_P in a commercial laboratory where a limited number of tests are made on any stratum of soil. Few laboratories have the resources to make as many tests as were possible to produce Table 4-1. From published sources, analyses by the author, and the very large series of tests published by the Waterways Experiment Station (Hammitt, 1966), it appears the standard deviation for carefully run tests is on the order of

$$w_L = \pm 3 \text{ percent} \qquad w_P = \pm 4 \text{ percent}$$

Table 4-1 Distribution of liquid and plastic limits for two student laboratory sections

Student no.	Section 1		Section 2	
	w_L	w_P	w_L	w_P
1	32.8	21.0	34.0	19.8
2	32.0	21.0	35.8	23.5
3	30.0	22.4	29.8	21.6
4	30.9	19.1	32.5	19.9
5	29.6	22.2	29.4	19.4
6	32.6	21.5	35.2	19.2
7	32.5	21.0	36.3	20.6
8	31.8	22.3	35.5	22.8
9	32.1	21.8	33.4	16.1
10	31.8	22.2	37.1	16.4
11	—	—	32.8	22.7

These values depend on soil type and may be only about half that shown above for soils of low plasticity.

The principal sources of error which affect reproducibility (between laboratories) and repeatability for the liquid and plastic limits are:

1. Care in preparing the soil to obtain all of the (−) No. 40 material and prewetting prior to the test.
2. Careful adjustment of the liquid limit machine to obtain a cup fall of 10 mm. Even small deviations from 10 mm can affect the liquid limit as much as 10 points.
3. Carefully controlling the quantity of soil in the liquid limit cup.
4. Careful attention to rolling the plastic limit thread to 3 mm and to a water content of "just crumbling." Use of a 3-mm rod for visual comparison is of considerable help.

Items 2 and 3 above are essentially limitations on the liquid limit machine and additionally it appears that the base (on which the cup impacts) and table on which the machine is set also affect w_L. The fall cone test has been used in Europe to increase the reproducibility of the liquid limit. This method purports to eliminate all the shortcomings of the current test as just itemized. Basic essentials of the fall cone test are shown in Fig. 4-2.

To use the fall cone test, one obtains the (−) No. 40 soil, just as for the liquid limit test. The difference is that it is a larger quantity and is placed in the cup shown in Fig. 4-2. The cone tip is brought into contact with the soil surface

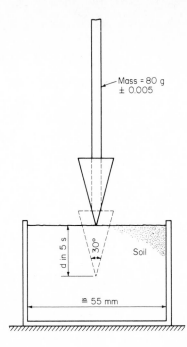

Figure 4-2 Fall cone test for liquid limit. Standard 30° cone is brought into contact with soil and clamped. At time zero cone is released and penetration at end of 5 seconds is recorded. Water content of soil is determined, cup emptied, and test repeated as necessary at other water contents to produce a semilog curve of w versus penetration.

and time-initialized. The cone is released for a free-fall into the soil and the penetration at the end of 5 seconds is recorded. Since the water content for a penetration depth of 20 mm (defines w_L) would be a chance occurrence, several trials at different water contents on both sides of 20-mm penetration are used to construct a semilogarithmic plot of penetration versus water content. One enters this curve at 20 mm (log scale) and reads the liquid limit on the water content scale similarly to obtaining the w_L from a blow count of 25 with the regular test.

The liquid limit can be statistically determined for a soil stratum, but in practice this is seldom done. In small building sites only a few values are determined in any layer—perhaps 1 to at most 5. This number of values is too few for any valid statistical conclusions so the values are either individually used or at most simply averaged.

Sieve Analysis

This test (ASTM D422, AASHTO T88) was briefly described in Sec. 2-12. The test has a number of limitations, including:

1. It is neither reproducible nor repeatable except by accident.
2. It is difficult to obtain truly representative test samples. Quantity is small and depends on grain size, i.e., larger samples for gravel than for sand or soils with large amounts passing the No. 200 sieve.

3. When more than about 10 percent (−) No. 200 material is present, it is particularly significant whether the sample is washed on the No. 200 sieve to determine the percent passing (the fines fraction).

Basically the grain size analysis consists of:

1. Obtaining a representative sample and reducing it to elemental particles by pulverizing with a mortar and pestle or washing on the No. 200 sieve.
2. Sieving the sample through a stack of four to six sieves and weighing the quantity retainined on each of the sieves. Common sieve sizes in use are illustrated in Table 4-2.
3. Computing the percent passing (or finer) for each sieve based on the cumulative weight retained on any sieve and the total sample weight.
4. Plot the percent passing versus the sieve opening (loosely termed the *grain diameter*) to a semilog plot.

The semilog plot (as Fig. 4-3) is used so that the grain diameter scale is extended to give all grain sizes approximately equal emphasis. This type of plot makes it easier to compare two or more soils plotted on the same graph since the curves will generally be more separated than using the arithmetic plot shown in Fig. 4-4.

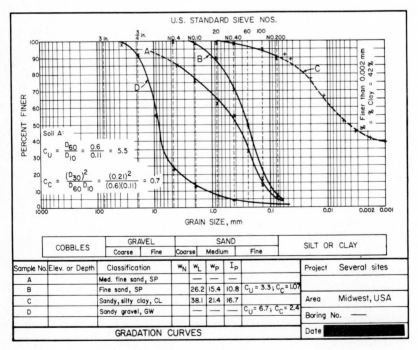

Figure 4-3 Grain-size distribution curves for several soils using a semilogarithmic plot. Here the plotting is made with grain sizes increasing from right to left.

Table 4-2 Standard U.S., British, French, and German sieves. U.S. sieves are all available in 20-cm diameters and most are available in 30.5-cm diameters

U.S.†		British standard‡		French§		German DIN¶	
Size or no.	Opening, mm	No.	Opening, mm	No.	Opening, mm	Designation, μm	Opening, mm
Size 4″	101.6						
3″	76.1						
2½″	64.0						
2″	50.8						
1¾″	45.3						
1½″	38.1						
1¼″	32.						
1″	25.4						25.0
¾″	19.0						20.0
							18.0
⅝″	16.0						16.0
½″	12.7						12.5
⅜″	9.51						10.0
5⁄16″	8.00						8.0
¼″ No. 3	6.35						6.3
No. 4**	4.76			38**	5.000		5.0
5	4.00			37	4.000		4.0
6	3.36	5**	3.353				
7	2.83	6	2.812	36	3.150		3.15
8	2.38	7	2.411	35	2.500		2.5
10	2.00	8	2.057	34	2.000		2.0
12	1.68	10	1.676	33	1.600		1.6
14	1.41	12	1.405	32	1.250		1.25
16	1.19	14	1.204				
18	1.00	16	1.003	31	1.000		1.0
20	0.841	18	.853				
25	0.707	22	.699	30	.800	800	.800
30	0.595	25	.599	29	.630	630	.630
35	0.500	30	.500	28	.500	500	.500
40††	0.420	36††	.422	27††	.400	400††	.400
45	0.354	44	.353	26	.315	315	.315
50	0.297	52	.295				

(*continued*)

Table 4-2—*Continued*

U.S.†		British standard‡		French§		German DIN¶	
Size or no.	Opening, mm	No.	Opening, mm	No.	Opening, mm	Designation, μm	Opening, mm
No. 60	0.250	60	.251	25	.250	250	.250
70	0.210	72	.211	24	.200	200	.200
80	0.177	85	.78	23	.160	160	.160
100	0.149	100	.152				
120	0.125	120	.124	22	.125	125	.125
140	0.105	150	.104	21	.100	100	.100
170	0.088	170	.089			90	.090
				20	.080	80	.080
200	0.074	200	.076			71	.071
230	0.063	240	.066	19	.063	63	.063
						56	.056
270	0.053	300	.053	18	.050	50	.050
325	0.044			17	.040	45	.045
400	0.037					40	.040

† ASTM E-11-70 (Vol. 14.02.).
‡ British Standards Institution, London BS-410.
§ French Standard Specifications, AFNOR X-11-501.
¶ German Standard Specification, DIN 4188.
** For standard compaction test.
†† For Atterberg limits.

An indication of the gradation may be numerically computed from the grain size curve for (+) No. 200 sieve sizes using the *coefficient of uniformity* C_U defined as

$$C_U = \frac{D_{60}}{D_{10}} \qquad (4\text{-}1)$$

The shape of the curve between the D_{60} and D_{10} grain sizes is somewhat defined with the *coefficient of concavity* C_C defined as

$$C_C = \frac{D_{30}^2}{D_{60} \cdot D_{10}} \qquad (4\text{-}2)$$

It is conventional procedure with grain size analyses to identify the grain size as D with the subscript identifying the percent finer—thus $D_{60} = 2.00$ is interpreted as 60 percent of the soil grains are smaller than 2.00 mm.

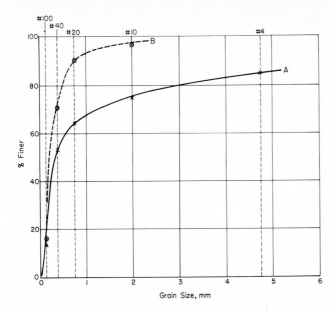

Figure 4-4 Replot of soils *A* and *B* of Fig. 4-3 using an arithmetic scale. Note how much detail is lost on the distribution of soil grain sizes less than 1 mm.

The values C_U and C_C are only used with the USC system and have no meaning when more than about 10 percent of the soil is $(-)$ No. 200 sieve. When C_U is large, we can say that there is a large difference between the D_{60} and D_{10} sizes as might occur with a distribution of sizes between these two percentages. The value C_C tends to identify the shape of the curve between the D_{60} and D_{10} sizes—a sag will produce a small value and a linear or convex downward shape will produce a larger value. A very large or small value of C_C indicates that some grain sizes are missing between D_{60} and D_{10}, so that the grading is not as desirable.

When several tests are run on the same soil the "percent finer" is given as a range from the lack of reproducibility and repeatability. For example, a specification may state the following:

Sieve No.	Percent passing
4	80–90
100	15–25

An individual sieve analysis on this soil producing 82 percent passing the No. 4 and 18 percent passing the No. 100 would meet specifications.

Hydrometer Analysis

The hydrometer analysis may be used to extend the range of the grain size distribution curve and to estimate sizes in that fraction smaller than the No. 200 sieve. An estimate of the "clay fraction" as the percent smaller than 0.002 mm may also be made. The hydrometer analysis is not directly used in any of the soil classification systems. Details of this test may be found in ASTM D422 or in Bowles (1978).

4-3 THE UNIFIED SOIL CLASSIFICATION SYSTEM

This system, originally developed for use in airfield construction, was reported by Casagrande (1948). It had already been in use since about 1942, but was slightly modified in 1952 to make it apply to dams and other construction.

The principal soil groups of this classification system are given in Table 4-3. As shown in the table under the column heading "Group Symbols," the soils are designated by group symbols consisting of a prefix and a suffix. The prefixes indicate the main soil types and the suffixes indicate the subdivisions within groups as follows:

Soil type	Prefix	Subgroup	Suffix
Gravel	G	Well graded	W
Sand	S	Poorly graded	P
		Silty	M
		Clay	C
Silt	M		
Clay	C	$w_L < 50$ percent	L
Organic	O	$w_L > 50$ percent	H
Peat	Pt		

A well-graded gravel is GW; a poorly graded sand is SP; a well-graded sand is SW; a silty sand is SM; a clay with a liquid limit > 50 percent is CH, etc. A verbal description† should accompany the classification symbols, e.g., brown, coarse, well-graded sand with trace of gravel, SW. Problems 2-16 through 2-19 illustrate the use of visual and verbal soil descriptions.

A soil is well graded or *nonuniform* if there is a wide distribution of grain sizes present, i.e., if there are some grains of each possible size between the upper and lower gradation limits. This can be ascertained by plotting the grain-size curve as illustrated in Fig. 4-3 and either observing the shape and spread of sizes or computing the coefficient of uniformity C_U and the coefficient of concavity C_C, as given by Eqs. (4-1) and (4-2), respectively.

---

† The ASTM (D2487) should be consulted for any requirements for classifying the soil where this standard is required.

A soil is poorly graded, or *uniform,* if the sample is mostly of one grain size or is deficient in certain grain sizes. A beach sand is an example of a uniformly graded soil.

Table 4-3 illustrates that only the sieve analysis and the Atterberg limits are necessary to classify the soil. A sieve analysis is performed and a plot of the grain-size distribution curve is made. When *less than 12 percent passes the No. 200 sieve,* it is necessary to obtain C_C and C_U to establish whether the soil is well or poorly graded. When more than 12 percent of the material passes the No. 200 sieve, the uniformity coefficient C_U and the coefficient of curvature C_C have no significance and only the Atterberg limits are used to classify the soil.

The Unified Soil Classification System defines a soil as:

1. *Coarse-grained* if more than 50 percent is retained on the No. 200 sieve
2. *Fine-grained* if more than 50 percent passes the No. 200 sieve

The coarse-grained soil is either:

1. *Gravel* if more than half of the coarse fraction is retained on the No. 4 sieve
2. *Sand* if more than half of the coarse fraction is between the No. 4 and No. 200 sieve size

The coarse-grained soil is:

GW, GP or SW, SP	≤ 5 percent passes No. 200 sieve
GW-GM, GP-GM, GW-GC, GP-GC or SW-SM, SP-SM, SW-SC, SP-SC	$5 <$ percent passing No. 200 sieve ≤ 12
GM, GC or SM, SC	> 12 percent passes No. 200 sieve

Classification of coarse-grained soils depends primarily on the grain-size analysis and particle size distribution. Note carefully that it is possible for a granular soil to have, say, 59 percent pass the No. 4 sieve and be called a "gravel" depending on the percent passing the No. 200 sieve. Add a little sand so that the percent that passes the No. 4 sieve increases to, say, 61 percent and the soil may be classified as a sand. The engineering properties would be about the same in both cases, but the classification is considerably different. A dual classification symbol could be used, such as GW-SW, but these were not included in the original classification to keep the system as simple as possible; also, others might not know what this means. A major classification change with a small increase or decrease in the percent passing the No. 4 or No. 200 sieve is another reason why it is necessary to include a verbal description along with the symbols, i.e., very sandy gravel, very gravelly sand, etc. *Peat* is a woody, fibrous material classified solely on the basis of visual appearance.

Classification of fine-grained soils also requires the use of the plasticity or A chart shown in Table 4-3. Each soil is grouped according to the coordinates of the plasticity index and liquid limit. On this chart an empirical line (the A line) separates the inorganic clays (C) from silts (M) and organic (O) soils. Although the silty and organic soils overlap areas, they are easily differentiated by visual examination (dark color, presence of organic materials) and odor.

Organic clay is now recognized and is identified as "organic" and may plot either above or below the A line. It may be determined visually or by color and odor. Alternatively, ASTM suggests that the suspect material be oven dried and Atterberg limits performed. If the oven-dry limits are less than 75 percent of the original limits, the soil is "organic."

The terms "lean" for $w_L < 50$ and "fat" for clays with $w_L > 50$ are suggested. Also, borderline cases should be classified on the basis of the poorer material.

Most inorganic clays lie fairly close to the A line. Kaolin clays (described in Chap. 5) tend to plot below the A line as inorganic silts (ML or MH) because of the similarity of engineering properties. Some active clays such as bentonite may lie above the A line and/or very close to the U line even though the liquid limit may be several hundred percent. In general, however, soils that have the same classifications tend to have the same engineering behavior.

The upper limit line (U line) shown in the A chart represents approximately the upper range of plasticity index and liquid limit coordinates found thus far for any soils. Any soil plotting to the left of the U line should be suspect and the limits rechecked as the first step in the classification sequence.

Example 4-1 Given the classification data for the following three soils, classify the soils using the Unified Soil Classification System.

Percent passing	Soil		
	A	B	C
No. 4 sieve	42	72	95
10	33	55	90
40	20	48	83
100	18	42	71
200	14	38	55
w_L, percent	35	39	55
w_P, percent	22	27	24
Visual observation	Dark tan, very gravelly	Grayish brown, some odor	Blue-gray, traces of gravel

Table 4-3 Unified soil classification chart

Including identification and description

COARSE-GRAINED SOILS More than half of material is larger than No. 200 sieve size (The No. 200 sieve size is about the smallest particle visible to the naked eye)				Field identification procedures (Excluding particles larger than 75 mm and basing fractions on estimated weights)		
	GRAVELS More than half of coarse fraction is larger than No. 4 sieve size.		CLEAN GRAVELS (Little or no fines)	Wide range in grain size and substantial amounts of all intermediate particle sizes.		
				Predominantly one size or a range of sizes with some intermediate sizes missing.		
			GRAVELS WITH FINES (Appreciable amount of fines)	Nonplastic fines (for identification procedures see ML below).		
				Plastic fines (for identification procedures see CL below).		
	SANDS More than half of coarse fraction is smaller than No. 4 sieve size. (For visual classifications, the 6 mm size may be used as equivalent to the No. 4 sieve size.)		CLEAN SANDS (Little or no fines)	Wide range in grain sizes and substantial amounts of all intermediate particle sizes.		
				Predominantly one size or a range of sizes with some intermediate sizes missing.		
			SANDS WITH FINES (Appreciable amount of fines)	Nonplastic fines (for identification procedures see ML below).		
				Plastic fines (for identification procedures see CL below).		
FINE-GRAINED SOILS More than half of material is smaller than No. 200 sieve size				Identification procedures on fraction smaller than No. 40 sieve size		
		SILTS AND CLAYS Liquid limit less than 50		Dry strength (Crushing characteristics)	Dilatancy (Reaction to shaking)	Toughness (Consistency near plastic limit)
				None to slight	Quick to slow	None
				Medium to high	None to very slow	Medium
				Slight to medium	Slow	Slight
		SILTS AND CLAYS Liquid limit greater than 50		Slight to medium	Slow to none	Slight to medium
				High to very high	None	High
				Medium to high	None to very slow	Slight to medium
HIGHLY ORGANIC SOILS				Readily identified by color, odor, spongy feel, and frequently by fibrous texture.		

(continued)

Table 4-3—*Continued*

Including identification and description

Group symbols	Typical names	Information required for describing soils
GW	Well-graded gravels, gravel-sand mixtures; little or no fines.	Give typical name; indicate approximate percentages of sand and gravel, max. size, angularity, surface condition, and hardness of the coarse grains; local or geologic name and other pertinent descriptive information; and symbol in parentheses.
GP	Poorly graded gravels, gravel-sand mixtures; little or no fines.	
GM	Silty gravels, poorly graded gravel-sand-silt mixtures.	
GC	Clayey gravels, poorly graded gravel-sand-clay mixtures.	For undistorted soils add information on stratification, degree of compactness, cementation, moisture conditions, and drainage characteristics.
SW	Well-graded sands, gravelly sands; little or no fines.	
SP	Poorly graded sands, gravelly sands; little or no fines.	EXAMPLE: Silty sand, gravelly; about 20 percent hard, angular gravel particles 12 mm maximum size; rounded and subangular sand grains coarse to fine; about 15 percent nonplastic fines with low dry strength; well compacted and moist in place; alluvial sand; (SM).
SM	Silty sands, poorly graded sand-silt mixtures.	
SC	Clayey sands, poorly graded sand-clay mixtures.	
ML	Inorganic silts and very fine sands, rock flour, silty or clayey fine sands with slight plasticity.	Give typical name; indicate degree and character of plasticity, amount and maximum size of coarse grains; color in wet condition, odor if any, local or geologic name, and other pertinent descriptive information; and symbol in parentheses.
CL	Inorganic clays of low to medium plasticity, gravelly clays, sandy clays, silty clays, lean clays	
OL	Organic silts and organic silt-clays of low plasticity.	
MH	Inorganic silts, micaceous or diatomaceous fine sandy or silty soils, elastic silts.	For undistorted soils add information on structure, stratification, consistency in undisturbed and remolded states, moisture and drainage conditions.
CH	Inorganic clays of high plasticity, fat clays.	
OH	Organic clays of medium to high plasticity.	EXAMPLE: Clayey silt, brown; slightly plastic; small percentage of fine sand; numerous vertical root holes; firm and dry in place; loess; (ML).
Pt	Peat, muck, peat-bog, etc	

Table 4-3—*Continued*

Including identification and description

	Laboratory classification criteria

Use grain-size curve in identifying the fractions as given under field identification

Determine percentages of gravel and sand from grain-size curve. Depending on percentage of fines (fraction smaller than No. 200 sieve size), coarse-grained soils are classified as follows:

GW, GP, SW, SP,

GM, GC, SM, SC,

<u>Borderline</u> cases requiring use of dual symbols.

Less than 5 percent

More than 12 percent

5 to 12 percent

$C_U = \dfrac{D_{60}}{D_{10}}$ Greater than 4 $C_C = \dfrac{(D_{30})^2}{D_{10} \times D_{60}}$ Between 1 and 3

Not meeting all gradation requirements for GW

| Atterberg limits below "A" line, or I_P less than 4 | Above "A" line with I_P between 4 and 7 are <u>borderline</u> cases requiring use of dual symbols. |
| Atterberg limits above "A" line with I_P greater than 7 | |

$C_U = \dfrac{D_{60}}{D_{10}}$ Greater than 6 $C_C = \dfrac{(D_{30})^2}{D_{10} \times D_{60}}$ Between 1 and 3

Not meeting all gradation requirements for SW

| Atterberg limits below "A" line or I_P less than 4 | Above "A" line with I_P between 4 and 7 are <u>borderline</u> cases requiring use of dual symbols. |
| Atterberg limits above "A" line with I_P greater than 7 | |

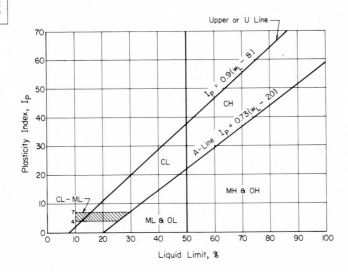

Plasticity or *A* chart for use in the Unified Soil Classification system. Extend as necessary for $w_L > 100$.

133

SOLUTION Since more than 12 percent passes the No. 200 sieve, we can immediately eliminate GW, GP, SW, and SP as possible classifications for all three soils.

(a) *For soil A:*

    (1) Less than 50 percent passes the No. 4 sieve; therefore, the soil must be predominantly gravel, thus *G*.

    (2) Considering the location of the $w_L = 35$ percent and $I_P = 13$ (computed) on the *A* chart, we find a CL.

    (3) From the preceding two observations and the visual description of this soil, soil *A* is: Dark tan, clayey gravel, GC.

(b) *For soil B:*

    (1) Less than 50 percent passes the No. 200 sieve; therefore, the soil is coarse-grained (either sand or gravel).

    (2) Compute the percent passing the No. 4 sieve and retained on the No. 200 sieve as

$$72 - 38 = 34 \text{ percent sand}$$

$$100 - 72 = 28 \text{ percent gravel}$$

Therefore, of the coarse fraction more than half is sand.

    (3) More than 12 percent passes the No. 200 sieve, and from the Atterberg limits, the soil plots below the *A* line ($w_L = 39$ and $I_P = 12$); thus, the (−) No. 40 fraction is an ML. Noting that the percentage of sand and gravel are nearly equal, soil *B* is grayish-brown, very gravelly, silty sand with trace of organic material, SM.

(c) *For soil C:*

    (1) With 55 percent passing the No. 200 sieve, the soil is fine-grained.

    (2) Using $w_L = 55$ percent and $I_P = 31$, the soil plots above the *A* line and also above the line of $w_L > 50$; therefore, soil *C* is blue-gray, fat, sandy clay with a trace of gravel, CH.

////

It was originally suggested by Casagrande (1948) that the fine-grained soils might better be classified as:

Soil	$w_L =$	0		35		50	
Clay			CL		CI		CH
Organic			OL		OI		OH
Silt			ML		MI		MH

This intermediate plasticity classification is not used at present and is not recommended.

With the coarse-grained soils, an additional subgrouping of GU and SU is sometimes used. For example, in Great Britain the "U" indicates a uniform gravel or sand, such as certain gravel deposits and some dune and beach sands which consist of primarily one or two sizes.

4-4 THE AASHTO SOIL CLASSIFICATION SYSTEM

The original BPR classification system of the late 1920s has been revised several times. It classifies soils into eight groups, A-1 through A-8, and originally required the following data:

1. Grain-size analysis
2. Liquid and plastic limits and the calculated I_P
3. Shrinkage limit
4. Field moisture equivalent—the maximum moisture content at which a drop of water placed on a small surface will not be immediately adsorbed
5. Centrifuge moisture equivalent—a test to measure capacity of soil to hold water (Dry soil is soaked in water for 12 hours and then centrifuged for 1 hour; the final water content thus obtained is the CME.)

The revised (Proc. 25th Annual Meeting of Highway Research Board, 1945) system retained the eight basic soil groups but added two subgroups in A-1, four subgroups in A-2, and two subgroups in A-7. Soil tests (4) and (5) were deleted, so that the only soil tests required are the *grain-size* analysis and the *liquid* and *plastic limits*. This revised classification system was adopted by AASHTO as standard M-145. Table 4-4 illustrates the current AASHTO soil classification system. Soil group A-8 is not shown, but is peat or muck based on a visual classification. Shown are groups A-1 through A-7 with two subgroups in A-1, four subgroups in A-2, and two subgroups in A-7, for a total of 12 soil subgroups in this classification system (exclusive of peat and/or muck). Figure 4-5 can be used to obtain the graphic ranges of w_L and I_P for the A-4 to A-7 groups and for the subgroup classification in the A-2 subgroup. A group index can be computed using Eq. (4-3) or Fig. 4-6 to further compare soils within a subgroup.

In general, this classification system rates a soil as:

1. Poorer for use in road construction as one moves from left to right in Table 4-4, i.e., A-6 soil is less satisfactory than A-5 soil
2. Poorer for road construction as the group index increases for a particular subgroup, i.e., an A-6(3) is less satisfactory than an A-6(1)

Table 4-4 AASHTO soil classification system

Note that A-8, peat or muck, is by visual classification and is not shown in the table.

General classification	Granular materials (35% or less passing No. 200)							Silt-clay materials (More than 35% passing No. 200)			
	A-1		A-3	A-2				A-4	A-5	A-6	A-7
Group classification	A-1a	A-1b		A-2-4	A-2-5	A-2-6	A-2-7				A-7-5; A-7-6
Sieve analysis: Percent passing:											
No. 10	50 max.										
No. 40	30 max.	50 max.	51 min.								
No. 200	15 max.	25 max.	10 max.	35 max.	35 max.	35 max.	35 max.	36 min.	36 min.	36 min.	36 min.
Characteristics of fraction passing No. 40:											
Liquid limit:				40 max.	41 min.	40 max.	41 min.	40 max.	41 min.	40 max.	41 min.
Plasticity index	6 max.		N.P.	10 max.	10 max.	11 min.	11 min.	10 max.	10 max.	11 min.	11 min.
Group index	0		0	0	0	4 max.		8 max.	12 max.	16 max.	20 max.
Usual types of significant constituent materials	Stone fragments, gravel, and sand		Fine sand	Silty or clayey gravel and sand				Silty soils		Clayey soils	
General rating as subgrade	Excellent to good							Fair to poor			

136

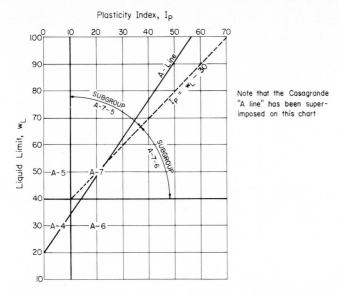

Figure 4-5 Liquid limit and plasticity index ranges for silt-clay soils (*A*-4 through *A*-7).

4-5 GENERAL DESCRIPTION OF AASHTO SOIL CLASSIFICATION SUBGROUPS

The *A*-1 through *A*-3 soils are *granular* with not more than 35 percent of the material passing the No. 200 sieve.

A typical material in group *A*-1 is a well-graded mixture of gravel, coarse sand, fine sand, and a binder [(−) No. 200] material which has little to no plasticity. Subgroup *A*-1*a*, which may contain appreciable gravel, is a coarser-graded material than *A*-1*b*, which is predominantly coarse sand. The binder material in this group may have a small amount of plasticity ($I_P \leq 6$).

The *A*-3 soil is fine, relatively uniform sand, typically a fine beach sand or desert blown sand. The *A*-3 subgroup may also include stream-deposited mixtures of poorly graded fine sands with traces of coarse sand and gravel. The silt or rock flour fractions, if any, passing the No. 200 sieve are *nonplastic* (NP).

Group *A*-2 is also granular but with appreciable (but not more than 35 percent) material passing the No. 200 sieve. These materials are on the borderline between the materials falling in groups *A*-1 and *A*-3 and the silt-clay materials of groups *A*-4 through *A*-7. Subgroups *A*-2-4 and *A*-2-5 include various materials in which not more than 35 percent is finer than the No. 200 sieve and which have the plasticity characteristics of the *A*-4 and *A*-5 groups. Subgroups *A*-2-6 and *A*-2-7 are similar to *A*-2-4 and *A*-2-5 except that the plasticity characteristics of the (−) No. 40 sieve fraction are those of the *A*-6 and *A*-7 groups. For example, if the soil has

$$w_L \leq 40 \text{ percent} \qquad I_P \geq 11 \qquad \text{Group index } GI \leq 16$$

and not more than 35 percent passes the No. 200 sieve, the soil is an *A*-2-6 since the plasticity characteristics are those of an *A*-6 soil as shown in Table 4-4. If more than 35 percent of the material had passed the No. 200 sieve, the soil would contain "appreciable" fines and be classified as an *A*-6.

Groups *A*-4 through *A*-7 are considered to be fine-grained soils, and all have more than 35 percent of the material passing the No. 200 sieve.

Soil group *A*-7 is further subdivided to

$$A\text{-}7\text{-}5 \text{ if } I_P < (w_L - 30)$$

$$A\text{-}7\text{-}6 \text{ if } I_P > (w_L - 30)$$

Figure 4-5 can be used to quickly classify the *A*-7 subgroups.

Soil group *A*-8 is peat (very organic) or muck (thin, very watery, and with considerable organic material) and is identified by inspection of the deposit.

4-6 THE AASHTO GROUP INDEX

To establish the relative ranking of a soil within a subgroup, the group index *GI* was developed. The group index is a function of the percent of soil passing the No. 200 sieve and the Atterberg limits. The group index can be obtained as the

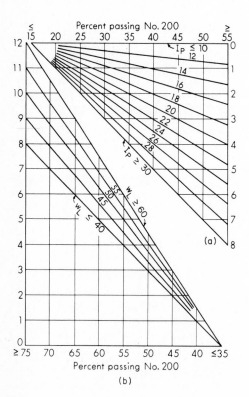

Figure 4-6 Charts to obtain group index of a soil. Group index equals the sum of readings from vertical of parts *a* and *b*.

sum of the values from Fig. 4-6a and b, a graphical presentation of the following equation:

$$GI = 0.2a + 0.005ac + 0.01bd \qquad (4\text{-}3)$$

where a = that part of the percent passing the No. 200 sieve greater than 35 and not exceeding 75, expressed as a whole number (range = 1 to 40)

b = that part of the percent passing the No. 200 sieve greater than 15 and not exceeding 55, expressed as a whole number (range = 1 to 40)

c = that part of the liquid limit greater than 40 and not greater than 60, expressed as a whole number (range = 1 to 20)

d = that part of the plastic index greater than 10 and not exceeding 30, expressed as a whole number (range = 1 to 20)

The group index should be rounded to the nearest whole number and placed in parentheses, as

$$A\text{-}2\text{-}6(3)$$

In general, the larger the group index value, the less desirable the soil for highway construction use within that subgroup.

Example 4-2 Given the same soil classification data as in Example 4-1.

REQUIRED Classify the three soils using the AASHTO classification system.

SOLUTION

(a) Classifying soil A:
(1) Proceeding from left to right in Table 4-4, the soil will be either an A-1, A-3, or A-2, since only 14 percent passes the No. 200 sieve.
(2) Based on $I_P = 13$ (computed), we eliminate A-1 and A-3.
(3) With $w_L = 35$ percent and $I_P = 13$, the soil fits the A-2-6 classification.
(4) The group index can be computed as

$$GI = 0.2(0) + 0.005(0)(0) + 0.01(0)(3) = 0.0$$

The group index is more conveniently obtained as the sum of the values from Fig. 4-6a and b:

$$\text{Fig. 4-6}a \cong 0$$

$$\text{Fig. 4-6}b \cong 0$$

$$GI = \overline{0}$$

(5) From inspection of the sieve analysis data and the classification data, soil A is dark tan, *silty* or *clayey sandy gravel*, A-2-6(0).

(b) *Soil B:*

(1) Proceeding from left to right in Table 4-4, the soil can only be an *A*-4, *A*-5, *A*-6, or *A*-7, since 38 percent passes the No. 200 sieve.

(2) Based on $I_P = 12$, the soil can only be an *A*-6 or *A*-7.

(3) With $w_L = 39$ percent, the soil is an *A*-6.

(4) The group index is

$$\text{Fig. 4-}6a = 0.5$$

$$\text{Fig. 4-}6b = 0.4$$

$$GI = \overline{0.9} \rightarrow 1.0$$

(5) From inspection of the sieve analysis data (31 percent gravel, 33 percent sand) and data just obtained, soil *B* is a grayish brown, *very gravelly sandy silt* or *clay* with trace of organic material, *A*-6(1).

(c) *Soil C:*

(1) With 55 percent passing the No. 200 sieve, the soil is an *A*-4, *A*-5, *A*-6 or *A*-7.

(2) With $w_L = 55$ percent and $I_P = 31$, the soil is an *A*-7-6 since $I_P > w_L - 30$ (also from Fig. 4-5).

(3) The group index is

$$\text{Fig. 4-}6a = 8$$

$$\text{Fig. 4-}6b = 5.8$$

$$GI = \overline{13.8} \qquad \text{say 14}$$

(4) Soil *C* is a blue-gray *sandy clay with trace of gravel*, *A*-7-6(14).

/////

4-7 FIELD IDENTIFICATION TESTS

Coarse-grained soils such as gravel, sand, sandy gravel, gravelly sand, etc., can be readily identified by inspection. Traces of silt and/or clay are somewhat more difficult to identify when mixed with these materials but may not be of much importance unless the quantity is over 5 to 10 percent. The sedimentation test described for fine-grained soils may be used to determine if significant quantities of silt, very fine sand, or clay are present.

Fine-grained soils can be identified using some, or all, of the following tests performed on approximately the ($-$) No. 40 sieve sizes (remove the larger particles by hand rather than actually sieving).

1. *Dilatancy* (or pore water mobility reaction to shaking). Prepare a pat of moist soil with a volume of 1 to 3 cm^3, using enough water to make the soil soft but not sticky. Place the pat in the open palm of one hand and jar the hand vigorously with the other. If the soil is fine sand, silt, or silty fine sand, the inertia forces due to jarring will force the water to the surface of the soil pat, and it will appear wet or glossy. When the sample is manipulated, this surface

water will disappear. In soils with substantial clay this test produces no reaction.

2. *Dry strength* (resistance of dry lumps to crushing). Mold a pat of soil to about the consistency of putty by adding water as necessary. Allow the pat to completely dry and then test the crushing strength by breaking or crumbling between the fingers. The dry strength increases with increasing plasticity. High dry strength is characteristic for clays of the CH group, lesser dry strength for CL and MH soils, and very low to nonexistent strength for OL and ML soils. Fine sand, silt, and sand-silt mixtures possess almost no dry strength. Note that one may do this test approximately on naturally dried in situ soil.

3. *Toughness* (consistency near the plastic limit). Take a specimen of about 1 cm^3 and mold it to the consistency of putty. Proceed to roll the soil in the palm (or on a smooth surface) into a thread about 3 mm in diameter. When the pat of soil crumbles and loses its plasticity, the plastic limit has been reached. The higher the resistance of the 3-mm thread to pulling apart, the higher the position of the soil on the plasticity chart with respect to the *A* line. A weak thread which is easily crumbled indicates silts or inorganic clays of low plasticity. Highly organic clays are also very weak but may feel spongy at the plastic limit.

4. *Sedimentation.* Place about 50 g (more for gravelly soils) in a glass jar, such as a beaker, test tube, glass graduate, or other jar on the order of 150 mm deep, with water to fill the jar. Vigorously shake for several minutes and allow to stand. Gravel and coarse sand will settle almost instantly. Medium to very fine sand will take not more than 1 to 3 minutes; silt will take not more than 15 minutes. Clay will take only slightly longer unless a deflocculating agent (see Sec. 5-8) is added. The relative thickness of the sediments is an indication of percentages of various grain sizes.

5. *Color.* In general, dark colors such as black, gray, and dark brown indicate organic soils.

6. *Odor.* Organic soils usually have a distinctive smell of decaying materials. This test should be applied to fresh samples which are still wet. Roots, pieces of weeds, wood, plants, etc., may be visually present as further aid.

7. *Feel.* Sands and silts dry rapidly and can be dusted from the hands easily. Clay tends to leave considerable discoloration after drying and the hands may have to be washed to remove all traces. Clay tends to be smooth to the touch or to leave a smooth streak when a spatula blade is moved across a wet mass. Silts and sands are rough and gritty and leave grain marks when a spatula blade is moved across a wet lump.

4-8 SUMMARY

This chapter has presented the two common systems of soil classification and a set of field identification tests widely used in foundation engineering. The Unified system is used both in the United States and, with only minor modification (if any), abroad. The AASHTO system is used by most of the state highway departments in the United States and some abroad.

Table 4-5 Comparison of the AASHTO and Unified soil classification groups

AASHTO	Unified
A-1a	GW, GP, SW, GM
A-1b	SW, SP, SM, GC
A-3	SP
A-2-4	CL, ML
A-2-5	CL, ML, CH, MH
A-2-6	CL, ML
A-2-7	CL, ML, CH, MH
A-4	CL, ML
A-5	CL, ML, CH, MH
A-6	CL, ML
A-7	CL, ML, CH, MH
A-8	Peat and muck or organic

Both classification systems require:

1. A sieve analysis using at least the Nos. 10, 40, and 200 sieves. If the Unified system is used and less than 12 percent passes the No. 200 sieve, enough additional sieves must be used to plot a reasonable grain-size curve.
2. Determination of the liquid and plastic limits.
3. Applying the process of elimination to classify the soil using the grain size and plasticity data.

Table 4-5 compares the soil classifications of the two systems used in this chapter, and it is readily seen that the systems overlap considerably.

Limitations on making reproducible (between laboratories) or repeatable (same operator) soil tests for classification have been shown to illustrate that it is possible to obtain alternate classifications on the same soil even with carefully performed tests. Only minor inattention to test details can result in major misclassification of some soils. Since there is a somewhat gradual transition in construction properties with the systems, in most cases classification errors are not serious. Exceptions are when carelessness results in too little soil passing the No. 200 sieve or obtaining a $w_L < 50$ when it is greater, or vice versa.

Note that there is not complete agreement on the particle size division between gravel, sand, silt, and clay. The following division, however, seems to be the most widely accepted and is satisfactory for geotechnical work.

Particle size	4.76 mm		0.074 mm		0.002 mm	
Sieve No.	4		200		—	
	Gravel		Sand		Silt		Clay

These arbitrary definitions are based on particle size, and it is possible for clay mineral particles, or plateiets, to be slightly larger than 0.002 mm. Likewise, it is possible for particle sizes of 0.002 mm or less to be rock flour or colloids instead of clay minerals, as further discussed in Chap. 5.

Another problem in soil classification occurs between countries (refer to Table 4-2). The following stacks of sieves might be used for grain-size analysis in the United States and Great Britain:

United States			Great Britain		
No. 4	4.76 mm	← sand and gravel division →	No. 7	2.411 mm
10	2.00		14	1.204
20	0.841		25	0.599
40	0.420	← for Atterberg limits →	36	0.422
60	0.250		72	0.211
100	0.149		100	0.152
200	0.074		200	0.076
	Pan			Pan

These typical sieve stacks also illustrate the small differences between the two countries for the sand and gravel size division and the sieve numbers used to obtain the soil fraction for determining the Atterberg limits.

PROBLEMS

4-1 For the assigned set of liquid and plastic limit data in Table 4-1:

(a) Compute the arithmetic mean for w_L, w_P and I_P.

(b) Compute the standard deviation for w_L, and w_P.

(c) Using the above information, compute I_P and give an estimate of the reliability of this computation. Be sure to comment on significance of values.

4-2 A sieve analysis performed on two soils produced the following data:

	Percent passing	
Sieve no.	Soil A	Soil B
4	85.0	100.0
10	72.2	94.0
30	53.6	77.1
60	34.4	43.8
100	26.1	19.6
200	5.1	8.8

Required: plot the grain-size curves for both soils on the same graph. Find D_{10} and D_{85} and compute C_C and C_U.

Ans: Soil A: $D_{10} = 0.09$ mm; $C_C = 0.6$; $C_U = 8$ (student answers may differ slightly because of plotting accuracy).

4-3 A grain-size analysis was performed on two soils as follows:

Sieve no.	Percentage passing, soil A	Percentage passing, soil B	
4	98.0	100	
10	88.2	100.0	
20	72.3	84.5	
40	54.1	61.3	
60	32.7	53.5	
100	17.1	39.2	
200	7.7	35.1	
		0.05 mm	23.2 Hydrometer
		0.01	15.8
		0.005	9.7
		0.001	3.4

Required: plot the two grain-size curves on the same graph sheet, find D_{10} and D_{85} for both soils, and compute C_C and C_U as appropriate. Estimate the clay fraction of soil B.

Ans: Soil B: C_C and C_U, no meaning; $D_{10} = 0.007$ mm; $D_{85} = 0.84$ mm, percent clay = 4.

Use the following data for Probs. 4-4 through 4-6.

In all classification problems, in addition to the group classification symbols, give a description of the soil (sandy clay, sandy clay with trace of gravel, etc.) as appropriate for the sieve analysis data.

	Percent passing				
	Soil				
Sieve	A	B	C	D	E
4	49	76	82	97	—
10	35	67	71	92	98
40	26	40	33	85	91
100	20	36	22	76	75
200	14	27	4	54	62
w_L, percent	32.6	41.3	NP	53.4	48.3
w_P, percent	21.5	22.3	—	31.6	23.1
Visual:	Dark brown	Dark tan	Light brown	Dark gray with woody odor	Reddish brown

4-4 Classify soils A, B, D, and E from the above groups using the Unified Soil Classification system. Do not plot a grain-size curve unless necessary.

4-5 Classify the assigned soils from the above group using the AASHTO classification system.

4-6 Plot a grain-size distribution curve for soil C, compute C_U and C_C, and classify in the Unified Soil Classification system.

4-7 Classify soils A and B of Prob. 4-2 using both the Unified and AASHTO systems.

4-8 Classify soils A and B of Prob. 4-3 using both the Unified and AASHTO systems. Take $w_L = 32.6$ and 46.5 percent, respectively, and $w_P = 16.3$ and 26.5 percent for soils A and B, respectively.

SOIL STRUCTURE AND CLAY MINERALS

5-1 SOILS AND SOIL FORMATION

Chapter 4 considered soil in terms of whether it was cohesionless (coarse-grained) or cohesive (fine-grained). These classifications used a visual grain description together with consistency index properties to describe the soil as gravel, sand, silt, clay, or a mixture, as sandy clay, silty sand, etc.

Chapter 3 considered soil formation as a cyclic geological process involving the weathering of rocks. The soil was then considered to be either a *residual* or a *transported* soil, and the agents of transportation and resulting deposits were considered in some detail.

This chapter will consider soil in terms of its composition, structure, and, in the case of clay, the clay minerals and their very important influence on soil behavior. Soil will at this point be further defined as follows:

1. The unconsolidated material of the earth's crust used to build upon or used as a construction material.
2. The loose, or unconsolidated, materials overlying bedrock produced by rock weathering.

In addition to soil, bedrock at shallow depths is often of interest, particularly in terms of quality, to the geotechnical engineer.

5-2 SOIL STRUCTURE AND FABRIC

Soil structure is both the geometric, skeletal arrangement of the particles, or mineral grains, and the interparticle forces which may act on them. Soil structure includes gradation, arrangement of particles, void ratio, bonding agents, and associated electrical forces. The structure is the property which produces a response to external changes in the environment, such as loads, water, temperature, and other factors.

Soil fabric is a more recently introduced (mid-1960s) term to describe the "structure" of clays. Fabric denotes the geometric arrangement of the mineral particles in a clay mass as observed by optical or electron microscopes. The geometric arrangement includes particle spacing and pore size distributions.

5-3 SOIL COHESION AND FRICTION

A measure of the attraction between cohesive soil particles is *cohesion* (symbol c). Similarly resistance to relative particle displacement in cohesionless soils is friction. Friction in the sense used here has the usual meaning in that the friction force F_f is

$$F_f = vN$$

where v = friction coefficient
N = normal force between particles

Many soils (including "cemented" cohesionless soils) exhibit both cohesion and friction resistance to interparticle displacement.

The friction coefficient in geotechnical work (refer to Fig. 5-1) is taken as

$$v = \tan \phi$$

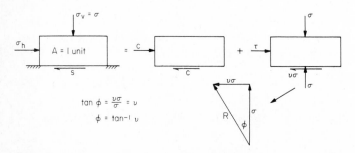

Figure 5-1 Relationship between angle of internal friction ϕ and normal pressure. Note use of superposition of cohesion and friction resistance to obtain total shear strength.

Table 5-1 Typical values of e_{max}, e_{min}, unit weight and angle of internal friction (used in Chap. 13) for several soils

	Loose			Dense		
	Dry unit weight	e_{max}†	ϕ‡	Dry unit weight	e_{min}†	ϕ‡
Gravel	16.0–18.0	0.62–0.44	32–36	18.0–20.0	0.44–0.30	35–50°
Coarse sand	15.0–17.5	0.73–0.50	32–38	17.5–19.6	0.50–0.33	35–48
Clayey sand	14.0–16.5	0.86–0.58	28–32	16.5–18.5	0.58–0.40	35–40
Silty sand	12.6–15.5	1.05–0.68	28–32	15.5–17.5	0.68–0.49	32–38
Fine sand	14.0–18.5	0.86–0.40	27–33	15.5–18.0	0.68–0.44	33–39
Sandy gravel	15.0–18.0	0.73–0.44	30–38	18.0–22.0	0.44–0.18	36–45
Gravelly sand	15.0–18.0	0.73–0.44	30–38	18.0–22.5	0.44–0.16	36–50
Silt	14.0–15.5	0.86–0.68	20–30	15.5–17.5	0.68–0.49	25–32

† Depends on G_s generally ranging from 2.65–2.72.
‡ Use higher values for angular particles.

where ϕ is termed the *angle of internal friction* of the soil. A lower limiting value for a dry sand, gravel, or sand-gravel mixture may be estimated as the angle formed from carefully pouring a quantity into a pile and measuring the angle of the resulting slope. The pile of soil is in about a minimum dense state e_{max} and for a sand will have an angle of repose (angle of internal friction) of about 30° (see Table 5-1).

In soil, either (or both) ϕ and c may be greater than zero, or zero—but not less than zero. If the soil state is such that both are zero, a very dangerous condition exists. These states will be considered in more detail in Chap. 13.

Since resistance to particle displacement, or shear strength, is also dependent on how well the particles are packed (density), particle interlocking (shape), degree of confinement, degree of saturation, and other factors, both cohesion and friction are statistical accumulations of these apparent effects for the instant soil state and under the conditions of measurement or test procedure. They are also subject to interpretation by the engineer. In any case they are widely used parameters which are state-dependent and not unique.

5-4 GRANULAR SOIL STRUCTURE

The arrangement of the individual soil particles in a granular soil may be termed *packing*. Packing of grains of soil, or any other particulate medium, is strongly influenced by both particle size distribution and particle shape.

Figure 5-2a illustrates ideal packing of spheres in a volume which is one sphere thick. In Fig. 5-2b the same number of spheres have been rearranged into a more dense configuration termed *rhombic packing*. According to theoretical

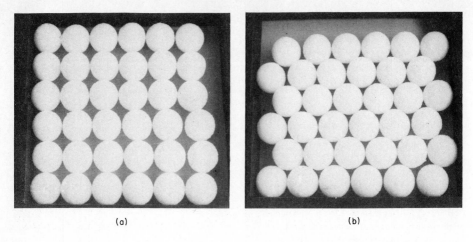

Figure 5-2 Photograph illustrating ideal packing of spheres: (*a*) simple cubical packing; (*b*) rhombic packing with the same number of spheres occupying a smaller volume along top and sides.

considerations—which are simple for the simple cubic packing of Fig. 5-2*a*—if we fit balls inside a cube of side $= nR$, we obtain

	Volume	Unit weight	Void ratio e
Simple cubic	$8R^3$	$\dfrac{\gamma'\pi}{6}$	0.91
($n = 2$)			
Rhombic	$4\sqrt{3}R^3$	$\dfrac{\gamma'\pi}{3}\sqrt{3}$	0.65
Pyramidal	$4\sqrt{2}R^3$	$\dfrac{\gamma'\pi}{4}\sqrt{2}$	0.35

$n = $ integer as 2, 4, 6, etc;. $\gamma' = G\gamma_w$ where $G = $ specific gravity of sphere.

This range represents the theoretical maximum and minimum void ratios of any particulate mass consisting of equal spheres of radius R.

Figure 5-3*a* illustrates the ideal particle size distribution for optimum packing. Approximations to this condition are desirable in many geotechnical engineering problems where stability is a concern. Typically these problems include highway and railroad fills, levees, and dams where optimum packing (translate: maximum density) tends to develop maximum shear resistance and minimum subsidence.

Ideal particle size distributions seldom exist in real soils; however, the packing of Fig. 5-3*a* establishes the upper limit, while the packing of Fig. 5-3*b* establishes the lower limit as obtained by a "one" size particle distribution. It should be noted that with equal spheres the variation of unit weight depends on the specific gravity G but is independent of sphere radius; with real soil the

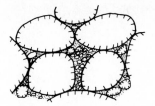 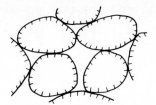

(a) Ideal packing with particles ranging from large to small (well graded).

(b) Same large grain structures as (a) but small sizes removed (poorly graded).

Figure 5-3 Particle packing.

variation of γ will depend on both the "one" size and the numerical value of that size as well as G_s. This is due both to "size" not being uniform and to "size" being a range of particle sizes which have passed one sieve opening and been retained on a smaller sieve size.

Geotechnical considerations, as stated previously, require optimum particle packing. In optimum packing, shear strength is increased because there is more particle contact, providing additional lateral support. Compression and/or subsidence is reduced since there are fewer available soil voids to allow volume change. Further, with packing there is less tendency for particles to roll, slip, and slide to new equilibrium positions under stress. Where a site is founded on a loose granular deposit, it will be necessary to increase the packing via compaction. Where the deposit is shallow, a vibratory compactor may be used. For deep deposits, vibroflotation (a vibratory device) may be used or piles may be driven and/or withdrawn. In either case vibration energy is used to displace the soil particles into a more dense configuration.

5-5 OTHER CONSIDERATIONS OF GRANULAR SOIL STRUCTURE; RELATIVE DENSITY

Cohesionless soils tend to form a *single-grained* structure, as illustrated in Fig. 5-4, which may be either loose (Fig. 5-4a) or dense (packed). Single-grained soil structures are formed when the soil grains independently settle out of a soil-water suspension, as opposed to "floc" settling (see Sec. 5-8). Generally grains larger than about 0.01 mm will form single-grained structures. This size is large enough that the interparticle forces and ionic forces in the water are not sufficient to overcome gravity forces acting on the soil grains. Piles of pure silt, sand, or gravel or silty sand, sandy gravel, etc., mixtures are single-grained structures. Very small soil particles on the order of 0.001 mm and smaller are *colloids*. Colloids are sufficiently small to be more affected by both interparticle forces and ions in soil-water suspensions than by gravity forces. Clay minerals are particles smaller than about 0.002 mm, but their behavior is such that they will be separately considered.

(a)   (b)

Figure 5-4 Single-grained soil structure of real soil. Note, however, that in the field geologically aged deposits often contain significant interparticle growths at the contact points (cementation) from precipitated calcium, iron, magnesium, aluminium, etc., salts or oxides. (*a*) Loose structure; (*b*) soil of part *a* rearranged into a more dense configuration.

True cohesionless soils can only be found in transported soil deposits where the wind or water has removed the colloidal and/or clay mineral contaminants. Typical cohesionless deposits include sand and gravel bars in streams, select glacial outwash deposits, some drumlins and eskers (discussed in Chap. 3), sand dunes and beach sands, and similar deposits. Other deposits can sometimes be washed to remove the cohesive materials and produce sands and gravels.

Certain conditions of deposition can produce a very loose (metastable) soil structure. This type of structure may be able to support a substantial static load but may collapse under relatively small dynamic, or vibratory, loads. This type of soil structure is most likely to have been formed geologically by an upward flow of water in the deposit which has since gradually lessened or disappeared altogether. The diminished flow of water allowed the soil grains to settle out of the flow into a very loose structure. Soil exploration may not detect this state, nor does a laboratory sample built to the same void ratio necessarily produce the same behavior. A careful examination of geologic evidence may give an indication of the potential problem. Careful observation of the penetration test may also give an indication, since the first blows of the test would cause a soil collapse and relatively large penetration per blow where the later penetration per blow rate would be less.

Where the soil grains are from about 5 to 0.05 mm (coarse to fine sand), the presence of a small amount of water can alter the engineering behavior considerably. The surface tension of water at conditions of $S \ll 100$ percent is sufficient to restrict particle movement in this range, producing an "apparent cohesion" which disappears when the soil dries out. Practically, apparent cohesion allows near vertical sand cuts or greater mobility of rubber-tired vehicles over damp sand. Apparent cohesion inhibits soil packing or produces what is commonly termed *bulking*. Figure 5-5 qualitatively illustrates the effect of bulking on unit

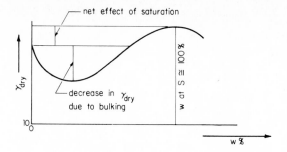

Figure 5-5 Qualitative effect of bulking on resulting dry unit weight. By placing water content at $S = 100$ percent, the maximum dry unit weight is obtained.

weight. At a water content of zero the unit weight is nearly the maximum possible for some energy input. At intermediate water contents, bulking, or surface tension accumulations, restricts particle movement and the unit weight decreases. As more water is added, some of the pores become saturated, with loss of surface tension and increased packing. At $S = 100$ percent all the surface tension effect is lost, and together with some particle lubrication effect, the maximum unit weight is obtained. In field densification operations, where adding water does not affect the surrounding soil detrimentally, flooding a sand (to ensure $S = 100$ percent) will assist considerably in increasing the unit weight.

Cementation may occur in cohesionless deposits as percolating water precipitates various carbonates and other materials. This may be a very important in situ feature which may be nearly indetectable on small soil quantities in any exploration program. In situ, the larger cumulative effects beneath a footing or similar may be of major importance. In situ particle cementation over geologic time periods cannot be easily duplicated in the laboratory on reconstituted samples—even to the same void ratio. In many cases, of course, the cementation is sufficient to produce easily detected effects (stratum is hard, dense, visually rocklike, etc), but in many others the cementation is difficult to detect directly.

Optimum packing of a granular soil results in the greatest unit weight and minimum void ratio $e_{\min}$. Conversely, minimum packing results in the loosest state, minimum unit weight, and maximum void ratio $e_{\max}$. The loosest state is approximately obtained by pouring dry sand (Bowles, 1978) into a calibrated mold. Sometimes the sand is allowed to fall through water to produce a known weight and volume. With G_s, the volume of the mold, and the weight of soil in the mold, the void ratio is easily computed. The densest state is obtained by vibrating a confined weight of sand and measuring the volume. Table 5-1 gives some ranges of void ratios and other data for several soils as an indication of the values one might expect to obtain. Tabulated values such as these are acceptable for preliminary design but should never be used for any final design.

The *relative density* is a measure of the in situ void ratio e_n as an index property related to the laboratory values of the maximum and minimum void ratios as

$$D_r = \frac{e_{\max} - e_n}{e_{\max} - e_{\min}} \tag{5-1}$$

Relative density can also be expressed in terms of the maximum (γ_{max}), minimum (γ_{min}), and in situ (γ_n) dry unit weights as

$$D_r = \frac{\gamma_{max}}{\gamma_n} \frac{\gamma_n - \gamma_{min}}{\gamma_{max} - \gamma_{min}} \tag{5-2}$$

This equation is preferable to Eq. (5-1) because of the greater ease of determining unit weights and because it does not require a determination of the specific gravity. Considerable importance was attributed to the relative density by early proponents, who attempted to relate various soil properties such as void ratio, angle of internal friction, and thus, indirectly settlement and strength characteristics to this index property. Relative density is sometimes used at present in liquefaction studies (Sec. 14-7) as a field compaction specification requirement, and to assess the competence of in situ granular materials for foundations.

The major reason for using relative density is that undisturbed sampling of in situ cohesionless sands and gravels is nearly impossible, and, as a consequence, penetrometer testing is widely used. A large data base presently exists—albeit with considerable scatter—relating penetration tests to relative density [e.g., see Eq. (6-3)]. Table 5-2 gives some simple field identification tests which may be used to estimate D_r.

Considerable research, with the latest reported in ASTM (1973), indicates that D_r is not a very reliable soil index property. It is quite possible for two sands with identical values of in situ void ratios e_n and D_r to have significantly different engineering behavior due to grain shape, cementation, confinement, and stratification resulting from deposition and stress history.

Since D_r depends on laboratory determination of γ_{max} and γ_{min}, or the corresponding void ratios, a large error may result in not accurately determining both of these values. Generally, a statistical determination of γ_{max} will produce a rather consistent (average or standard deviation) value about 0.45 kN/m³ too small, and conversely for the minimum unit weight a value about 0.45 kN/m³ too large. This is illustrated in Ex. 5-1.

Table 5-2 Terms and field identification in relative density

Soil state†	D_r	Field identification
Very loose	0–0.20	Easily indented with finger, thumb, or fist
Loose	0.20–0.40	Somewhat less easily indented with fist. Easily shoveled
Medium compact	0.40–0.70	Shoveled with difficulty
Compact	0.70–0.90	Requires pick to loosen for shoveling by hand
Very compact	0.90–1.00	Requires blasting or heavy equipment to loosen

† Not all authorities agree on either the terminology describing the soil state or the value of D_r to which it applies. These values are as good as any proposed and with the subjective nature of this index property may be used with reasonable confidence.

Example 5-1 A medium coarse, gravelly sand was tested in the soil laboratory by a group of 10 students. Each student did three tests each for the maximum and minimum unit weights and reported the extreme (not the average) values obtained. The sand was returned to the source container and well mixed for further use. The data were as follows (the Δ values were later computed for obtaining the standard deviation $\bar{\sigma}$):

Test	1	2	3	4	5	6	7	8	9	10
γ_{min}	14.36	14.58	15.02	15.12	14.78	14.51	14.40	15.08	15.15	14.43
Δ	0.0	0.22	0.66	0.76	0.42	0.15	0.04	0.72	0.79	0.07
γ_{max}	18.57	18.70	18.86	18.39	18.50	18.32	18.45	18.42	18.49	18.75
Δ	0.29	0.16	0.0	0.47	0.36	0.54	0.41	0.44	0.37	0.11

REQUIRED Compute the standard deviation and assess the error in D_r if the in situ value of $\gamma_n = 16.0 \text{ kN/m}^3$.

SOLUTION The standard deviation is computed based on the maximum and minimum values of unit weight, not on the average, since the definition of D_r is based on the extreme values. With this concept, the Δ values are obtained as the difference between the smallest unit weight of 14.36 and the other nine values; thus

$$14.58 - 14.36 = 0.22$$

Similarly, with the maximum γ, we obtain

$$18.86 - 18.57 = 0.29$$

The standard deviation is computed as

$$\bar{\sigma} = \sqrt{\frac{\sum (\Delta^2)}{9}} = 0.52 \quad \text{for } \gamma_{min}$$

and 0.37 for γ_{max}. Thus, one would expect that on the average the minimum unit weight would vary as

$$\gamma_{min} = 14.36 + 0.52 \text{ kN/m}^3 \quad \text{(and not } \pm 0.52\text{)}$$

(range from 14.36 to as much as 14.88.)

Since the range of values of γ_{min} and γ_{max} must be investigated, it will be useful to make a plot of D_r versus γ as shown. The scale is adjusted so that the curve is linear, since Eq. (5-2) directly plots a curve. The unit weight scale is adjusted, once and for all, by obtaining the range of γ and computing D_r for the scale increment; the spacing of this increment is made such that the slope is constant, using a constant increment for the D_r scale. For example, for the unit weight increment 11 to 12, the spacing may be 30 mm; then from 12 to 13, use 25 mm; from 13 to 14, use 22.9 mm; etc. This is shown on Fig. E5-1.

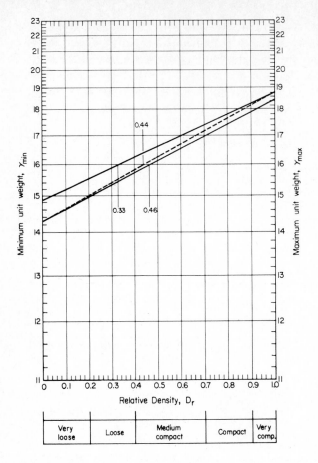

Figure E5-1

From Fig. E5-1 obtain the "true" value for D_r (assuming $\gamma_n = 16.00$ is exact) of 0.44 and extreme values of 0.46 and 0.33, respectively. The probable maximum error is

$$\text{Probable error} = \frac{0.44 - 0.33}{0.44} \times 100 = 25 \text{ percent}$$

////

5-6 STRUCTURE OF COHESIVE SOILS

A cohesive soil may be defined as an aggregation of mineral particles which has a plasticity index defined by the Atterberg limits and which forms into a coherent mass on drying such that force is necessary to separate the individual microscopic grains. The ingredients necessary to give a soil deposit cohesion are *clay minerals*, sometimes termed *argillaceous materials.* The amount of cohesion depends on the relative sizes and amounts of various soil grains and argillaceous materials present. Generally when over 50 percent of the deposit consists of particles 0.002 mm and smaller, the deposit is termed "clay." With this relative percentage the larger soil particles are suspended in a fine-grained soil matrix. Where 80 to 90 percent of the deposit material is smaller than the No. 200 sieve (0.075 mm), as little as 5 to 10 percent clay can give the soil a cohesive label.

Seldom does a pure clay deposit exist naturally; it is nearly always contaminated with silt and/or fine sand particles as well as colloidal ($<$0.001) sizes. Colloids, sometimes called *rock flour,* are the byproduct of rock abrasion and do not possess clay mineral properties (Sec. 5-7) even though the size range is similar.

A complete description of the structure of a fine-grained cohesive soil requires a knowledge of both the interparticle forces and the geometrical arrangement, or fabric, of the particles. It is nearly impossible to measure the interparticle force fields surrounding clay particles directly; therefore, the fabric is the principal focus in studies of cohesive soils. From the fabric studies, estimates are theorized or postulated of the interparticle forces. The interparticle forces appear to be developed from three different types of electric charge:

1. *Ionic bonds.* Bonds due to a deficiency in electrons in the outer shells of atoms making up the basic soil units
2. *Van der Waals bonds.* Bonding due to alternations in the number of electrons at any one time on one side of an atomic nucleus
3. *Others.* Includes hydrogen bonds and gravitational attraction between two bodies

Recent studies of clay soils with the scanning electron microscope (SEM) show the individual clay particles to be aggregated or flocculated together in submicroscopic fabric units which are called *domains* by numerous recent researchers (Collins and McGown, 1974; Yong and Sheeran, 1973). Domains in turn group together to form submicroscopic groups called *clusters.* These groupings are due to the interparticle forces acting on the small basic units. Clusters group together to form *peds* and groups of peds of macroscopic size. Nonscientific terms for the peds include soil "crumbs" and soil "aggregations." Peds and other macrostructural features such as *joints* and *fissures* constitute the macrofabric soil structure. A sketch of this system is shown in Fig. 5-6. Photographs of various clays using the SEM technique indicate a very complex structure. This may in part account for the complex engineering behavior of clay soils.

Macrostructure, including the stratigraphy, of fine-grained soil deposits has an important influence on soil behavior in engineering practice. Joints, fissures, root holes, varves, silt and sand seams and lenses, and other discontinuities often control the behavior of the entire soil mass. The strength of a soil mass is significantly less along a crack, or fissure, than through intact material, particularly in laboratory tests where the defect may extend completely through a small

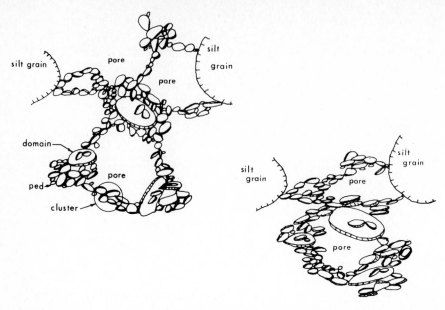

Figure 5-6 Structure of a clay soil: (*a*) a porous, flocculated sediment interspersed with silt grains; (*b*) the sediment after it has been subjected to overburden or other stresses which have resulted in a reorientation of the domains, clusters, and peds into a more parallel (dispersed) state.

sample. If an in situ defect happens to be unfavorably situated with respect to applied stresses, failure or instability may occur unless the surrounding material provides sufficient confinement. The drainage of a clay layer can be markedly affected by the presence of even a very thin layer(s) of silt or sand. Therefore, in any engineering problem involving stability or settlements, the geotechnical engineer must carefully investigate the clay macrostructure.

Microstructure is more important from a fundamental than from an engineering viewpoint, but it is useful as an aid in the general understanding of soil behavior. The microstructure of a clay is the complete geologic history of that deposit, including both stress changes and environmental conditions during deposition. These geologic imprints tend to affect the engineering response of the clay very considerably. Recent research on clay microstructure suggests that the greatest single factor influencing the final structure of a clay was the electrochemical environment existing at the time of deposition. Flocculated structures, or aggregations, of varying degrees of packing and interconnections result during sedimentation. The general degree of packing appears to be particularly sensitive to whether the deposition took place in a marine, brackish, or freshwater environment. The ion concentrations in these three waters would range from high in the case of marine to low in freshwater. The degree of packing also appears to be influenced to a large degree by the clay mineralogy as well as by the amount and angularity of the fine sand or silt grains present. Silt particles in a cohesive deposit have been observed to have thin skins of apparently well oriented clay

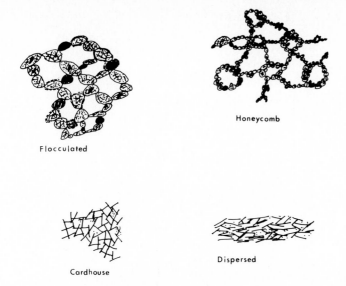

Flocculated

Honeycomb

Cardhouse

Dispersed

Figure 5-7 Structure of clay soil using earlier terms of structure orientation. The flocculent structure might be obtained from sedimentation in water with a low salt content. The honeycomb structure could be obtained from sedimentation in a marine (high-salt-content) environment. The "card-house" description was widely used prior to SEM studies. The "dispersed" state is a convenient description for reorientation from compaction as in Chap. 7.

particles. Both silt and clay particles/aggregations often contain thin films of amorphous materials (organic, silica, or iron compounds) on their surfaces. Leaching into or out of these fine-grained deposits may change the soil characteristics considerably as the peds and clusters become coated or the amorphous material is leached away.

Early descriptions of cohesive soil included honeycomb, flocculent, and dispersed structures. These terms, illustrated in Fig. 5-7, are still widely used to describe the total cohesive soil structure. The honeycomb structure may very well be a situation where the clusters form particular groupings during sedimentation, and the flocculent structure may be a situation where either silt grains attract coatings of clay minerals or peds form and produce the porous and random flocculent structure.

Present evidence tends to the theory that the cluster arrangement between peds of the flocculent structure and the somewhat analogous cluster arrangement between honeycomb cells produce the initial distortions (settlements) under stresses. Some research evidence also indicates that water flow through these soils may dislodge domains and/or clusters, which increases the pore spaces and flow. Deposition downstream, however, may plug other pores in the same mass and result in a decreased flow rate. This concept is of particular importance in permeability studies and may produce some governing factors as considered in Sec. 8-5.

5-7 CLAY AND CLAY MINERALS

Clay minerals are predominantly silicates of aluminum and/or iron and magnesium. Some of them also contain alkalies and/or alkaline earths as essential components. These minerals are predominantly crystalline in that the atoms composing them are arranged in definite geometric patterns. Most of the clay minerals have sheet or layered structures. A few have elongated tubular or fibrous structures. *Clusters* are books of sheetlike units or bundles of tube or fiber units. Soil masses generally contain a mixture of several clay minerals named for the predominating clay mineral with varying amounts of other nonclay minerals.

Clay minerals are very small (less than 2 μm) and very electrochemically active particles which can be seen using an electron microscope only with difficulty. In spite of their small size, however, the clay minerals have been studied extensively (Grimm, 1968; Mitchell, 1976) because of their economic importance, particularly in ceramics, metal molding, oil field usage, and engineering soil mechanics. The clay minerals exhibit characteristics of affinity for water and resulting plasticity not exhibited by other materials even though they may be of the clay size or smaller. For example, finely ground quartz does not exhibit plasticity when wetted. It should be particularly noted that any fine-grained "clay" deposit is likely to contain both clay minerals and a wide range of particle sizes of other materials which are essentially "filler."

There are two fundamental building blocks for the clay mineral structure. One is a silica unit (Fig. 5-8), in which four oxygens form the tips of a tetrahedron

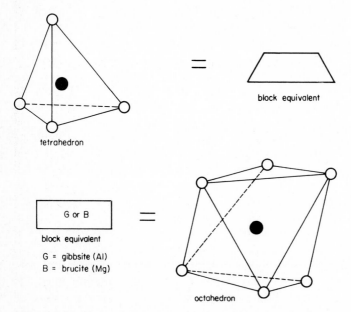

Figure 5-8 Simplified diagrams of the silica tetrahedron and octahedral units. Note carefully the orientation of the diagram for the tetrahedron to match points on the octahedron.

and enclose a silicon atom, producing a unit approximately 4.6 Å high.† The other unit is one in which an aluminum or magnesium (and sometimes Fe, Ti, Ni, Cr, or Li) atom is enclosed by six hydroxyls having the configuration of an octahedron which is about 5.05 Å high. Figure 5-8 gives the block diagram equivalents of these two units. The octahedral unit is called *brucite* if the metallic atom is mainly magnesium and *gibbsite* if the atom is aluminum. All the possible combinations of these basic units to form clay minerals produce a net negative charge on the exterior of the clusters. A soil-water suspension will thus have an alkaline reaction (pH > 7) unless the soil is contaminated with an acidic substance.

The principal source of clay minerals is the chemical weathering of rocks which contain

Orthoclase feldspar
Plagioclase feldspar
Mica (muscovite)

all of which may be termed *complex aluminum silicates* (see Sec. 3-2 for chemical equations). However, according to Grimm (1968), clay minerals can be formed from almost any rock as long as there are sufficient alkalies and alkaline earths present to effect the necessary chemistry. The decomposition of orthoclase feldspar to give a clay mineral was given in Sec. 3-9.

Weathering action on rocks produces a very large number of clay minerals with the common property of affinity, but in widely differing amounts, for water. Some of the most common clay minerals are the following.

Kaolinite

The name "kaolinite" is a modification of "Kauling," meaning high ridge of a hill near Jauchau Fu, China, where a white kaolinite clay was obtained several centuries past. The term kaolin actually describes several distinct clay minerals. Engineers use the term to describe a clay group characterized by low activity.

The kaolinite structural unit consists of alternating layers of silica tetrahedra with the tips embedded in an alumina (gibbsite) octahedral unit as in Fig. 5-9. This alternating of silica and gibbsite layers produces what is sometimes called a 1 : 1 basic unit. The resulting flat sheet unit is about 7 Å thick and extends infinitely (relative to 7 Å) in the other two dimensions. The kaolinite cluster is a stacking of 70 to 100 or more of these 7-Å sheets as a book with hydrogen bonds and van der Waals forces at the interface. The resulting formula is approximately

$$(OH)_8Al_4Si_4O_{10}$$

The bonding combination of hydrogen and van der Waals forces results in considerable strength and stability with little tendency for the interlayers to take on

† Angstrom unit Å = 10^{-10} m.

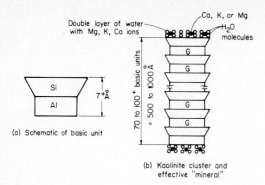

(b) Kaolinite cluster and effective "mineral"

Figure 5-9 The kaolinite clay mineral.

water and swell ("activity") and results in the mineral book being on the order of 500 to 1000 Å thick. Kaolinite is the least active of the clay minerals so far observed. Kaolinite can be produced by weathering of certain of the more active clay minerals as well as being directly formed as a byproduct of rock weathering.

Another 1:1 mineral of the kaolinite "family" is *halloysite*. It differs from kaolinite by being more randomly stacked, so that a single molecule of water may enter between the 7-Å units, giving the equation

$$(OH)_8Al_4Si_4O_{10} \cdot 4H_2O$$

Halloysite further differs from kaolinite in that the elemental sheets are rolled into tubes, as illustrated by the SEM of Fig. 5-10*b*. Dehydration by heat on the order of 60 to 70°C, and even air drying, will often permanently alter halloysite, either reducing it to $2H_2O$ or completely removing the water molecules and producing approximately kaolinite. The engineering properties of halloysite are considerably different from those of kaolinite, and since even air drying may affect the chemistry which is indirectly measured by the Atterberg limits, great care is necessary in obtaining realistic samples for the Atterberg limits and hydrometer analysis.

Kaolinite and halloysite clays are widely used for chinaware due to the absence of iron and subsequent iron discoloration on firing at high temperature. Kaolin clay is widely used as an intestinal absorbent to combat intestinal infections, i.e., in antidiarrheal medicines and for digestive disorders.

Kaolinite tends to be found in regions of heavier rainfall, as in the southeastern United States, China, parts of Europe, South America, and other more local areas.

Illite

Illite is a general term for a clay group first identified in Illinois. These clay minerals are of the general equation

$$(OH)_4K_y(Si_{8-y} \cdot Al_y)(Al_4 \cdot Mg_6 \cdot Fe_4 \cdot Fe_6)O_{20}$$

where y is between 1 and 1.5. Illite is derived principally from muscovite (mica) and biotites and is sometimes called mica clay.

Figure 5-10 SEMs of clay minerals in the kaolinite clay group: (*a*) kaolinite *(SEM courtesy of Dr. R. J. Kriezek, Northwestern University)* (read 5 μ as 5 μm); (*b*) Halloysite clay minerals *(from Tovey and Yan, 1973)*.

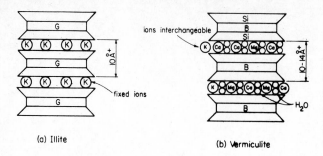

Figure 5-11 (*a*) Illite and (*b*) vermiculite clay minerals. Note, as shown, that seldom will only potassium ions, in the case of illite, or magnesium and/or calcium ions, in vermiculite, be the sole ions found.

The illite clay mineral (Fig. 5-11) consists of an octahedral layer of gibbsite sandwiched between two layers of silica tetrahedra. This produces a 1 : 2 mineral with the additional difference that some of the silica positions are filled with aluminum atoms and potassium ions are attached between layers to make up the charge deficiency. This bonding results in a less stable condition than for kaolinite, and thus the activity of illite is greater.

Vermiculite is a clay mineral in the illite family which is similar except for a double molecular layer of water between sheets interspersed with calcium and/or magnesium ions and substitution of brucite for gibbsite in the octahedral layer.

Illite and vermiculite clays and clay shales are widely used in making lightweight aggregates (sometimes called "expanded shale" or "vermiculite"). The vermiculite in particular expands considerably on high heating because the water layers quickly turn to steam with resulting large expansions.

Illite clays tend to be found in areas of moderate rainfall, as in the central United States, England, and Europe.

Montmorillonite

Montmorillonite was the name given (ca. 1847) to a clay mineral found at Montmorillon, France, with the general formula

$$(OH)_4Si_8Al_4O_{20} \cdot nH_2O$$

where nH_2O is the interlayer (*n* layers) of adsorbed water. The term "smectite" is also used for this clay mineral group.

The montmorillonite clay mineral is made of sheetlike units ordered, also as a 1 : 2 unit, as schematically illustrated in Fig. 5-12. The intersheet bonding is due mainly to the van der Waals forces and is, thus, very weak relative to hydrogen or other ion bonding. Various substitutions take place, including Al for Si in the tetrahedra and Mg, Fe, Li, or Zn, for Al in the octahedral layer. These exchanges result in a relatively large net unbalanced negative charge on the mineral, with resulting large cation exchange capacity and affinity for water with H^+ ions in the absence of metallic ions.

Bentonite is a montmorillonitic clay found in partially weathered volcanic deposits in Wyoming, Switzerland, and New Zealand. This clay mineral is particularly active in terms of swelling in the presence of water and has been widely

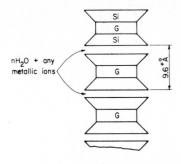

(a) Schematic of montmorillonite

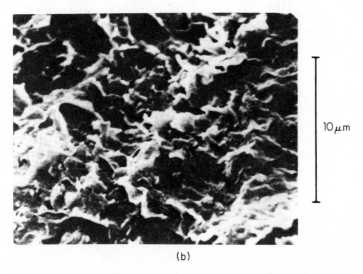

$10\,\mu m$

(b)

Figure 5-12 Montmorillonite and bentonite clay minerals: (*a*) schematic of montmorillonite; (*b*) SEM of bentonite *(from Matsuo and Kamon, 1973)*.

used in drilling oil wells and in soil exploration as a drilling mud and as a clay grout. Bentonite is somewhat variable in its properties depending on the source and amount of weathering of the parent volcanic material. At present, bags of clays of high activity, commercially available for drilling and grouting, are loosely called "bentonite" although they are merely montmorillonite clays.

Weathering of montmorillonite clay minerals often produces kaolinite clay, and in areas where weathering has progressed both minerals are usually present.

Montmorillonite is found in the more arid regions of the world, as in the western United States, Australia, New Zealand, and southern Africa.

5-8 GENERAL CLAY MINERAL PROPERTIES

Several characteristics are similar for all the clay minerals.

Hydration

Clay particles are almost always hydrated, i.e., surrounded by layers of water molecules called "adsorbed water." This layer is often at least two molecules thick and is called the "diffuse layer," the "double diffuse layer," or simply the "double layer." This water is firmly attracted and/or contains metallic ions. A diffusion of the adsorbed cations from the clay mineral extends outward from the surface of the clay into the adsorbed water layer. The effect of this is to produce a net ($^+$) charge near the mineral particle and a ($^-$) charge at a greater distance. This diffusion of cations is a phenomenon very similar to the diffused interface between a free water surface and the atmosphere where the diffused material is water molecules. This water is often so firmly attracted it behaves more as a solid than as a liquid, and some researchers report the density $\rho_w \rightarrow 1.4$ g/cm^3.

This layer of water may be lost at temperatures higher than 60 to 100°C and will reduce the natural plasticity (reduce w_L, say, 6 to 10 percent) of the soil. Some of this water may be lost by air drying. Generally, if the double layer is dehydrated at low temperatures, the plasticity properties may be recovered by mixing with sufficient water and "curing" for 24 to 48 hours. If dehydration occurs at higher temperatures, the plasticity properties are permanently lowered.

Clay minerals have sufficient attractive potential to H^+ ions that a layer of water up to about 400 Å can surround the particle as illustrated in Fig. 5-13. This qualitatively illustrates the difference between kaolinite and montmorillonite clays in terms of in situ water content and possible liquid-limit values.

Activity

The edges of all the clay minerals have net negative charges. This results in attempts to balance the charges by cation attraction. The attraction will be in proportion to the net charge deficiency and may be related to the *activity* of the clay. The activity may be defined as

$$\text{Activity} = \frac{\text{plasticity index } I_p}{\text{percent clay}} \tag{5-3}$$

where the percent clay is taken as the soil fraction <2 μm. Activity is also related to relative potential water contents as illustrated in Fig. 5-13. Typical activity values based on Eq. (5-3) are as follows:

Kaolinite	0.4–0.5
Illite	0.5–1.0
Montmorillonite	1.0–7.0

Although activity is numerically defined in Eq. (5-3), a better practical indicator of activity is the shrinkage limit. The shrinkage limit (defined in Sec. 2-5) is the threshold water content to initiate volume change. Activity in terms of volume change is a principal concern in evaluating the soil for use in earthworks and foundations.

Plate plan	0.1 to 1 μm	10 μm	0.3 to 4 μm
Specific surface	800 m²/g	80 m²/g	15 m²/g (statistical average)

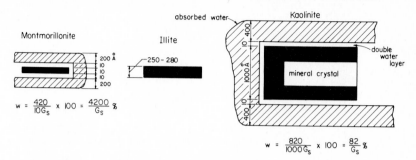

Figure 5-13 Relative sizes, absorption potential, relative range in water content, grain size, and specific surface for montmorillonite, illite, and kaolinite clay minerals.

The cation exchange capacity in terms of millequivalents per 100 g of clay is also used as an indication of activity. For example, 1 meq of $Na_2/100$ g $= 0.031$ percent Na_2O. The exchange capacity of several clay minerals is as follows:

Clay	Exchange capacity, meq/100 g
Kaolinite	3–15
Halloysite ($4H_2O$)	10–40
Illite	10–40
Vermiculite	100–150
Montmorillonite	80–150

In practical terms, clays can have their activity in terms of the plasticity characteristics altered by substitution of metallic ions of higher order as in the following substitution scale:

$$Li < Na < NH_4 < K < Mg < Rb < Ca < Co < Al$$

According to this scale, Ca will replace Na or Mg more easily than either Mg or Na will replace Ca. Also, from a practical standpoint, the higher the exchange capacity, the more cations (in some form of admixture) will be required to effect a change in activity.

Many clays, particularly in the southwest part of the United States, tend to swell large amounts when they become saturated. This swelling can be considerably reduced by cation exchange where the usual cation is calcium, furnished by mixing lime with the clay. This process is termed *soil stabilization*. Other additives such as cement and fly ash (a byproduct of burning coal in steam power plants) will produce similar results due to the high concentration of Ca and Al ions in these materials.

A chemical analysis is not generally used to determine the chemical formula of a clay because of the presence of large numbers of different clay minerals. Instead, a trial mix procedure is used where the soil is mixed with one or more additives in varying percentages to obtain the optimum percent of the particular additive for field use.

Flocculation and Dispersing

Clay minerals almost always produce soil-water suspensions which are alkaline (pH > 7) as a result of the net negative charges on the mineral units. A few exceptions may occur when the minerals are contaminated with amorphous substances. Because of these charges, the H^+ ions in water, the van der Waals forces, and the small size of the particles, they tend to become attracted together on collision (or even in very close proximity) in a solution. Several particles thus attracted form a randomly oriented *floc* or structure of larger size which will settle out of suspension very rapidly to form a very loose sediment. In the laboratory, a 50- or 60-g clay sample will settle out of a 1000-milliliter suspension in about 30 minutes unless floc formation is controlled. To avoid flocculation a dispersed soil-water suspension may be neutralized by adding additional H^+ ions as furnished by acidic materials. The most common material for laboratory work is sodium hexametaphosphate (trademark Calgon, $NaPO_3$), which produces an acidic solution when the dry material is mixed with water. The solution can be checked for acidity by using blue litmus paper (which should turn pink). When the solution is neutralized, the clay particles do not form flocs on collision in the water. Addition of an alkaline material such as sodium hydroxide (NaOH) or alum $[KAl(SO_4)_2]$ will cause very rapid flocculation.

A freshly flocculated clay, in these circumstances, can be easily dispersed back into solution by shaking, indicating that the interparticle attractions are far less than the shaking forces. After the clay has been standing some time, however, dispersion is not easily accomplished, indicating a *thixotropic* (strength gain with aging) effect. Piles driven into saturated soft clays will remold the soil structure in a zone around the pile. Initial load capacity is often extremely low, whereas after 30 or more days of aging the design load may be developed by adhesion between the clay and the pile.

Water Effects

The water phase of clay soils is not very likely to be chemically pure water. This water accounts for the plasticity properties of clay. In laboratory tests for the Atterberg limits it is specified by ASTM that distilled water be added as required. The use of distilled water, which is relatively ion free, may produce results somewhat different from those obtained using a more contaminated water such as may enter the soil in situ.

A particular phenomenon of clay is that a clay mass which has dried from some initial water content forms a mass which has considerable strength. If these lumps are broken down to elemental particles, the material behaves as a cohesionless particulate medium. When water is again added, the material becomes plastic with some strength intermediate to the dry lump strength. If the wet clay is again dried, it forms hard, strong lumps. The role of water in this phenomenon is not fully understood, although in drying, surface tension certainly pulls the particles into maximum contact with the very minimum of interparticle spacing so that the interparticle forces are a maximum. It appears that the higher density resulting from packing and the close spacing resulting in the maximum effect of interparticle force attraction give this very high strength. We can readily observe that the strength of the clay varies from a very low value at $S \to 100$ percent to a very high value at $S = 0$. It is of interest to note that the use of water, which is a dipolar agent, will produce this effect, whereas a nonpolar agent such as carbon tetrachloride (CCl_4) does not. A dipolar agent is one which tends to develop a $+$ and $-$ charge on opposite sides of the molecule. The $+$ charge on one side of a dipole tends to attract the $-$ charge of any material present—including both clay particles and the negative side of other water molecules.

5-9 SUMMARY

This chapter has considered the definition of soil as commonly used for engineering purposes. We have looked at soil in terms of both structure and fabric. Fabric is an appropriate term for describing cohesive soils, but structure, more specifically single-grained structure, is preferable in describing the geometric arrangement of cohesionless soils.

The concept of *cohesion c* and *angle of internal friction ϕ* as soil parameters to measure resistance to interparticle slip (strength) was introduced. It was also presented that these are state parameters rather than unique values for a given soil.

We note that the smallest soil unit of cohesive soils is the *domain*, which in turn forms *clusters* and clusters form *peds* or visible soil crumbs.

The effect of water on granular packing to produce bulking of sands was noted. The concept of relative density D_r to describe packing of cohesionless soils was given, together with observation that large errors may be associated with this index property.

The clay mineral was considered in some detail. It is noted that there are three main groups of clay minerals of particular geotechnical engineering importance. These three groups are:

Kaolinite—least active
Illite—intermediate activity
Montmorillonite—most active

It was also given that the clay mineral develops the plasticity in cohesive soils and that water has a significant effect on clay minerals. It was noted that seldom is a single type of clay mineral found in a soil deposit and that because of this a chemical analysis is seldom directly used in soil stabilization studies. Rather soil stabilization using additives is based on obtaining the optimum amount of material by trial mixes.

Finally we have learned why sodium hexametaphosphate is used as a dispersing agent in the hydrometer test.

PROBLEMS

5-1 Verify that the void ratio for simple cubic packing is 0.91, based on $n = 6$.

5-2 Derive Eq. (5-2) using Eq. (5-1).

5-3 Make a master graph sheet for a D_r versus unit weight graph as used in Fig. E5-1.

5-4 Referring to Example 5-1, redo and obtain the range of possible error in D_r using a standard deviation based on the *average* test results. Comment on this procedure vs. using the procedure based on the extreme values of the example.

5-5 Referring to Example 5-1, what is the percent chance (approximate) that a single test would give the correct value of D_r? Note that you also have to define what the correct value is! How many tests would be required to give a 95 percent reliability on D_r?

5-6 Given the following data:

$$\gamma_{max} = 21.3 \text{ kN/m}^3 \text{ but the average of 20 tests is } 21.0$$

$$\gamma_{min} = 15.7 \text{ kN/m}^3 \text{ but the average of 20 tests is } 16.0$$

$$\gamma_n = 19.8 \pm 0.3 \text{ kN/m}^3$$

what is D_r and what is the maximum error which might be obtained on a single test?

5-7 Given the following data obtained by two student laboratory groups on the same soil:

Trial	Group 1		Group 2	
	γ_{max}	γ_{min}	γ_{max}	γ_{min}
1	18.08	15.94	18.19	15.83
2	18.36	15.83	17.78	16.04
3	17.45	15.36	17.97	15.75
4	18.15	16.00	17.92	15.83
5	18.17	15.50	18.09	15.93
6	17.97	15.79	17.97	16.13
7	18.09	15.83	18.23	16.05
8	18.14	15.77	18.17	16.00
9	17.90	15.61	18.19	15.75
10	18.11	16.08	18.03	16.00
11	18.19	15.83	17.98	15.68
12	18.22	15.82	17.98	15.90

Required:

(*a*) Compute the standard deviation of each group and of the total of 24 tests and compare.

(*b*) Compute the relative density of this soil based on $\gamma_n = 17.8\,\text{kN/m}^3$.

5-8 Explain how you would obtain the Atterberg limits on a soil where air drying sufficiently to sieve through the standard sieve would alter the limits too much to tolerate. What is the sieve number for this test?

5-9 A 4 percent by weight addition of lime will adequately alter the Atterberg limits and swell characteristics of a clay soil for an airfield. For a treatment to a depth of 1 m, how large an area will a 45-kg sack cover?

Hint: You will have to do some estimating!

SIX

SOIL EXPLORATION AND SAMPLING

6-1 INTRODUCTION

Anyone who has driven along a highway in an area where the roadway has cut through hills has observed the many types, or at least colors, of soils which are exposed. One who has visited building sites where excavation for basements has taken place has probably observed the several types and colors of the exposed soils. It may be inferred from even these limited observations that soil occurs in nature in a highly variable manner. On one hand the soil may be relatively homogeneous over an area several hundred meters horizontally and at least several meters vertically; alternatively, the soil may vary considerably in distances of 1 m or less both horizontally and vertically. It may contain a number of strata vertically which contain lenses (local globs) or be a mixture of "soils" with large amounts of roots and other fibrous (woody or peaty) materials. Consistency may range from almost bricklike in firmness to very soft. This latter can occur in a "homogeneous" cohesive deposit where the color indicates no discernible stratification over the depth of interest and where the principal variables are the clay and water content.

Field exploration provides data on surface and underground conditions at the proposed site. Samples may be obtained for visual inspection and to determine physical and index properties. Depending on the proposed site use, relatively undisturbed samples may be obtained to make estimates of engineering properties for strength, stability, and water flow.

There are very few situations where some site exploration is not required for engineered design. Because of the potential for large variations in soil types and properties in short distances—both lateral and vertical—exploration data from an adjacent property may be substantially in error.

Other major factors requiring site exploration include:

1. Necessity of using marginal land as the more desirable sites have already been used near urban areas.
2. Necessity of using reclaimed land such as old landfill sites and urban renewal areas. These latter often contain rubble-filled basements with little to no quality control over the fill.
3. Local building code requirements.
4. Potential for a lawsuit if substantial redesign is required when the site is opened for construction. Subsequent building damages from settlement or other foundation problems which could have been avoided by a soil investigation (properly interpreted) are also recoverable claims against geotechnical consultants and designers.

Of necessity, only a limited discussion of soil exploration and sampling can be presented in a textbook. The interested reader is referred to Broms (1980) and in particular to the substantial volume prepared by Hvorslev (1949); the latter is sufficiently applicable that several recent reprintings have been made.

6-2 SITE INVESTIGATION

The field exploration program will include inspection of any available topographic maps of the area and viewing of aerial photographs, if available, and may include an on-site inspection to observe surface features and conditions of adjoining construction. Discussion of the site and area with local residents may be valuable in determining locations of filled-in areas or recent topographical changes.

The field exploration will ultimately involve drilling holes into the ground to determine the stratigraphy and to obtain disturbed and/or undisturbed soil samples for testing. This operation, coupled with an on-site inspection, study of maps, etc., should enable the geotechnical engineer to:

1. Determine the source and nature of the deposits (recent history of filling or excavation, flooding, and local geology).
2. Determine the depth, thickness, and, with testing, composition of the several strata making up the soil profile.
3. Determine the location of bedrock. The quality of bedrock may also be determined, but this is done only when necessary due to the excessive costs of rock, compared with soil, drilling.

4. Determine the location and variation in the ground water table or determine that the water table is not in the zone of design interest.
5. Obtain the quantity and types of soil samples needed to determine the engineering properties for design. Samples may range from bags of disturbed soil used for index properties to undisturbed samples obtained in thin-wall tubes for strength and settlement tests.

In some cases the exploration program is broken into phases as follows:

Phase 1   Initial reconnaissance (field trip and/or office study of available maps and other information)
Phase 2   Preliminary borings—just sufficient to obtain the general character of the subsoil, obtain index properties, and possibly perform some strength-deformation tests
Phase 3   Detailed program with additional borings and/or more carefully performed strength/deformation tests

For smaller routine projects, phases 1 and 2, or many times just phase 2, complete the exploration. In some cases phase 2 determines the feasibility of using the site. The time lag between phases 2 and 3 may be several years as the client uses the initial data in a preliminary design to run project cost estimates so that money can be obtained for doing the project.

The cost of an adequate site exploration, including laboratory testing and submitting a report to the client which outlines what was found and the recommendations of the geotechnical engineer, runs from about 0.1 to 0.3 percent of the total cost for structures except bridge and dam sites, where the exploration costs may exceed 1 percent. This cost must be weighed against overdesign (for the designer's protection) costs plus the additional contract bid cost to cover either contractor uncertainty or the cost of hiring a geotechnical expert to check the site. There are numerous cases where overdesign did not work, i.e., where site conditions discovered during construction required a redesign.

6-3 SITE GEOLOGY

A geologic study is often very useful in planning a site investigation. This study may enable a prediction of the type and properties of the site materials so that the best methods and equipment may be selected in advance of the actual operations. The study may aid in interpreting the data subsequently found in the exploration as well. An excellent summary of the beneficial effects of making a geologic study is given by Legget (1979).

The geologic history may reveal old filled-in stream channels and lakes, sinkhole activity, rocks, and areal uplifts. Rock quality may be estimated on the basis of outcrops, uplifts (producing fractures, etc.), and general existing topographical features.

Low areas, stream plan and profiles, and erosion patterns may indicate problems with transported soil deposits. General topography may indicate the extent of weathering of residual deposits.

Mineral (including coal and water) deposits may be significant in future site use. Coal and other underground mining activity may require special design precautions to avoid mine roof collapses beneath important structures.

Seismic potential is a major design factor in many parts of the world. A few regions (California, Chile, Italy, Japan, Turkey, etc.) are well known for seismic activity, but the potential exists in other areas. Since seismic records have been accumulated for such an insignificant time span relative to geologic time, there is no great certainty that an earthquake cannot occur anywhere at any time. Admittedly, the likelihood increases of having more events in "active" areas. Interestingly, however, no dam failures have occurred from being located on or very near a fault of the some 25 000 dams of record (Sherard et al., 1974).

Earthquakes are believed to be caused by accumulations of stress which exceed the ultimate rock strength at depths in the crust overlying the molten magma. The resulting rock failure termed an "earthquake" is sudden and produces cracking, shattering, and relative movement (either vertical, lateral, or both) along the failure surface. It would appear reasonable that once an earthquake occurs, less stress is required to initiate further slip than in sound rock. If this assumption is valid, the "active" zones are those where the rock crust is already fractured. Surface identification of these zones may be difficult unless slippage is recent or of a magnitude such that later erosion has not removed the surface evidence. This is the case with the San Andreas fault in California (which can easily be seen from the air in many places), but many faults are beneath oceans or occurred so long ago that surface erosion has removed most to all of the visible effects. Also some earthquakes produce such small relative movements that no surface evidence occurs.

The seismic potential must always be a factor in any risk analysis for important structures (high-rises, dams, nuclear facilities, and similar). Encountering fractured rock in a boring is not absolute evidence that an earthquake has occurred. After all the available information is assembled, one may make some kind of risk analysis, but there is always substantial uncertainty involved.

Geologic information may be obtained from the U.S. Geological Survey (USGS) in the United States. Data may also be available from certain other countries which contracted with the USGS for survey work. Additionally, State Transportation Departments and Mining Departments often have maps outlining significant geologic features. Aerial photographs are widely available from both local and the U.S. Department of Agriculture which display many useful features. Satellite photographs are also sources of aerial geologic data but may not be as readily available as other aerial photographs. Where the project warrants the cost, aerial photographs may be contracted from firms specializing in this work.

6-4 DETERMINING SUBSURFACE CONDITIONS

There are two general types of subsurface exploration:

1. Samples are not taken.
2. Samples are recovered—usually used to correlate "1" when "samples not taken" is the principal method used.

The determination of engineering soil properties using in situ testing may be considered a combination of soil exploration and testing. The general mechanics of in situ testing will be briefly described in those chapters where the soil properties are required in the analysis method being considered.

In most soil exploration, samples of the soils are recovered, but in some cases it may not be necessary. In these cases it is only necessary to ascertain the underground conditions or the exploration is to supplement recovered samples. Common methods with sample recovery include:

1. Probing (or sounding)—a rod or probe is driven into the soil with the driving resistance related to the soil state.
2. Seismic—measuring the speed of shock waves from blasting or hammering from source to pick-up unit.
3. Electrical resistivity—measuring the voltage drop (resistivity) between two electrodes inserted a known distance apart.

All three of these methods require equipment calibration and verification information obtained from borings and sample recovery.

The seismic method is particularly useful to profile lengthy projects (such as a roadway or airstrip) so that fewer intermediate soil borings are required. It is also useful to obtain dynamic engineering properties in situ.

Exploration where samples are recovered is slower, more costly, but more reliable than probing or using seismic or electrical resistivity methods. These exploration methods may be categorized as follows.

Test Pits

Test pits may be opened using a backhoe or front-end loader, but these are generally limited in depth, require a large site, and are rather costly. Test pits are seldom dug by hand because of the time and labor costs involved. Test pits are most suitable for borrow sites where the pit can be later used for initial access and where large representative samples can be visually obtained. Test pits may be used to obtain carefully hand-trimmed "undisturbed" samples for laboratory testing. On occasion test pits are used to provide sites for in situ tests—usually load tests using flat metal plates. These plates are placed on the ground and loaded in increments to simulate footings. This is termed a "plate-load test," and the ultimate plate load capacity is related to the allowable soil pressure recommended to the designer for footing loads.

Drilling

The most common method of determining subsurface conditions and obtaining samples is to drill holes at selected points over the site (or foundation area). Any type of well drilling equipment with adequate depth capacity can be used to advance the borehole. The two most common methods are wash boring and augering.

The details of wash boring are illustrated in Fig. 6-1. Essentially, the method requires equipment to lift, rotate, and drop the bit and a water supply with a pump. Dropping of the bit (on the end of a string of drill rods) chisels the soil loose into fragments. Water is continuously pumped through the drill rods and out the end of the bit to carry the soil cuttings to the surface where the water-soil suspension is caught in a sump. The cuttings settle (and segregate into sizes) and can be saved for visual inspection and very disturbed samples collected for laboratory testing. The water is recirculated to bring up additional cuttings.

The method of auger drilling requires rotary drill equipment to provide both a push against the drill head and to rotate it at a speed on the order of 75 revolutions per minute (rpm). Commonly, solid or hollow-stem continuous flight auger rods with an outer diameter of 102 to 152 mm (OD) are used. As the drill advances, additional auger flights (1.5-m or 5-ft lengths) are added. Soil is

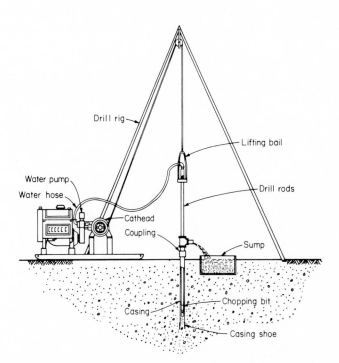

Figure 6-1 Schematic of a wash boring set-up. Sump may be a portable container or a pit dug on site.

(a)

(b)

Figure 6-2 Auger boring. (*a*) Driller observing auger spill and depth while assistant in foreground is assembling the drill rods and split spoon for a penetration test. (*b*) Making the penetration test. The auger drive has been disconnected from the auger and shifted back and the split spoon inserted into the hollow stem. The driller is pulling on the rope wrapped around the p.t.o. (cathead), and the hammer is being raised to free fall on the drill rods when driller slacks the rope. The assistant is observing the penetration and counting the blows N.

…mple quantity (which is large for most projects) is roughly 0.01 m³ …otal of 400 000 m³. This is the basis on which the mass properties …determined" for whatever use. The percentage sampled is too small …prehend. The application of statistical methods to assess the …lity" here would be mostly an exercise in mathematics.

////

…example (greatly idealized) is typical of sampling operations. At most …al strata would be encountered in a 20-m depth with usually one or two …collected from each borehole in each stratum (depending on thickness, …spacing of samples, field inspection, and type of sample needed). It would …re occasion when one could always recover a sample of constant (here …) length.

…should be evident that the stratum will be reliably determined if the small …e quantities are truly representative. Whether they are is likely to be a …what random occurrence—but probably biased toward being representative …a conscious effort is usually made.

…sturbed Samples

…isturbed samples are obtained where physical and index properties are required …nd preservation of the in situ soil structure (and stress history) is not necessary. …o put this in perspective, soil in situ at a borrow pit before it is removed and …placed in a truck is "undisturbed" and has a distinctive, unique skeleton struc- …ture, and some quantity of soil water. When it is removed from the ground and …placed in the truck, the structure is largely destroyed (except in the large lumps). …When the soil is brought to the fill site, dumped, spread, and graded, the initial …structure is completely lost and a new one formed. The new state is termed …"remolded." Auger cuttings from a bore hole are similarly remolded or disturbed.

…We may obtain disturbed samples from (1) pick and shovel operations, (2) …auger cuttings, and (3) the penetration testing described in Sec. 6-9. Samples from …sources 1 and 2 are commonly placed in bags and returned to the laboratory. If …the natural water content is to be preserved, the bags may contain plastic liners …or friction top cans are used instead of bags. Samples from source 3 are com- …monly placed in glass jars with screw caps which are marked to show boring …number, depth, and other data. This sample collection is shown in Fig. 6-3.

Undisturbed Sampling

Undisturbed samples are obtained when effort is made to minimize changes in …the existing soil structure. These samples are required to estimate engineering …properties for strength and stability analyses as well as for water flow studies.

Undisturbed samples are commonly obtained from:

1. Test pits. These probably produce the best quality.

brought to the surface and spills off the flights around the borehole where it can be visually inspected and "disturbed" samples put in a bag for later testing if desired. Figure 6-2 illustrates the sequence involved in continuous flight auger boring using a hollow-stem auger.

The continuous flight auger is almost exclusively used at present since it has a number of advantages over other methods, such as the following:

1. It is very rapid.
2. Samples are less disturbed than from wash borings.
3. It is easier to visually detect stratum changes from spoil from the flights—but recognizing that current spoil is from strata already penetrated.
4. Rock drilling can be done with the same drill rig by changing bits.
5. The hole does not require casing when the hollow stem auger is used since additional testing and sample recovery can be made through the stem.
6. Penetration testing and undisturbed samples can be recovered by either pulling the auger (solid stem) or through the hollow stem after removal of the bit plug. Care is necessary in pulling the auger (or the plug) that a differential water head does not cause a "quick" state in the soil at the bottom of the hole.

6-5 LOCATION, SPACING, AND DEPTH OF BORINGS

Boring location is site-subjective; thus hard, fast rules are not possible. Location generally depends on site topography and/or proposed location of the structure. The depth and spacing depends on:

1. Purpose or intended use (size, type of building, weights, etc.)
2. Needed information (physical properties of the soil, strength, water flow)
3. Site conditions discovered as the boring project progresses
4. Geotechnical consultant experience—total and in this area

In general, borings should be located to obtain maximum information for the minimum meterage of holes since this is usually how the client is charged. Enough borings should be taken to provide a reasonably confident assessment of the subsurface conditions for preliminary studies. This may be as few as two or three to as many as 15 or 20 (on large sites). Confidence is subjective in nature so that while one geotechnical consultant might accept three or four borings, another might insist on six or eight for the same site.

The purpose (nature of the construction) often both determines the number of borings and their depth. Buildings require sufficient borings to delineate "poor" and "good" soil zones by interpolation of information from adjacent holes. Often the building corners, if known, and a middle point are drilled. Additional borings under heavily loaded columns, machinery pads, and similar may also be required.

Roads, airfield, and water and sewer line work usually specifies borings spaced along the centerline (or lines for divided highways). Other areas such as taxiways, parking stands, etc., require additional borings for airports. Road and airport borings both locate rock and soil stratification, as well as materials which are either suitable or unsuitable for fill. Water and sewer work often has a principal objective of locating the water table and rock line.

Generally at least one boring for a building should extend to a sonsiderable depth—often to bedrock if practical. The remaining borings may extend to a depth where the building load stresses are insignificant if the deep boring has not disclosed any stratum which may dictate the type of foundation. The depth of significant stress is commonly taken as two times the least width of the building (or typical footing for warehouses and similar) and is based on stress profiles as displayed in Fig. 1-12.

It is generally good practice not to terminate any borings for buildings in soft strata or in strata with significant amounts of organic material. Both of these types of materials are subject to large time-dependent settlements and may control the foundation design. Caution is also necessary to ensure that borings to bedrock actually terminate on that material and not on a suspended boulder.

Borings for roads and similar may range from every station to perhaps 100 to 150 m (300 to 500 ft). Borings are commonly 1.5 to 3 m below the proposed grade line except in deep fill sections where poor soils at greater depths may produce long-term settlements (and pavement bumps) from the fill weight. Borings beneath large culverts and bridge piers are taken to substantially greater depths so that the foundations can be adequately designed for stability and settlement.

Borings in river channels for bridge piers should attempt to ascertain both scour and competent soil depth. Scour is the erosion of the river bed to greater depths during floods which later fills or "silts" to the low water stream bed.

Number of Borings

We have not actually defined the "number" of borings required for a project. We have discussed considerations which enter into the equation. It is evident that the variables in the equation are heavily project- and site-dependent as well as on the professional judgment of the geotechnical engineer. The number must be sufficient to give the geotechnical engineer reasonable confidence that the underground conditions have been identified well enough to make a recommendation at a reasonable risk level. Since any recommendation carries some risk, the lower the confidence level, the more conservative will be the recommendation. If these are overly conservative, the owner client often incurs additional design costs which can easily exceed the cost of making several additional borings. It should be evident that site exploration, in these circumstances, is a poor place to shave project design costs.

This does not mean that the geo___ the uncertainty. The following are not ___

1. Making large numbers of unnecessar___ client has essentially said "do what is ___ of borings to delineate "poor" soils wh___ that the entire site is underlain by erratic s___
2. Making an excessively conservative recom___ an adequate exploration program was done ___ mend a pile foundation when spread footings ___

6-6 SAMPLING

Sampling or collection of small quantities of soil for p___ classification, strength correlations, or for engineering ___ categorized as either disturbed sampling or undisturbed ___ samples should be representative. This is often very dif___ through a number of strata are the material source. Th___ illustrates the small quantities of soil used in site studies.

Example 6-1 Given: a site plan 100×200 m and a single st___ (visual observation during boring).

REQUIRED Boring coverage and percentage of sample coverag___ soil is recovered and inspected every 0.75 m of boring depth. ___ sample recovered is bagged. The sample diameter is 35 mm.

SOLUTION After the first boring disclosed a uniform stratum, proba___ two additional borings would be taken (total of three).

The vertical profile will produce

$$\frac{20}{0.45 + 0.75} = 16.67 \qquad \text{say, 17 samples}$$

With one sample in three bagged, there are six samples per hole bagged. Since the length per sample is $0.45 + 0.75 = 1.20$ m, we have the

$$\frac{\text{Sampling percent}}{\text{hole}} = \frac{0.45 \times 100}{1.2} = 37.5 \text{ percent}$$

The sample quantity at six samples per hole is

$$V = 3 \times 6 \times (0.7854 \times 0.035^2)(0.45) = 0.00779 \text{ m}^3$$

The percent samples for the site is

$$\text{Percent} = \frac{0.00779 \times 100}{100 \times 200 \times 20} = 1.9 \times 10^{-6} \text{ percent}$$

(a)

(b)

Figure 6-3 A recovered sample. (*a*) The split spoon shown in Fig. 6-5*a* has been recovered from Fig. 6-2*b*, and the coupling and drive shoe (behind screwdriver) have been removed and the spoon halves separated to display the sample. The assistant is taking a pocket penetrometer test to determine the unconfined compression strength. Shown in the background are sample bags and several spare thin-walled sample tubes with plastic end covers for "undisturbed" tube samples if needed. (*b*) A portion of the sample has been placed in a sample jar which has been identified with the label placed on the lid. Also shown is a close-up of the pocket penetrometer being used in part *a* and the field log being kept as the boring progresses.

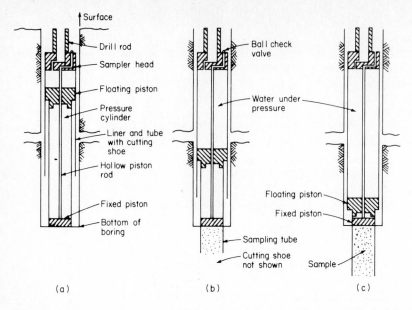

Figure 6-4 Piston sampling. (*a*) Sampler is lowered to bottom of boring and fixed piston is seated on soil. (*b*) Increasing pressure in the pressure cylinder advances the sample tube. (*c*) At full advance the sample is recovered on withdrawing the sampler unit. Note that many details are omitted for greater clarity.

2. Thin-wall tube samples with nominal diameters of 50 to 150 mm. The 50- to 75-mm tubes are most commonly used. Thin-wall tube samples is the most common source of "undisturbed" samples.
3. Various types of piston sampler (one type is shown in Fig. 6-4).
4. Various types of foil samplers [see Hvorslev (1949) for a typical description].

   There is, of course, no such thing to date as an "undisturbed" soil sample. At the very least any sample recovered will be unloaded of the in situ overburden pressure (which may result in some expansion or swell). Samples below the water table will undergo a loss of pore pressure and/or pore fluid. Samples recovered by means of 2, 3, and 4 above are disturbed some amount from the soil displacement occurring from the sampler volume. The use of thin-wall tubes reduces the sampler volume but does not eliminate the disturbance completely.

   The amount of sampler disturbance is dependent on several factors as follows:

1. Shape of cutting edge. This should be tapered and swaged to a slightly smaller diameter than the tube (to reduce inside friction).
2. Area of container relative to area of soil. This area ratio A_r is computed as

$$A_r = \frac{D_o^2 - D_i^2}{D_i^2} \tag{6-1}$$

where D_o and D_i = outside and inside (swaged) tube diameters, respectively. Ideally, A_r should be under 0.10.

3. Friction between soil and inside tube wall. This will compress the sample so that the recovery length may be less than the pushed length. Large stones, roots, and similar may tear the sample, causing it to lengthen—also unloading of the overburden pressure. A measure of this disturbance is the *recovery ratio* L_r, defined as

$$L_r = \frac{\text{Length recovered}}{\text{Length pushed}} \qquad (6\text{-}2)$$

Obviously values less than 1.0 indicate compression; greater than 1.0 is an expansion. Samples with L_r much different from 1.0 are substantially disturbed.

4. Soil character—whether the soil is gravelly, contains appreciable roots and other solid organic material, or has cracks and fissures from shrinkage.

Piston samplers represent efforts to control sample loss by using a valving arrangement to retain the sample via suction. The piston can also prevent soil flow from any pressure differential by limiting the soil recovery length to the tube advance. This may not improve the sample quality but will certainly prevent a recovery ratio $L_r > 1$.

Foil samplers are tube devices with an inner foil system or liner which surrounds and unreels with sample recovery to hold the soil as it advances into the container. Since the foil holds the sample away from the tube wall, friction is greatly reduced so that greater lengths of sample can be recovered at a time without either substantial friction disturbance or compression of the lower soil from the column above. These devices are most commonly used in very soft cohesive deposits where it is desired to obtain nearly 100 percent vertical recovery. Since great lengths can be obtained at a time, the sampling operation is rather rapid, in comparison to obtaining 0.5 m or less samples at a time.

While "undisturbed" cohesive soil samples of good quality can be obtained, it is nearly impossible to obtain cohesionless samples of similar quality. Structure is nearly always lost in these materials. Cementation (or aging) effects, which may be significant in mass performance but are so small in sample sizes that minor remolding breaks the particle contacts, are often lost. Remolding from perimeter gravel is likely to penetrate nearly through sand samples compared to cohesive samples of the same diameter.

Efforts to inject grout (cement or other materials) or asphaltic material meet with mixed success but may be suitable for certain analyses. Freezing has also been used to recover samples.

Fortunately, truly undisturbed samples of cohesionless materials are seldom necessary. Other indirect methods are commonly used to estimate engineering properties, and the physical properties such as grain size analyses can be performed on disturbed samples.

After obtaining "undisturbed" field samples, there is the problem of preserving them in this state while transporting them to the laboratory, storing them as necessary, removing them from the recovery container, trimming to size, and inserting them into the testing device. It is evident that few "undisturbed" samples can survive all of these hazards without degradation.

In situ testing for engineering properties is gaining in popularity because of the great difficulty in recovering and testing undisturbed samples. Principal among the in situ tests is the cone penetration test (CPT), borehole pressuremeter, and vane shear. Other methods include borehole shear, hydraulic fracturing, and various types of pressure-sensitive devices which are pushed into the stratum; the resistance (either push or lateral on the device) is related to the soil property of interest. Most of these devices do not test an "undisturbed" soil since any volume inserted into the soil produces disturbance and no device to date has been found which has zero volume.

6-7 WATER TABLE

When the water table is near the ground surface, its position may influence the foundation type or construction method. The allowable bearing capacity tends to be reduced when the water table is close to or at the footing level. Ground water table (GWT) fluctuations may reduce the stability or even float the structure out of the ground.

Exploration to site landfills and hazardous waste disposals requires a very accurate location of the GWT and often the direction of flow.

The most accurate method of GWT location is to directly measure it by lowering a tape or other device down the boring after the water level has stabilized. This is usually done by bailing the boring below the GWT and allowing it to fill back to the GWT. It may take several days to several weeks for this to occur, and the hole should be plugged so that surface water and contaminants are excluded. Hvorslev (1949) provides a computational method to estimate the GWT when the soil is fine-grained so that it takes a very long time for the GWT to stabilize. In the computational method the hole is bailed and the inflow level measured at selected time intervals.

Generally, borings should be plugged if there is any possibility of surface contamination of the groundwater. Borings through clay layers into granular subsoils which contain the GWT should also be plugged since this state represents a confined aquifer which may develop artesian pressure under certain conditions (such as river flood stages), and the boring is converted into an artesian well (refer to Fig. 1-2b).

6-8 SAMPLE QUALITY AND ENGINEERING PRACTICE

Disturbed, but representative, samples are quite adequate for physical properties and index-consistency tests.

Good-quality undisturbed samples are required for engineering soil properties for cohesive soils. In most cases the engineering properties are obtained by using substantially disturbed samples from the standard penetration test of the next section. Supplementary thin-wall tube samples for laboratory testing may also be obtained where the soils are very soft or the project is of a substantial nature.

The standard penetration test with some material recovered for visual inspection and possibly sieve analyses tends to be the norm for most cohesionless soils.

6-9 THE STANDARD PENETRATION TEST (SPT)

The SPT is currently the most widely used method for determining soil conditions (worldwide). The SPT has been extensively studied and has been (and is) subject to much criticism (e.g., De Mello, 1971; Schmertmann, 1975) but is still widely used. It has a number of shortcomings, or sources of errors, but has the particular advantage of being rapid. Another distinct advantage is the large data base of success; its simplicity is also a major consideration. Both these factors allow making a large number of tests, rapidly, at a relatively low cost.

The SPT has been standardized by ASTM (D1586) and consists in:

1. Driving the standard split barrel (also called "split spoon") sampler shown in Fig. 6-5 a distance of 460 mm (18 in) into the soil at the bottom of the boring. The first 152 mm (6 in) is to seat the driving shoe onto undisturbed soil.
2. Counting the blows to drive the sampler the next 305 mm (12 in). The counts are usually obtained for each of the 152-mm (6-in) increments and summed to obtain the blow count N.

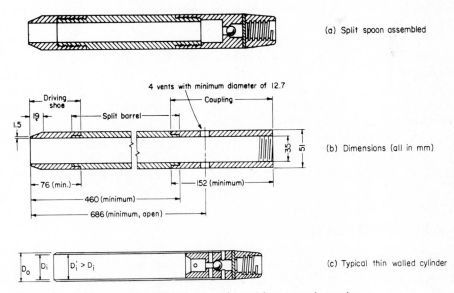

Figure 6-5 Dimensions of the standard split spoon used for penetration testing.

3. Using a 63.5-kg (140-lb) mass falling free from a height of 760 mm (30 in).

Since the sampler is driven by the impacting of the falling weight onto the drive rods connected to the sampler and extending to the ground surface, the procedure is also called a *dynamic penetration test*. The free fall and height of drop are loosely interpreted since a large number of drill rigs use a rope wrapped around a power take-off unit. The driller pulls on the rope which then tightens on the pulley so that friction winds (and raises) the rope until the weight is visually 760 mm above the reference mark, at which time the rope is released with the weight unwinding it for the drop. Any coil congestion around the power pulley will inhibit free fall so that an increased blow count will be obtained. Several manufacturers attempt to avoid this by using a mechanical hoist-trip device. This is, however, not the only source of error—other factors such as pushing a rock, damaged driving shoe, missing a count, and pore pressures also contribute error (or at least variations so that N may not be reproducible in situ).

The boring log shows "refusal" when the blow count $N \geq 100$. The log may show a ratio, such as 70/150, 50/100, and so on, indicating 70 blows for 150 mm of penetration or 50 blows for 100 mm of penetration. Large blow counts both greatly reduce the daily borehole meterage and cause rapid equipment wear and damage. Standardization of the blow count for "refusal" by ASTM at 100 assists both the client and the geotechnical engineering firm to better identify drilling costs.

The SPT was originally developed to investigate the status of cohesionless deposits for pile installation. At present it is widely used in both cohesionless and cohesive deposits for routine exploration for all types of foundations. In loose sands and very watery cohesive soils, inserts are available to aid in retaining the sample so that it can be brought to the surface without falling out of the sampler end.

At the surface, the sampler is removed from the string of rods, the sampler unscrewed, and the split barrel separated to display (see Fig. 6-3a) the recovered sample for inspection. Cohesive samples are usually immediately tested for strength by using either a pocket penetrometer (Fig. 6-3a) or a portable field compression tester. Parts of the sample are usually stored in small glass jars which are marked with the job, boring number, sample depth, and blow count N. These are returned to the laboratory and used as necessary for sieve analyses, natural water content determination, and Atterberg limits. The boxes of samples may be given to the client or saved in the soil laboratory for a stated period of time.

6-10 SPT CORRELATIONS

A number of correlations between N and soil properties have been proposed. Many are useful, whereas others are little better than guesses. For example, in Table 6-1, giving the unit weight as shown is almost meaningless. The estimate

Table 6-1 Standard penetration test (SPT) correlations

Strength correlations will be given in later chapters as needed. Values shown are primarily for " order of magnitude."

	Cohesionless Soil			
N	0–10	11–30	31–50	>50
Unit weight γ, kN/m^3	12–16	14–18	16–20	18–23
Angle of friction ϕ	25–32	28–36	30–40	>35
State	Loose	Medium	Dense	Very dense

Relative density D_r see Eq. (6-3) and Eq. (6-4) since depends on $p_0 = \gamma y$

	Cohesive Soil				
N	<4	4–6	6–15	16–25	>25
Unit weight† γ, kN/m^3	14–18	16–18	16–18	16–20	>20
q_u, kPa†	<25	20–50	30–60	40–200	>100
Consistency	Very soft	Soft	Medium	Stiff	Hard

† Values heavily dependent on water content.

for angle of internal friction ϕ is generally conservative, and (as noted in Chap. 13) it is common to estimate ϕ as 30 to 32° for many projects.

The relative density D_r is often related to N but is often a very poor correlation. This results from N being somewhat project- and site-dependent and from D_r being rather tenuous to define (or reliably compute). As a consequence of this and some recent work which seems promising, it was decided not to include D_r in Table 6-1, but rather provide the current "best estimate" equations.

According to Marcusson and Bieganousky (1977)

$$D_r = 0.086 + 0.0083(2311 + 222N - 711(OCR) - C_1\sigma'_v)^{1/2} \tag{6-3}$$

and according to Fardis and Veneziano (1981), who applied much of the data used to develop Eq. (6-3), the relationship is

$$\ln N = C_2 + 2.06 \ln D_r + C_3 \ln \sigma'_v \tag{6-4}$$

where $C_1 = 7.7$ for σ'_v in kPa; 53 for psi units

    $C_2 = $ depth function which should be determined at a site by measuring N and D_r†

    $C_3 = 0.222$ for σ'_v in kPa; 0.442 for psi units

    $OCR = $ overconsolidation ratio defined by Eq. (11-2)

Both of these equations are based on regression analyses. Equation (6-3) is based on four dissimilar soils and a large number of tests and claims a 78 percent reliability with a ± 0.075 standard deviation.

Example 6-2 Given: the SPT blow count at a depth of 4 m is 12. The soil is very sandy with traces of gravel and has an estimated unit weight $\gamma = 17.9$ kN/m^3. The soil is damp but above the water table.

---

† If no correlation is made for C_2, use the value of $C_2 = 2.67$ obtained from the data base used for the equation.

REQUIRED Estimate the relative density D_r.

SOLUTION We will use both Eqs. (6-3) and (6-4), which for this case should give nearly identical values. We will take $OCR = 1$ (for a normally consolidated sand:

$$\sigma_v' = 4(17.9) = 72 \text{ kPa}$$

By Eq. (6-3):

$$D_r = 0.086 + 0.0083(2311 + 222N - 711(OCR) - 7.7\sigma_v')^{1/2}$$

$$D_r = 0.086 + 0.0083(2311 + 222(12) - 711(1) - 7.7(72))^{1/2} = 0.59$$

By Eq. (6-4) (use $C_2 = 2.67$ since we have no site correlation data):

$$\ln N = 2.67 + 2.06 \ln D_r + 0.222 \ln \sigma_v'$$

Rearranging and inserting values, we obtain

$$-2.06 \ln D_r = 2.67 - \ln 12 + 0.22 \ln 72$$

$$-\ln D_r = \frac{1.134}{2.06}$$

$$D_r = 0.577$$

Best estimate of $D_r = 0.58$

////

6-11 BORING LOGS

Information obtained during the field drilling operations is recorded on a field log. A typical log is shown in Fig. 6-6a, which should be carefully studied for the type of information which is collected at this stage.

An office log is prepared as a composite of the field log, supplementary laboratory tests, and sample (jar) inspection as in Fig. 6-6b. These data may also be shown on a line drawing which includes a site plan and the borings in elevation. In most cases, however, the boring logs shown in Fig. 6-6b are sent to the client along with a location plan of the several borings, and recommendations for soil quality control, suggested earthwork specifications (how to fill, compaction procedures, required density, dewatering methods, etc.) and bearing capacity, location of basement floor, footing depth, or pile lengths—depending on the purpose of the project.

A transcribed copy (the field original is usually dirty) of the field log may also be furnished. The jars of soil samples may be sent to the client if they were requested during contract negotiations, but usually they are retained in case additional verification tests are required or other questions or problems develop.

6-12 CONE PENETRATION TEST (CPT)

A method of in situ exploration where no samples are taken is the cone penetration test (CPT). In this test a cone with standard tip (or cone) geometry as in Fig. 6-7 is pushed into the ground. Since the cone is pushed rather than driven, the test is also termed a *static penetration test*. Various cone configurations exist which allow measuring of point resistance as well as resistance from a portion of shaft near the cone. Some cones allow measuring the excess pore pressure which will exist in cohesive soils in the vicinity of the cone tip since it is commonly pushed at a rate of 15 to 20 mm/s, which is too fast for drainage to occur in many fine-grained soils.

Some of the advantages of the cone method are that it:

1. Is very rapid—particularly when electronic data acquisition equipment is used to record the tip pressure and/or side resistances
2. May allow a nearly continuous record of resistance in the stratum of interest
3. Is useful in very soft soils where recovery of " undisturbed " samples would be very difficult
4. Allows use of a number of correlations between cone resistance and the desired engineering property

Some of the major disadvantages are that:

1. This method is applicable only in fine-grained deposits (clay, silt, fine sands) where the material does not have massive resistance to cone penetration.
2. Interpretation of soil type producing the cone resistance requires either (*a*) considerable experience or (*b*) recovery of samples for correlation testing.

The CPT has been used in Europe for many years, particularly in sedimentary deposits along the North Sea. It has only been used substantially in the United States since the 1970s and mostly in sedimentary deposits along the Gulf Coast.

Several cone correlations will be given in subsequent chapters as appropriate. The cone does not seem to have been used to correlate with the relative density so far.

6-13 ROCK DRILLING

Where rock (or refusal) is encountered and it is desired to investigate the quality of the material, corings are undertaken. In rock coring, special bits are attached to a core barrel (see Fig. 6-8) and the bit rotated at high speed (order of 400 to 1000 rpm) to make the cut by abrasion. It is rare that an intact core can be recovered, particularly in the smaller sizes, say, under 50 mm. Considerable deg-

SOIL TESTING SERVICES, INC. 111 PFINGSTEN ROAD NORTHBROOK, ILL. 60062

Chicago Phone 273-5440
Northbrook Phone 272-6520

TECHNICIAN JN
DRILLER MED
HELPER MED
RIG NO. 4

SURFACE ELEV. _____
BORING STARTED 4-29-81
BORING COMPLETED 4-30-81
STATION Plan
OFF SET 0

CASING USED 10' SIZE 4"

Sheet 1 of 1

WATER LEVEL OBSERVATIONS
WL: 10' WS OR WD
WL: _____ BCR 65' ACR
WL: _____ AB _____ Hr. AB
WL: _____ 24 Hr. AB

ABBREVIATIONS
F.T.-Fish Tail
W.O.-Wash Out
S.S.-Split Spoon
D.B.-Diamond Bit
P.A.-Power Auger
R.B.-Rock Bit
W.S.-While Sampling
W.D.-While Drilling
B.C.R.-Before Casing Removal
A.C.R.-After Casing Removal
A.B.-After Boring

DRILL CREW CHECK LIST
Topsoil Thickness 6"
Fill Thickness None

CAVE IN LEVEL:
While Drilling and Sampling None
After Boring Completion _____

WATER LOSS:
At 15' To 20'
Percent Loss 50%
At _____ To _____
Percent Loss _____

BOULDERS OR OBSTRUCTIONS
At None To _____
At _____ To _____

ARTESIAN PRESSURE:
Depth None
Height of Soil Rise In Casing

FB: BL 9/71

JOB NO. 221-----   BORING NO. 1   CLIENT ____ Co.   WEATHER ____

Sample No.	Depth From	To	Sampling Method	Split Spoon Blows (6" 6" 6" / 2 Feet)	R Length Recovered in Feet	Qp Penetrometer Test in TSF	Sample Description
1	0	2.0	ST		18"	2.0	Red Br. Silt some CL -- Tr. Sand
	0	2.5	PA				
2	2.5	4.0	SS	7 15 20	18"	2.25	Red Br. Sandy CL
	2.5	5.0	PA				
3	5.0	7.0	ST		12"	4.5*	Br. Sand Med
	5.0	7.5	PA				
4	7.5	9.0	SS	15 20 7	16"	-	Br. Sand Med - Tr. Grav.
	7.5	10.0	PA				
5	10.0	11.5	SS	8 10 11	18"	-	LG Sand Med & Med Grav BR
	10.0	12.5	PA				
6	12.5	14.0	SS	8 12 14	18"		"
	12.5	15.0	PA				
7	15.0	16.5	SS	6 11 10	16"		"
							Strata Change 18'
8	15.0	20.0	PA				Shale BL -- Layers fine Gr Sand
	20.0	21.0	SS	25 125/5"	8"		
	20	25	RB				
9	25.0	25.5	SS	125/5"	5"		Shale BL
							EOB

(a)

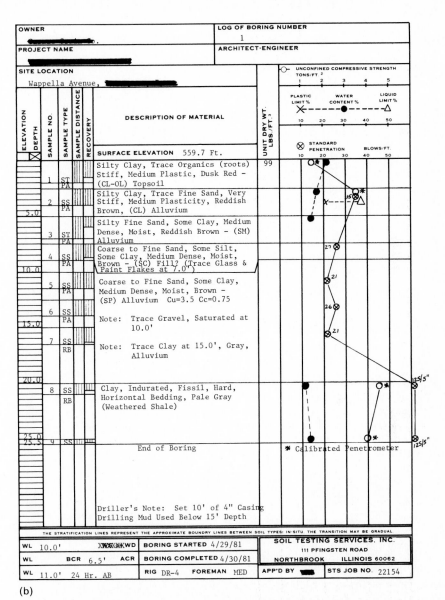

(b)

Figure 6-6 Boring logs. (*a*) Transcribed field log so that it could be reproduced. (*b*) Office log which included data from the field log shown in part *a* as well as any supplementary laboratory work. Note that the soil description has been substantially improved from that used by the drill crew in part *a*.

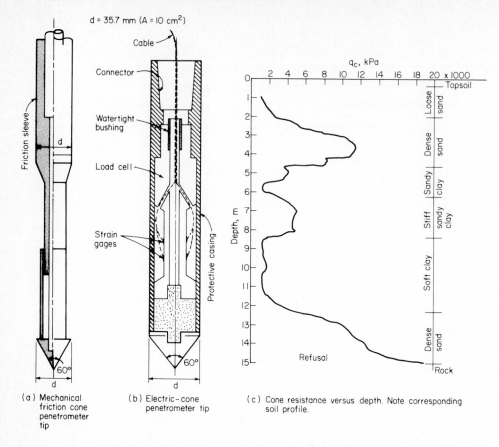

Figure 6-7 Cone penetrometers and output. Typical dimensions of cone *a* given in ASTM D3441. Test consists in pushing cone then pushing friction sleeve. Cone *b* may also be called a "Fugro cone." Versions are available as shown and also with a friction sleeve and with facilities to measure pore pressure at the cone tip. Cone data *c* are presented as a plot of cone resistance q_c versus depth, usually requiring a supplemental boring to identify the several strata penetrated. Otherwise, about all that is known is whether the resistance is large or small.

radation of the core also occurs when the core fractures and the pieces interact with others inside the core barrel.

A measure of the core quality is the recovery ratio [Eq. (6-2)], but the rock quality designation (RQD) proposed by Deere (1968) is currently more commonly used. The RQD is computed on the basis of the sum of the rock fragments that are 100 mm or larger in length divided by the core advance ($\times 100$ to produce a percentage). An intact length of rock core would have RQD = 100 and a recovered barrel of gravel would have RQD = 0. Refer to Fig. 6-9 for a general computation of RQD and qualitative rock rating.

Figure 6-8 Rock bits. *(Courtesy of Soiltest, Inc.)*

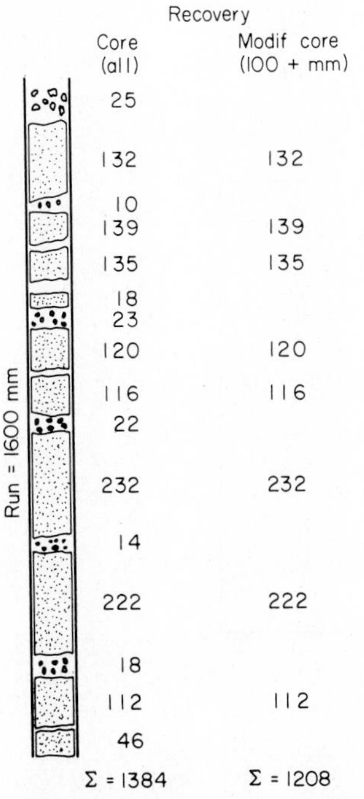

Recovery

	Core (all)	Modif core (100 + mm)
	25	
	132	132
	10	
	139	139
	135	135
	18	
	23	
	120	120
	116	116
	22	
	232	232
	14	
	222	222
	18	
	112	112
	46	
	Σ = 1384	Σ = 1208

Run = 1600 mm

$$\% \text{ Recovery} = \frac{1384}{1600} \times 100 = 86.5\%$$

$$\text{RQD} = \frac{1208}{1600} \times 100 = 75.5\%$$

Rock quality = "fair"

Rock quality[a]	RQD, %
Very poor	0 - 25
Poor	25 - 50
Fair	50 - 75
Good	75 - 90
Excellent	90 +

[a] From Deere (1968)

Figure 6-9 Rock quality designation (RQD) index. Method of computation and qualitative determination of rock quality.

6-14 SEISMIC EXPLORATION

Seismic exploration using shock waves refracted from lower and more dense strata may be used in extended sites to supplement more expensive borings. Seismic methods are also used to obtain in situ dynamic engineering properties for vibration analyses.

In seismic exploration a small explosive charge or a heavy hammer blow on a special plate provides the shock wave. Geophones are placed on the line of interest at increasing distances from the shock source as shown on Fig. 6-10a. The time required for the shock wave to reach each geophone unit is recorded by a pickup unit called a seismograph (Fig. 6-11), which is connected to both the shock source and geophone so the travel time can be accurately obtained. The

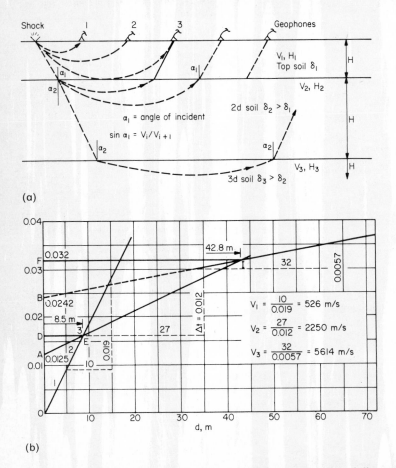

Figure 6-10 A seismic survey. Scaled values are used in Example 6-3, and distances *DE* and *FI* are those on Fig. 6-13. (*a*) General orientation of shock and pickups (geophones). May have only one geophone which is moved along line of 1, 2, 3, etc. with shock repeated as needed. (*b*) Plot of arrival time from corresponding numbered (below) geophones.

Fig. 6-11 Seismic surveying. Shock is being provided in background to be picked up by geophone and transmitted to seismograph. Tape on ground orients survey and locates shock source and pickup (at arrow in foreground). *(Courtesy of Soiltest, Inc.)*

results can be plotted as shown on Fig. 6-10*b*. Interpretation is fairly simple if the stratum is relatively uniform in thickness and the density of the layers increases with depth. It is more difficult if the layers are of variable thickness, slope, or a more dense layer overlies a less dense one. It is assumed that shock waves travel faster in more dense materials. The thickness of sloping strata can be approximately obtained by making the survey in both directions.

Assuming that the upper layer is less dense, the shock waves arriving first to the several geophones travels in a direct path through the upper layers. Eventually the distance is such that a shockwave travels down through the top layer, is refracted into the lower more dense layer, travels through it at greater speed, and is refracted back into the upper layer and to the geophone. This refracted wave will arrive before a wave traveling only through the less dense upper layer. A plot of distance versus arrival time directly gives the apparent velocity in any layer which defines the inverse of the slope of the time-versus-distance plot since

$$v = \frac{\Delta d}{\Delta t} = \frac{d_2 - d_1}{t_2 - t_1}$$

In Fig. 6-12 the time for travel $OC > OABC$. Use of a rectangular travel path (not a refracted one) gives the approximate thickness of the upper stratum as

$$\frac{DE}{V_1} = \frac{2H_1}{V_1} + \frac{DE}{V_2}$$

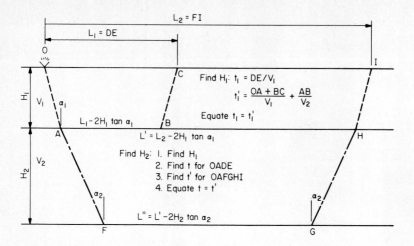

Figure 6-12 Exact method for finding thickness of strata using the seismic method. Obtain DE and FI from a plot such as that in Fig. 6-10b.

based on equating equal travel times for a direct traverse (DE) versus the rectangular path. The distance DE must be scaled from a plot such as that in Fig. 6-10b. Solving for H_1 we obtain

$$H_1 = \frac{V_1}{2}\left(\frac{DE}{V_1} - \frac{DE}{V_2}\right)$$

(6-5)

More exactly, angles of incidence and refraction can be used, and one can obtain (Shepard and Haines, 1944) the following (refer to Fig. 6-10):

Top layer (may have more than one layer) and with shock source (dynamite blast) at depth d_b below surface is

$$H_1 = \frac{OA(V_1)}{2\cos\alpha_1} + \frac{d_b}{2}$$

(6-6)

For the second layer

$$H_2 = \frac{AB(V_2)}{2\cos\alpha_2}$$

(6-7)

where

$$\cos\alpha_i = \left[1.0 - \left(\frac{V_i}{V_{(i+1)}}\right)^2\right]^{1/2}$$

(6-8)

The time values OA and AB are scaled from the plot (Fig. 6-10). It is evident that the accuracy of the resulting computations is dependent on the plot scale.

An alternative expression for H_i using the scaled distances between successive breaks (such as $C_1 = DE$ or FI) on the time-versus-distance plot and assuming that direct and refracted travel times are equal gives

$$\frac{C_1}{V_1} = \frac{C_1}{V_2} + \frac{2H_1 \cos \alpha_1}{V_1}$$

Substitution for $\cos \alpha_1$ and solving for H_1 obtain

$$H_1 = \frac{C_1}{2} \left(\frac{V_2 - V_1}{V_2 + V_1} \right)^{1/2} \tag{6-9}$$

This gives the approximate stratum thickness of any layer but may be in some error for strata other than the first and more exact computations (as in the following example) may be required.

Where wave velocities are desired for a particular stratum it is possible to drill two holes a known distance apart and apply a shock in one at the point of interest and hang a pick-up unit in the other at the corresponding elevation (Stokoe and Wood, 1972).

Finally, we should note that seismic surveying is not quite as simple as this discussion would indicate. There are three types of ground waves (see Sec. 14-6) generated when a shock is made and two of these travel near the surface where the geophone is located with nearly the same speed. We have a problem of identifying the type of wave which has arrived first so that we can identify the wave which actually traveled down into the ground and back up. Since very small time intervals are involved any interpretation errors are significant and nearly always require corroboration via borings. Wave velocities shown in Table 6-2 illustrate the difficulties in making velocity measurements.

Table 6-2 Velocity and electrical resistivity of soils and rocks from several sources

Values are shown to indicate the wide range in values which might be expected from a particular site

Material	V, km/s	R, $\Omega \cdot$m
Igneous rocks	4–6	10 000[+]
Metamorphic rocks	4–6	1000[+]
Limestone	3–6	10 000[+]
Sand:		
Loose (dry)	0.2–0.5	1000–50 000
Dense (dry)	0.5–2	2000–100 000
Glacial till	1.5–2.5	200[+]
Clays and silts	0.5–2	100–500
Frozen soil	1–2	
Water	1.5	

Example 6-3 Using the data shown on Fig. 6-10 and Fig. 6-12 find the thickness of both strata.

SOLUTION The stratum velocities are computed as shown on Fig. 6-10b.

For top stratum: scale $C_1 = 8.5$ m; $t_1 = OA = 0.0125$ second as shown in Fig. 6-10. Substituting into Eq. (6-9), we obtain

$$H_1 = \frac{8.5}{2} \left(\frac{2250 - 526}{2250 + 526}\right)^{1/2} = 3.35 \text{ m}$$

Alternatively compute H_1 using Eq. (6-6):

$$\cos \alpha_1 = \left[1.0 - \left(\frac{526}{2250}\right)^2\right]^{1/2} = 0.9723 \qquad \alpha_1 = 13.52°$$

and

$$H_1 = \frac{0.0125(526)}{2(0.9723)} = 3.38 \text{ m}$$

Scaling and round-off has produced these small differences. If we average values, we obtain $H_1 = 3.36$ m, say, 3.4 m

For the second layer (also referring to Fig. 6-12):

$$\cos \alpha_2 = \left[1.0 - \left(\frac{2250}{5614}\right)^2\right]^{1/2} = 0.9162 \qquad \alpha_2 = 23.63°$$

Now for $L_1 = 42.8$ m we have

$$t_{OAHI} = \frac{2 \times 3.4}{526(\cos 13.52)} + \frac{42.8 - (2 \times 3.4) \tan 13.52}{2250}$$

$$t_{OAHI} = 0.013296 + \frac{41.16}{2250} = 0.03159 \text{ second}$$

$$t_{OAFGHI} = \frac{2 \times 3.4}{526 \cos 13.52} + \frac{2H_2}{2250 \cos 23.63} + \frac{41.16 - 2H_2 \tan 23.63}{5614}$$

$$= 0.013296 + 0.0009702H_2 + 0.0073317 - 0.0001559H_2$$

Equating times as indicated on Fig. 6-12, we obtain

$$t_{OAFGHI} = t_{OAHI}$$

$$0.013296 + 0.0009702H_2 + 0.0073317 - 0.0001559H_2 = 0.03159$$

$$0.0008143H_2 = 0.0109617$$

$$H_2 = 13.5 \text{ m}$$

Approximately, H_2 can be found using Eq. (6-7) as

$$H_2 = \frac{AB(V_2)}{2 \cos \alpha_2} = \frac{(0.0242 - 0.0125)(2250)}{2(0.9162)} = 14.4 \text{ m}$$

And using Eq. (6-9) we obtain H_2 as

$$H_2 = \frac{42.8}{2} \left(\frac{5614 - 2250}{5614 + 2250} \right)^{1/2} = 14.0 \text{ m}$$

We might note that $H_2 = 13.5$ m is "exact" but depends on the scaled value of 42.8 m. The true value is probably between 13.0 and 14.5 m but can only be reliably established by borings, and even this will be subject to the regularity of both ground surface and stratum profile.

/////

6-15 RESISTIVITY EXPLORATION

One other type of exploration sometimes used is the electrical resistivity method. In this procedure, illustrated in Fig. 6-13, a pair of electrodes is placed in the ground a distance d apart. A second pair is placed in line also a distance d from the first electrodes. The voltage drop is measured together with the current I generated by the power source. The assumption made is that the voltage drop is in a hemispherical volume of soil of radius d so that the resistance can be computed (Griffiths and King, 1965) as

$$R = 2\pi d \, \frac{V}{I} \qquad (6\text{-}10)$$

where R = resistivity, $\Omega \cdot d$ (i.e., units of d as $\Omega \cdot$ m, $\Omega \cdot$ ft, etc.)
      d = common electrode spacing, m or ft (or other)
      V = voltage from power source
      I = current, amperes (or, usually, milliamperes)

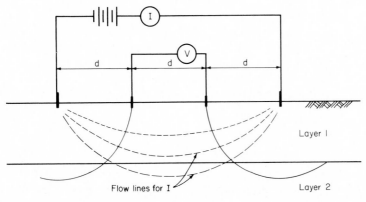

Figure 6-13 General schematic of a resistivity measurement. Values of I and V are taken, and probe spacing d is incremented or all four probes are moved to a new grid point.

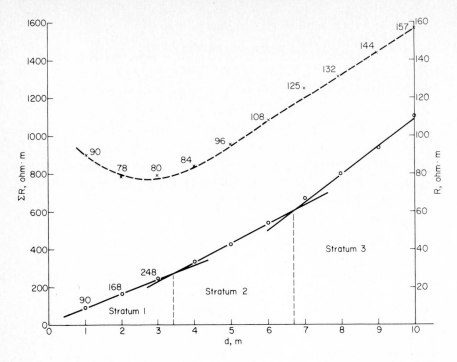

Figure 6-14 Plotting results of a resistivity survey to obtain stratum depths as indicated above. The first stratum is about 3.4 m, and so forth.

The battery (or other) voltage should be converted to alternating current unless some means to avoid polarizing (H^+ ions attracted to cathode) the electrodes is used.

In locating vertical strata, the electrode spacing is incremented from some initial spacing in constant increments (say, 0.5, 1, or 1.5 m) as illustrated in Fig. 6-14. To locate an area of shallow rock or the water table, the electrode spacing is held constant and the four electrodes are placed at several points making a surface grid. The results are formed into a contour map of resistivity to identify the desired areas.

Typical resistivities of several soil materials are given in Table 6-2 to indicate orders of magnitude which may be measured at any given site.

6-16 THE TOTAL STRESS CELL AND DILATOMETER

Two other methods which are recent introductions and provide a combination of in situ testing and exploration are the total stress cell and dilatometer. These two methods are very similar in equipment in that a thin plate is pushed into the soil.

(a)   (b)

Figure 6-15 Total stress cell: (*a*) cell in protective insert; (*b*) cell extended out of protective casing. *(Courtesy of K. R. Massarsch, Royal Institute of Technology, Stockholm, Sweden.)*

The total stress cell method (Massarsch and Broms, 1976) uses a thin cased plate which is pushed to about 0.3 m above the test point. The measuring cell (Fig. 6-15) is advanced out of the protective casing to the point of interest. The cell is about 3 mm thick so that volume displacement is not large and reportedly its effect dissipates in about 3 days so that at the end of this time the cell can measure the in situ lateral pressure.

The flat dilatometer (Fig. 6-16) is an even more recent introduction (Marchetti, 1980) that is similar to the total stress cell. This spade-shaped stainless steel device is 14 mm thick with a circular pressure membrane on one side. In use it is pushed (or driven) to a point and the pressure membrane activated. The pressures to reference the membrane to zero and to push it 1 mm into the soil are both recorded. An estimate of in situ pore pressure and overburden pressure is needed which, together with the pressure measurements are used to estimate soil properties. Estimates of grain size distribution, in situ lateral pressure and the lateral stress-strain modulus of Chap. 14 can be made. The device is sufficiently rugged that it can penetrate fairly hard soils. The penetration rate is from 20 to 40 mm/s, and tests may be made every 0.2 to 0.5 m of depth. Noting that the total stress cell is 3 mm thick and requires about 3 days to allow the stresses from volume increase to dissipate, it is evident that the dilatometer requires some interpretation to obtain the in situ lateral pressure from measurements taken immediately after an even larger volume change.

Figure 6-16 Flat dilatometer kit. Round expansion cell is clearly visible on spade shaped device. Meter to read expansion pressure also shown. *(Courtesy of Williams and Associates, Clearwater, Florida.)*

6-17 SUMMARY

This chapter has introduced the reader to some of the considerations, methods, and limitations in site exploration to estimate the underground conditions at a site to limit the risk factor to a tolerable amount.

We find that there are no hard and fast rules for the number of borings but that this depends rather on site geology and topography, the type of project, conditions found as the exploration commences, and the professional competence of the geotechnical engineer.

We see that boring depths are on the order of two times the building or footing width but that all (or at least one) may go to competent stratum. Borings for important structures are never terminated in soft (poor) soil.

It was given that the most common exploration method is to use auger borings together with the standard penetration test (SPT). Another method that is fairly common for fine-grained soils is the cone penetration test (CPT). The SPT has a number of correlations with some shown here (several others will be introduced later as needed).

We have briefly looked at seismic and electrical resistivity methods as well as the total stress cell and dilatometer. These latter several methods include references which the interested reader should consult if a more in-depth study is needed.

PROBLEMS

6-1 Compute the area ratio A_r for the standard split spoon (Fig. 6-5b) used in the SPT.

Ans.: 1.12

6-2 What is the area ratio of a thin-walled tube of outer diameter (OD) = 51 and inner diameter (ID) = 46 mm?

6-3 Three thin-walled tube samples were recovered as follows:

Sample	Length pushed	Length recovered, mm
1	610	608
2	610	583
3	590	610

Compute the recovery ratios and comment on overall sample quality.

Ans.: $L_{r(2)} = 0.96$

6-4 Referring to Fig. 6-9, if the core run is 1800 mm with the recovery shown, what is the percent recovery and RQD?

6-5 Referring to Fig. 6-9, if the 116 mm fragment falls apart after minimal handling, what is the revised RQD?

6-6 What is the relative density in Example 6-2 using Eq. (6-3) if $\gamma = 18.9$ kN/m³ instead of 17.9 given? What if N increases to 14 with the increased unit weight?

Ans.: 0.59; 0.62

6-7 Using the data of Prob. 6-6 and Eq. (6-4), compute the two values of D_r and compare with the above answers. Make any appropriate comments.

6-8 What would you suggest for a value of q_c for the dense sand layer (-2 to -4m) in Fig. 6-7c?

6-9 What would you suggest for q_c for the dense sand at -12.5 m to rock in Fig. 6-7c?

6-10 Referring to Figs. 6-10 and 6-12, what would the strata thicknesses be if $V_1 = 833$ m/s instead of 526 given?

Ans.: $H_1 \cong 5.8$ m

6-11 Make a literature search and find at least three *different* correlations between SPT N and the modulus of elasticity E. Do not use correlations given later in this book. Cite author, date, source title, page number, the equation, an example, and limitations given for the equation.

6-12 Make a literature search as in Prob. 6-11 above but for the CPT to find equations for E based on q_c.

6-13 Using the data shown on Fig. 6-14 for the individual measurements (upper curve), replot the figure to a suitable scale, draw a smooth curve, and obtain the values from the curve to replot the cumulative curve and determine the stratum depths. Does obtaining the "smoothed" data greatly affect the computed stratum depths? Which method gives the most reliable depth values?

SEVEN

COMPACTION AND SOIL STABILIZATION

7-1 GENERAL CONCEPT OF SOIL STABILIZATION

When the soils at a site are loose or highly compressible, or when they have unsuitable consistency indices, too high permeability, or any other undesirable property making them unsuitable for use in a construction project, they may have to be stabilized. *Stabilization* may consist of any of the following:

1. Increasing the soil density
2. Adding inert materials to increase the apparent cohesion and/or friction resistance
3. Adding materials to effect a chemical and/or physical change in the soil material
4. Lowering the water table (soil drainage)
5. Removal and/or replacement of the poor soils

Any alteration of the physical or engineering properties of a soil mass will require investigation of economic alternatives such as relocation on the site or using an alternative site. At present most of the more desirable building sites near urban areas have been used, so that an alternative location may not be practical. Currently, sites such as abandoned sanitary landfills (garbage dumps), swamps, bays, marshes, hillsides, and other poor areas are being used for construction sites, with this trend expected to both continue and accelerate. When alternative sites are not available or environmental considerations, citizen opposition, and zoning regulations severely limit the options available, it becomes more and more necessary to modify or stabilize the site soil to obtain the needed properties. Economically feasible solutions may severely tax the ingenuity of the geotechnical engineer.

In cases such as earth dams, embankments, levees, or other fills, where select materials in sufficient quantities may not be available, selective use of the available materials, and understanding of both the function of the earth structure and the mechanics of the earth mass, can produce a satisfactory solution via use of zoned construction.

7-2 SOIL AS A CONSTRUCTION MATERIAL

Soil is one of the most readily available of construction materials at a site, and, when it can be used, it is usually the most economical. Earth dams, river levees, and highway and railway embankments are all economical uses of earth as a construction material; however, as with any other construction material, it must be used with quality control. If soils are dumped or otherwise placed at random in a fill, the product will be a fill of low unit weight with a resulting low stability and high subsidence. Subsidence is used to describe the mechanism of vertical movements within a fill due to self-weight; settlement is the vertical movement of the underlying soil caused by the weight of the fill.

Early road fills were usually constructed by end-dumping fill from wagons or trucks with very little attempt to compact or densify the soil and no quality control via specifications. Failures of high embankments were common. Many earthworks, such as levees and earthen dams, are almost as old as humankind, but where the structures have survived it was a happy combination of inadvertent quality control and luck. For example, in ancient China and India, the structures were constructed by workers carrying small baskets of earth and dumping them in the embankment. Laborers bringing additional fill walked over the previously dumped loose earth, compacting it. In some cases herds of goats and sheep were driven across the loose earth for additional compaction with (presumably) the organic matter resulting being picked up by hand and removed. Even today, in some areas, elephants are used to compact earth embankments; however, Meehan (1967) reported that, although the elephant has sufficient mass, he refuses to distribute his walking pattern sufficiently to achieve uniform compaction.

7-3 SOIL STABILIZATION

Soil stabilization may be any, or a combination of one or more, of the following:

Mechanical—densification with various types of mechanical equipment as rollers, falling weights, explosives, static pressure, fabrics, freezing, heating, etc.

Additives—gravel, to cohesive soils; clay, to granular soils; and chemical additives such as portland cement, lime, fly ash (byproduct from coal burning)—often with lime and/or portland cement, asphalt cement, sodium and calcium chlorides, paper mill wastes, and others (sodium silicate, polyphosphates, etc.)

The use of rollers, falling weights, explosives, static pressures, and fabrics will be considered in separate following sections; however, an in-depth treatment of these several topics is well beyond the scope of this text. The interested reader should refer to ASCE (1968, 1978, 1982) for additional background.

Addition of gravel layers to be partially worked into a clayey road surface is common for stabilizing unpaved roads. Traffic continues to work the gravel into the underlying soil; periodic grading replaces the gravel displaced by wheels to the road edges. Gravelly road areas can be somewhat similarly stabilized with clay-sand additives, which tends to prevent large gravel displacements under traffic. In the former case the friction resistance of the soil is increased; in the latter the cohesion is altered.

A common stabilization procedure in fine-grained soils is to excavate to a depth and blend the excavated soil with

Portland cement
Portland cement and fly ash
Lime } with sufficient water
Lime and fly ash

The stabilized soil is replaced and compacted with rolling equipment; the cured result is a low-grade concrete. When portland cement is used, the stabilized soil is called "soil-cement."

Asphalt cement admixtures are used like portland cement in soil stabilization.

Chloride stabilization is usually based on the hygroscopic (attraction for water) properties of these materials to produce a damp soil to increase cohesion and to reduce dust problems on roads from traffic.

Addition of lime, fly ash, and sometimes portland cement is made in clayey deposits—particularly those subject to large volume changes—to effect a Ca^{2+} ion exchange to reduce the activity of the clay minerals. Soils treated in this manner can have a substantial reduction in I_P and shrinkage and/or swell depending on the amount of lime used. The plasticity index reduction is primarily due to an increase in w_P, although in some soils there may also be a substantial reduction in the liquid limit w_L.

Heating or freezing the soil may affect permanent strength increases but cheaper and more reliable methods should be investigated first.

Grouting is the injection of a fluid mixture (slurry) of

Clay and other soil such as fine sand
Cement or cement and clay
Sodium silicates, and similar chemicals

into the soil to reduce the porosity n (fill voids, cracks, etc.) and increase strength on setting.

Bentonite clay is a common grout material used to create a water barrier to halt wet weather seepage into poorly constructed basements. Bentonite is injected in a series of closely spaced holes around the perimeter or leaking zone. During dry periods the clay dries out; when water migrating toward the basement walls during wet weather encounters the clay, its extreme activity is such that the water is absorbed and the clay swells to close the soil voids to halt further flow. Success depends on sufficient clay being placed that a clay barrier (during swelling) is fully developed. Any holes in the barrier will have to be "patched" or the basement may continue to take water.

Grouting is commonly used in dam abutments and beneath dams to fill cracks in rock and reduce the porosity of the soil to control seepage. It is also used to strengthen the soil (when the grout cures) beneath footings. Beneath footings subject to machine vibrations, grout may stiffen the soil so that vibration amplitudes (displacements) are within equipment tolerance.

Grouting is a highly specialized branch of soil stabilization. The interested reader should consult ASCE (1978) for a more in-depth coverage and reference sources.

7-4 SOIL COMPACTION

Compaction is the densification of soils by the application of mechanical energy to produce particle packing. The soil may be pretreated by drying; addition of water; aggregates; or a stabilization agent such as portland cement, lime, fly ash, or other material. Additional mechanical pretreatment via discing, harrowing, blading, or using a blending machine may be used depending on the soil state.

Compaction energy in the field may be obtained from rollers, vibratory devices, and falling weights. In the laboratory, test samples to establish quality control are compacted using impact (or dynamics), kneading devices or static pressure using a piston and a compression machine.

The objective of densification is to improve the engineering properties of the soil mass. Several advantages accrue from compaction including:

1. Reduction of subsidence (vertical movement within the mass) from the reduced void ratio.
2. Increase in soil strength.
3. Reduction in shrinkage—decrease in volume resulting as the water content reduces from the reference value during drying.

The principal disadvantages are that swell (increase in water content from the reference value) and frost heave potential are often increased.

The following several sections indicate that compaction control may be used to produce selected engineering properties in the soil mass.

7-5 THEORY OF COMPACTION

A control specification for the compaction of cohesive soils was developed by R. R. Proctor while constructing dams for the Los Angeles Water District during the late 1920s. The original method was reported through a series of articles in the Engineering News Record (Proctor, 1933). For this reason the standard laboratory dynamic procedure is commonly called the " Proctor " test.

Proctor defined the four variables of soil compaction as:

1. Compaction effort (or energy)
2. Soil type (gradation, cohesive, or cohesionless, particle size, etc.)
3. Water content
4. Dry unit weight (Proctor used the void ratio)

Compactive effort and energy (CE) is a measure of the mechanical energy applied to the soil mass. In the field compactive effort is related to the number of passes of rolling equipment, number of drops of a falling weight, energy in a blast, and similar on a given volume of soil. Compaction energy is seldom a part of earthwork specifications because of the extreme difficulty in measuring it. Rather, the use of a particular piece of equipment, the number of passes, or more commonly, the end product of dry unit weight is specified.

In the laboratory, CE is developed by impact (usually), kneading, or by static compression. During impact compaction a hammer is dropped from a known height a number of times on the several layers in a mold to produce a sample of known volume. The size and shape of hammer, number of drops, number of layers, and mold volume are specified in tests standardized by ASTM and AASHTO. These are given in Table 7-1.

Kneading tests are similar except a punching type device is used to produce a kneading action on the soil. The CE from the impact hammer is readily computed and is shown for the standard tests in Table 7-1. The CE is not readily computed for kneading and static compaction procedures.

Table 7-1 Elements of the standard compaction tests

	Standard (ASTM D698)	Modified (ASTM D1557)
Hammer	24.5 N (5.5 lb)	44.5 N (10 lb)
Height of hammer fall	305 mm (12 in)	457 mm (18 in)
Number of layers	3	5
No. of blows/layer†	25	25
Mold volume	0.000 942 2 m ($1/30$ ft³†	
Soil	($-$) No. 4 sieve	
Compaction energy (CE)	595 kJ/m³ (12 400)	2698 kJ/m³ (56 250 lb · ft/ft³)

† Using the 102-mm (4-in) diameter mold.

Note: Larger molds are used when soils with ($+$) No. 4 particles are used. Author research using a 1000-cm³ mold with a diameter = 10.3 cm and height 12 cm requires 3 layers at 26 blows per layer to duplicate the standard compaction test. There will be a number of advantages in using a 1000-cm³ mold for SI computations.

In both impact and kneading tests, several samples of the soil are mixed with increasing quantities of water and compacted in a mold and weighed. Knowledge of the weight of wet soil in the mold of known volume, the wet unit weight is directly computed as

$$\gamma_{wet} = \frac{\text{Weight of wet soil in mold}}{\text{Volume of mold}}$$

Water content samples are taken from the compacted soil and the dry unit weight obtained as

$$\gamma_{dry} = \frac{\gamma_{wet}}{1 + w}$$

The data from the several compacted samples are used to plot a curve of dry unit weight versus water content as shown in Fig. 7-1.

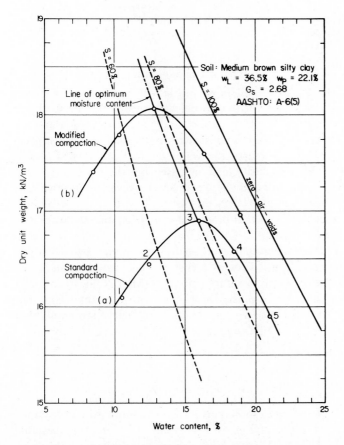

Figure 7-1 Standard and modified compaction test curves for a clayey glacial soil from near Peoria, Illinois.

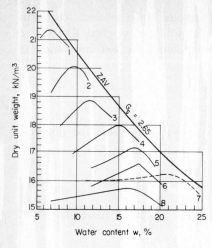

Soil No.	Description	w_L	I_P
1	Well-graded loamy sand	16	NP
2	Well-graded sandy loam	16	NP
3	Medium-graded sandy loam	22	4
4	Lean sandy silty clay	28	9
5	Lean silty clay	36	15
6	Loessial silt	26	2
7	Heavy clay	67	40
8	Poorly graded sand	—	NP

Figure 7-2 Standard compaction curves for several soils. Note that as the percent passing the No. 200 sieve (soils 4 through 7) increases the dry unit weight drops rapidly. The low unit weight for soil 8 indicates the difficulty in compacting sand by impact. (*After Johnson and Sallberg, 1960.*)

All soils on which compaction tests have been performed display a weight versus water content curve similar to that shown in Fig. 7-1. Sometimes the curve has more or less concavity as shown for the several soils in Fig. 7-2. From these curves it is evident that compaction effort, water content, and soil type are all significant parameters. Figure 7-1, curve *b*, illustrates another feature of compaction curves, namely, as the CE increases the water content at maximum dry density moves to the left. The dry density also increases, but there is no predictable relationship between dry density and CE. There are, however, limiting values of dry density ranging from about 11 kN/m³ for soil with no compaction (dumped in a pile) to a value less than that of a solid ($e = 0$) as obtained from the product of $G_s \gamma_w$.

The peak value of dry unit weight is termed "maximum dry density," although some technicians call it "Proctor density." The water content at maximum dry density is the optimum moisture content (OMC). A line of zero air voids can be drawn and is always above the compaction curve (at any CE) if the correct value of G_s has been used. The zero air voids (ZAV) curve represents the dry density when saturation is 100 percent ($S = 100$) and is readily computed using Eq. (2-13) as

$$\gamma_{ZAV} = \frac{G_s \gamma_w}{1 + wG_s} \tag{7-1}$$

In most cases OMC occurs at $S = 75$ to 80 percent for the standard compaction test. This is illustrated with the soil shown in Fig. 7-1. Note that several "saturation" lines, using Eq. (2-13), have been drawn in addition to the line through the two OMC values.

A compaction curve may be interpreted (refer to point numbers for curve *a* in Fig. 7-1) as follows:

1. At low water content and at some CE the lumps are not sufficiently slaked (water softened) and broken down. Note that a higher CE does break more soil lumps at the same or lower water content.
2. As the water content increases the lumps are slaked somewhat and water films are partially developed around clay minerals. More particle orientation takes place and density increases.
3. At the OMC this process is optimized for this particular CE.
4. Additional water produces free water in the voids so that the CE causes excess pore pressures to develop. This liquefies the soil temporarily so that particles float about with additional dispersion of the clay minerals but with a reduction in particle packing.
5. At large water contents the compaction efficiency falls off rapidly. Note, however, that the process does not produce a saturated soil because of the continual particle movement and floating from the excess pore pressures that are being developed from application of the CE.

One may check the accuracy of the compaction data by plotting the ZAV on the compaction curve plot. If the ZAV intersects the compaction curve some kind of error has been made; wrong G_s, data from two soils, weighing or computation errors and the data should be rechecked and/or the compaction test rerun.

The locus of points through the maximum dry densities of several compaction curves on the same soil at different CEs produces the "line of optimums" in Fig. 7-1. This curve approximately parallels the ZAV curve and is useful to interpolate intermediate values of OMC.

Compaction tests are reasonably repeatable and reproducible. The standard deviation should be on the order of ± 0.4 kN/m^3 for carefully performed tests. The following are the major factors affecting reproducibility:

1. Scale of plot. Note the scale of Fig. 7-1 versus Fig. 7-2 allows the unit weight to the nearest 0.1 with little difficulty.
2. Number of tests and the spacing on the water content scale. Five tests are usually sufficient if one is near OMC and the water content spread between points is not more than 5 (preferably 2 to 3) percent.
3. Mixing time. The more the soil is mixed, the higher the maximum density since the additional mixing produces a more dispersed clay structure.
4. Whether the test is done on a given quantity of soil which is broken down (e.g., using the Bowles pulverizer shown in Fig. 7-3) or fresh samples are used for each test point. It should be noted that ASTM requires fresh samples for each curve point.

Literature reports that reusing the sample versus using fresh samples can produce unit weight differences as high as 0.3 to 1.3 kN/m^3. During development

Figure 7-3 Bowles Soil Pulverizer used to reduce compacted samples where soil is reused in compaction test. Required finding correct blade advance and blade configuration to reduce soil efficiently without particle degradation.

of the soil pulverizer shown in Fig. 7-3, the author did a large number of compaction tests (unpublished) and found the following:

1. If the mixing time in a mechanical mixer is on the order of 15 minutes and the soil for the initial test was mixed with some water (5 to 6 percent for most soils) and cured overnight, it made little difference whether the soil was reused or fresh samples were used.
2. If the soil was not very cohesive, the effect of mixing was not pronounced and again it made little difference whether the soil was reused or fresh samples used. In passing, it is noted that one instance in the reported literature found just the opposite for this type of soil.
3. Plotting scale can easily produce differences of up to 0.3 kN/m^3.

7-6 COMPACTION OF COHESIONLESS SOILS

Cohesionless soils cannot be readily compacted using either impact or kneading methods. These soils can be compacted by using confined static compression, but not very efficiently. The usual method for compacting these soils is to use a combination of confinement and vibration. In the field this is accomplished by using smooth wheel rollers with an internal vibrator. The soil is confined vertically along a strip the width of the wheel in instant contact with the ground. The soil is confined laterally by the soil in front and behind the contact strip. Vibration shakes the grains into a more packed state.

Another method sometimes used is to flood (saturate) the soil and roll it—preferably using vibratory rollers. This method is most suitable for very sandy soils where the excess water is not a problem (creating mud, softening adjacent soil, etc.). Saturation ensures breaking of any surface tensions from just damp soil and the rapid application of the roller pressure creates excess pore pressure. The excess pore pressure temporarily liquefies the soil so that the grains are easily displaced into a more packed state during pore drainage.

In the laboratory, cohesionless soils are compacted by confining layers of dry soil in a compaction mold and sharply rapping the sides with a rubber mallet. Alternatively, compaction may be produced by filling a mold, applying a surcharge, and placing the system onto a vibrating table for a period of time. In any case, several trials may be necessary to determine the maximum unit weight value to use. A curve is usually not drawn to obtain a maximum dry density.

7-7 STRUCTURE AND PROPERTIES OF COMPACTED COHESIVE SOILS

The structure and engineering properties of compacted cohesive soils will depend greatly on the method of compaction, the CE, and the water content at compaction. Usually the water content during compaction of a soil is referenced to the OMC as dry of optimum, at optimum or wet of optimum. Research on laboratory compacted clay samples has shown that when they are compacted dry of optimum the structure of the soil is essentially independent of whether impact or kneading methods are used (Lambe, 1958; Seed and Chan, 1959). Wet of optimum, however, the compaction method has a significant effect on the soil fabric and thus the strength and compressibility of the soil.

At the same CE with increasing water content the soil fabric becomes increasingly oriented (dispersed) as shown qualitatively in Fig. 7-4. Dry of optimum, the soil tends to produce a flocculated (card-house) fabric. If the compactive effort is increased the fabric tends to become more dispersed even though the water content remains constant as point E of Fig. 7-4. Note also that the fabric is very considerably more dispersed at point C than at point A for the same CE but large change in water content from dry to wet of optimum.

Laboratory tests indicate that swelling is greater and shrinkage is less for clay compacted on the dry side of optimum. This is attributed to a combination of the flocculated fabric, sensitivity of additional water at the contact points, and the lower reference water content for swelling. For soil wet of optimum, the reference water content is already high so that little further water is required for $S = 100$ percent—thus some limitation on swell. Shrinkage from this higher water content reference would logically be more since the percent change in water content will be greater if the soil "dries out."

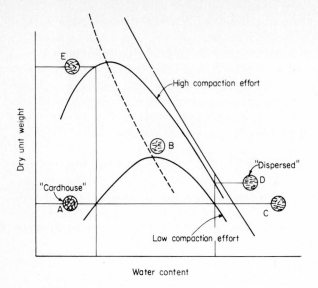

Figure 7-4 Qualitative effect of compaction on soil fabric and structure. (*After Lambe, 1958.*)

Flow of water through soil (permeability as described in Chap. 8) depends on both the void ratio and fabric orientation. Laboratory tests indicate that flow rates (permeability) decrease

1. At constant CE and increasing water content since a dispersed fabric is less permeable.
2. At increased CE and constant water content for the same reasons as " 1."

Compressibility (Chap. 11) of compacted clays seems to be a function of both the method of compaction and stress level subsequently imposed on the soil mass. At relatively low stress levels, clays compacted wet of optimum appear to be more compressible. At high stress levels, clay compacted wet of optimum are less compressible. An explanation put forth by Lambe (1958) is that the flocculated structure produced on the dry side of optimum has a large accumulation of interparticle bonds from a water deficiency compared to the dispersed fabric on the wet side of OMC. Applications of low stresses on the dry side are not sufficient to overcome the particle bonding to (1) first reorient the fabric to a more dispersed state and then (2) squeeze particles closer together, so compressibility (reduction in void ratio) is low. On the wet side, step 1 has already been done in the compaction process so that step 2 effects predominate and the compressibility is larger.

At high stresses such that interparticle bonds are broken both steps 1 and 2 occur for the flocculated (dry side) fabric and result in a large amount of compressibility. For the wet side fabric factor (2) predominates with a dispersed fabric so that less compressibility is produced.

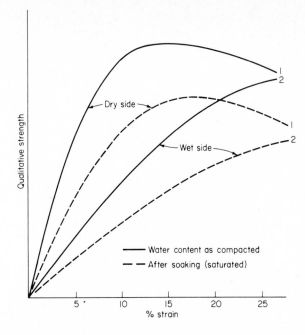

Figure **7-5** Qualitative stress-strain curves for clay compacted with respect to OMC as shown and tested. Also shown is strength after soaking to produce saturation. At high strains most soils converge to a limiting strength as qualitatively shown above. At low strains the compacting water content is a significant parameter.

The strength of cohesive samples is also dependent on the water content at compaction (and at test time). Strength is also dependent on the strain level defining "strength." For example, if we define the strength to occur at a large strain (e.g., 20+ percent), the strength of compacted clay is relatively independent of molding water content. Apparently the large amount of sample remolding during strain produces the same ultimate fabric in the failure zone.

At a strength defined at, say, 5 percent strain, clays compacted on the dry side of optimum produce larger strengths. When clays compacted on the dry side of OMC are later saturated, the strength is also higher than for clays compacted on the wet side (and saturated), but at high strains strengths are about equal (see Fig. 7-5).

This means that where strength at low deformations is critical (as a pavement subgrade) and swell is not a problem, the soil should be compacted dry of optimum. For a dam core (which will later saturate) which relies on a shell and mass for stability rather than strength, compaction should be on the wet side of optimum so that settlement deformations may be accommodated without cracking.

7-8 STABILIZATION OF EXPANSIVE CLAYS

Expansive clay is a very common occurrence. Clay expansion results when the water content increases from the reference value. Shrinkage occurs when the water content falls below the reference value and down to the shrinkage limit.

Generally a clay is suspect of being subject to large volume changes (expansive) if the plasticity index $I_p \geq 20$. Several procedures are available to stabilize (reduce volume change in) this type of soil.

1. Add an admixture such as hydrated lime. Usually 2 to 4 percent will reduce I_p to less than 20.
2. Compact the soil well on the wet side (3 to 4 percent) of optimum. This ensures a reasonably dispersed clay structure and at the same time produces a low dry density (see point 5 in Fig. 7-1). It appears that the dry density of expansive clay is a particularly significant parameter.
3. Control the change in water content from the reference value (w at time clay is finally used as a foundation support).

The economic importance of coping with expansive clays can hardly be underestimated since these materials are so prevalent and often produce damage. Four international conferences on the volume change problem have been held to date. The latest was sponsored by ASCE (1980) and is a suggested reference source for current methodology and bibliographical source.

7-9 OTHER COMMENTS ON SOIL COMPACTION

It has been shown that soil properties can be obtained by exercising selective compaction control. Compaction control will be considered in some detail in following sections.

Unfortunately, it is often difficult to obtain direct comparison between soils compacted by field and laboratory methods. This is well illustrated in Fig. 7-6, where a soil was compacted by the several methods shown. Note that there is a different OMC for each method—in fact, dry unit weights are also method-dependent. Fortunately in many cases we are concerned with relative results; thus we might specify OMC + 3 percent for a particular project and realizing that the actual water content will vary $\pm$ several percentage points. This specification may produce an adequate representation of the laboratory impact procedure versus using a sheepsfoot roller (kneading compaction) on site. If we want the roller to compact the soil wet of optimum we might require +5 percent above the laboratory OMC. Of course, we would want to observe the project to ensure that this does not make the soil so wet it will not compact.

Often field densities are simply stated as some percent of the maximum dry density obtained from the laboratory test. Ranges from 90 to 105 percent are common—particularly for building foundations (around and beneath footings and beneath floor slabs)—and water content is not specified.

Considering these several factors as well as test procedures, it is suggested that great refinement in compaction tests is not necessary. In particular, the original method of reusing the soil to produce a compaction curve seems prefer-

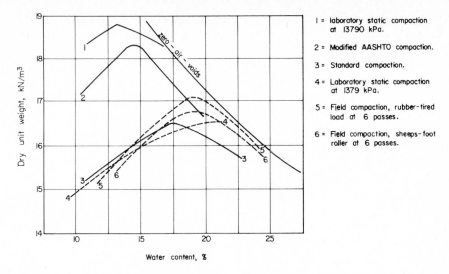

1 = laboratory static compaction at 13790 kPa.

2 = Modified AASHTO compaction.

3 = Standard compaction.

4 = Laboratory static compaction at 1379 kPa.

5 = Field compaction, rubber-tired load at 6 passes.

6 = Field compaction, sheeps-foot roller at 6 passes.

Figure 7-6 A comparison of field and laboratory compaction unit weights on same soil. *(Turnbull, 1950; see also Lambe and Whitman, 1969.)*

able to the use of fresh samples for the curve points. The following are some additional laboratory considerations in using the original " Proctor " method:

1. Requires less soil from the field, less storage, and reduces quantity to process; also less soil to dispose of at end of testing
2. Technician is likely to have a better feel for the next point on the compaction curve
3. Tends to reduce the number of test points
4. Reduces substantially the cost of making a compaction curve

Note that if a somewhat higher density is obtained for the job specification, a better in situ product is generally produced at about the same compaction cost.

7-10 EXCAVATION AND COMPACTION EQUIPMENT

Soil to be used in a compacted fill is excavated from a borrow area. The borrow may be either on or off site. Power shovels, draglines, and self-propelled scrapers are used to excavate the borrow material. These devices may cut through layers of different materials, allowing the several soils to become mixed. The power shovel mixes the soil by cutting along a vertical surface, while the scraper mixes the soil by blading across a sloping surface where different layers may be exposed.

The borrow area may be near or several kilometers from the fill site. Various types of equipment are used to transport and spread the soil on the fill area. These include scrapers, conventional trucks, and trucks especially made for this purpose. It is preferable to spread the material when dumping to save spreading time. Often, however, the borrow requires processing prior to compaction, which may include running the soil through a processor which thoroughly mixes it, simply spreading the dumped earth and using a farm disc to pulverize it, adding water, or drying the soil. The layer of loose earth placed over a previously compacted layer is called a *lift*. Lifts range from 15 to 50 cm depending on the size and type of equipment and the type of soil to be compacted.

We should carefully note that the following occurs in obtaining a soil from a borrow (or cut area within a project) and placing it in a fill:

1. In situ soil is at some in place density or void ratio e_i.
2. On excavation the in situ soil increases in volume—termed *swell*. Swell values may range from 15 percent for a gravelly material to 40 percent for clay; for rock swell is on the order of 60 to 65 percent.
3. The soil is dumped at the site, processed as required, and compacted to a new void ratio e_f (or final density). For soil e_f is usually less than e_i; for rock $e_f > e_i$ since broken rock can never be compacted back to its initial void ratio.

These concepts can produce some interesting computations to relate final compacted quantities to borrow (for royalty payments) or cut quantities used as fill as illustrated in the following example.

Example 7-1 Given: 2000 m^3 of fill is placed to $e_f = 0.353$. The borrow source has $e_i = 0.60$.

REQUIRED Number of cu m of borrow to produce the fill.

SOLUTION The borrow owner is paid on the basis of before and after surveys taken at the borrow pit. We must relate the in place fill volume to the measured quantity of soil removed from the borrow pit.

For final conditions [and fundamentals from Chap. 2: $e = V_v/V_s$; $V_v + V_s = V_t$ and $V_s = W_s/(G_s\gamma_w)$], we obtain the following:
Let:

$$V_s = 1.0$$

Therefore,

$$V_v = e = 0.353$$

Also let total weight of solids in 2000 m^3 of fill = W'_s. By proportion

$$\frac{W_s}{1 + e} = \frac{W'_s}{2000}$$

From which

$$W'_s = \frac{2000 W_s}{1 + e} = \frac{2000 G_s \gamma_w}{1.353} \qquad (a)$$

For the initial conditions and using X m^3 as the quantity of borrow, we obtain similarly

$$W''_s = \frac{(X)G_s \gamma_w}{1.6} \qquad (b)$$

Since $W'_s = W''_s$ (weight in place is same as weight removed from borrow pit) we equate (a) and (b) to obtain

$$\frac{(X)G_s \gamma_w}{1.6} = \frac{2000G_s \gamma_w}{1.353}$$

Canceling terms, this borrow volume is

$$X = 2365 \text{ m}^3$$

////

The compacting equipment used depends on the type of soil (and on the equipment available to the contractor). Equipment is available which applies pressure, vibration, vibration and pressure, and kneading energy. The equipment may be self-propelled or towed. Figure 7-7 illustrates several types of compaction equipment common to fill projects. Particular features of several types of compaction equipment include:

1. *Smooth-wheel rollers.* These rollers give 100 percent coverage beneath the wheel with contact pressures up to about 400 kPa. These may be used on all types of soil except when large boulders are present. These rollers are also called *steelwheel* rollers. They are particularly suited for cohesionless soil with and without vibration devices attached. They are very efficient when used with flooding for sand compaction. These rollers are also commonly used for finish rolling of subgrades and base courses and compacting asphalt pavements; rollers are self-propelled.
2. *Pneumatic or rubber-tired rollers.* These rollers give about 80 percent coverage; tire pressures go up to about 700 kPa. Several rows of four to six closely spaced tires with front and rear spacing alternated are used for the higher percent coverage cited. They may be towed but are generally self-propelled. These rollers may be used for either cohesive or cohesionless soil. The tires are sometimes misaligned vertically (wobbly wheel), to produce a kneading action on the soil.
3. *Segmented wheel compactors.* These produce about 60 percent coverage and generate contact pressures up to 1000 kPa. They are designed for cohesive soils, and the lift is generally restricted to 15 cm.
4. *Tamping foot rollers.* These rollers produce about 40 percent coverage and contact pressures from 1400 to about 8500 kPa depending on the diameter of the roller and whether the drum is filled for added weight. The tamping foot roller has small, rectangular feet similar to the sheepsfoot roller. Compaction begins in the soil below the foot projection (about 15 to 20 cm depth), and the

Figure 7-7 Compaction equipment of several types: (*a*) smooth-wheel rollers—the roller on the left is equipped with a vibratory device; (*b*) grid roller; (*c*) sheepsfoot rollers towed by large agricultural-type tractor; (*d*) steel-wheel compactor in sanitary landfill; (*e*) self-propelled tamping foot roller; (*f*) using a disc to process fill soil prior to compacting.

    depth of penetration is successively less on subsequent passes (called "walking out"). These rollers are only suitable for cohesive soils. They are commonly self-propelled, with either two rollers for front wheels or four rollers for both sets of "wheels."

5. *Sheepsfoot rollers.* These rollers produce about 8 to 12 percent coverage due to the small "sheepsfoot" projections of 3 to 8 cm² area. The drum can be filled with water or sand to increase the weight. Contact pressures range from about 1400 to 7000 kPa. Sheepsfoot rollers are commonly towed in parallel or two in parallel with a trailing roller to cover the gap between the front rollers. Six to eight passes (producing 40 to 60 percent coverage) will usually obtain the required density due to the spreading of the contact pressure from the sheepsfoot tips. The roller is only suitable for cohesive soil.

Table 7-2 Compaction equipment and soil compaction applications

Soil group	Soil type		Degree of compaction	Typical compaction equipment	Shallow fills and backfills No. of passes or coverages	Lift thickness, mm	Placement w, %	γdry, kN/m³	Field control	Deep foundations Comp. method	Field control
Pervious or free-draining	GW GP SW SP	Compacted	95 to 105 percent of standard compaction test or 70 to 85 percent of D_r	Steel-wheeled roller or vibratory compactor / Rubber-tired roller / Crawler-type tractor / Hand tampers (mass > 45 kg)	As required / 2–5 / 2–5 / As required	As required / 300 mm / 200 / < 150	Saturate by flooding	17–21	Field density tests at random locations	None available except for near surface as listed at left	
		Semi-compacted	90 to 95 percent of standard compaction test or 60 to 70 percent of D_r	Rubber-tired roller / Crawler tractor / Hand tampers / Controlled routing of construction equipment	2–5 / 1–2 / As required	350 / 250 / < 200	Saturate by flooding	16–20	Field density tests at random locations	Vibroflotation, compaction piles, sand piles, explosives, and surface methods	Undisturbed samples from borings or test pits; SPT before and after compaction
Semipervious to impervious	GM GC SM	Compacted	95 to 105 percent of standard compaction test	Rubber-tired roller / Sheepsfoot roller / Hand tampers	2–5 / 4–8 / As required	200 / 150 / < 100	OMC based on lab compaction test	16–20	Field density tests at random locations to determine RC	Preload fills Lower water table Generally consolidation theory applies	
	SC ML CL OL MH CH OH	Semi-compacted	90 to 95 percent of standard compaction test	Rubber-tired roller / Sheepsfoot roller / Crawler tractor / Hand tampers / Controlled routing of construction equipment	2–4 / 4–8 / 2–4 / As required / As required	250 / 200 / 150 / < 150 / < 200		14–19			

Notes: 1. Rubber-tired rollers with tire pressures of 550 to 700 kPa.
2. Sheepsfoot rollers with foot pressures on order of 1700 to 3500 kPa.
3. Crawler tractors weighing over 85 kN and track pressures greater than 45 kPa.
4. A *pass* is made with a sheepsfoot roller; *coverage* is made with rubber-tired rollers, steel-wheel rollers, and crawler tractors. Coverage includes 100 percent of surface, whereas sufficient compaction may be obtained with 3 to 5 passes which includes only 45 to 50 percent of surface area.

(a)

(b)

Figure 7-8 Small compactors for limited areas. (*a*) Wacker Vibro Plate compactor being used to compact backfill adjacent to wall footing for floor slab. (*b*) Wacker Rammer compactor used for narrow areas where vibro plate is too large. Here compacting base for a wall footing. (*Courtesy of Wacker Corp., Milwaukee, Wisc.*)

6. *Mesh or grid pattern rollers.* These produce about 50 percent coverage, and contact pressures range from 1400 to about 6000 kPa. This roller is useful for compacting rocky soil, gravels, and sands.

Vertical vibrators are attached to the steel-wheel rollers to densify granular soil more efficiently. The principle is that of confining the soil across most of the contact area, applying pressure from the weight of the device, and using vibration to break or dislodge the particle contact points so that particle slip can occur. Small hand-towed vibrating plates are available ranging from about 0.23 to 1.2 m square with masses from 450 to 5000 kg. These devices are for use in small areas such as culvert, pipe, or trench backfill; behind retaining walls, and adjacent to foundation walls for buildings. Effective compaction depth ranges from about 7 to 20 cm. Table 7-2 tabulates recommended compaction equipment for several soils based on the Unified Soil Classification system.

Figure 7-8 illustrates two small compactors which may be used for limited areas such as wall footings, basement or floor slab backfill, and behind retaining walls. Backfill compaction around culverts and sewer pipes in trenches may also be readily accomplished with these devices. Best results are obtained by using a lift thickness under 150 mm.

7-11 COMPACTION SPECIFICATIONS

The objective of compaction is to improve the engineering properties of the soil. The dry unit weight should be specified to accomplish this goal and not simply specify, "the soil shall be compacted to 95 percent of the unit weight obtained in the standard compaction test." Many standard soil specifications contain clauses such as this for ease of specification writing. Unfortunately, in many cases these standard clauses are used without any real concern for engineering properties. Assume that an engineering analysis has determined that the field compaction should be some percent of that obtained on the laboratory compacted sample after inspection of the moisture-dry unit weight curve to see what the OMC and maximum dry density value is. Presumably enough classification tests or strength tests have been performed so that judgment now applies, and it is decided that the relative compaction (percent of that obtained in the laboratory) is defined as

$$RC = \frac{\text{field dry unit weight}}{\text{maximum dry unit weight from laboratory test}} \times 100 \quad \text{percent}$$

$$(7\text{-}2)$$

which sets the value of the field dry unit weight to be obtained. The value of RC is typically from 105 to about 90 percent, usually based on γ, index properties, classification, and previous performance. Note that it is defined from the laboratory test and may be the standard, modified, or some other amount of compaction energy. Coupled with RC may be a compaction water content requirement, say, 3 percent wet of optimum, optimum ± 1.5 percent, 3 percent dry of optimum, etc., based on climatic factors, fill use considerations, and other factors such as those already considered.

The reader should note that there is a considerable difference between RC and relative density as defined by Eq. (5-1). Relative density (if used) applies only to cohesionless soils with little (−) No. 200 fines; however, the author has successfully used a unit weight comparison instead of D_r on numerous occasions. The compaction test is generally used if the soil contains more than 12 percent fines, but there is no reason that some type of unit weight test cannot be used for all soils to establish compaction specifications.

A relationship between relative density and relative compaction is shown in Fig. 7-9. A statistical study by Lee and Singh (1971) on 47 different granular soils indicated that the relative compaction corresponding to zero relative density is about 80 percent. It should be further noted that relative compaction can never be less than about 80 percent, since soil always has a unit weight and simply dumping it in a pile will produce unit weights on the order of 11 to 14 kN/m^3.

Unit weight and/or water content (end product) specifications are used for most highways and buildings. As long as the contractor is able to obtain the specified relative compaction, it does not matter how it is obtained or what equipment is used. Project economics should ensure that the contractor will utilize the most efficient compaction procedures. The most economical compaction conditions are illustrated in Fig. 7-10, which shows three qualitative compaction curves for the same soil at different compactive efforts. A study of these curves, assuming that curve 1 represents the curve which can be obtained with existing equipment, indicates that to achieve, say, 90 percent relative compaction, the placement water content of the fill must be greater than water content a and less than water content c. These points are found where the 90 percent RC line intersects curve 1. If the fill water content is not in the range a to c as in zone A of Fig. 7-10, it will be difficult, if not impossible, to obtain the required RC without increasing the compactive effort by increasing the number of passes or using heavier equipment. This is why it is necessary to wet (by sprinkling) or dry (by discing) the soil prior to rolling.

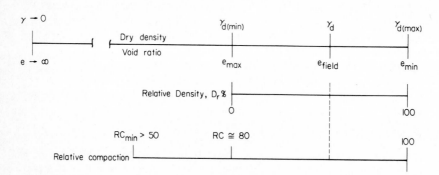

Figure 7-9 Relationship between relative density, density and relative compaction. (*After Lee and Singh, 1971.*)

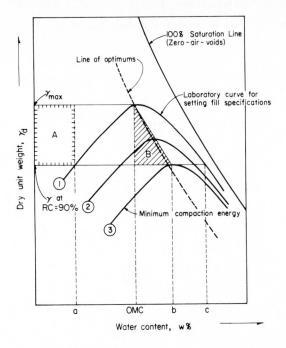

Figure 7-10 Dry unit weight versus water content illustrates most economical field compaction conditions. Assuming that the placing of water content is not a significant parameter, any w in zone B will result in maximum compaction efficiency.

Referring again to Fig. 7-10, from a purely economical standpoint, the most economical water content would be at b, where the minimum compactive effort is needed to attain 90 percent RC. However, consistently achieving the minimum RC for a project requires the use of a slightly higher compactive effort, as in curve 2. With this consideration (noting that there will be some inevitable fluctuation in water content due to environmental factors), the most economical fill water contents exist in the hatched zone B between the OMC and b and are "wet of optimum."

7-12 FIELD CONTROL OF COMPACTION

Field tests are required to determine if the specifications for relative compaction have been met. The test location(s) should be representative or typical of the compacted lift and borrow material. Typical job specifications may specify a field test for unit weight† for every 1000 to 3000 m^3 of fill or when the borrow material changes significantly. It may be necessary to make the field test on the lower lift if for some reason inspection was temporarily suspended, especially if sheepsfoot rollers are used. Likely sites for tests may be randomly selected, determined by probing with a 12- to 15-mm bar (for soft spots), or based on judgment.

---

† This is usually called a field density test. This terminology is likely to continue and will be one of the "problems" in using SI.

Table 7-3 Table of random numbers. Use entire number, first digit, second digit, first two digits, etc., of each entry as a random number

	1	2	3	4	5	6	7	8	9	10
1	16057	43688	44334	73470	36029	98628	65873	95981	16409	39478
2	90781	31691	58816	84699	42103	30877	64837	18638	74650	69832
3	16912	50831	53883	46964	28355	97607	96167	47999	21406	63691
4	83985	19760	93957	19489	91200	77460	93463	20248	29432	53652
5	23887	61081	78790	94252	92003	68007	95276	89500	74668	48696
6	58466	88644	27123	90111	31217	38330	30489	54761	83117	93031
7	29299	82138	53630	92397	59271	17913	25445	23067	26689	17551
8	49598	95030	77669	43385	74670	62512	78271	45552	12081	21582
9	33896	30936	83344	28733	14580	13850	40619	65704	32284	71527
10	56994	30339	10399	81934	15452	20487	39270	84252	84969	34925
11	36894	36092	56141	33005	61489	98271	70052	14176	19258	65804
12	84181	48602	27486	60619	29858	52449	66877	26789	50822	80501
13	18461	59131	59651	66667	25752	47027	70791	33144	31610	90766
14	35004	17366	10528	54429	85783	34960	22855	57764	32264	20082
15	43873	72014	26305	26562	82931	86516	44909	48288	90932	36015
16	48204	75791	63079	69512	25532	44530	94331	69846	64175	40864
17	94364	66167	41721	78013	68174	48069	65946	84485	32139	12772
18	32341	57576	82616	42470	11107	48341	46697	78290	41703	88563
19	44772	85906	20207	52697	52904	57420	86667	93784	21296	38184
20	77562	97250	77693	33952	31771	94514	99730	51652	52445	49397
21	62417	10349	84296	29440	43012	39920	18373	25506	23004	52726
22	76442	86840	65332	71065	72133	54460	12671	88171	66015	71785
23	96900	27897	80531	61432	45690	67476	10666	99470	89055	56170
24	93419	85823	24387	29005	24944	14714	49948	32762	40196	90955
25	88405	63859	21544	21142	38450	81002	36981	24422	39675	40194
26	89660	10750	97814	65802	62022	53586	72648	17047	76282	22567
27	42314	89765	64787	45291	91159	92491	26070	82484	61974	73654
28	46944	41836	69151	34955	91135	72669	61806	41477	17172	38816
29	83740	43131	21062	68972	26378	35988	47376	65070	97459	70006
30	82523	82256	20151	99233	38053	67565	28000	65743	30203	70695
31	87602	53569	31347	74647	90912	73002	41613	35895	40808	94110
32	82906	91833	67205	54785	29921	29927	97673	23682	52653	28407
33	90161	27905	58192	56878	99741	22540	11519	19039	79195	23449
34	79382	90238	88670	48204	71521	65893	90058	10930	12892	85028
35	66484	16313	82920	77322	31492	92498	36672	35206	94974	99123
36	27462	85739	51713	30093	45021	61949	95123	14020	58246	29479
37	66414	67612	35651	84658	87061	39212	27017	91553	94703	84771
38	51325	34019	48488	37111	17171	77727	76492	36703	44002	43310
39	10219	95606	70067	60742	19477	13451	15798	96236	11690	96916
40	99391	67465	84013	30452	62321	81155	63610	47089	74115	91706

On large projects a table of random numbers (Table 7-3) can be used to obtain test location coordinates to avoid technician bias in selecting the locations (Sherman et al., 1967).

Example 7-2 Given a fill area 16 × 150 m and two lifts.

REQUIRED Use a table of random numbers to obtain coordinates for four field test locations in each of the two lifts.

SOLUTION Use Table 7-3 and interpret the first two digits of column 1 and the first two digits of column 2 of every fifth line to give the X and Y coordinates from the left end of Fig. E7-2. Obtain:

Lift	1		2	
Test	X	Y	X	Y m
1	16	4.3	77	9.7
2	23	6.1	88	6.3
3	56	3.0	82	8.2
4	43	7.2	6.6	16

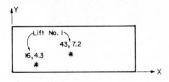

Figure E7-2

Obviously other combinations of random numbers could be used, but these coordinates give a reasonable coverage of the fill area for each lift.

////

Field testing to determine the in situ density may be either *destructive* or *nondestructive*. In destructive testing the density may be determined by driving a cylinder into the ground to recover a sample of known volume. More commonly, these tests involve excavating a hole, recovering all the soil, and measuring the hole volume V_h. Usually V_h is measured using the sand cone method (Fig. 7-11a–d) or with balloon equipment (Fig. 7-11e–f). Specific test details are given in ASTM or Bowles (1978). The wet soil removed from the hole is weighed to determine W_T and water content samples taken to determine w_n. The wet and dry unit weights can then be computed as

$$\gamma_{wet} = \frac{W_T}{V_h} \qquad \gamma_{dry} = \frac{\gamma_{wet}}{1 + w_n}$$

Problems encountered with destructive testing include:

1. Time lag in determining the water content using oven drying. Special ovens and devices using chemicals to produce a gas pressure related to water content are available to reduce the time to determine w.
2. Backfilling the hole. Often this is not critical, but in dam cores and similar it may be desirable to carefully backfill and compact the test holes.
3. Inattention to detail so that the unit weight is incorrect. Only small technician errors can produce a soil not meeting compaction specifications since the hole volume is so small.

Nondestructive testing involves using radiation devices which can be placed on the ground to directly measure density and water content. Figure 7-12 is typical of these field units. Particular advantages include

1. Making a large number of tests rapidly. A better statistical quality control may be made when the number of tests is large.
2. Directly obtaining both γ_{wet} and w.

Figure 7-11 Field unit weight equipment. (*a*) Sand-cone equipment; jug; cone, template, digging tools, and can to recover wet soil. (*b*) The site is smoothed and the template positioned. (*c*) The hole is dug as large as practical; the can should be about 80 percent filled. (*d*) Cover the can and position the sand jug to fill hole with sand. (*e*) Balloon equipment; volumeasure, digging equipment, template, and can to recover wet soil. (*f*) The hole is dug and the volumeasure positioned on the template—note that the water level compared to that in part *e* indicates the hole is nearly filled.

Figure 7-12 Nuclear density meter used for testing density of soil to support a floor slab. *(Courtesy of Soiltest, Inc.)*

The principal disadvantages include:

1. Equipment cost
2. Careful calibration required on the soil of interest

Example 7-3 Given field unit weight (sand cone) test with the following data obtained:

Weight of wet soil removed from test hole = 1942 g
Weight of sand used to fill hole and cone = 2744 g
Density of sand = 1.60 g/cm³
Weight of sand to fill cone = 1289.7 g (by calibration)

REQUIRED Find the in situ wet and dry unit weights if the entire sample is oven-dried to 1708.7 g.

SOLUTION By proportion the volume of the hole is

$$V_{\text{hole}} = \frac{2744 - 1289.7}{1.60} = 908.9 \text{ cm}^3$$

The wet unit weight is immediately computed as

$$\gamma_{wet} = \frac{1942}{908.9}\,(9.807) = 20.95 \text{ kN/m}^3$$

The dry weight is also directly obtained, since the entire sample was dried, to give

$$\gamma_{dry} = \frac{1708.7}{908.9}\,(9.807) = 18.44 \text{ kN/m}^3$$

The in situ water content is

$$w = \frac{1942 - 1708.7}{1708.7}\,(100) = 13.7 \text{ percent}$$

////

7-13 STATISTICAL FIELD UNIT WEIGHT CONTROL

Seldom do 100 percent of the field tests for unit weight meet specifications, just as steel or concrete quality control tests also do not. Since it would be unfair to the contractor to require all the test locations not meeting specifications to be excavated and replaced, it is necessary to establish how many tests have to fail to meet specifications before

1. The work is considered unacceptable and the soil must be removed and replaced.
2. The contractor must accept a lower unit price for the work.

Formerly the acceptance criteria depended considerably on a "feel" for the project, which could be somewhat arbitrary, depending on the project engineer. Now it is more appropriate—especially for large projects—to quantify the "feel" using statistical quality control. This method is described in some additional detail by Turnbull et al. (1966) and will be briefly introduced here. Basically the procedure is as illustrated in Fig. 7-13. One develops the normal frequency distribution curve using either unit weight directly or relative compaction and estimates a reasonable standard deviation $\bar{\sigma}$. If this is estimated as, say, 2 percent relative compaction, then 98 percent of the field tests would be equal to or greater than the specified unit weight. Alternatively, the standard deviation might be, say, ± 0.80 kN/m^3 (about 5 pcf), and any values of unit weight less than this value are unacceptable. In statistics terminology the 2 percent value is the coefficient of variation $\bar{C}$ for the unit weight. For a large project, a plot of all the field tests would result in a normal frequency distribution curve unless bias were introduced. Bias might develop from technician error, poorly calibrated equipment, or changes in compaction equipment. It might also develop from unnoticed changes in fill material and poor control of fill moisture content. The

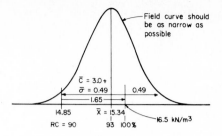

Figure 7-13 Normal distribution curve as would be obtained from plotting a number of field tests. A qualitative curve can be made to determine initial testing and for predetermining general compaction control requirements. Numbers shown correspond to text discussion.

normal frequency distribution curve is, of course, symmetrical as in Fig. 1-3 and as in Fig. 7-13, which shows the curve that might be obtained from a reasonably large number of tests—more than 20 to 30.

Now if we assume that the standard deviation $\bar{\sigma} = 0.80$ kN/m³ is to be based on this large number of tests, how many (or what percentage) can be allowed which fail to meet specifications? Note that unit weights that are too large are acceptable; it is only those too low which are "not to specification." The contractor might be concerned if a large number of tests are too large, since this represents a profit reduction. Whatever the statistical criteria, they must be reasonable, or earthwork costs would rapidly escalate because the contractor would lose money on the first job on which the criteria were applied, but word would get around and on the next job the earthwork costs would reflect the additional compaction effort. One report of what is "reasonable," reported by Sherman et al. (1967), indicates that a standard deviation[†] of 0.4 is adequate, with somewhat larger values, around 0.5 to 0.6, for clayey soils with considerable gravel. A study of 29 projects gave a field average $\bar{\sigma} = 0.5$ for all the projects. Thus, it appears that a standard deviation $\bar{\sigma}$ of 0.5 ± 0.05 will generally be adequate for all projects. Figure 7-13 illustrates the application of a standard deviation of 0.49 ($\bar{C} = 0.49 \times 100/15.34 = 3.19$ percent) to an example with numbers so that the principle can be more easily understood.

Shown on Fig. 7-13 is a situation where RC = 90 percent and the soil has a laboratory-determined standard compaction test dry unit weight of 16.50 kN/m³. The mean value $\bar{X}$ of a large number of field tests is determined (or estimated) to be 15.34 kN/m³. The relative locations of these several values are shown (in both RC and unit weight). Now the question is how many tests out of this large number could be less than 90 percent RC for "acceptable" work. We will assume that the project is one involving several hundred cubic meters of soil, so that at least 30 tests will be performed (in any case an estimate of the total number of tests should be made). Now from Fig. 7-13 we obtain the required standard deviation as $15.34 - 0.9(16.50) = 0.49$ as shown. From general statistics considerations this is

$$\bar{X} \pm 1.000\bar{\sigma} = 15.34 \pm 0.49$$

[†] A standard deviation $\bar{\sigma} \cong 0.4$ corresponds to a coefficient of variation $\bar{C}$ of about 2.0 percent.

We need now to determine for these 30 (or other assumed number) tests the reliability which will produce the coefficient of 1.000 for $\bar{\sigma}$. From Table 1-2 at $N = 30$, we interpolate (noting that 1.0 is between 80 and 85 percent) in the table for the coefficient:

$$\begin{array}{cccc} \text{Reliability percent} & 80 & P & 85 \\ \text{Coefficient} & 0.854 & 1.000 & 1.055 \end{array}$$

By proportion P is obtained as

$$P = 80 + \frac{1 - 0.854}{1.055 - 0.854}(5) = 83.6 \text{ percent}$$

With an 83.6 percent reliability, the number of failing tests (out of 30) would be

$$30\frac{100 - 83.6}{100} = 4.92 \qquad \text{or, say, 5 tests}$$

This means that if we do 30 tests, we would expect 1 test in 6† to be either larger than $15.34 + 0.49 = 15.83$ kN/m^3, which is also acceptable, or less than $15.34 - 0.49 = 14.85$ (14.85 is the value for RC = 90 percent), which is unacceptable. The project engineer would now have to decide if this is satisfactory after consideration of all factors. These will include the fact the poorly compacted soil may be surrounded by satisfactorily compacted soil, the test locations, and the intended use of the fill. On some projects, particularly small ones such as basement floors and around building walls, all the tests are required to meet specifications.

To obtain a smaller percentage of tests "failing" on any project, one can:

1. Increase the number of tests (but this is not of very much help. Inspection of Table 1-2 for $N = 100$ tests shows a reliability of only 84 percent).
2. Use a smaller RC, say 0.85 in this example (but we do not need statistics to tell us this!).
3. Reduce difference between $\bar{X}$ and γ_{max} by exercising close field compaction control.

7-14 DEEP COMPACTION OF IN SITU SOILS

In many cases a thick layer of loose soil overlies the site where it is not practical to excavate and replace it with compaction control. In these cases in situ dynamic compaction procedures may be used. Quality control for cohesionless deposits is usually on the basis of relative density using the SPT blow counts before and after densification together with relationships relating SPT blow count N to D_r [e.g., Eq. (6-3) or (6-4)].

---

† Strictly, this is 1 in 12 since a bell curve implies every other "bad" test is a (+) which is satisfactory.

Compaction by blasting or dropping weights may be termed dynamic compaction. In the case of blasting, the soil at a depth below the explosion is compacted. The soil in the immediate vicinity of the blast may be rather loose and additional compaction by other means will be required. Blasting seems to be most effective in saturated cohesionless soils where the blast produces an instantaneous liquefaction from pore pressure increases. The large pore pressure increase produces drainage and volume reduction so that the overburden weight causes the soil to resettle to a more dense state.

Dropping weights is an ancient method that has recently been patented and called dynamic consolidation. Since ideas are not patentable the specific equipment configuration used by the patent holder may be the only valid claim the patent holder can exercise. Indeed the diesel powered tamper of Fig. 7-8b is conceptually identical except for the mechanical configuration. The diesel explosion provides the impact to the ground instead of the falling weight so that the equipment is more compact and portable. This is also analogous to using the diesel hammer versus the drop hammer in driving piles.

Dropping a weight from a height densifies the soil at the point of impact as well as radiating out and downward. Site compaction proceeds by dropping the weight, which may vary from a steel wrecking ball to a specially designed mass weighing anywhere from 1 to about 20 Mg (or more) depending on the capacity of the lifting device. Drop heights may range from 6 to 20 m. The surface may become cratered from the drops, particularly if the soil is very loose, and must be backfilled and compacted with other equipment.

Density is determined by trial and field procedure is to establish a grid on which the weight is dropped. One or more drops is made on each grid point. Preliminary work requires establishing the grid spacing, number of drops, and drop height and weight.

Dynamic consolidation (compaction) is applicable to both cohesionless and cohesive soils. In cohesionless soils Leonards et al. (1980) suggest the depth of influence D is approximately

$$D = 0.5(Wh)^{1/2} \text{ (m)} \tag{7-3}$$

In cohesive soils Menard and Broise (1975) suggest that D is approximately

$$D = (Wh)^{1/2} \text{ (m)} \tag{7-4}$$

where W = weight of falling mass in metric tonnes (1000 kg)
  h = drop height, m

Other dynamic compaction procedures involve inserting vibrating devices into the soil. One such device is the vibroflot, a cylindrical element with an internal vibrator and water jet. The device is sunk by a combination of vibrations and jetting and slowly withdrawn. This is done on a grid pattern to densify the site. Another procedure is to attach a pipe pile to a vibrating hammer (terraprobe) and insert and withdraw the pile on a grid pattern. Both these methods work best in loose sand deposits.

7-15 STATIC STABILIZATION

Static stabilization may be used to increase soil density and may be accomplished by:

Lowering the water table. Increasing the intergranular pressure, causing settlement, and also possibly altering the water content of "soft" soils to produce adequate strength.
Preloading site.

Often only small changes in density as from lowering the water table may be sufficient (coupled with lower water content) to produce a stable mass. Lowering water tables, however, may be environmentally unacceptable.

Preloading the site with one or more meters of soil depth, diking, and filling with water, or similar methods is sometimes used. Preloading is particularly suitable where settlements under loads must be limited. By loading the site to produce pressures larger than those anticipated under service loads, settlements are greatly accelerated. When the preload is removed and service loads applied, further settlements (if any) may be very small amounts.

Preloading is often used with soil drainage methods such as installing a sand layer beneath the preload and/or installing columns of sand, gravel, or porous geotextiles into the underlying soil. These materials being porous allows water to freely move and by controlling the spacing the length of lateral water travel in the soil to the vertical drain is made small so that drainage (producing settlement) is much more rapid than otherwise.

Negative preloading by removing soil so that the service load is less than the initial soil load is another alternative to reducing soil settlements.

7-16 FABRIC STABILIZATION

Improving the strength of soil by mixing straw with mud to form hut walls and make sun-dried bricks easier to handle is a very old practice. Currently this procedure has been extended to the use of metal or synthetic fabrics in the form of strips on top of some—or all—of the lifts in fills to improve the strength and stability. This strength improvement translates into the soil being stable at steeper slopes or being able to carry larger surface loads without large deformations. Soil with reinforcement strips installed is called *reinforced earth.* Two common reinforced earth configurations are shown on Fig. 7-14.

Certain reinforced earth configurations are patented—but as with any patent, the fundamental idea is not patentable. Only the particular strip configuration might produce a valid claim for the patent holder. The U.S. Patent Office checks to see only if a similar patent has been issued; it does not check as to whether the applicant has filed on a basic idea.

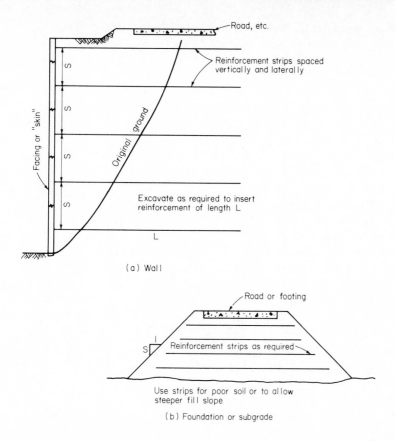

Figure 7-14 Reinforced earth applications. (*a*) Allows a vertical wall in a limited area. Facing may be thin metal sheets or precast concrete plates. (*b*) Use strips spaced vertically and longitudinally (usually by trial) in areas of low capacity or to limit fill slope.

Metal and fabric strips, cables and wire mesh have all been used (as well as straw, bamboo, saplings, etc. in remote areas) but metal tends to have a short service life in the hostile soil environment. Short life coupled with not requiring high tensile strength in the strips, suggests that alternative materials may be better to use. Synthetic fibers in particular seem to have a long service life in the soil environment and there are at present a very large number available. Synthetic fabrics produced for use in soil installations are collectively called *geotextiles*.

Another area of geotextile use is in isolating soil from water. An excavation beneath the road surface or building foundation is made to some depth, a sheet of fabric is laid and the soil backfilled and compacted. Care is taken that any fabric joints are well sealed. Water vapor migrating upward from the water table is stopped at this vapor barrier so that condensation and soil saturation is prevented in the critical zone just beneath the surface element. This both increases the load capacity of the base soil and controls volume change in swell-susceptible soils. Obviously, if surface water can get into the protected zone from above, the vapor barrier is useless and may even be detrimental.

7-17 SOIL-CEMENT AND LIME–FLY-ASH SOIL STABILIZATION

Soil-cement and lime-fly ash (LFA) are very commonly used in soil stabilization. Laboratory procedures are specified by ASTM, AASHTO, PCA (1959) and for LFA in a U.S. Department of Transportation publication (DOT, 1976). These stabilization procedures are used worldwide with select modifications for local climate and soil peculiarities.

Basically the field stabilization criteria for density and percent of admixture (cement, lime, fly ash, etc.) based on dry weight of soil is established in the laboratory as follows:

1. Establish the standard compaction curve for reference.
2. Mix several compaction samples at possible percentages of admixture. Admixture percentages are usually in the following ranges:

    Cement: 3 to 12 percent for A-1 through A-4 soils
              8 to 16 percent for A-4 through A-7 soils
    LFA: 12 to 30 percent with larger percentages for poorer soils.
3. Produce compaction curves for these several percentage mixes and obtain:

    OMC
    Optimum percent of admixture(s) based on maximum dry unit weights obtained using the several percentages.
4. Prepare several additional samples at the one or more selected percentages of admixture and using the corresponding OMC, compact samples to obtain approximately the previously determined maximum dry density.
5. Test these samples for freeze-thaw (durability), strength, etc., as desired. The percentage of admixture producing the best overall response in the several performance tests is used for the job specification.

Strength tests may be made on samples directly extruded from the standard compaction mold and cured (often 2, 7, and 28 days). Since the L/d ratio is near 1 for the compaction mold, some organizations compact samples in smaller molds of greater depth to produce a larger L/d ratio. Relative strength is usually satisfactory so that compaction mold samples are adequate for most pavement subgrades.

If fly ash is used with lime (or cement), it is usually held to a constant ratio, for example, a 30 percent LFA mixture with $1:3$ lime–fly ash (1 part lime to 3 parts fly ash).

The weight of the admixture(s) can be readily computed from the ratio (if used) and the percent of admixture as follows:

$$W_a + W_s = \gamma_{dry}$$

$$W_a = FW_s$$

Making the indicated substitution we obtain

$$W_s + FW_s = \gamma_{dry}$$

and

$$W_s = \frac{\gamma_{dry}}{1 + F}$$

and finally obtain

$$W_a = \gamma_{dry} - W_s$$

where γ_{dry} = dry unit weight of soil-admixture
W_a, W_s = weights of admixture and soil, respectively in units of γ_{dry}
F = percent admixture (as decimal)

Once the weight of admixture is found the sacks of cement (94 lb or 45 kg) or lime (50 lb or 25 kg) or simply the weight of material can be readily computed as in the following example.

Example 7-4 Given the dry unit weight of soil–lime–fly-ash mixture = 19.06 kN/m^3; percent F by weight of admixture = 0.15 at 1 : 3.

REQUIRED Sacks of lime and tons (metric) of fly ash to treat a section of roadway 2 km $\times$ 8 m $\times$ 200 mm thick (final compacted).

SOLUTION The weight of admixture in 1 m^3 is:

$$W_a = 19.06 - \frac{19.06}{1.15} = 2.486 \text{ kN/m}^3$$

at 1 : 3 the lime $= \dfrac{2.486}{4} = 0.6215$ and fly ash $= 3 \times 0.6215 = 1.8645$ kN/m^3;

$$\text{Total} = 0.6215 + 1.8645 = 2.486 \text{ (checks)}$$

For the project (total weight = total volume $\times$ admixture)

$$\text{Volume} = 2 \times 1000 \times 8 \times \frac{200}{1000} = 3200 \text{ m}^3$$

No. of sacks of lime at 25 kg/sack (50 lb); also use 1000/9.807 to convert kilonewtons to kilograms:

$$\text{Lime} = \frac{3200 \times 0.6215 \times 1000}{9.807 \times 25} = 8111.7 \text{ use } 8120 \text{ sacks}$$

$$\text{Fly ash} = \frac{3200 \times 1.8645 \times 1000}{9.807} = 608\ 382 \text{ kg}$$

$$= 608.4 \text{ tons (metric)}$$

Note that sacks of lime and kilograms of fly ash per square meter can be readily computed for spreading and mixing on site. The values shown are, however, more suitable for plant or machine mixing on site.

////

7-18 SPECIAL PROBLEMS IN SOIL COMPACTION

Care must be taken that the fill being used is that specified. Stratified borrow should be used in fill in the same manner it was classified and laboratory tested, i.e., blended or on selected strata.

Compaction of sanitary landfills is difficult. Figure 7-6d illustrates a compactor used for landfill operations which can crush boxes, small metal cans, etc. Compaction is very haphazard in these sites. The primary purpose is to cover the material at the end of each day's operations with 15 to 30 cm of earth which has been compacted; this is primarily for rodent, insect, and odor control. It is recognized that over a period of time decay will produce a fill with large voids even if the earth layers are heavily compacted. Some advantage might be obtained by segregation of fill: papers and other organics in one location; tires, old refrigerators, hot water tanks, building rubble, etc., in another.

Placing logs, stumps, and large boulders in the bottom of fills as a means of disposal is not recommended. Unless wood is permanently below the water table, it will decay. It is difficult to compact the soil adjacent to any of these large objects, and local subsidence may be a problem after a period of time.

Replacing soil in roadway trenches such as those for water, power, or telephone replacement/repair is very difficult. Almost always the trench settles because the soil has not been backfilled properly. The result is an expense to road users from tire and vehicle damage as well as being a nuisance. This problem can be avoided over 90 percent of the time by carefully backfilling in layers and compacting for the full depth. The compaction density may be determined by using a pocket penetrometer (in fine-grained soils) to compact the replaced soil to the same penetration resistance as the trench walls. Hand compactors are usually used due to the limited work area. It should be noted that slight overfill may be required, based on local experience, to allow for subsidence within the fill. Note, however, overfill should not be allowed where the trench was simply backfilled with a tractor to the full depth and then compacted. In this case only the top

150 mm is likely to meet any compaction density. This latter is, however, an extremely common method of backfill if there is no quality control inspector on site.

Frozen soil should not be compacted, as the lumps will thaw and produce soft lenses. Additionally, even if the soil lumps containing ice are broken down, compaction at temperatures below 0°C produces lower unit weights than above 0°C. Where it is absolutely essential to compact soil at temperatures below 0°C the soil should be treated with 0.5 to 1.5 percent calcium chloride to reduce the freezing point of the water. Even with the addition of $CaCl_2$, it is difficult to compact the soil to an equivalent above 0°C density. It appears that a combination of cold soil and water reduces compaction efficiency starting about 5°C on down with substantial effects occurring below 0°C.

7-19 SUMMARY

This chapter has introduced the reader to a number of methods to stabilize (or improve) a soil using both mechanical means and admixture and/or additives. The economic advantage, and often the necessity, of improving the site soil was pointed out.

In depth coverage was limited to the several common compaction methods (both rolling and dropping weights) with a number of pieces of equipment described which give the desired results. It was noted that additive stabilization, particularly grouting, are highly specialized procedures. These procedures depend on trial mixes for optimum quantities of materials rather than sophisticated chemical analyses.

The common laboratory impact compaction test was described in some detail together with the concept of optimum moisture content (OMC), maximum dry density and test limitations. Plotting the zero-air-voids (ZAV) curve as a check on laboratory data was noted. Compaction procedures for cohesive soils to obtain selected soil response/properties (as measured in the laboratory) was considered.

It was found that different procedures are necessary to compact cohesive and cohesionless soils. We may summarize these compaction procedures as follows:

Method	Roller Equipment	Soil
Pressure-confinement-vibration	Smooth-wheel, rubber-tired	Cohesionless
Kneading	Sheepsfoot type, rubber-tired	Cohesive

It was also noted that selective compaction to produce a low dry unit weight was most helpful in controlling volume change with expansive soils.

Quality control of field compaction, including using a random number table to locate check points and a statistical method to estimate the number of failing check tests to allow in "acceptable" work, was also considered.

PROBLEMS

7-1 For the compaction curves of Fig. 7-1:
  (a) Estimate the maximum dry density and OMC for both curves.
  (b) Indicate the range of w for 95 percent RC for both curves.

7-2 The natural moisture content of a borrow material is 8 percent. Assuming 3000 g of moist soil for a standard compaction test, how much water is to be added to bring the sample to 11, 13, 15, 17, and 20 percent water contents?

7-3 For the soil shown in Fig. 7-1, a field unit weight test gave the following information:

$$w = 13.5 \text{ percent}$$

$$\text{Wet unit weight} = 20.09 \text{ kN/m}^3$$

Compute RC for both curves.
  Ans.: Curve 1—approx. 105 percent.

7-4 Redo Example 7-1 if $e_f = 0.40$.

7-5 Given 4000 m³ of fill is placed at a dry unit weight of 19.5 kN/m³. The borrow source has a dry unit weight of 16.5 kN/m³ and $G_s = 2.69$. What volume of borrow is required to make the compacted fill?
  Ans.: 4728 m³

7-6 Approximately how many sacks of cement at 45 kg/sack are required to produce a soil-cement road base using 5 percent by dry weight of cement. The road is 1.6 km × 8 m × 150 mm thick compacted to $\gamma_{dry} = 19.5$ kN/m³.

7-7 How much water must be added to Prob. 7-6 and allowing 2 percent loss for low humidity. The in situ water content is 6.5 percent and OMC is 12.5 percent.

7-8 A lime–fly-ash stabilization is necessary for a fat clay soil used as a subgrade for an airport runway. Laboratory tests have determined that a 20 percent admixture (based on dry unit weight) at a 1 : 3 lime–fly-ash ratio is satisfactory. The dry unit weight of the compacted soil admixture is 18.2 kN/m³. How many sacks of lime and tonnes of fly ash are required for a 2750 × 23 × 1 m thick runway?

7-9 Set up station numbers and distances (X and Y coordinates) from the left side of the runway as field unit weight test locations to randomly sample the in situ compaction *and* cement contents for what you feel will be enough tests to obtain a 95 percent confidence in the work of Prob. 7-8.

7-10 Sixty field unit weight determinations were made for an earth dam as follows:

No. of tests:	16	5	4	6	2	10	7	6	5
RC:	98	102	99	101	100	97	96	95	90

  (a) Is the job satisfactory if RC = 95 percent per specifications?
  (b) What, if anything, will the contractor need to do to improve the efficiency of the operation?

7-11 Referring to Fig. 7-13 and Sec. 7-13, if all data are the same but $\bar{\sigma}$ is computed as 0.39, what percent reliability do we obtain for 30 tests?
  Hint: The new n is 0.49/0.39 = 1.256.
  Ans.: Approximately 89 percent.

7-12 Referring to Prob. 7-11, what $\bar{\sigma}$ will give only 1 "bad" test out of the 30? What RC would this correspond to?
  Ans.: $\bar{\sigma} \cong 0.24$.

EIGHT

SOIL HYDRAULICS, PERMEABILITY, CAPILLARITY, AND SHRINKAGE

8-1 WATER IN SOIL

One of the most important considerations in soil mechanics is the effects of water in the soil on its engineering properties. The Atterberg limits test displays how soil may vary from a solid to a viscous fluid with water content. Individual observations of dry and wet soil on and around excavations, along roadways, and elsewhere indicate a considerable range in state. Cohesive soils are very hard, brittle, and tend to shrink when dry and are very soft, plastic, and tend to swell when wet. Cohesionless soils range from moldable to crumbly for the wet and dry states, respectively.

Chapter 3 introduced the concept of the ground water table and water flow from a higher to a lower energy potential. Wells as a source of water supply are intimately concerned with the flow of water through soils. Highway subdrainage is a water flow problem. Frost action in soils is a flow problem and also depends on capillary action. Chapter 2 introduced the concept of the buoyant, or submerged, unit weight and the loss of effective pressure which occurs due to soil pore water pressure. Seepage flow quantities, to be considered in Chap. 9, and consolidation settlements, in Chap. 11, are water-in-soil problems or conditions.

The following sections will present the general concepts and theory involving the effects of water on, and water flow in, soil. Many situations will occur where proper use of these concepts produces an adequate solution so that the soil and/or site can be used.

8-2 PERMEABILITY

The facility of fluid flow through any porous medium is an engineering property termed *permeability*. For geotechnical engineering problems the fluid is water and the porous medium is the soil mass. Any material with voids is porous and, if the voids are interconnected, possesses permeability. Thus, rock, concrete, soil, and many other materials are both porous and permeable. Materials with larger void spaces generally have larger void ratios, and thus, even the densest soils are more permeable than materials such as rocks and concrete. Materials such as clays and silts in natural deposits have large values of porosity (or void ratio) but are nearly impermeable, primarily because of their very small void sizes, although other factors may also contribute. The terms "porosity" n and "void ratio" e are both used to describe the voids in a soil mass.

The permeability of a soil mass is important in:

1. Evaluating the amount of seepage through or beneath dams and levees and into water wells
2. Evaluating the uplift or seepage forces beneath hydraulic structures for stability analyses
3. Providing control of seepage velocities so that fine-grained soil particles are not eroded from the soil mass
4. Rate of settlement (consolidation) studies where soil volume changes occur as water is expelled from the soil voids as a rate process under an energy gradient (Chap. 12).
5. Controlling seepage from sanitary landfills and hazardous liquid waste dumps.

To provide an appreciation of the flow of water through a soil mass, we will first develop the general equation of laminar flow through a capillary tube, as illustrated in Fig. 8-1a. During laminar flow the velocity varies across the tube diameter from zero at the tube walls—because of friction or viscosity effects—to a maximum value at the center.

From Fig. 8-1a, the velocity gradient at a radial distance r from the tube center is $-dv/dr$ and the unit shear force at this distance r is $-\eta(dv/dr)$, as shown in Fig. 8-1b. The forces on a free body of a cylinder of water with dimensions as shown in Fig. 8-1b at a radial distance r from the center of flow are also shown. Equating these forces, we obtain

$$-\eta(2\pi rL)\frac{dv}{dr} = \pi r^2 h_1 \gamma_w - \pi r^2 h_2 \gamma_w$$

Separating variables, combining terms, letting $i = (h_1 - h_2)/L$ be defined as the hydraulic gradient, and integrating, we obtain

$$v_r = -\frac{\gamma_w}{4\eta} r^2 i + C \qquad (a)$$

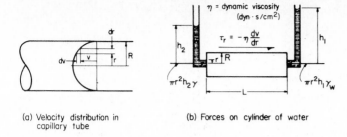

(a) Velocity distribution in capillary tube

(b) Forces on cylinder of water

Figure 8-1 Fluid flow (right to left) through a capillary tube [η = dynamic viscosity (dyne·s/cm^2)].

At $r = R$, $v_r = 0$, from which Eq. (a) becomes

$$v_r = i \frac{\gamma_w}{4\eta} (R^2 - r^2) \qquad (b)$$

The total flow in a unit time is obtained as the following integral:

$$q = i \frac{\gamma_w}{4\eta} 2\pi \int_0^R (R^2 - r^2) r \, dr$$

Integrating, we obtain

$$q = \frac{\gamma_w}{8\eta} \pi R^4 \qquad (c)$$

If the area of the tube is taken as $A = \pi R^2$, the average velocity of flow is

$$v = \frac{q}{A} = \frac{\gamma_w}{8\eta} R^2 i \qquad (8\text{-}1)$$

This expression for velocity is termed *Hagen-Poiseuille's law* since Hagen and Poiseuille, working independently, obtained experimental results almost simultaneously in 1839 and 1840.

Darcy (ca. 1856), in considering the flow of water through sand filters in France, proposed that flow of water through a soil could be expressed as

$$v = ki \qquad (8\text{-}2)$$

where $i = \Delta h/L$ = the head loss in a length of filter bed L and is commonly referred to as the hydraulic gradient.

k = coefficient of permeability with units of velocity.

Darcy's law is a statistical representation of the average flow conditions in a porous medium. This equation is considered to be one of the most important equations in soil mechanics. It is generally considered valid for *laminar* flow but, as will be seen later, is applicable for any flow.

From a comparison of Eqs. (8-2) and (8-1) it is evident that

$$k \cong \frac{\gamma_w}{8\eta} R^2$$

thus depends on the unit weight and viscosity of the fluid—which is temperature-dependent—and on the tube radius. The tubes through a soil mass are of irregular shape, both in diameter and in the direction (longitudinal) of flow, and are dependent on the void ratio and, in particular, on the effective grain size. For this reason, k in coarse sands is larger by many orders of magnitude than in silts and clays even though the void ratio in the silts or clays may be as large as in sand, or often much larger.

Laminar flow occurs in smooth, straight pipes with Reynolds numbers N_R up to about 2100. The Reynolds number is defined as

$$N_R = \frac{v\rho\,d}{\eta} \quad \text{(dimensionless)} \tag{8-3}$$

where v = velocity, cm/s

d = tube diameter, cm

ρ = mass density = $\dfrac{\gamma}{g}$ (g/cm^3)

η = dynamic viscosity (dynes $\cdot$ s/cm^2) = g/cm $\cdot$ s

In soil it appears that turbulent flow occurs at much lower numbers, perhaps as low as 300 to 600. Darcy's law has been found experimentally not to be valid at N_R values smaller than this (see the discussion by Rumer, 1964), because of the discrepancy attributed to inertia forces developed in the water due to abrupt changes in direction of flow resulting from the irregularities in pores, pore sizes, and interconnections. Since these factors are not readily determinable, it follows that the best (and most valid) procedure is to obtain the permeability coefficient using a hydraulic gradient which is as close as practicable to the prototype. If this is done, it is academic whether the flow is laminar, turbulent, etc., since the k determined will be representative of flow conditions to be expected. A further mitigating factor is that for many fine-grained soils the flow velocity is so low (under the field hydraulic gradient) that the flow is laminar and the inertia forces are insignificant.

Inertia forces from near turbulent flow velocities may be a critical factor in causing internal soil erosion. Inertia forces large enough to dislodge small silt or fine sand grains, or overcome interparticle attractive forces in clays, will result in material loss as these small particles are washed out of the soil matrix, with flow channel enlargement and further increased erosion. This phenomenon is called *piping*, and control measures are considered in Chap. 9.

8-3 SOIL WATER FLOW AND THE BERNOULLI ENERGY EQUATION

The Bernoulli equation is commonly used in pipe flow but also applies to flow of water through a soil mass. Figure 8-2 illustrates application of the Bernoulli equation for flow conditions at two points in a soil mass a length L apart. As in any fluid mechanics text, refer all heads to an arbitrary, but as convenient as possible, datum line.

At location B the total energy available, described as a measurable distance (called head) above the reference datum, is†

$$h_1 = Z_1 + \frac{p_1}{\gamma_w} + \frac{v_1^2}{2g} = \text{total energy} = \text{constant} \qquad (a)$$

At location C, a distance L downstream from point B,

$$h_2 = Z_2 + \frac{p_2}{\gamma_w} + \frac{v_2^2}{2g} \qquad (b)$$

The head loss between these two points is defined as $\Delta h = h_1 - h_2$.

For a constant area A of *saturated*, percolating soil from B to C, continuity of flow must exist, from which

$$q_\text{in} = q_\text{out} = Av‡$$

This requires that $v_1 = v_2 = \text{constant}$, and equating Eqs. ($a$) and ($b$) for total energy, we have

$$\Delta h = Z_1 - Z_2 + \frac{p_1 - p_2}{\gamma_w} \qquad (8\text{-}4)$$

A critical evaluation of Eq. (8-4) shows that if $\Delta h = 0$, no flow can possibly take place.

---

† This distance can be measured by inserting a piezometer tube into the pipe as shown; the water level will rise to a level representing the energy head available at that point.

‡ This equation represents a *steady state* condition and, thus, the requirement of a saturated soil condition, since with a nonsaturated soil condition the voids could retain some of the entering water.

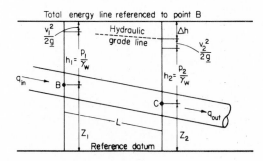

Figure 8-2 The Bernoulli energy equation for pipe flow.

In soil masses, p_1 and p_2 will be atmospheric pressures, unless we are investigating artesian conditions, resulting in the head loss being simply the change in free water surface elevation between any two points B and C. Thus

$$\Delta h = Z_1 - Z_2 \tag{c}$$

This elevation change is generally a very small value for in situ groundwater conditions. Since $i = \Delta h/L$ in Darcy's equation, if i is very small, it necessarily follows that the soil-water velocity v will also be very small.

With the rationale established for small velocities of flow in soil, it also follows that the velocity head ($h_v = v^2/2g$) will be very small and can nearly always be neglected, e.g., for a velocity of 0.3 m/s the velocity head is only

$$h_v = \frac{0.3^2}{2(9.807)} = 0.0045 \text{ m (about 4.5 mm)}$$

In most soils the velocity is well under 0.3 m/s, resulting in a velocity head so small there would be some difficulty in accurately measuring it. For steady state flow, and a velocity head of zero, the piezometric head that is measured can only be the static or position head

$$h = \frac{p_{\text{static at piezometer point}}}{\gamma_{\text{water}}}$$

Considering Fig. 8-3, which represents a homogeneous, isotropic, confined soil mass, if piezometers are inserted at points A, B, C, and D, the water levels will be observed as shown. The piezometers shown are understood to be tubes of sufficient diameter that capillary effects (Sec. 8-7) are negligible. They are positioned to obtain the water level corresponding to the water pressure and would include any $v^2/2g$ effects (but we have already concluded that these are nearly zero) at the tip.

In geotechnical field work, a simple piezometer may consist of a 3- to 5-cm-diameter tube with the tip placed in a sand filter device and into a drill hole. The upper portion of the hole is plugged with a clay seal so that surface water is excluded and so that water pressures at the tip are observed. The piezometric head is measured by lowering a weighted tape or an electrical indicator which completes its circuit on contact with water in the piezometer pipe. Some piezo-

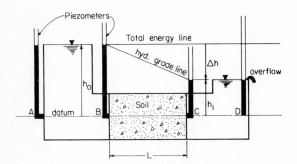

Figure 8-3 Flow of water through an isotropic, homogeneous soil mass of length L with the head conditions shown.

meter installations may use bourdon tube pressure gages or electronic pressure indicators, but these require a considerable installation expense.

The piezometric heads of Fig. 8-3 are obtained as follows:

1. Take the datum along line $ABCD$ so that elevation head does not need to be considered. Also take $v^2/2g = 0$ at all locations.

2. At A: $h_A = h_0$ *(d)*

3. At B: $h_B = h_0$ (with velocity head $= 0$ is essentially a static head)   *(e)*

4. At C: $h_C = h_1$ *(f)*

5. At D: $h_D = h_1$ *(g)*

With $h_A = h_B = h_0$, and $h_C = h_D = h_1$, it follows that the head loss Δh shown occurs across the soil sample of length L. As we can only assume a uniform loss since the soil is considered to be homogeneous and isotropic, the slope of the energy line is as drawn on the figure. In this case the hydraulic gradient is

$$i = \frac{h_0 - h_1}{L} = \frac{\Delta h}{L}$$

The hydraulic gradient is defined as the slope of the energy line defined by the free surface of flowing water in open channels, or the slope of the piezometric heads between two points in confined flow. It represents the head loss or energy loss per unit of length. It should again be noted that if $i = 0$ in Fig. 8-3, $h_1 = h_0$ and no flow can take place.

The low velocity of the pore fluid in a soil mass means that after sudden changes, as for example increasing h_0 or decreasing h_1 in Fig. 8-3, there will be a certain time lag before a new steady state condition is obtained. In earth dams or levees which become saturated during impoundment, or high water during floods, and then undergo a sudden to very rapid emptying, the time lag represents a time during which soil stability may be critical.

It may be apparent from Fig. 8-3 that the approach velocity v_i and the discharge velocity v_d are different from the actual velocity of the water through the soil pores v_a. From continuity, however, the flow rate must be constant; thus

$$q_B = Av_i = Av_{s(\text{apparent})} = Av_d = q_C$$

And in the soil mass, referring to Fig. 8-4, we have

$$q = Av_s = A_v v_a \qquad (h)$$

But also from Fig. 8-4 and Eq. (2-3), we have

$$\frac{A_v}{A} = \frac{V_v}{V} = \text{porosity } n \qquad (i)$$

and substituting into Eq. (h),

$$Av_s = nAv_a$$

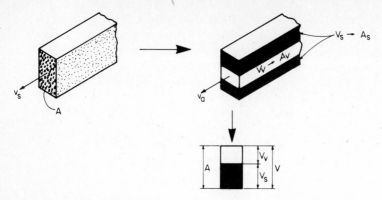

Figure 8-4 Distribution of the cross-sectional area of a percolating soil mass as voids and solids to approximate the true seepage velocity.

and since

$$n = \frac{e}{1 + e}$$

the actual pore velocity in the soil mass, after substitution for n and canceling A values, is

$$v_a = \frac{1 + e}{e} v_s \tag{8-5}$$

This equation indicates that the actual seepage velocity in a soil mass may be substantially larger than the apparent seepage velocity v_s, which is usually computed when using Darcy's law.

Example 8-1 Given: a permeability test using a loose, coarse sand. $Q = 1650$ cm^3 in a time of 15 minutes; void ratio $e = 0.65$; area of sample $= 45.4$ cm^2.

REQUIRED What is the actual water velocity through the sand in centimeters per second?

SOLUTION The nominal discharge velocity v_s is

$$v_s = \frac{1650}{45.4(15)(60)} = 0.0404 \text{ cm/s}$$

Using Eq. (8-5), the actual velocity is approximately

$$v_a = \frac{1.65}{0.65} (0.0404) = 0.102 \text{ cm/s} \text{ (more than double the nominal value)}$$

This computation clearly illustrates that the void ratio or porosity of a soil affects the permeability [k as used in Eq. (8-2)], or how water flows through it.

////

8-4 DETERMINATION OF THE COEFFICIENT OF PERMEABILITY

The determination of the coefficient of permeability may be made in one of several ways. An approximate value may be obtained in the laboratory using either a *constant-head* or a *falling-head* permeability test. The falling-head test is more economical for tests of long duration, while the constant-head test is preferred for soils, such as sands or gravels, which have large void ratios and for which a large flow quantity is desired to improve computational precision. Figure 8-5 gives a schematic test setup for both tests. The test has been standardized for a temperature of 20°C as a convenience. Since the viscosity of water varies from 0.0157 dyne·s/cm² at 4°C to 0.008 97 at 25°C (the possible range of in situ soil temperatures of interest), the factor is 1.75; thus, a 175 percent difference in k can be obtained from two tests at 4 and 25°C.

For the constant-head permeability test (use the right side of Fig. 8-5):

$$Q = Avt = Akit$$

and by rearranging for k as the only unknown, obtain

$$k = \frac{QL}{Aht} \qquad \text{cm/s usual units} \tag{8-6}$$

where Q = total discharge volume, cm³, in time t, seconds
$\quad A$ = cross-sectional area of soil sample, cm²
$\quad h$ = differential head across sample, cm

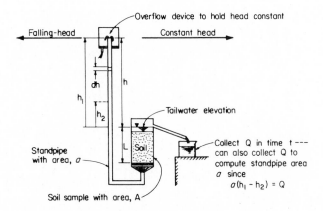

Figure 8-5 Line details of the laboratory test for determining the coefficient of permeability. Use the right side of the figure to identify terms used for the constant-head test and the left side for the falling-head test.

For the falling-head permeability test (use the terms on the left side of Fig. 8-5), in the standpipe of cross-sectional area a cm^2:

$$v = -\frac{dh}{dt} \quad \text{(use minus since the head is decreasing)}$$

The resulting flow into the sample from the standpipe is

$$q_{in} = -a\frac{dh}{dt}$$

and the flow through and out of the sample is

$$q_{out} = Av = Aki$$

From continuity we can equate q_{in} and q_{out} to obtain

$$-a\frac{dh}{dt} = Ak\frac{h}{L}$$

Separating variables, integrating, and starting at time $t_1 = 0$ (which results in the constant of integration being 0) gives

$$k = \frac{aL}{At}\ln\frac{h_1}{h_2} \quad \text{cm/s} \tag{8-7}$$

when all dimensional units are in centimeters and seconds.

The coefficient of permeability may be estimated from a consolidation test (Chap. 11) using Eqs. (12-3) and (12-1) with the coefficient of consolidation c_v.

Example 8-2 A laboratory falling-head permeability test is performed on a light grey, gravelly, well-graded sand with the following test data obtained:

$$a = 0.96 \text{ cm}^2 \qquad A = 45.4 \text{ cm}^2 \qquad L = 20.0 \text{ cm}$$

$$h_1 = 160.2 \text{ cm} \qquad h_2 = 43 \text{ cm} \qquad \begin{array}{l} t = 65 \text{ s for head to} \\ \text{fall from } h_1 \text{ to } h_2 \end{array}$$

$$\text{Water temperature of test} = 20°C$$

REQUIRED Compute k.

SOLUTION Make a direct substitution into Eq. (8-7) to obtain

$$k = \frac{0.96(20.0)}{45.4(65)}\ln\frac{160.2}{43.0} = 0.0085 \text{ cm/s}$$

Note:

1. The hydraulic gradient is very large in this test and may be totally unrealistic (it also may produce turbulent flow).
2. If the water temperature had differed from 20°C, a temperature correction would be required.

////

When the test temperature differs from 20°C, the coefficient of permeability should be corrected to 20°C by recognizing that the value of k is inversely proportional to viscosity to obtain

$$k_{20} = k_T \frac{\eta_T}{\eta_{20}} \qquad (8\text{-}8)$$

where k_T is the coefficient of permeability at any test temperature T. Table 8-1 gives several values of η versus T.

Figure 8-6 gives the approximate range of k values one may expect to obtain. The coefficient of permeability is plotted on a log scale, since the range of permeabilities is so large. No other engineering property of any material exhibits such a large range of values as does the permeability of soil.

An empirical equation relating the coefficient of permeability to the effective grain size (D_{10}) from a sieve analysis was reported by A. Hazen (ca. 1892) based on work with rapid sand filters in water treatment plants. He found that for sands with D_{10} sizes between 0.1 and 3.0 mm, the coefficient of permeability could be expressed approximately† as

$$k = C(D_{10}^2) \qquad \text{cm/s} \qquad (8\text{-}9)$$

Table 8-1 Temperature versus dynamic viscosity and surface tension for water

T, °C	γ, kN/m^3	η, dyne · s/cm^2	Surface tension, dynes/cm
4	9.807	0.015 67	75.6
16	9.7969	0.011 11	73.4
18	9.7935	0.010 56	73.1
20	9.7896	0.010 05	72.8
22	9.7854	0.009 58	72.4
24	9.7808	0.009 14	72.2
26	9.7758	0.008 74	71.8
28	9.7704	0.008 36	71.4
30	9.7646	0.008 01	71.2

† Hazen (1911), which is commonly cited as the reference of origin, does not give this form of the equation; instead, $v = cd^2h/L(0.73 - 0.03T)$ is given, where T is the temperature of the water.

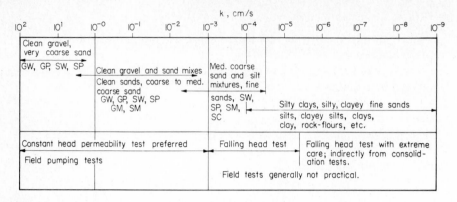

Figure 8-6 Typical ranges of permeability coefficients and suggested test methods.

Is this equation D_{10} is the effective grain size in centimeters, with C such that k is in centimeters per second. The coefficient C varies, according to Hazen, from about 40 to 150 and the values may be taken as follows:

C	Sand (any or all of the following applies)
40–80	Very fine, well graded or with appreciable fines [(−) No. 200]
80–120	Medium coarse, poorly graded; clean, coarse but well graded
120–150	Very coarse, very poorly graded, gravelly, clean

One would expect that poorly graded sand would have a larger coefficient than well-graded materials, since the void spaces would be more ordered and larger with poorly graded soil.

An estimate of the permeability k_2 at a void ratio of e_2 when a test was performed with results of k_1 at void ratio e_1 may be made as

$$k_2 = k_1 \left(\frac{e_2}{e_1} \right)^2$$

Other equations more complicated than this have been suggested, but in the range of void ratios (0.5 to 1.1) likely to be used, this equation is considerably simpler and the results are sufficiently precise considering the precision with which k_1 can be determined.

8-5 LIMITATIONS AND OTHER CONSIDERATIONS IN DETERMINING k

The laboratory test for determination of k is very unreliable, with considerable attention to test procedure and equipment design necessary to obtain even the correct order of magnitude of the permeability coefficient. Some of the factors producing this unhappy condition are:

1. Soil in situ is generally stratified, and it is hard to duplicate in situ conditions in the laboratory test. The horizontal value k_h is usually needed, but tube samples are likely to be tested, with vertical values k_v obtained.
2. In sand the values of k_v and k_h are considerably different, often on the order of $k_h = 10$ to $1000k_v$, due to the sedimentation process of soil deposit formation. The field soil structure is invariably lost in the laboratory because an undisturbed sample cannot be tested (even if it were possible to obtain one), since it would have to be transferred from the recovery device to the permeameter.
3. The small size of laboratory samples leads to effects of boundary conditions, such as smooth sides of the test chamber affecting flow and air bubbles either in the water or trapped in the test sample affecting the test results.
4. No method is available to evaluate k for other than saturated steady state soil conditions, yet many flow problems will involve partially saturated soil-water flow. Where k is very small, as in clays and fine silts, it may be difficult to determine when a steady state has been obtained.
5. When k is very small, say from 10^{-5} to 10^{-9} cm/s, the time necessary to perform the test will cause evaporation and equipment leaks to become very significant factors. Leaks which are so small that no visible accumulation of water occurs (because of evaporation) can produce a computed k that is too large by several orders of magnitude.
6. In the interests of time, the laboratory hydraulic gradient $\Delta h/L$ is often made 5 or more, whereas in the field more realistic values may be on the order of 0.1 to less than 2.
   (a) In clays a threshold i of 2 to 4 may be necessary to produce any flow (and an apparent k); thus, a totally unrealistic flow may be obtained if the field hydraulic gradient is not as large as the threshold gradient.
   (b) In sands, unrealistically high i values may produce turbulent conditions, indicating a flow condition different from the field i, which may indeed produce laminar flow.
   (c) Unrealistically high i values may produce sample packing and a void ratio different from that in the field. This may be very critical when testing loose sands.

In the field, the permeability may be evaluated on the basis of observing the length of time it takes dye, salt, or radioactive tracers to travel between two wells, wellpoints, or borings in which the differential head can also be measured. Field pumping tests and the well equations given in Chap. 9 can be used to compute the coefficient of permeability. The permeability may be computed from the rate of water fall in a borehole filled with water (above the water table) or from its rising some height in the borehole after bailing (below the water table). Procedures and computations to obtain the coefficient of permeability using the several methods cited above can be found in references such as Cedergren (1977), USBR (1968), and NAFAC (1971). Almost any well-performed field test gives a more reliable value of the coefficient of permeability than a laboratory test, except on remolded soil used for compacted fills, but is considerably more expensive and time-consuming.

8-6 EFFECTIVE COEFFICIENT OF PERMEABILITY OF STRATIFIED SOILS

Figure 8-7 illustrates a stratified (an *anisotropic* soil mass) soil condition. It is usually convenient to replace this system by an equivalent soil mass with a single effective thickness $L = \sum H_i$ and a single value of k—either k'_v or k'_h, depending on the direction of flow being considered. There is a direct analogy between a simple electric circuit containing only resistors in either series or parallel and water flow through the stratified soil mass of Fig. 8-7. Flow perpendicular to parallel-bedded strata is analogous to series resistor networks. When resistors are in series, the *largest* resistance controls the current flow (analogy = fluid), and with flow perpendicular to the bedding planes, as k_v of Fig. 8-7, the stratum with the *smallest* permeability essentially controls the flow quantity. When flow is parallel to the bedding planes, as k_h of Fig. 8-7, the flow quantity is controlled by the stratum with the largest coefficient of permeability (and in electric circuits by the smallest resistor in parallel). With these concepts in mind, we will now develop equations for the equivalent k's for the stratified deposit. For equivalent k'_v, we have

$$q_{in} = q_{out}$$

from continuity; therefore $v = $ constant, and

$$v = k'_v i = k_1 \frac{h_1}{H_1} = k_2 \frac{h_2}{H_2} = k_3 \frac{h_3}{H_3} = \cdots = k_n \frac{h_n}{H_n}$$

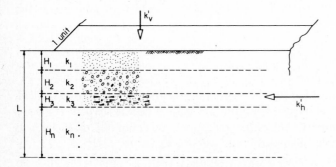

Figure 8-7 Stratified soil system with permeability as shown.

Rearranging, we have

$$\frac{H_1}{k_1} = \frac{h_1}{v}$$

$$\frac{H_2}{k_2} = \frac{h_2}{v}$$

$$\frac{H_3}{k_3} = \frac{h_3}{v}$$

$$\cdots\cdots\cdots$$

$$\frac{H_n}{k_n} = \frac{h_n}{v}$$

Adding, we obtain

$$\frac{h_1}{v} + \frac{h_2}{v} + \frac{h_3}{v} + \cdots + \frac{h_n}{v} = \frac{H_1}{k_1} + \frac{H_2}{k_2} + \frac{H_3}{k_3} + \cdots + \frac{H_n}{k_n}$$

Factoring the left side and recognizing that $H_1 + H_2 + H_3 + \cdots + H_n = L$ and $v = k'_v(h/L)$, with some rearranging to solve for k'_v we obtain

$$k'_v = \frac{L}{H_1/k_1 + H_2/k_2 + H_3/k_3 + \cdots + H_n/k_n} \tag{8-10}$$

The equivalent k'_h can be obtained as

$$q = Av_{\text{average}} = L(k'_h)i$$

which is also the sum of the flow in each stratum:

$$L(k'_h)i = k_1 H_1 i + k_2 H_2 i + k_3 H_3 i + \cdots + k_n H_n i$$

Canceling i and solving for k'_h, we obtain

$$k'_h = \frac{k_1 H_1 + k_2 H_2 + k_3 H_3 + \cdots + k_n H_n}{L} \tag{8-11}$$

Example 8-3 Given the stratified soil system shown in Fig. E8-3.

REQUIRED What is the head drop across each stratum, and what is the corresponding flow through a unit area of soil? For the flow computation, use both basic concepts and Eq. (8-10).

SOLUTION From Fig. E8-3

$$\Delta h_1 + \Delta h_2 + \Delta h_3 = h = 8 \text{ units} \tag{a}$$

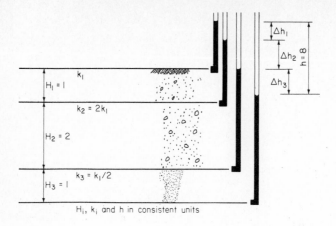

H_i, k_i and h in consistent units

Figure E8-3

From continuity of $q_{in} = q_{out}$ through any stratum we have

$$q_1 = \frac{k_1 \, \Delta h_1}{H_1} = \frac{k_2 \, \Delta h_2}{H_2} = \frac{k_3 \, \Delta h_3}{H_3}$$

or

$$q_1 = \frac{k_1 \, \Delta h_1}{1} = \frac{2k_1 \, \Delta h_2}{2} = \frac{k_1 \, \Delta h_3}{3 \cdot 1}$$

Now find Δh_1 and Δh_3 in terms of Δh_2 (canceling k_1)

$$\Delta h_1 = \Delta h_2; \qquad \Delta h_3 = 2 \, \Delta h_2$$

Substituting these values into Eq. (*a*) above, we obtain

$$\Delta h_2 + \Delta h_2 + 2 \, \Delta h_2 = 8$$

$$\Delta h_2 = \frac{8}{4} = 2$$

From which $\Delta h_1 = 2$; $\Delta h_3 = 2 \times 2 = 4$

The flow quantities are:

$$q_1 = \frac{k_1(2)}{1} = 2k_1$$

$$q_2 = \frac{2k_1(2)}{2} = 2k_1$$

$$q_3 = \frac{k_1(4)}{2 \cdot 1} = 2k_1$$

Thus the flow in each stratum is the same as required from continuity.

By Eq. (8-10) the equivalent

$$k'_v = \frac{4}{1/k_1 + 2/2k_1 + (1/k_1)/2}$$

$$k'_v = \frac{4k_1}{1 + 1 + 2} = k_1$$

The flow quantity $q = k'_v h/L = k_1(8)/4 = 2k_1$ (as by direct computation above)

////

8-7 CAPILLARITY AND CAPILLARY EFFECTS IN SOIL

All materials possess intermolecular forces. These may be termed *cohesion* for the internal molecular forces and *adhesion* for the attraction between molecules of dissimilar materials, such as water and glass. If the adhesion forces between a liquid and any other material are larger than the intermolecular attraction of the liquid, the surface of the dissimilar material will be "wetted" by the liquid. Mercury, for example, has a very considerable cohesion; thus it will wet only a limited number of dissimilar materials. Water, on the other hand, with little internal cohesion, will wet almost all materials it contacts. Wetting agents may be used to increase the adhesion effects between liquids and solids.

Any quantity of liquid will behave as though the surface were a tightly stretched skin due to the intermolecular attractive forces in the interior. This phenomenon is termed *surface tension*. This liquid material property accounts for the spherical shape of water drops on oily dust and the nearly true spheres of mercury drops on glass plates. Wetting agents (soaps and detergents) tend to reduce surface tension (the internal cohesion) and increase adhesion of the material to the foreign surface (or make it wet).

Since surface tension is a material property of liquids and depends on intermolecular attraction, it will be temperature-dependent (below some critical temperature, which for water is 100°C, the material is a liquid, above it is a gas). Table 8-1 also gives the surface tension value for water at several values of temperature.

Surface tension allows a razor blade or needle carefully placed on the surface of water to float and produces the rise above the static surface level of water in a small glass tube placed in a container of water.

Since surface skin tension effects are caused by intermolecular attraction, it follows that at the interface with air the skin is in vertical equilibrium and that over the surface the pull must be equal in all directions, or perpendicular to any line of interest as in Fig. 8-8a. With a skin thickness of one molecule, the units of surface tension must be force/length.

Now let us investigate the effects of surface tension on a curved surface, as shown in Fig. 8-8b.

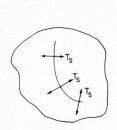

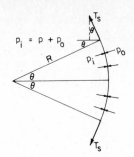

(a) Surface tension perpendicular to any line of interest on surface of liquid.

(b) Surface tension on a portion of 3-dimensional curved surface.

Figure 8-8 Surface tension.

For equilibrium, $\sum F_h = 0$, or

$$T_s \sin \theta (2\pi R \sin \theta) = \frac{p}{4} \pi (2R \sin \theta)^2$$

from which

$$p = \frac{2T_s}{R} \tag{8-12}$$

This equation states that the pressure difference $(p_i - p_o)$ inside the curved surface is directly proportional to the surface tension T_s and inversely proportional to the radius of curvature R. It is well known that water vaporizes (or boils) at temperatures less than 100°C when the pressure is less than atmospheric (101.3 kPa or 1 bar). Equation (8-12) indicates that if R is sufficiently small, a higher temperature will be necessary to cause vaporization due to the increased pressure.

Height of Water Rise in Capillary Tubes

When a hollow, open-ended tube is inserted into a container of liquid, and if the liquid wets the contact surface, it will climb the inside walls of the tube because of surface tension effects, producing a concave spherical upper surface, as shown in Fig. 8-9a. This is both a theoretical and readily observed phenomenon.

For clean glass, the angle α of the concave surface film with the tube walls is:

Liquid to clean glass	α, degrees
Water	0
Mercury	139

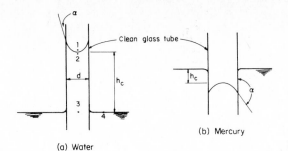

(b) Mercury

(a) Water

Figure 8-9 Capillary rise.

The height of rise h_c can be computed as $\sum F_v = 0$, from which in Fig. 8-9a

$$\frac{\pi d^2}{4} h_c \gamma = \pi d T_s \cos \alpha$$

and solving for h_c,

$$h_c = \frac{4 T_s \cos \alpha}{\gamma d} \qquad (8\text{-}13)$$

where T_s = surface tension
γ = unit weight of fluid
d = tube diameter, all in consistent units

For water at $20°C$; $T_s \cong 72.8$ dynes/cm; $\gamma = 9.7896$ kN/m^3. Since 10^8 dynes = 1 kN and 10^6 cm^3 = 1 m^3, we obtain

$$h_c = \frac{4(72.8 \text{ dynes/cm})(10^6 \text{ cm}^3/\text{m}^3)}{9.7896 \text{ kN/m}^3(10^8 \text{ dynes/kN})(d \text{ cm})}$$

From which the following is obtained:

$$h_c = \frac{0.297\,46, \text{ cm}^2}{d, \text{ cm}} \qquad (8\text{-}14)$$

Example 8-4 Given $d = 0.001$ cm, $T = 20°C$.

REQUIRED Height of capillary rise h_c.

SOLUTION Using Eq. (8-14),

$$h_c = \frac{0.297\,46}{0.001} = 297.46 \text{ cm} = 2.97 \text{ m}$$

////

Example 8-5 Given $d = 0.05$ cm:

$$\text{Mercury with } \alpha = 139°$$

$$\gamma_{Hg} = 13.6 \times 9.807 = 133.4 \text{ kN/m}^3$$

$$T_s = 473 \text{ dynes/cm}$$

REQUIRED Height of capillary rise.

SOLUTION Use Eq. (8-13).

$$h_c = \frac{4(473)(10^{-2})(\cos 139°)}{133.4(0.05)} = -2.14 \text{ cm (below liquid surface)}$$

Note that the sign of the angle allows Eq. (8-13) to be used for all liquids, and that $\alpha > 90°$ will produce a depressed height of capillary rise.

////

Example 8-6 Referring to Fig. 8-9a, (1) compute the stress at points 2 and 3, and (2) derive a general expression for the stress at any height h_i in the capillary zone h_c.

SOLUTION

(1) Find stresses:
   (a) At point 3 the stress is equal to the stress at point 4 from fluid static principles. This pressure is atmospheric = 0 gage

$$\text{Atmospheric pressure} = 14.7 \text{ psi (absolute)} = 101.3 \text{ kPa}$$

   (b) The stress at point 2 can be computed from Eq. (8-12) as follows: Water angle on tube, $\alpha = 0°$; tube radius $R = d/2$. Also from Fig. 8-8b: $p_i = p + p_o$ but by inspection $p_i = 0$ (gage), from which

$$0 = p + p_o \qquad \text{which gives } p = -p_o$$

$$(-) = \text{tension stress}$$

Using Eq. (8-12):

$$-p_o = \sigma_2 = \frac{2T_s}{R} = \frac{4T_s}{d}$$

From Table 8-1 for 20°C the surface tension $T_s = 72.8$ dynes/cm, so that $4T_s = 291.2$, say, 300 dynes/cm for the temperature range 0 to 20°C.
At point 2 approximately:

$$\sigma_2 = \frac{300}{d, \text{ cm}} (10^{-8} \text{ kN/dyne})(10^4 \text{ cm}^2/\text{m}^2) = \frac{0.03}{d, \text{ cm}} \text{ kPa} \qquad (a)$$

For $d = 0.001$ cm obtain:

$$\sigma_2 = 0.03/0.001 = -30 \text{ kPa}$$

(2) Obtain a general expression for the stress at any height $0 \le h_i \le h_c$. Since the tension stress is just sufficient to pull a column of water to height h_c, we have

$$W_c = -\gamma_w h_c A_{\text{tube}} \qquad\qquad (b)$$

But $-\gamma_w h_c = \sigma_t$ and $W_c = \gamma_w h_c A_{\text{tube}}$, and equating and canceling A, we obtain

$$\sigma_t = \gamma_w h_c \qquad\qquad (c)$$

From Example 8-3, $h_c = 2.97$ m and $\gamma_w = 9.7896$ kN/m (Table 8-1) Inserting values, we obtain:

$$\sigma_t = -9.7897(2.97) = 29.1 \text{ kPa} \qquad (\text{vs. } -30 \text{ from part } b)$$

The discrepancy is due to using $4T_s = 300$ as an average value instead of the table value at 20°C.

From Eqs. (b) and (c) it is evident (or from using a free body) that the tension stress in the water at any height h_i in the capillary zone h_c is

$$\sigma_t = -\gamma_w h_i \qquad\qquad (8\text{-}15)$$

$$////$$

Referring to Eq. (a) of Example 8-6 and solving for d to produce a stress of 100 kPa (approximately 1 atm):

$$d = \frac{0.03}{100} = 0.0003 \text{ cm} \qquad (\text{very small})$$

From Eq. (8-15) the height of capillary rise is

$$h_c = \frac{100}{9.808} = 10.2 \text{ m} \qquad (\text{approximately})$$

Theoretically, we would have a negative pressure in the water of 1 atmosphere in a capillary tube of approximately

$$10.2 \text{ m high} \times 0.0003 \text{ cm in diameter}$$

At $(-)$ one atmosphere (a vacuum) the water would be expected to boil at normal ambient temperatures. This might occur in the laboratory under ideal conditions, but not visible to the eye (0.0003 cm = 0.003 mm); in the soil pore water contamination and other factors occur, thus greatly reducing the likelihood of any water "boiling."

Figure 8-10a illustrates the normal capillary supply in the soil as from a ground water table. When the water table fluctuates and surface water infiltration occurs, the capillary tubes may fill as in Fig. 8-10b. Note that a filled capillary tube is a saturated condition for that string of soil pores but does not necessarily mean the soil is "saturated."

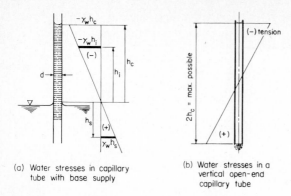

(a) Water stresses in capillary tube with base supply

(b) Water stresses in a vertical open-end capillary tube

Figure 8-10 Water stresses in capillary tubes with the boundary conditions shown.

Capillary Tubes with Variable Radius

Figure 8-11 illustrates the rise which can be obtained in capillary tubes of variable cross section. Figure 8-11a and b illustrates that the capillary rise h_c depends on tube diameter, and Fig. 8-11a further illustrates that the meniscus may not fully develop if the tube height is less than h_c. Figure 8-11c illustrates that a sudden enlargement may halt the capillary rise unless it is located at a height which corresponds to diameter d_2. Figure 8-11d illustrates how capillary rise can bypass a tube enlargement if a water supply from above flows downward and fills the tube to above the enlargement. The enlargement can be bypassed by raising the water supply so that h_c for diameter d_2 is above the enlargement; after the tube is filled, the water surface can be lowered. This condition corresponds to rainfall infiltration and/or a temporary rise in the ground water table. In Fig. 8-11e, an open based box with a capillary tube in the top is lowered into the water until the capillary tube has water in the zone h_c for that tube diameter. If the box is slowly raised, sufficient capillary tension stresses develop (at least in theory) throughout the water to hold it in the box. Figure 8-11f is a large tube with soil particles such that the pores make capillary tubes. Note that the large void area in the center at the top produces an irregular h_c level. In this case the tensile stresses in the water produce compressive interparticle stresses, resulting in an increase in the intergranular pressure.

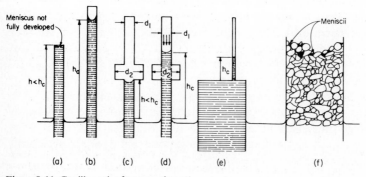

Figure 8-11 Capillary rise for several cases.

Capillarity in Soils

Figure 8-12 represents several soil particles with the voids filled with water and evaporation taking place. The soil "tubes" will vary in cleanliness and the water-soil interface will range from smooth to rough; however, the upper limit of intergranular stress due to compression of the tube walls as tension stresses develop in the pore water (refer to Fig. 8-12c) will be on the order of 80 to 100 percent of the maximum possible capillary tension stresses computed using Eq. (8-14). The pore tubes are quite irregular and variable in diameter, but it is still convenient, and a satisfactory approximation, to use Eq. (8-13a) to obtain h_c.

The pore diameter is impossible to measure and must be approximated. A commonly used approximation is

$$d \cong \tfrac{1}{5}D_{10} \qquad \text{mm}$$

Particle packing may influence the pore diameter(s) considerably, but this is not reflected in this equation. In clayey soils where D_{10} may be on the order of 0.0015 mm, the tube diameter is approximately

$$d = \frac{0.0015}{5 \times 10} = 0.003 \text{ cm}$$

and the corresponding height of capillary rise is

$$h_c = \frac{0.297}{0.003} \cong 100 \text{ cm} \qquad \text{(about 1 m)}$$

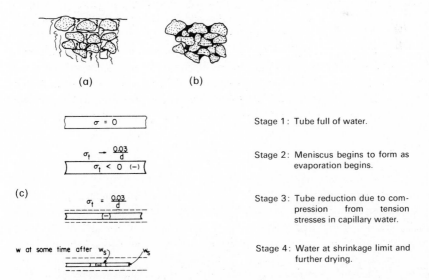

Figure 8-12 Capillary effects to produce soil shrinkage. (a) Saturated soil. (b) Soil after partial drying, with meniscii, water tension, and grain compressive stresses being developed. (c) Evaporation stresses in a horizontal capillary tube.

Capillarity and Shrinkage

The shrinkage limit was defined in Chap. 2 as the water content ($S = 100$ percent) below which no further volume change occurs. Figure 8-12c illustrates the general conditions of shrinkage; at stage 1 the tube is full of water and no meniscus radius can form. As evaporation takes place, the meniscus forms and water tension stresses develop; this squeezes water out of the pores, reducing pore sizes, tube diameter, and radius, and more water evaporates. The limiting case occurs when the soil matrix will reduce in volume no further from the water tension stresses. This is the shrinkage limit, and from this point on the water simply evaporates with no further soil structure change. As Eq. (8-12) indicates, very high compressive stresses can be developed during drying due to the large number of soil pores simultaneously being subjected to water tension. It is not possible to evaluate the magnitude of the compressive stresses developed; however, the effects have been observed. For many cohesive soils the drying (often several times over long geologic periods and perhaps aided by pore and pore-water contamination as in marine deposits) has altered the soil structure just as if large depths of overburden had been applied and later eroded. In many parts of the United States the drying has loaded the soil (termed *preconsolidated* in Chap. 11) to apparent overburden loads on the order of 200 to 800 kPa. Tschebotarioff (1936) and others found shrinkage values on the order of 900 kPa in the Nile valley of Egypt. Large values of shrinkage stresses have also been found in Australia, India, and the Middle East.

Capillarity is a significant factor in sands (gravel is too large to be much affected), especially in fine to medium-fine sands. When the sand is fully saturated or completely dry, capillarity is not present (Fig. 8-12c, stage 1 for the saturated case) and the sand grains are easily displaced. At intermediate water contents, capillary effects are present, and due to the many particles the cumulative effect is great enough to allow sand to stand on vertical cuts and to be molded when damp. It is well known and easily observed that it is much easier to walk or drive a wheeled vehicle on damp than on dry sand due to grain displacement. Another practical consideration with damp sand is *bulking*, which was considered in Sec. 7-6. When sand is dry, it is totally impossible to make a vertical excavation or to mold it. The capillary effect in damp sand is termed *apparent cohesion* since it disappears when the sand is either dry or saturated.

Slaking is the rapid, almost explosive, disintegration of dry, or nearly dry, cohesive soil lumps when they are immersed in water; it is a capillary phenomenon. Soil that is not very nearly dry will only swell when immersed in water; however, if the water content is well below the shrinkage limit, the soil capillaries contain air. When the soil is placed in water, surface tension pulls water into the capillary tubes, confining and compressing the air. As the air becomes more compressed, the interparticle tension stresses increase, and when they become larger than the particle attractions, the soil slakes. Only the clayey soils and some clayey shales, where the material in the elemental state has a shrinkage limit, will slake. Other soils and rocks without a shrinkage limit do not exhibit this phenomenon.

Shrinkage and Volume Change

Volume change is a serious problem in shrinkage-susceptible soils all over the world. It can involve almost any cohesive soil, but is more pronounced in arid to semiarid areas, or where the more active montmorillonite, or bentonite, clay minerals have not sufficiently weathered to a less active state. Large areas of the western and southwestern United States. Australia, India, the Middle East, and southern Africa are covered with soils subject to large volume changes which cause serious engineering problems. Cohesive soils in other areas are often susceptible to lesser volume changes, but these are usually more a nuisance than a serious problem. Jones and Holtz (1973) estimated that some 20 to 25 percent of the land area in the United States is covered with soils susceptible to volume change. At that time the economic damage estimate was over 2.25 billion dollars annually. Holtz (1980) notes that a number of authorities are of the opinion the economic damage is now closer to 5.5 billion.

Expansive clay soils are dense and very hard in the dry state due to shrinkage stresses. Even at small water contents the soil is quite dense and hard—so hard that obtaining thin-walled tube samples for laboratory testing is often difficult or nearly impossible. Often these soils will contain a maze of shrinkage cracks ranging from hairline to 2 or 3 cm wide. Figure 8-13 illustrates shrinkage cracks in a grass-covered area and in a dried lake bed, indicating that the phenomenon is not limited to freshly deposited water deposits. Shrinkage cracks similar to those shown for the grassy area can considerably influence the rate of soil saturation from surface infiltration. Laboratory strength test values are considerably affected by cracks and/or cracks which have "silted up" with foreign windblown

(a)   (b)

Figure 8-13 Shrinkage cracks. (a) Typical for clayey soil. The crack is about 2 cm wide and about 1.5 m deep, determined by inserting a small wire. (b) Cracks in a lake bed deposit. They are about 4 cm wide and at least 0.3 m deep.

or water-borne material. If the test samples are badly fissured, strength tests may not be practical.

The expansion problem can be avoided only in cases where the soil can be reliably protected against water infiltration by using surface or subsurface drainage, landscaping, or using impervious membranes (asphalt or plastic cloth). The only other recourse is altering the clay by chemical additives such as cement, lime, lime–fly ash, cement–fly ash, calcium chloride, etc. Simply covering an area with a floor slab or pavement does not control water infiltration in the zone of importance, since water vapor tends to condense on the under side over a period of several years and will saturate the soil. This can be readily observed by turning over a small rock in the summer; the underside will be damp, even if there has been a prolonged dry period. Unfortunately, near the edges of a floor slab or pavement the water content will be less than in the interior due to perimeter evaporation; thus the volume change will be differential rather than uniform.

Volume change is directly related to the shrinkage limit and somewhat related to the liquid and plastic limits. Table 8-2 gives an approximate relationship which has been found to be reasonably reliable in predicting the occurrence of volume change. Unfortunately, there is no reliable way to numerically quantify volume change. One would expect a "little" potential to be less than a "high" potential, but for design it would be of considerable benefit to be able to say that a "little" is, say, 2 or 3 percent or less, and "high" corresponds to 8 or 10 percent or more. At this point there is no means of quantifying the potentials given in the table, and considerable research needs to be done before they can be quantified. Factors such as type of clay, surcharge load, void ratio, method of saturation, and general environment produce too wide a range of parameters to provide a ready and direct answer. The most that can be said is that, as stated in Chap. 7, if $I_P \geq 20$, we are likely to have volume change problems which require some kind of precautionary measures.

We should note carefully that since volume change is the result of changes in water content in cohesive soils, a problem can develop in less arid areas as a result of tree and shrub growth. In the growing season tree roots may temporarily desiccate the soil, but when the tree is in the dormant season the water

Table 8-2 Relationship between Atterberg limits and volume change potential†

| Volume change potential | Plasticity Index I_p | | Shrinkage limit w_S |
	Arid area	Humid area	
Little	0–15	0–30	> 12
Moderate	15–30	30–50	10–12
High	> 30	> 50	< 10

† From Holtz and Gibbs (1956).

content again builds up, resulting in a seasonal volume change. This can develop into an environmental problem: how to avoid the volume change and at the same time avoid destroying the trees or shrubs.

Example 8-7 Given: the shrinkage limit of a very clayey soil is 9.6 percent. $G_s = 2.70$, for the clay. Compaction test data gives $\gamma_d = 16.70$ kN/m^3 at OMC = 16.5 percent.

REQUIRED Estimate the dry unit weight of the soil at the shrinkage limit and compare with the standard compaction test value.

SOLUTION By definition, at the shrinkage limit $w_s = 9.6$ percent, the degree of saturation $S = 100$ percent. From Eq. (2-12) we have

$$Se = wG_s \quad \text{and} \quad e = \frac{0.096(2.70)}{1.00} = 0.2592$$

From Eq. (2-11) the dry unit weight at the shrinkage limit is

$$\gamma_d = \frac{G_s \gamma_w}{1 + e} = \frac{2.70(9.807)}{1 + 0.26} = 21.02 \text{ kN/m}^3$$

The percent improvement is

$$\text{Percent improvement} = \frac{21.02}{16.70}(100) = 126 \text{ percent}$$

This computation indicates that the capillary stresses during drying produce very high compressive stresses in the soil structure and are considerably more efficient in producing a dense soil matrix than the standard compaction test. Note, however, that this unit weight was computed and not actually measured; a field value might be somewhat less, but could be even more if a large enough number of drying cycles had been applied.

////

8-8 SEEPAGE FORCES AND QUICK CONDITIONS

From Eq. (2-20) the intergranular or *effective* stress is

$$\sigma' = \sigma_t - u \qquad\qquad (2\text{-}20)$$

A careful examination of this equation shows that both the total pressure on the plane of interest and the pore water pressure can be measured and/or calculated. The pore water pressure can be measured using a piezometer with little difficulty (see discussion in Sec. 8-3). Note, however, that the effective stress can only be calculated.

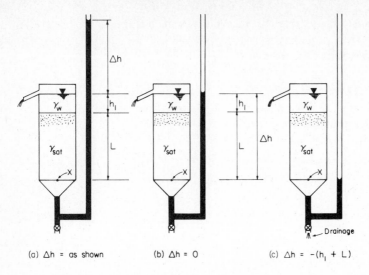

(a) Δh = as shown   (b) Δh = 0   (c) Δh = -(h₁ + L)

Figure 8-14 Intergranular pressure at point X for several differential heads.

When water flows in soil under a hydraulic gradient as considered in Secs. 8-3 and 8-4, the differential pressure head produces a force on the soil grains in the direction of flow. This pressure accumulation will be termed a *seepage force*. Consider the intergranular pressure at point X of Fig. 8-14 for the cases shown in (*a*), (*b*), and (*c*).

In Fig. 8-14, at the instant the differential head is as shown, the following are obtained for point X for all three cases:

$$\sigma_t = h_1\gamma_w + L\gamma_{\text{sat}} = h_1\gamma_w + \frac{G_s\gamma_w}{1+e} + \frac{\gamma_w e}{1+e} = h_1\gamma_w + \frac{(G_s + e)\gamma_w}{1+e}$$

$$\gamma' = \gamma_{\text{sat}} - \gamma_w$$

$$u = L\gamma_w + (h_1 + \Delta h)\gamma_w$$

In Fig. 8-14*a*, the intergranular pressure is

$$\sigma' = \sigma_t - u$$

$$\sigma' = h_1\gamma_w + L\gamma_{\text{sat}} - L\gamma_w - (h_1 + \Delta h)\gamma_w$$

or

$$\sigma' = L\gamma' - (\Delta h)\gamma_w \tag{a}$$

and flow is upward through the sample. The seepage force F_s based on the soil mass area A at point X is based on the differential head producing the flow (Δh), and we obtain

$$F_s = \Delta h\gamma_w A \tag{b}$$

In Fig. 8-14b, the intergranular pressure at point X is simply

$$\sigma' = L\gamma' \qquad (c)$$

There is no seepage force since there is no flow with $\Delta h = 0$.

In Fig. 8-14c, the intergranular pressure is computed as

$$\sigma' = L\gamma_{sat} + h_1\gamma_w - 0 \qquad (d)$$

We note that there is no pore pressure since the piezometric head is level with point X. We may also observe that the direction of flow is in the process of reversing from upward to downward. After some time elapse, of course, the sample will drain and only damp soil will remain above point X. At this instant when we have just opened the drain valve to drop the piezometric head as shown, the seepage force (downward) is obtained directly from fluid statics as

$$F_s = (h_1 + L)\gamma_w A \qquad (e)$$

Comparing this seepage force to Eq. (d), we see that the hydraulic gradient produces a seepage force which increases the effective soil pressure. This computation suggests that a means to stabilizing a soil mass might be to apply a vacuum at a level in the soil. A vacuum would both create a flow gradient and produce an atmospheric differential pressure that would increase the effective pressure—at least until sufficient pore drainage occurred to "break" the vacuum.

From inspection of the computations we obtain a seepage force only when there is a differential pressure head across the flow path of interest. This differential pressure head represents the energy loss in the pore fluid from viscosity, friction, and inertia effects as the fluid moves through the voids and around the soil grains making up the irregular and rough flow channels.

When the seepage force (or excess pore pressure) is sufficiently large, the individual soil grains will be suspended in the upward flowing water—a condition visibly resembling boiling. This phenomenon can be readily observed both in the laboratory and in field locations (on the land side of levees during floods) and can be approximately evaluated for sands. With cohesive soils, interparticle attractive forces produce a condition in which a mass of soil may be lifted rather than individual grains. These conditions will be considered separately.

Quick Conditions in Sands

When the intergranular pressure is zero for sand, the soil grains just touch with no friction resistance ($\sigma_f = f\sigma_n$) available. This state is termed a *quick* or *quicksand* condition. Intergranular stresses less than zero are not possible in sand, since this would be a state of tension. At a quick condition,

$$\sigma' = 0 = \sigma_t - u$$

Thus, when the pore pressure equals the total pressure on the plane, a quick condition exists, and the pore pressure can only equal the total pressure when $\Delta h > 0$, which is a flow condition.

The pore pressure necessary to produce a quick (or quicksand) state must be from an external source which produces a sufficient hydraulic gradient or from a shock or other suddenly applied load which temporarily increases the pore pressure. The increased pore pressure will produce a fluid flow so that it is not very likely that (as in the movies) one simply stumbles into a "quicksand" pit with no external signs or warning. Further with the sand grains suspended in the upward flow of water the effective unit weight is considerably more than the 9.807 kN/m^3 of ordinary water. It would be very difficult to sink into this fluid since a person will very nearly float in ordinary water.

Liquefaction

When a fine, or medium-fine, saturated, loose sand deposit is subjected to a sudden shock the mass will temporarily liquefy. This phenomenon is termed *liquefaction*. In the situation just described, four criteria were given: a particular sand, loose state, saturation, and a sudden shock.

The shock temporarily increases the pore pressure. The total pressure is not large when the soil is loose—also, the structure is somewhat unstable. The grain size is such that the pore pressure can "float" the grains. The result is a temporary liquefaction of the sand mass until pore drainage occurs. During this time lag the very viscous sand-water mixture has little shear strength to support any structures on it and, if not confined, may flow laterally. This phenomenon has been observed to occur in several fairly recent earthquakes. It also sometimes occurs during pile driving (noted when the pile has great penetration for several of the hammer blows). It may not be limited to sands as some "mudflows" are probably saturated cohesive (silty, sandy) soils where a local failure (root tear, animal hole, etc.) produces a local shock which liquefies the soil—further movements produce a progressive liquefaction, and the mass flows down hill.

Liquefaction can be readily observed in the laboratory by building a "quicksand" tank. The tank is about 0.6 m on a side. A diffuser should be placed in the bottom and several water entries below this (the diffuser spreads water uniformly over the base). Fill the tank with about 0.45 m of fine, and medium-fine sand. Turn on the water and visually allow the soil to loosen (not boil). After several minutes of flow, gradually close the entrance valve and allow the sand to settle loosely and the surface to drain to the soil surface. Now rap the side of the tank sharply with a rubber mallet and observe the surface to cover with 1 to 2 cm of water and the sand settle. If a small weight were placed on the surface prior to producing liquefaction, it would be observed to sink into the sand. This event takes place very rapidly—only a few seconds. Rapping the side of the box a second time will produce little to no effect.

The phenomenon of liquefaction has been extensively studied in the laboratory to avoid siting hazardous (nuclear and other) plants in locations susceptible to this during an earthquake. These studies indicate that liquefaction may also occur in other soils and at less than a saturation state. The necessary conditions being that the soil undergoes a larger series of shocks termed *cyclic loading* (as in an earthquake). In these cases liquefaction depends on: soil type and state, degree of saturation, load intensity and number of cycles of load. We will consider this latter case further in Chap. 14.

Example 8-8 Given void ratios of $e = 0.5$, 0.8, and 1.0 for a sand with $G_s = 2.67$.

REQUIRED What is the critical hydraulic gradient i_c for these void ratios?

SOLUTION Noting from the just completed discussion that a quick condition is a flow condition, the *critical hydraulic gradient* will be taken as that hydraulic gradient causing a quick condition,

$$\sigma_t = u$$

For any tailwater h_i and differential head Δh sufficiently large to produce i_c, we have

$$L\gamma_{\text{sat}} + h_i\gamma_w = (L + h_i + \Delta h)\gamma_w$$

and simplifying,

$$(\gamma_{\text{sat}} - \gamma_w)L = \Delta h\gamma_w$$

$$L\gamma' = \Delta h\gamma_w$$

Rearranging,

$$\frac{\Delta h}{L} = i_c = \frac{\gamma'}{\gamma_w}$$

or the critical hydraulic gradient i_c is

$$i_c = \frac{\gamma'}{\gamma_w} \tag{8-16}$$

Next we will develop a general expression for γ' using Eq. (2-10) and the definition of saturated conditions to obtain

$$\gamma_{\text{sat}} = \frac{\gamma_w(G_s + e)}{1 + e}$$

Using the definition for effective unit weight, we obtain

$$\gamma' = \frac{\gamma_w(G_s + e)}{1 + e} - \gamma_w$$

$$\gamma' = \frac{\gamma_w(G_s - 1)}{1 + e} \tag{8-17}$$

Now from Eqs. (8-16) and (8-17) the critical hydraulic gradient is

$$i_c = \frac{G_s - 1}{1 + e} \tag{8-18}$$

and for this example,

$$e = 0.5: \quad i_c = \frac{2.67 - 1}{1 + 0.5} = 1.11$$

$$e = 0.8: \quad i_c = 0.93$$

$$e = 1.0: \quad i_c = 0.835$$

The maximum range of i_c for any sand is on the order of

e	i_c	
0.3	1.3	(No sand is likely to have a void ratio much less than this value)
1.2	0.76	(No sand is likely to have a void ratio much larger than this value)

The likely range of e for sand deposits is from about 0.45 to 0.7; thus $i_c \cong 1.0$ for any practical G_s.

$////$

Seepage Uplift Pressure on Clay Strata

The interparticle attractive forces in clays are such that it is more likely that the mass, rather than individual particles, may float or isolated channels may erode (caused by defects from worm or animal burrows and root decay). Consider Fig. 8-15 as a typical condition where seepage forces would require investigation of stability. Assume that a piezometer located on the bottom of the silty clay indicates the piezometric head shown. One would be concerned with the amount of

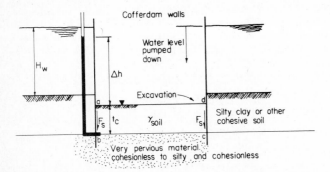

Figure 8-15 Excavation into a silty clay deposit inside a cofferdam.

excavation to produce t_c which can be safely done. Neglecting any side friction resistance τ_s, and with the excavation pumped dry as shown, $\sum F_v = 0$ gives the critical thickness t_c where the clay is on the verge of floating (safety factor $F = 1$). There are no seepage forces to consider within the clay, since the coefficient of permeability is so low that the seepage quantity is very nearly zero. With $\sum F_v = 0$ on block *abcd*, we obtain

$$t_c \gamma_{\text{soil}} = (\Delta h + t_c)\gamma_w$$

from which

$$t_c(\gamma_{\text{soil}} - \gamma_w) = \Delta h \gamma_w$$

where γ_{soil} is probably (but not necessarily) the saturated unit weight. Rearranging, we obtain

$$t_c = \frac{\Delta h \gamma_w}{\gamma_{\text{soil}} - \gamma_w} \qquad F = 1 \qquad (8\text{-}19)$$

and any $t > t_c$ gives a safety factor $F > 1$.

Example 8-9 Given the sheet pile conditions shown in Fig. E8-9.

REQUIRED

(a) What is the safety factor F against the soil in zone A rising under the seepage force from the unbalanced head Δh?

(b) What thickness of the silty clay t_c is required to maintain an $F = 1.10$?

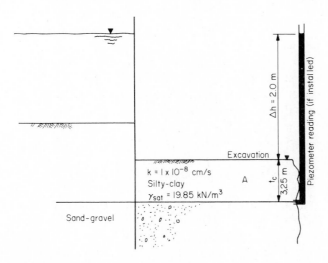

k = 1 x 10⁻⁸ cm/s
Silty-clay
γ_{sat} = 19.85 kN/m³

Sand-gravel

Excavation

A

$\Delta h = 2.0$ m

Piezometer reading (if installed)

t_c
3.25 m

Figure E8-9

SOLUTION Note that the low coefficient of permeability k will produce a full head drop across the thickness t_c until uplift occurs.

(a) The current safety factor F. The seepage force/unit area is (including position of piezometer tip)

$$F_s = (2 + 3.25)9.807(1) = 51.48 \text{ kN}$$

The weight of the block resisting this force and neglecting side friction since adjacent blocks are subject to same uplift, is

$$W_r = t_c \gamma_{sat}(1) = 3.25(19.85) = 64.51 \text{ kN}$$

The safety factor is defined as the ratio of the resisting forces to the driving forces; thus

$$F = \frac{W_r}{F_s} = \frac{64.51}{51.48} = 1.25$$

(b) Finding t_c which reduces F to 1.10. From the equation for F above rearrange to obtain

$$W_r = F(F_s)$$

$$t_c(19.85) = 1.1(51.48)$$

$$t_c = 2.85 \text{ m}$$

Note that this example illustrates the "total" stress concept to avoid a block uplift. The concepts illustrated with Fig. 8-14 and Eqs. (b) and (e) are based on the "effective" stress concept.

////

8-9 SUMMARY

This chapter has shown several reasons why soil water is one of the most important considerations in geotechnical engineering. The concept of the coefficient of permeability to relate the flow quantity through a soil mass was introduced. It was shown that this coefficient depends on a number of factors, including:

Pore size, shape, and roughness
Pore fluid (viscosity, temperature effects)
Hydraulic gradient (use as close to field value as practical)

It was noted that the horizontal k_h value may be several times larger than the vertical k_v field value and that it is difficult to obtain reliable values of laboratory k.

The analogy between fluid flow and electrical current flow was illustrated in developing equations for obtaining a mass equivalent k in stratified deposits.

The similarity of pipe and soil fluid flow was given, and it was noted that both the nominal and apparent flow velocities in soils are small but that the apparent velocity together with the irregular channel shapes and bends may well produce "turbulent" rather than "laminar" flow. With turbulent flow the smaller soil particles may become eroded to increase the flow channels (and flow rate).

Capillary effects in soil were given as the reason that soils subject to shrinkage on drying would produce compaction effects (increase density) and shrinkage cracks. It was noted that the capillary rise h_c in soils could be quite large. It was also noted that the effective pore diameter for capillary activity might be estimated by using $0.2D_{10}$.

The concept of effective pressure was used to define "quick" conditions for water flow in sand deposits. The concept of a seepage force based on the hydraulic gradient across a soil mass was used to evaluate uplift forces in cohesive soil deposits. Equation (8-16) displayed the critical hydraulic gradient being defined as that gradient producing an effective stress $= 0$. It was given as the hydraulic gradient which is equal to the effective (or buoyant) unit weight divided by the unit weight of water (γ'/γ_w). It was also shown that the unit weight ratio producing the critical hydraulic gradient is near 1 for the sands to which the concept of critical gradient applies. Note again that for cohesive soils, the seepage force which is on the verge of displacing a soil mass may be critical, depending on the resultant safety factor.

PROBLEMS

8-1 Estimate the Reynolds number for a soil under the following conditions:

$$k = 1 \times 10^{-1} \text{ cm/s}$$

$$\text{hydraulic gradient } i = 3$$

$$D_{10} = 0.15 \text{ cm (use } d = \tfrac{1}{5}D_{10})$$

$$T = 20°C \text{ (of pore water)}$$

Ans.: $N_R \cong 0.9$.

8-2 A sample of soil for a constant-head permeability test provides the following data (refer to Fig. P8-2): diameter $= 7.6$ cm; $L = 20.0$ cm; $\Delta h = 15.0$ cm; $e = 0.55$; $\gamma_{sat} = 2.00$ g/cm³; time of test duration $= 6$ min; $Q = 1200$ cm³.

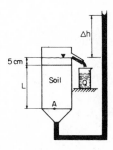

Figure P8-2

Required: sketch the test setup and compute

(a) Coefficient of permeability k, cm/s

(b) Nominal seepage velocity, cm/s

(c) Approximate actual discharge velocity

Ans.: (c) $v_a = 0.206$ cm/s.

8-3 Assume the sample of Prob. 8-2 is vertical as in the sketch.

Required: What differential head Δh will cause a quick condition at point A?

Ans.: 20 cm.

8-4 Given are the following data from a falling-head permeability test: Diameter $= 7.62$ cm.; $L = 20.3$ cm; $h_1 = 50$ cm; $h_2 = 20$ cm; time $= 82$ s; $Q = 36$ cm^3, and $T = 26°C$.

Required: sketch the test setup and find k both for test temperature and for 20°C.

Partial ans.: 5.97×10^{-3} cm/s at 28°C.

8-5 Refer to Fig. 8-7, and take:

Stratum	k, cm/s	H_i, m
1	1.5×10^{-2}	1.6
2	2×10^{-4}	2.5
3	8×10^{-2}	1.5
4	6×10^{-3}	0.9

Required: find the equivalent k_v and k_h for the soil mass.

Ans.: $k_v = 5.09 \times 10^{-4}$; $k_h = 2.31 \times 10^{-2}$ cm/s.

8-6 Referring to Example 8-3, if $\Delta h_1 = 5$ and $k_2 = 10k_1$, find Δh_2, Δh_3, and k_3.

Ans.: $\Delta h_3 = 2$.

8-7 A falling-head permeability test was performed with data as follows: $h_1 = 150$ cm; $h_2 = 50$ cm; $L = 15$ cm; $A = 100$ cm^2; time $= 10$ min; area of standpipe $a = 0.959$ cm^2; $T = 25°C$.

Required: compute the coefficient of permeability at the temperature of the test and at 20°C.

Ans.: $k_{20} = 2.3 \times 10^{-4}$ cm/s.

8-8 What is the expected height of capillary rise of the soil of curve D in Fig. 4-3?

8-9 Assuming the shrinkage limit of the soil of Fig. 7-1 $w_s = 9.4$ percent, what is the density at the shrinkage limit and how does this compare to the density values shown?

8-10 A sand soil has $G_s = 2.65$. Make a plot of critical hydraulic gradient i_c versus void ratio for the likely range of realistic void ratios.

8-11 Referring to Fig. 8-15, if $\gamma_{soil} = 19.2$ kN/m^3 and $\Delta h = 6.3$ m and t must be excavated to a thickness of 3.0 m, draw a neat working drawing of the problem conditions and compute the safety factor F; if it is less than 1.15, state what can be done to allow the excavation to safely proceed.

SEEPAGE AND FLOW NET THEORY

9-1 INTRODUCTION

The concept of using a coefficient of permeability k to describe the facility of fluid flow through a porous medium was introduced in Chap. 8. We will now use this coefficient to estimate the amount of seepage through a soil mass.

The seepage estimate is important when using wall barriers to limit inflow into excavations. Walls may be constructed using wood or precast concrete plank, steel sheeting, clay, concrete, a combination of steel sheeting and soil (called *cofferdam* cells), or other materials. The wall is seldom required to be completely impermeable to produce a dry work site. A dry site can be obtained if the reduced inflow can be pumped out as it enters.

Seepage estimates are important in dam construction—both earth and concrete types. Most dams allow seepage (or leaks) either through (earth types) the dam or through the base (both concrete and earth types) material. If the base and side material is rock, it is often grouted to fill cracks and reduce the permeability. Grout is sometimes used to reduce the permeability when the base material is soil.

Disallowing structural failures or washouts, we may define a successful dam as one in which the net retention (total inflow − outflow, seepage, and evaporation) is sufficient to satisfy the design requirements.

Several methods to reduce seepage other than grouting will be briefly introduced as well as elements of simple well hydraulics.

9-2 SEEPAGE FLOW THROUGH SOIL— THE LAPLACE EQUATION

Seepage analysis is not a very exact procedure, with computed results often not much more than estimates. This unhappy situation may be improved somewhat if steady state flow is considered. Steady state flow is obtained when the soil is fully saturated, the pressure gradient is unchanging, a constant soil mass is involved, and the flow rate is constant. We will consider only two-dimensional flow (in the XY plane), since a very large number of problems are, or can be, treated as two-dimensional flow by a transformation (or rotation) of axis. The soil is often assumed to be homogeneous but may be anisotropic ($k_x \neq k_y$).

With these ground rules established, and referring to Fig. 9-1, for any typical differential element,

$$q_{in} = q_{out}$$

At the entrance faces of the element and parallel to the X and Y axes, the hydraulic gradients are

$$i_x = \frac{\partial h}{\partial x} \qquad i_y = \frac{\partial h}{\partial y}$$

respectively, and at the exit faces the hydraulic gradients are

X axis:
$$i_x + \frac{\partial i_x}{\partial x}\,dx = \frac{\partial h}{\partial x} + \frac{\partial}{\partial x}\left(\frac{\partial h}{\partial x}\right)dx = \frac{\partial h}{\partial x} + \frac{\partial^2 h}{\partial x^2}\,dx$$

Y axis:
$$i_y + \frac{\partial i_y}{\partial y}\,dy = \frac{\partial h}{\partial y} + \frac{\partial^2 h}{\partial y^2}\,dy$$

For the entering flow quantities, using Darcy's law of $v = ki$ and a flow rate of $q = Av$, we obtain the following:

$$q_{x,\,in} = dy\,dz\,k_x \frac{\partial h}{\partial x}$$

$$q_{y,\,in} = dx\,dz\,k_y \frac{\partial h}{\partial y}$$

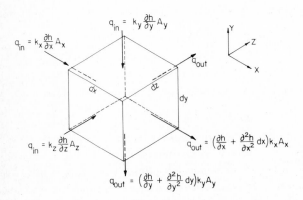

Figure 9-1 General flow of water through a soil element of dimensions dx, dy, and dz as shown.

The exit flow quantities are based on the exit hydraulic gradients, giving

$$q_{x,\,out} = dy\,dz\,k_x\left(\frac{\partial h}{\partial x} + \frac{\partial^2 h}{\partial x^2}\,dx\right)$$

$$q_{y,\,out} = dx\,dz\,k_y\left(\frac{\partial h}{\partial y} + \frac{\partial^2 h}{\partial y^2}\,dy\right)$$

Equating q_{in} to q_{out} and canceling appropriate terms, we obtain the Laplace equation for two-dimensional flow as

$$k_x\frac{\partial^2 h}{\partial x^2} + k_y\frac{\partial^2 h}{\partial y^2} = 0 \tag{9-1}$$

By direct analogy, three-dimensional flow is

$$k_x\frac{\partial^2 h}{\partial x^2} + k_y\frac{\partial^2 h}{\partial y^2} + k_z\frac{\partial^2 h}{\partial z^2} = 0 \tag{9-1a}$$

Equation (9-1) produces families of curves intersecting in the XY plane. Let us investigate whether these curves have any special properties. Figure 9-2 is a physical interpretation of two of these curves. One of the curves is the *flow path* of a particle of water from A to B, and the other curve is a line of constant pressure head h, termed an *equipotential line*. At point C the slope of curve AB is α, computed as follows:

The velocity vectors are (noting that a sign will go with the derivatives)

$$v_x = k_x\frac{\partial h}{\partial x} \qquad v_y = k_y\frac{\partial h}{\partial y}$$

By inspection of the figure,

$$\tan\alpha = \frac{v_y}{v_x} = \frac{k_y\,\partial h/\partial y}{k_x\,\partial h/\partial x} \tag{a}$$

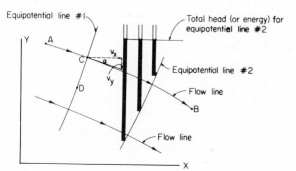

Figure 9-2 Intersection of a flow line with an equipotential line.

Now along any line of constant pressure (or head), say from C to D, h = constant and therefore the derivative $dh = 0$, but the derivative dh is

$$k_x \frac{\partial h}{\partial x} dx + k_y \frac{\partial h}{\partial y} dy = dh = 0$$

Dividing by dx and solving for dy/dx, we obtain

$$\frac{dy}{dx} = -\frac{k_x \, \partial h/\partial x}{k_y \, \partial h/\partial y}$$

which is the negative reciprocal of tan α; thus, the families of curves defined by the Laplace equation always intersect at right angles.

The lines along which the velocity vectors were considered are called stream or *flow lines*. The lines along which the total energy or head = constant are called *equipotential lines*. Note that along any equipotential line

Total head h = static (elevation) head + pressure head

+ velocity head (negligible quantity)

Both the static and pressure heads can vary with the Y coordinate but the total head is a constant. The total head can be measured in the field with a piezometer.

9-3 FLOW NETS

Figure 9-3 represents a layer of soil one unit in width confined between two impervious plates (or soil layers). The soil will be divided into sections by drawing a series of flow lines and equipotential lines which intersect at as close to right angles as it is practicable to draw. The channel formed by any two adjacent flow lines will be called a *flow path*. The difference in energy head represented by any two adjacent equipotential lines is a head loss, defined by Δh, and it is evident that the total head loss between two points, say, B and C, is $\sum \Delta h_i$. The

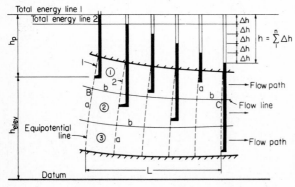

Figure 9-3 A percolating soil mass subdivided into a grid or flow net using flow lines and equipotential lines.

dimensions of the segments in Fig. 9-3 are a and b as shown on the figure. From this, the area of element 1 though which water flows is

$$A = a \times \text{width} = a(1)$$

The total area is obtained from the three flow paths ($n_f = 3$) as

$$A_{\text{total}} = n_f(a)$$

where n_f = number of flow paths and may be an integer or decimal, as 5, 7, 9, 9.2, 10.7, 4.3, etc.

The length L of soil across which a total head loss h occurs is, by inspection of Fig. 9-3.

$$L = n_d(b)$$

where n_d = number of equipotential drops (spaces, not lines) between the two boundary points where h is developed and must be an integer since $\Delta h = \text{constant}$

The flow quantity per unit of width can be computed as

$$Q = kiA = k\frac{h}{L}A = k(h)\frac{n_f}{n_d}\frac{a}{b}$$

Now if squares are drawn, then $a = b$ for any possible quadrilateral, and one obtains

$$Q = k(h)\frac{n_f}{n_d} \qquad (9\text{-}2)$$

Squares can be drawn when the soil is isotropic ($k_x = k_y$), as is evident by inspection of the Laplace equation [Eq. (9-1)]. By use of the method to be considered in Sec. 9-9, squares may be obtained when $k_x \neq k_y$.

Equation (9-2) can be used to obtain seepage quantities by using a graphical solution of Eq. (9-1). This is especially useful since an analytical solution of the Laplace equation for seepage can be obtained only for very simple cases, and even then the mathematics generally is prohibitive. Flow nets can be rapidly sketched, and they provide answers well within the precision with which the coefficient of permeability is likely to have been determined.

Approximate flow nets can be obtained for very complicated boundary conditions for which analytical solutions would be very nearly impossible, and these approximations will provide seepage quantities with an accuracy that is compatible with the determination of the coefficient of permeability.

One of the principal drawbacks of evaluating seepage quantity is that k_x is very seldom equal to k_y. Laboratory tests often do not reveal this situation—it usually requires field testing to obtain reliable values for the two values.

9-4 FLOW NETS FOR SHEET-PILE CUT-OFF WALLS

Using the principles just discussed, a flow net can be drawn for a sheet-pile cut-off wall as might be used in constructing a pier along a river or lake front or similar (refer to Fig. E9-1). In Fig. E9-1, one side of the wall has been excavated and pumped to produce the total head difference h shown. We may assume that over the interval from point X to Y (flow length L) the head drop is a series of equal drops Δh such that $\sum \Delta h = h$. Note that the flow length L is a minimum along the faces of the sheet pile and larger in all other areas within the flow zone. Water particles traveling from X to Y will travel a series of smooth flow paths. If we draw a smooth line through the locus of points producing any Δh, we have an equipotential line which intersects the flow paths at right angles and forms squares if $k_x = k_y$ according to Eq. (9-1). We note that a *singular* point exists at the sheet pile tip since it is impossible to draw a square here.

The flow net is drawn by trial and visually adjusted until a mesh is produced that generally intersects at right angles and makes "squares" as shown. Squares and other features will be taken up in following sections.

Several important flow and stability computations that can be made from flow nets will be illustrated in the following four examples using the sheet pile cut-off wall.

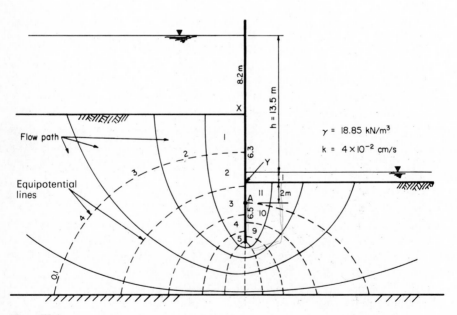

Figure E9-1

Example 9-1 Given the flow net shown for a sheet-pile cutoff wall. Assume the sheet-pile cutoff is impervious (although in practice it seldom is).

REQUIRED Find the effective pressure at point A (2 m below the downstream ground surface) and compute the seepage quantity per meter of wall width per day based on $k = 4 \times 10^{-2}$ cm/s.

SOLUTION

Step 1 Draw the sheet-pile wall-soil system to scale and sketch an acceptable flow net, as shown in Fig. E9-1.

Step 2 Scale the distance from the ground surface to point A, which is the next-to-last equipotential line downstream. The downstream ground line is the last equipotential line. For the scale of this plot we obtain 2 m as shown.

Step 3 Find the remaining seepage head at point A. Since Δh_i is defined as constant between any two consecutive equipotential lines and the total head loss for 11 drops is 13.5 m, by proportion the remaining seepage head at A is

$$\Delta h = \tfrac{1}{11}(13.5) = 1.23 \text{ m}$$

Step 4 The total static (as shown by a piezometer) pressure head at A is

$$h = 1.23 + 1.0 + 2.0 = 4.23 \text{ m}$$

See Fig. E9-4 of Example 9-4 if this computation is not clear. Total vertical pressure due to the weight of material overlying a plane through point A is

$$\sigma_{\text{total}} = 1.0(9.807) + 18.85(2) = 47.51 \text{ kPa}$$

Step 5 Compute the effective pressure as

$$\sigma' = \sigma_{\text{total}} - u$$

$$\sigma' = 47.51 - 4.23(9.807) = 47.51 - 41.48 = 6.03 \text{ kPa}$$

Since $\sigma' > 0$, point A is not "quick," but 6.03 kPa is not large enough to be very confidence-inspiring.

Step 6 Compute the seepage quantity. Use Eq. (9-2) and count $n_f = 4.1$, $n_d = 11$; 1 day = 86,400 s, and 1 m = 100 cm. Substituting into Eq. (9-2),

$$Q = k(h) \frac{n_f}{n_d} \times \text{width} = (4 \times 10^{-2})(13.5)\frac{4.1}{11}\frac{86,400}{100}$$

$$= 173.9 \text{ m}^3/\text{day/m of wall width}$$

////

Example 9-2 Given the flow net of Example 9-1 (Fig. E9-1).

REQUIRED What depth of water outside the sheet-pile wall will cause instability at point A, i.e., a "quick" condition?

SOLUTION

Step 1 Equate the vertical pressures at point A to 0 as follows:

σ_{total} downward = 47.51 kPa   from Step 4 in Example 9-1

Finding the piezometric head of water at A which gives 47.51 kPa, we have $h' = 47.51/9.807 = 4.84$ m

Since 3.0 m of water is caused by tailwater (1 m) and location of A (2 m), it follows that

$$\Delta h' = 4.84 - 3.0 = 1.84 \text{ m}$$

Step 2 Convert $h = 1.84$ m to the equivalent depth of water. Since n_f is unchanged, we can back compute the h' to produce 1.84 m as

$$h' = 1.84 \frac{11}{1} = 20.24 \text{ m}$$

Since there is $6.3 - 1 = 5.3$ m of free excavation depth, the new water depth is

$$20.24 - 5.3 = 14.94 \text{ m} \qquad \text{(versus 8.2 m previously)}$$

////

Example 9-3 Given the flow net in Example 9-1 (Fig. E9-1).

REQUIRED What is the effective pressure at A if the water level stabilizes on both sides of the wall at 8.2 m, as shown in Fig. E9-3?

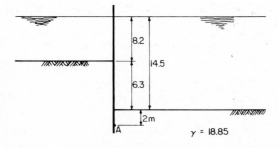

Figure E9-3

SOLUTION The total pressure on A is

$$2(18.85) + 14.5(9.807) = 179.90 \text{ kPa}$$

$$u = (14.5 + 2)9.807 = \overline{161.81}$$

$$\text{Effective pressure } \sigma' = \overline{18.09 \text{ kPa}}$$

$////$

Example 9-4 Given the problem conditions and the flow net in Example 9-1.

REQUIRED What surcharge W, in kilopascals, will provide a safety factor $F = 2.0$ at point A for the unbalanced water head of 13.5 m?

SOLUTION When $F = 1$, Eq. (8-19) rearranged applies, giving $\sum \sigma_v = 0$. To obtain this condition, refer to Fig. E9-4 and obtain the following:

$$W + 2(18.85) + 1.0(9.807) - (2 + 1.0 + 1.23)9.807 = 0$$

Solving, we obtain for $F = 1$

$$W = -6.02 \text{ kPa}$$

For $F = 2$, we must have

$$\sigma_{\text{total(down)}} = 2\sigma_{\text{up(pore pressure)}}$$

The reader should verify that this is correct, not the alternative, $2\sigma_{\text{total}} = \sigma_{\text{up}}$.

$$\sigma_{\text{total(down)}} = W + 2(18.85) + 1(9.807)$$

$$= W + 47.51$$

$$2\sigma_{\text{up}} = 2(4.23)9.807 = 82.97 \text{ kPa}$$

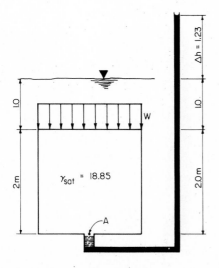

Figure E9-4

Equating, we obtain for $F = 2$

$$W + 47.51 = 82.97$$

$$W = 35.46 \text{ kPa}$$

The reader can readily verify that this value of W is correct.

////

9-5 FLOW NETS FOR EARTH DAMS

Flow through earth dams was one of the first applications of flow net theory. Consider the idealized earth dam in Fig. 9-4 with a *phreatic*, or *saturation*, line representing the upper flow boundary. A wet capillary zone exists above the phreatic line.

Note that this idealized dam section may be the entire dam or only the relatively impermeable clay core used as the primary water barrier. If the shell surrounding the clay core has a k larger than the core (say, $k_{\text{shell}}/k_{\text{core}} > 100$), the flow boundary and phreatic surface will be about as shown as the energy drop in the shell will be negligible relative to the core.

Phreatic Line Entrance Geometry

Model studies have shown that the apparent origin of the phreatic surface is upstream from the dam face a distance of approximately $0.3S$ as shown in Fig. 9-4. The distance S depends on the slope of the upstream face of the dam. Since the upstream face of the dam is an equipotential line, the actual phreatic line

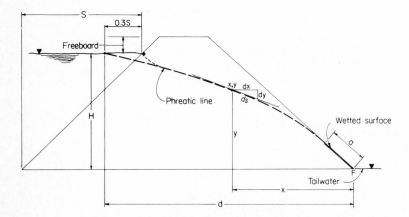

Figure 9-4 Developing the phreatic line for an idealized dam cross section. Note that point F is always taken at the intersection of the tailwater and the downstream face. In the case shown here, the tailwater is coincident with the "original" ground line.

enters and intersects the face at right angles [according to Eq. (9-1)] and at the intersection of the pool elevation with the face. The phreatic line is then faired into a smooth intersection with the main part of the phreatic line as shown.

The Phreatic Line

At a point of coordinates (x, y) measured from the downstream toe as shown we have

$$i = \frac{dy}{ds} \qquad v = ki = k\frac{dy}{ds} \qquad A = y \times \text{width}$$

For cases of small β angles (generally $\beta < 30°$), one may replace dy/ds with $i = dy/dx$. Making this substitution and solving for the flow rate per unit width, we obtain

$$q = Av = k\frac{dy}{dx}(y)(1) \qquad\qquad (a)$$

Separating variables, we have the following differential equation:

$$q(dx) = k(y)\,dy$$

This equation assumes that either $k = k_x = k_y$ or, if $k_x \neq k_y$, a linear transformation, as in Sec. 9-9, has been used. Integrating this differential equation, one obtains

$$q(x) = k\frac{y^2}{2} + C \qquad\qquad (b)$$

At $x = d$, $y = H$ and we obtain

$$C = q(d) - k\frac{H^2}{2}$$

and substituting this value of C into Eq. (b), we obtain

$$q(x - d) = \frac{k}{2}(y^2 - H^2) \qquad\qquad (9\text{-}3)$$

Equation (9-3) shows that the equation of the phreatic (or saturation) line is a parabola. A discontinuity in the parabola occurs at the downstream exit producing the wet zone a shown in Fig. 9-4.

Angle of Exit of the Phreatic Line at "a"

The angle of exit of the phreatic line can be developed by referring to Fig. 9-5 as follows—at point 1 along the downstream face:

$$\frac{\Delta h}{c} = \sin(\beta - \alpha) \qquad\qquad (c)$$

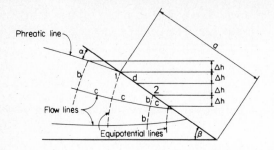

Figure 9-5 Exit of phreatic line at downstream face of an earth dam.

Since we will always use squares, inspection of the figure gives $b = c$; thus

$$\Delta h = b \sin (\beta - \alpha) \qquad (d)$$

Also from the figure,

$$\frac{\Delta h}{d} = \sin \beta \qquad (e)$$

At point 2 we have

$$\frac{c}{d} = \cos \alpha \qquad (f)$$

Now dividing (e) by (f), we obtain

$$\frac{\Delta h}{c} = \frac{\sin \beta}{\cos \alpha} \qquad (g)$$

Finally, equating Eqs. (c) and (g),

$$\sin (\beta - \alpha) = \frac{\sin \beta}{\cos \alpha} \qquad (h)$$

This equality can only be obtained if $\alpha = 0$, since $\cos \alpha = \cos 0 = 1$ and $\sin (\beta - 0) = \sin \beta$. With the angle of exit of the phreatic surface $= 0$, the exit is parallel and coincident with the downstream face of the dam at the top of the wet zone a of Fig. 9-6.

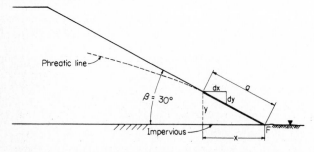

Figure 9-6 Computation of the wetted downstream face distance a and the direct computation of seepage quantity for small β angles.

Computing Length of Wet Zone "a"

The downstream wetted surface a in Fig. 9-6 can be computed as follows. Rearrange Eq. (9-3) to obtain

$$q = \frac{k}{2} \frac{y^2 - H^2}{x - d} \qquad (i)$$

The flow quantity q can also be obtained as follows:

$$i = \frac{dy}{ds} \cong \frac{dy}{dx} \qquad \text{for } \beta \leq 30°$$

Define $\tan \beta = dy/dx$ at the exit of the phreatic line. From Fig. 9-6, $y = a \sin \beta$, and since fluid flow $q = Av$, we have (per unit of width):

$$q = kiA = k(\tan \beta)(a \sin \beta)$$

or, rearranging,

$$q = k(a) \sin \beta \tan \beta \qquad (9\text{-}4)$$

Now if we equate Eq. (9-4) and Eq. (i) and substitute for $y = a \sin \beta$ and $x = a \cos \beta$ in Eq. (i) and solve the resulting quadratic, we obtain

$$a = \frac{d}{\cos \beta} - \sqrt{\frac{d^2}{\cos^2 \beta} - \frac{H^2}{\sin^2 \beta}} \qquad \text{for } \beta \leq 30° \qquad (9\text{-}5)$$

where all terms are identified on Fig. 9-4. This equation allows a direct computation of the wet distance a for a graphical solution or for an analytical solution using Eq. (9-4) with the limitations on β indicated.

9-6 METHODS FOR OBTAINING THE PHREATIC LINE FOR EARTH DAMS

The phreatic surface should be considered separately for cases of $\beta < 30°$ and for larger downstream slope angles.

Case I: $\beta \leq 30°$

Draw the earth dam to scale as in Fig. 9-7. Calculate the wetted downstream face using Eq. (9-5), and lay off as shown. Also locate the apparent origin of the parabola at a distance of $0.3S$ upstream, as shown in the figure.

Since the phreatic line is a parabola, we may use the simplest form of the equation,

$$y = Kx^2$$

and at x_0, $y = y_0$, which gives $K = y_0/x_0^2$.

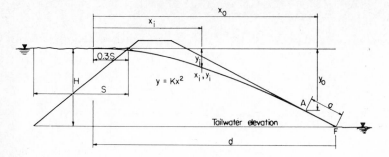

Figure 9-7 Locating the phreatic line when $\beta \le 30°$.

Scale several distances x_i, compute the corresponding offsets y_i, and, using a french curve, draw a smooth curve through the locus of points thus obtained. Note that the parabola is tangent to the downstream face at the top of the wetted face (point A) and is faired into a perpendicular to the upstream face at the water line. This is necessary, since the upstream face is an equipotential line and the phreatic line is a flow line.

Case II: $\beta > 30°$

An alternative means of establishing the phreatic line may be required when $\beta > 30°$. The reason is that the method of Case I is only valid as long as $dy/ds \cong dy/dx$, with the break point occurring at approximately 30°. How much larger than 30° one may continue to use Case I is an exercise in engineering judgment, taking into account the order of magnitude of k and how accurately k can be determined.

To obtain the phreatic line when it is necessary to account for $dy/ds \ne dy/dx$, it is necessary to obtain the parameter distance p of a parabola as measured from the focus F (see Fig. 9-8 and any text on analytical geometry). Once the half-parameter distance p is established, this locates a point on the phreatic line. An arc from F as shown on Fig. 9-8 locates a second point on the parabola along a perpendicular from F to the arc. Since the upstream starting point is known, this

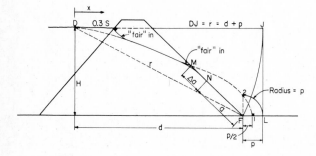

Figure 9-8 Locating the phreatic surface when $\beta > 30°$.

may give sufficient points to draw the parabolic phreatic line. If not, other offsets from the horizontal line DJ may be computed using

$$y = \frac{y_0}{x_0^2} x^2 = \frac{H}{r^2} x^2$$

as for Case I.

The procedure for obtaining the distance p for the parabola is as follows:

1. Let
   p = half-parameter of any parabola
   p = perpendicular distance from F both to the parabolic curve at point 2 and to the directrix (point K) of Fig. 9-8
   F = focus of the parabola, the intersection of the tailwater elevation and the downstream slope
2. From the dimensions d, H, and r of Fig. 9-8, one can compute

$$p + d = \sqrt{H^2 + d^2} = r$$

and rearranging, we obtain

$$p = \sqrt{H^2 + d^2} - d \tag{9-6}$$

3. Next let
   M = point where the theoretical parabola outcrops the downstream face using the p values computed from Eq. (9-6)
   N = point where the actual seepage line outcrops the downstream face a distance of Δa from M (to be found)
   FN = the wetted distance a (also to be found)

It has been found (Casagrande, 1937) that the ratio $\Delta a/(a + \Delta a)$ (refer to Fig. 9-8) is a special scalar which we may call ψ. From theoretical work cited by Casagrande, this scalar is related to the downstream slope angle β as follows:

β	ψ
30°	0.375
60	0.320
90	0.260
120	0.185
150	0.105
180	0.000

Steps in computing the seepage quantity for $\beta > 30°$ are as follows:

1. Compute p using Eq. (9-6), and note that d includes the distance of $0.3S$. Alternatively, initially locate the point D on the upstream surface, then with a compass scribe an arc to the horizontal DJ using DF as a radius, as shown in

Fig. 9-8. Drop a vertical JK as shown, measure p, and locate the two points (1 and 2) from F using the radius distances of p and $p/2$, also as shown.

2. Using any accepted procedure, construct the theoretical parabola from D through M to point 1.
3. Locate and scale the distance $FM = a + \Delta a$.
4. Compute Δa using the value of ψ from the table of values depending on the value of β. Interpolate for intermediate values of β. Then

$$\Delta a = FM(\psi)$$

5. Lay off Δa from point M and fair in the parabola so that it is *tangent* to the downstream face at N for all $\beta \le 90°$ and is *vertical* for all $\beta > 90°$.

Other Cases When $\beta > 30°$

The odd downstream face slopes shown in Fig. 9-9 are attempts to depress the seepage outcrop to the interior of the dam so that the downstream face is dry

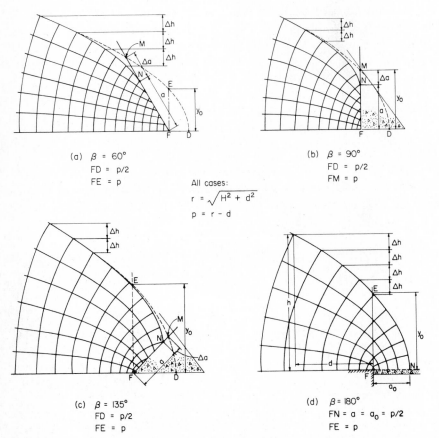

(a) $\beta = 60°$
FD = p/2
FE = p

All cases:
$r = \sqrt{H^2 + d^2}$
$p = r - d$

(b) $\beta = 90°$
FD = p/2
FM = p

(c) $\beta = 135°$
FD = p/2
FE = p

(d) $\beta = 180°$
FN = a = a_0 = p/2
FE = p

Figure 9-9 Downstream exit faces. In all these cases, the exit is into granular (or rock toe or underdrain) material with $k_2 \gg k_{dam}$. *(Casagrande, 1937.)*

(more confidence-inspiring), to control seepage pressures, and to reduce the possibility of piping (Sec. 9-12 will consider piping in some detail).

9-7 FLOW NET CONSTRUCTION

Figure 9-10 illustrates several flow nets for indicated flow geometry. The general shape of a flow net will be determined by the boundary conditions in most cases. Exceptions are at singular points, where the flow net may determine the boundary conditions. General requirements for the flow net boundary conditions are:

1. The flow net intersects with equipotential lines at right angles, except at singular points where the velocity is zero (stagnation) or $v \to \infty$, as occurs at corners or tips of impervious cutoff walls.
2. From the definition of q, Δh must be the same amount for each equipotential line.
3. The pressure head at the intersection of the phreatic line and any equipotential line is zero.
4. All flow paths must have continuity, so that $q_{in} = q_{out}$.

In constructing flow nets, the following guidelines may prove helpful.

1. Always draw squares which intersect at right angles as nearly as it is possible to draw them (exceptions are at singular points, such as corners). When conditions exist which require the use of rectangles, refer to Sec. 9-9 for modifications to allow the use of squares.
2. Use as few flow paths (and resulting equipotential drops) as possible while maintaining square sections. Generally four to six paths will be sufficient, used with a modest plotting scale so that the drawing does not tend to be more precise than soil data can justify.
3. Check the accuracy of squares by adding selected lines and observing if they subdivide large squares into recognizable smaller ones.
4. Use a pair of dividers to measure the square dimensions as illustrated in Fig. 9-11.
5. Always watch the appearance of the entire flow net. Do not make fine detail adjustments until the entire flow net is approximately correct.
6. Take advantage of symmetry where possible. Symmetrical geometry may result in portions of the net being exact squares; if so, develop this area first, then extend the net into the adjacent zones.
7. Use smooth transitions around corners and reentrant corners. Use gradual transitions from small to larger "squares."
8. A discharge face in contact with air is neither a flow line nor an equipotential line; however, such a boundary must fulfill the same conditions of equal drops in potential where the equipotential lines intersect it (a concept used in Fig. 9-5).

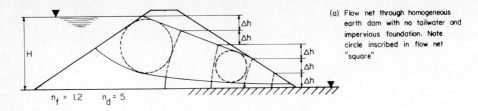

(a) Flow net through homogeneous earth dam with no tailwater and impervious foundation. Note circle inscribed in flow net "square"

$n_f = 1.2$ $n_d = 5$

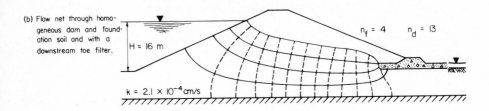

(b) Flow net through homogeneous dam and foundation soil and with a downstream toe filter.

$H = 16$ m

$n_f = 4$ $n_d = 13$

$k = 2.1 \times 10^{-4}$ cm/s

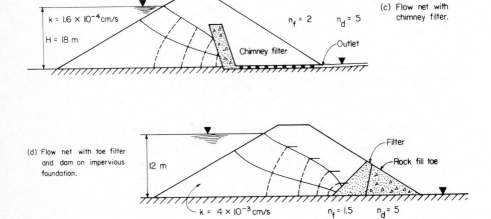

(c) Flow net with chimney filter.

$k = 1.6 \times 10^{-4}$ cm/s

$H = 18$ m

$n_f = 2$ $n_d = 5$

Chimney filter   Outlet

(d) Flow net with toe filter and dam on impervious foundation.

12 m

Filter

Rock fill toe

$k = 4 \times 10^{-3}$ cm/s $n_f = 1.5$ $n_d = 5$

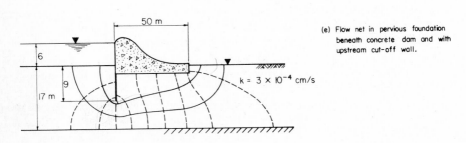

(e) Flow net in pervious foundation beneath concrete dam and with upstream cut-off wall.

50 m

6

9

17 m

$k = 3 \times 10^{-4}$ cm/s

Figure 9-10 Typical flow nets for conditions shown.

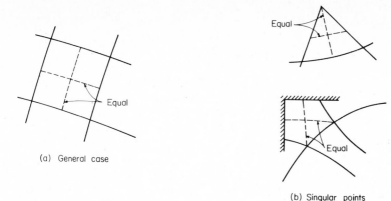

(a) General case

(b) Singular points

Figure 9-11 Definition of a "square" for constructing flow nets.

9. To get very good results a roughed-in flow net is adequate. While rough nets are usually not acceptable for academic work, the student should be aware that the ratio of n_f/n_d will be relatively unchanged in a very precise, right-angle-intersection flow net from a very rapidly drawn and uncorrected flow net. Keep in mind that the coefficient of permeability may be correct to the order of magnitude (power of 10), which is far less precise than the n_f/n_d ratio.

Example 9-5 Given the earth dam cross sections shown in Fig. E9-5. The dam section of Fig. E9-5b has a toe filter so that the phreatic line is entirely inside the dam.

REQUIRED Compute the expected seepage in cubic meters per second per meter of dam width.

SOLUTION Superimpose the flow nets directly on the profile section after drawing to scale to save text space and obtain the flow nets shown.

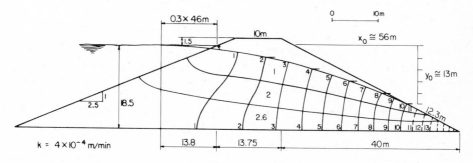

$k = 4 \times 10^{-4}$ m/min

Figure E9-5a

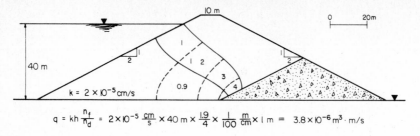

$$q = kh\frac{n_f}{n_d} = 2\times10^{-5}\ \frac{cm}{s} \times 40\ m \times \frac{19}{4} \times \frac{1}{100}\ \frac{m}{cm} \times 1\ m = 3.8\times10^{-6}\ m^3 \cdot m/s$$

Figure E9-5b

For Fig. E9-5a the flow rate is

$$q = (4 \times 10^{-4})(18.5)(1)\left(\frac{2.6}{15}\right) = 12.8 \times 10^{-4}\ m^3/min/m\ of\ width$$

The number of drops, 15, is averaged from the 2.6 flow paths. Alternatively, each flow path in turn could be computed and the results summed to obtain the total flow rate q.

////

9-8 DIRECT COMPUTATION OF SEEPAGE QUANTITY

Equation (9-4) can be used for a direct computation of seepage quantity and is generally recommended to obtain a rapid solution when $\beta \le 30°$. When $\beta > 30°$, the use of Eq. (9-4) may give satisfactory estimates of the seepage quantity in many cases.

Example 9-6 Compute the estimated seepage quantity for the dam shown in Example 9-5 (Fig. E9-5a).

SOLUTION Obtain dimensions and data from Fig. E9-5a as follows:

$$H = 18.5\ m \qquad d = 40 + 13.8 + 46(0.3) = 67.6\ m$$

$$k = 4 \times 10^{-4}\ m/min \qquad \beta = \tan^{-1} \frac{20}{40} = 26.6°$$

Step 1 From Eq. (9-5) compute the wet face distance a as

$$a = \frac{67.6}{\cos 26.6} - \sqrt{\frac{67.6^2}{\cos^2 26.6} - \frac{18.5^2}{\sin^2 26.6}}$$

$$= 75.6 - \sqrt{5715.7 - 1707.1}$$

$$= 75.6 - \sqrt{4008.6}$$

$$= 75.6 - 63.3 = 12.3\ m$$

Step 2 From Eq. (9-4),

$$q = k(a) \sin \beta \tan \beta$$
$$= (4 \times 10^{-4})(12.3)(\sin 26.6)(\tan 26.6)$$
$$= 11.03 \times 10^{-4} \text{ m}^3 \cdot \text{min/m of width}$$

This value compares very well to 12.8×10^{-4} obtained from using flow nets in Example 9-5.

////

9-9 THE FLOW NET WHEN $k_x \neq k_y$

When the coefficient of permeability in the horizontal direction (usually k_x) is not the same as in the vertical direction, which occurs most of the time as a result of the process of stratification resulting from sedimentation, the Laplace equation

$$k_x \frac{\partial^2 h}{\partial x^2} + k_y \frac{\partial^2 h}{\partial y^2} = 0$$

indicates that rectangles may result, the side ratio being $\sqrt{k_x/k_y}$. Therefore, let

$$u^2 = \frac{k_x}{k_y}$$

from which $k_x = u^2 k_y$, and substituting this into the Laplace equation obtain

$$u^2 k_y \frac{\partial^2 h}{\partial x^2} + k_y \frac{\partial^2 h}{\partial y^2} = 0$$

which can be rewritten after dividing by k_y as

$$\frac{\partial^2 h}{\partial (x/u)^2} + \frac{\partial^2 h}{\partial y^2} = 0 \qquad (9-7)$$

A similar approach using

$$v^2 = \frac{k_y}{k_x}$$

can also be made.

From Eq. (9-7) above, we have

$$\frac{x}{u} = \frac{x}{\sqrt{k_x/k_y}}$$

but the value of x/u is simply a transformed x distance we may call x'. Once the transformed distance x' is used, Eq. (9-7) indicates that the coefficient of permeability is no longer a factor. Since this is so, it follows that if we use transformed distances when $k_x \neq k_y$, then the resulting flow net is the same as when

$k_x = k_y$, and we always use squares for that case. The use of "squares" results in a net that is easy to check with a pair of dividers and easier to draw since the sides are the same.

Example 9-7 Given $k_x = 12.1 \times 10^{-3}$ cm/s; $k_y = 3.0 \times 10^{-3}$ cm/s.

REQUIRED Set up a table of several transformed x' distances for drawing the outlines of an earth dam.

SOLUTION Let

$$u^2 = \frac{k_x}{k_y} = \frac{12.1 \times 10^{-3}}{3 \times 10^{-3}} = 4.0$$

$$u = \sqrt{\frac{k_x}{k_y}} = \sqrt{4} = 2.0$$

$$x' = \frac{x}{u} = \frac{x}{2}$$

For:

x, m	x', m	y, m
100	50	25
50	25	10
25	12.5	5

These distances would be plotted to obtain the shape of the dam and various parts of the dam so that a flow net could be sketched using squares.

If this transformation had not been made, the resulting flow net would be rectangular with dimension $b = u(a) = 2a$. This would be hard to check using dividers, as they would have to be reset each time for each flow net square.

////

Computation of Seepage Quantity When $k_x \neq k_y$

When the flow net is completed using the transformed dimensions for x, the flow rate is computed as

$$q = k'(h) \frac{n_f}{n_d}$$

where k' = the effective coefficient of permeability, obtained as follows (refer to Fig. 9-12).

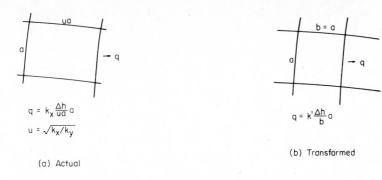

(a) Actual

(b) Transformed

Figure 9-12 Obtaining the effective coefficient of permeability when $k_x \neq k_y$.

Since the quantities of flow in Fig. 9-12a and 9-12b are the same, we may equate the flow quantities shown on the figure to obtain

$$k'(\Delta h)a/b = k_x(\Delta h) \frac{a}{a\sqrt{k_x/k_y}}$$

Rationalizing the denominator and simplifying, we obtain the equivalent coefficient of permeability as

$$k' = \sqrt{k_x k_y} \tag{9-8}$$

9-10 CONTROL OF SEEPAGE THROUGH DAMS

From Eq. (9-2), the flow net equation for seepage, it is seen that the quantity of seepage depends on

1. The coefficient of permeability, k
2. The differential head h across the flow path
3. The length of the flow path (or number of drops, n_d)
4. Number of flow paths (or area)

We may use grout to lower k, but other than rarely being able to use excavation and soil replacement in the critical areas, other means to control seepage are necessary. The use of a core or cut-off wall with a low coefficient of permeability are the principal control measures available that are practical.

The differential head may be controlled in some instances, but this is usually fixed for the project.

The length of the flow path can be increased using a clay blanket as shown in Fig. 9-13b. Upstream and downstream cut-off walls may be used as in Figs. 9-13c and 9-13d. Generally the downstream cut-off wall will be more efficient in reducing seepage flow but the upstream wall will reduce the seepage uplift pressure (refer to Fig. 9-14). Either an upstream or downstream clay blanket can be

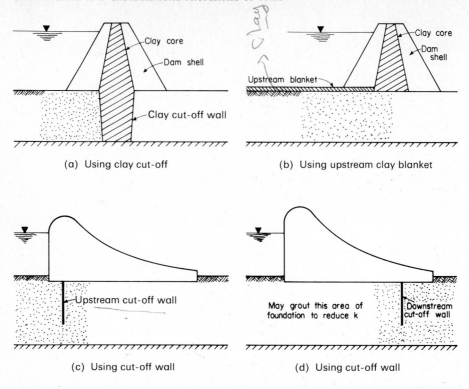

(a) Using clay cut-off

(b) Using upstream clay blanket

(c) Using cut-off wall

(d) Using cut-off wall

Figure 9-13 Several methods of controlling seepage. Note that foundation grouting may be used with parts *b* to *d*. Foundation grouting may be used in lieu of the cut-off walls shown in parts *c* and *d*.

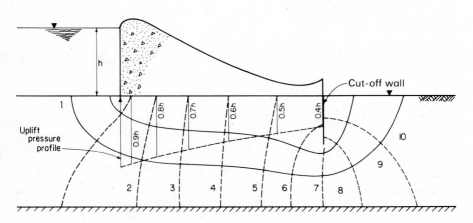

Figure 9-14 Flow net used to obtain seepage uplift forces beneath a concrete dam with a sheet-pile cut-off wall on the downstream end.

used but the upstream blanket will be more efficient in controlling seepage uplift since the pool elevation will have a canceling effect. As a matter of fact, filter drains are often used downstream of clay cut-offs or cores to reduce the seepage uplift as a side benefit.

Note that where clay exists in the reservoir pool area (a natural blanket) and is to be used for the dam core, care should be taken not to accidentally excavate through into any more pervious lower soil. After core excavation and prior to filling the pool, the pool area should be rolled and carefully inspected for any cuts (or animal holes) into the lower soil. Any such areas should be carefully covered and compacted. If holes are left the dam may leak so much that the elevation cannot be obtained. Leaks of this type usually get worse with time rather than silting over.

9-11 SUDDEN DRAWDOWN AND SEEPAGE FORCES

Figure 9-15 illustrates a condition where rapid drawdown occurs behind a dam or a water reservoir. Note that the loss of the stabilizing effect of water on the upstream face may create a temporary "quick" condition. The seepage forces are analyzed for stability as in Sec. 8-8.

Figure 9-14 illustrates the seepage force concept used to compute the uplift force acting on a concrete dam using pressures obtained from the equipotential lines of a flow net.

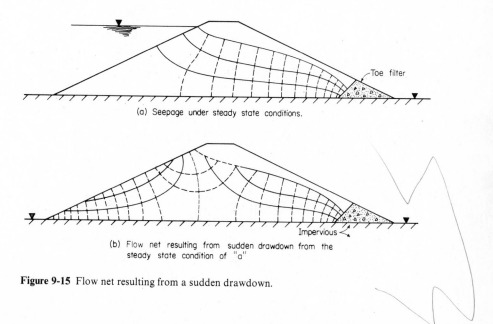

(a) Seepage under steady state conditions.

(b) Flow net resulting from sudden drawdown from the steady state condition of "a"

Figure 9-15 Flow net resulting from a sudden drawdown.

9-12 PIPING AND CONTROL OF PIPING

Soil grains will be dislodged and eroded away when the hydraulic gradient at the exit face of a percolating soil mass is large enough to overcome the interparticle bonding forces and/or cementing. This is a progressive phenomenon, with the smaller particles eroding first. As erosion of the smaller particles takes place, the resistance to flow decreases, with a corresponding increase in the hydraulic gradient. With the increase in hydraulic gradient, larger particles can now be eroded, and the process accelerates and tends to the formation of an underground flow channel or "pipe" which moves upstream, as in Fig. 9-16. Flow increases a tremendous amount when the pipe reaches the impounded water, rapidly enlarging the pipe cavity, and can cause a dam failure. The failure usually occurs through collapse of the pipe roof and overtopping in the channel thus formed, which, being of soil, rapidly erodes as a result of the large water velocity. This mode of failure may conceal the fact that piping was the real cause of the disaster.

Piping is a *progressive* failure, and monitoring of a site to observe excessive seepage flow will usually detect it in time for preventive measures. Piping can occur at a considerable time after a water retention structure has been built if an event occurs which increases the exit gradient sufficiently to cause soil erosion at a localized point. This may happen, for example, from

1. Cavities toward upstream formed when roots decay
2. Animal burrows (those of muskrats, gophers, etc., including snake burrows)
3. Excavations downstream, either pits or in some cases simply skinning the top surface to some critical depth

Piping often occurs during floods on the land side of protective levees. Here the direction of flow is initially vertical, as the increased flood level raises the hydraulic grade line and water in the underlying pervious soil attempts to seek piezometric level. Root holes, animal burrows, etc., can provide a less resistant localized flow path, and water seeps out—slowly and carrying fine sand and silt and clay particles at first, then more rapidly until it "boils." These local pipes can be halted at relatively early stages by placing a sandbag ring around the "sand boil" to develop sufficient water head that the back pressure equalizes the hydraulic gradient. Untreated "sand boils" can result in the levee collapsing into

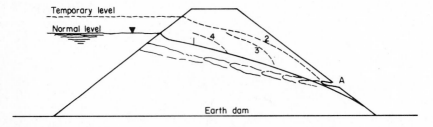

Figure 9-16 Formation of a "piping" condition.

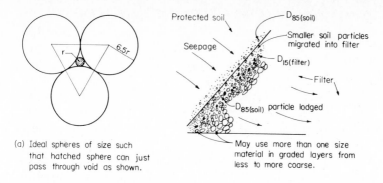

(a) Ideal spheres of size such
that hatched sphere can just
pass through void as shown.

Figure 9-17 Soil-filter concept: (*a*) relative diameters of ideal spheres so that a particle can just pass through filter spheres; (*b*) interface of exit face and filter.

the underground pipe cavity and being overtopped with flood water. Figure 1-2*b* illustrates a case where sheet piling was used in a futile attempt to halt the piping. The Teton dam (Fig. 1-2*d*) failure was probably caused by piping in the discontinuities between the dam and the rock abutment, which were apparently improperly grouted.

Piping can be controlled and/or eliminated in earth dams by using a *filter* or *graded filter* on the exit side of the impermeable element (core or cut-off wall). The filter should be graded to ensure that the protected soil cannot wash through the filter material (see Fig. 9-17).

Experience, supported by tests made by Bertram (1940), indicates that if the following filter criterion is met, piping will be adequately controlled:

$$\frac{D_{15(\text{filter})}}{D_{85(\text{protected soil})}} < 4 \text{ to } 5$$

This criterion states that the *piping ratio* of the D_{15} size of the filter soil is not more than four or five times the D_{85} size of the protected soil. A further criterion is:

$$\frac{D_{15(\text{filter})}}{D_{15(\text{protected soil})}} > 4 \text{ to } 5$$

This criterion states that the D_{15} size of the filter soil should be more than four or five times the D_{15} size of the protected soil. The U.S. Corps of Engineers also recommends that the ratio of the D_{50} filter and protected soil sizes should be as follows:

$$\frac{D_{50(\text{filter})}}{D_{50(\text{protected soil})}} \leq 25$$

It follows, also, from an earlier statement concerning interparticle forces and cementing, that the protected soil should be cohesive.

9-13 OTHER METHODS FOR SEEPAGE QUANTITIES

Flow Net Alternatives

Several methods of obtaining flow nets are available, including the building of sand models with dye injected at select points to trace the flow paths. The most popular and rapid method is to use an electrical analogy. This method requires the use of a controlled voltage electrical supply, silver paint or other highly conducting material for electrodes, and a probe. The shape of the percolating soil mass is cut out of some electrically conducting material, and electrodes are placed to appropriately simulate the differential of water head. The probe is used to find a locus of points corresponding to a constant voltage loss (value), which is by definition an equipotential line. The exact details and equipment needed are outlined in Bowles (1978) for the interested reader.

Special Problems

Radial flow nets may be constructed to model three-dimensional seepage problems such as for excavation dewatering. Details for constructing a radial flow net may be found in Taylor (1948) or Mansur and Kaufman (1962).

When the pervious soils are stratified, the flow net requires modification to account for flow crossing the boundary similar to light refraction. This problem is taken up in some detail by Casagrande (1937) and Cedergren (1977).

Analytical Models

The finite element of the elastic continuum can be used to compute pressures at the nodes of a finite element mesh. An elementary treatment of this method is in Desai (1979) and in further depth in Zienkiewicz (1977). The finite difference method may also be used but has the disadvantage over the finite element method of requiring a uniform grid and there is considerable difficulty in writing finite difference equations at boundary nodes. Some details of the finite difference method are given by McNown et al. (1955), and a short computer program limited to a rectangular flow area (but illustrates the programming methodology) is given in Dunn et al. (1980).

Harr (1962) presents a Russian (translation) method for analytically computing seepage for confined flow called the *method of fragments*. Confined flow would be obtained from a flow configuration as in Fig. 9-14.

In most cases a roughly sketched flow net will be adequate. This is likely to be at least as accurate as the coefficient(s) of permeability are known. Further a boundary condition such as a sheet wall may not be pervious—neither are any adjacent soil boundaries. If a computed seepage quantity is close to the measured value, it is likely to be a fortunate coincidence rather than any superiority of the computational method.

9-14 ELEMENTS OF WELL HYDRAULICS

Well flow can be approximated by flow nets; however, it may be more convenient to directly compute the quantity. Several cases exist, of which those shown in Fig. 9-18 will be analyzed in additional detail to obtain flow equations.

For the slot well (Fig. 9-18a) under *artesian* conditions with a single line source (approximately doubled for source on both sides),

$$Q = kDL \frac{dh}{dx}$$

Separating variables and integrating, we obtain

$$h = \frac{Qx}{kDL} + C$$

The limits are $h = H$ at $x = R$; $h = h_e$ at $x = 0$ (at slot face). This gives

$$Q = \frac{kDL}{R}(H - h_e) \tag{9-9}$$

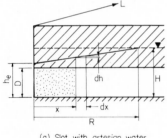

(a) Slot with artesian water

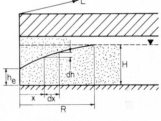

(b) Slot with gravity flow

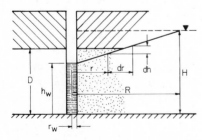

(c) Artesian well fully penetrating aquifer

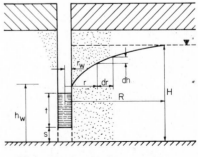

(d) Gravity well either fully or partially penetrating aquifer

Figure 9-18 Several cases of well flow.

For the slot under *gravity* flow and a single line source (see Fig. 9-18b),

$$Q = khL \frac{dh}{dx}$$

Integrating and inserting limits of $h = H$ at $x = R$ and $h = h_e$ at the slot, we obtain

$$Q = \frac{khL}{2R}(H^2 - h_e^2) \tag{9-10}$$

For an *artesian well* fully penetrating the aquifer (see Fig. 9-18c for identification of terms), we have

$$Q = kiA \qquad \text{and} \qquad i = \frac{dh}{dr} \qquad A = 2\pi rD$$

Substituting, we obtain

$$Q = k(2\pi rD)\frac{dh}{dr}$$

and integrating and inserting limits of $h = H$ at $r = R$ and $h = h_w$ at $r = r_w$, we obtain

$$Q = \frac{2\pi kD(H - h_w)}{\ln (R/r_w)} \tag{9-11}$$

For the *fully penetrating gravity well,*

$$i = \frac{dh}{dr} \qquad \text{and} \qquad A = 2\pi rh$$

and

$$Q = k(2\pi rh)\frac{dh}{dr}$$

Integrating and inserting limits, we obtain

$$Q = \frac{\pi k(H^2 - h_w^2)}{\ln (R/r_w)} \tag{9-12}$$

Research and observations show that the flow from partially penetrating gravity wells (the most common case encountered) is better described (Mansur and Kaufman, 1962) as

$$Q = \pi k \frac{[(H - s)^2 - t^2]}{\ln (R/r_w)}\left[1 + \left(0.3 + \frac{10r_w}{H}\right)\sin \frac{1.8s}{H}\right] \tag{9-13}$$

where terms are as identified in Fig. 9-18d. Note that Eq. (9-13) becomes Eq. (9-12) when the well fully penetrates the aquifer, since $s = 0$ and $t = h_w$.

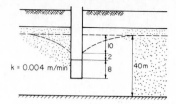

Figure E9-8

Example 9-8 Given are the gravity well conditions shown in Fig. E9-8.

REQUIRED Estimate the flow of the well when the drawdown is 10 m as shown.

SOLUTION With 10 m drawdown,

$$h_w = 30 \text{ m} \qquad s = 20 \text{ m}$$

$$t = 20 - 10 - 2 = 8 \text{ m}$$

$$r_w = 0.3 \text{ m} \qquad H = 40 \text{ m}$$

Using Eq. (9-13),

$$Q = 3.14(0.004)\frac{(40-20)^2 - 8^2}{\ln{(R/0.3)}}\left[1 + \left(0.3 + \frac{10(0.3)}{40}\right)\sin{\frac{1.8(20)}{40}}\right]$$

$$Q = 3.14(0.004)\frac{400 - 64}{\ln{(R/0.3)}}(1.29) = \frac{5.45}{\ln{(R/0.3)}}$$

Since R is unknown, we will assume several values of R and compute the corresponding flow quantity Q.

R, m	$\ln R/0.3$	Q, m^3/min
20	4.20	1.30
40	4.89	1.11
50	5.12	1.07
100	5.81	0.94

from which it appears that $Q \cong 1$ m^3/min, and it is not necessary to have an exact value of the radius of drawdown influence R.

/////

9-15 SUMMARY

This chapter has presented a means of obtaining seepage quantities and seepage forces for various geometrical configurations for which closed form solutions are not readily obtainable. The general procedure is to construct a flow net. The flow net consists of families of curves intersecting at 90° and forming squares, except at singular points, where judgment must be exercised. We note that if $k_x \neq k_y$, the intersection angle is still 90° but the quadrilateral is not square.

We may transform either the x or y dimension using a transformation variable

$$u = \sqrt{\frac{k_x}{k_y}} \qquad \text{or} \qquad v = \sqrt{\frac{k_y}{k_x}}$$

to obtain either $x' = x/u$ or $y' = y/v$, but not both simultaneously. With transformed dimensions, we can use squares. The seepage quantity is computed as

$$q = kh \frac{n_f}{n_d}$$

with k = coefficient of permeability or a transformed valued of $k' = \sqrt{k_x k_y}$.

It was shown that uplift pressures can be obtained from a flow net. The flow net can be used in the control of piping to obtain the hydraulic gradient. The flow net is also useful in determining the optimum location of cut-off walls.

Select elements of well hydraulics have been introduced. We also note that excavation dewatering is similar to well hydraulics and that a plan flow net can be used to determine flow quantities.

PROBLEMS

For Probs. 9-1 through 9-4, refer to Fig. P9-1.

9-1 Find the seepage in cubic meters per day per meter of dam width. Use a scale of 1 cm = 5 m.
  Approx. ans.: 1.1 m³/day/m.

9-2 Find the uplift force per meter of width and plot the pressure profile along the base of the dam. Use a scale of 1 cm = 5 m for dimensions and any convenient pressure scale.
  Approx. ans.: 1436 kN/m.

9-3 Install a cut-off wall 4.0 m downstream from the front of the dam as shown by the dashed lines in Fig. P9-1. The wall will be 9 m in length (the tip will be 13 m below the ground surface). Compute the seepage as in Prob. 9-1.

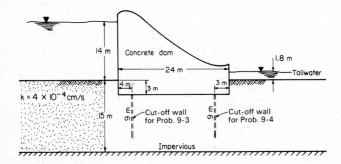

Figure P9-1

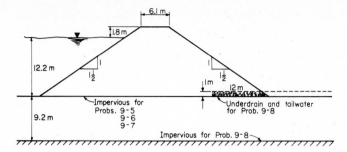

Figure P9-5

9-4 Install a cut-off wall 3 m from the downstream end of the dam as shown by the dashed lines and remove the cut-off wall from the upper end. The cut-off wall is 9 m in length. Compute the seepage as in Prob. 9-1.

Ans.: 0.66 m^3/day·m.

For Probs. 9-5 through 9-8, refer to Fig. P9-5.

9-5 Find the seepage if the dam rests on an impervious stratum and there is no underdrain. Take $k = 3.5 \times 10^{-3}$ cm/s and any convenient scale.

Approx. ans.: 6.5 m^3/day·m.

9-6 Redo Prob. 9-5 if there is an underdrain with the dimensions shown.

9-7 Redo Prob. 9-6 if $k_x = 8 \times 10^{-4}$ and $k_y = 2 \times 10^{-4}$ cm/s.

9-8 Redo Prob. 9-5 if there is an underdrain and the soil on which the dam rests is also pervious.

Approx. ans.: 15.0 m^3/day·m.

9-9 What would be the expected well flow of a well at point A of Fig. P9-9 (center of excavation) if r_w will compute the area of the excavation and assuming a fully penetrating well with water elevation in the well the same as it would be for the excavation (i.e., 0.5 m below the bottom of the excavation)?

9-10 Same as Prob. 9-10 except $r_w = 0.5$ m and fully penetrating.

Ans.: 47 m^3/day.

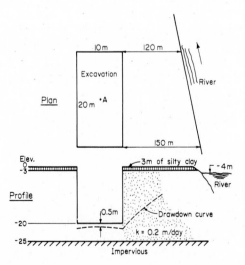

Figure P9-9

TEN

STRESSES, STRAINS, AND RHEOLOGICAL CONCEPTS

10-1 GENERAL CONSIDERATIONS

One of the most important functions of the study of soil mechanics is the prediction of the magnitude of the stresses under some loading which will produce excessive deformations, termed *failure stresses*. Any load will produce stresses and resulting strains which can be integrated over the stress zone of interest to obtain deformation. The deformation is usually termed *settlement*, and considerable effort is often expended to obtain a settlement prediction. This chapter and the several chapters following will be concerned in particular with stress and deformation "predictions" and some of the reasons for terming this activity a prediction.

Resisting stresses are developed when any material is subjected to a load. The study of strength of materials is devoted to evaluation of:

1. Magnitude of the resulting stresses σ
2. Whether the stresses result in a material failure
3. Magnitude of strains ε

The problem is considerably less difficult when dealing with an isotropic, homogeneous, linearly elastic (obeying Hooke's law), isolated body with clearly defined boundaries. We are concerned here with the stresses and strains developed in soil masses due to applied loads such as fills, building foundations, or negative loads such as from excavations. In soil masses the materials are generally:

Not	But may
Isotropic (elastic properties the same in all directions)	Be anisotropic (elastic properties not the same in different directions)
Homogeneous (constant material properties of e, w, γ, structure, etc., throughout)	Have lenses of different materials, be stratified, increase in density with depth, and contain varying amounts of moisture
Linearly elastic	Be nonlinear and be elastic only over a very limited stress range
An isolated body	Be a stressed zone in a semi-infinite medium

Since soil is a particulate material, failure is primarily by a rolling and slipping of grains and not via simple tension or compression. Because of this failure mode, the stresses of interest are *shear stresses*. The soil resistance or soil strength of interest is the *shear strength*. Conceptually, soil strength is quite different from the ultimate strength of materials such as steel or concrete. The soil locally subjected to a load (or stressed) is always surrounded by the remainder of the semi-infinite medium. Exceptions occur, of course, such as adjacent to excavations or over old mine tunnels or other subsurface cavities. With the local stressing, we have a situation analogous to a ship floating in the ocean. The stressed zone will "fail" if the stresses are too large. Failure is defined as a considerable alteration or state change in soil structure (or remolding), accompanied by substantial deformation and enlargement of the stressed zone until deformation eventually halts. The resulting total deformation is the deformation under stresses up to failure plus the larger deformation occurring after failure. The soil strength after "failure" is termed the *residual strength*.

When any viscous fluid, usually water, is present in the soil void spaces, the particle rolls and slips will be resisted by the pore fluid. The amount of resistance will be proportional to the quantity of pore fluid present in a range of saturation from 0 to 100 percent. The duration of pore fluid resistance will depend on the effective coefficient of permeability k.

Rheology is the study of materials in a fluid state as a rate process. Soil deformation which depends on the coefficient of permeability becomes a rate process, and some effort has been directed toward using rheological methods for analysis. Several rheological models will be presented in Sec. 10-20.

Most efforts to predict soil response to applied loads have used theory of elasticity methods. A few researchers have also used plasticity theory. The major problem has been that both elasticity and plasticity theories are for an elastic continuum, whereas soil is an aggregation of particulate material. An additional major problem with soil is that it is *state-dependent*; that is, it changes in volume (1) with changes in water content and (2) under stress. Either of these state changes produces a different material from what was started with. Proper allowance—not always done—should be made for any state changes.

In addition to state changes, soil has several additional very significant factors which require consideration, such as:

1. Any new stress condition starts from an initial nonzero stress state. Soil is always subjected to in situ all around stresses from existing overburden, changes in water content, or an overburden which has since eroded. This is somewhat different from the usual condition of zero† stresses in structural analysis.
2. Initial in situ conditions of stress, density, and similar are generally products of previous (or geologic) stress history. This means that stress history is a permanent imprint on the soil structure or skeleton and any remolding or other disturbance destroys, or greatly alters, any properties resulting from this effect.
3. Stress-strain conditions in a soil mass are three-dimensional. This differs from many structural problems which are treated as two-dimensional. It is common, however, to analyze many soil problems using a two-dimensional model.
4. Soil strength is heavily dependent on water content and pore pressures.

Considering these several factors, it should be readily apparent that a high degree of success in predicting soil deformations is not possible. This and the next several chapters will focus on making design estimates and some of their limitations.

10-2 GENERAL STRESSES AND STRAINS AT A POINT

Figure 10-1 illustrates the stresses on a six-sided differential element in a soil mass. Not shown on this element are the stresses on the three planes away from the reader, for additional clarity and because they would simply be equal and

---

† But it is analogous to residual stresses often in rolled structural shapes.

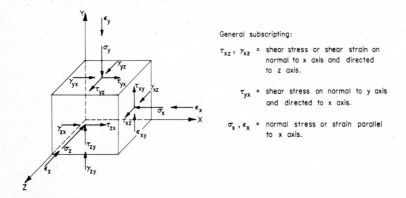

General subscripting:

τ_{xz}, γ_{xz} = shear stress or shear strain on normal to x axis and directed to z axis.

τ_{yx} = shear stress on normal to y axis and directed to x axis.

σ_x, ϵ_x = normal stress or strain parallel to x axis.

Figure 10-1 Stresses and strains on a soil element with coordinate axes arbitrarily shown.

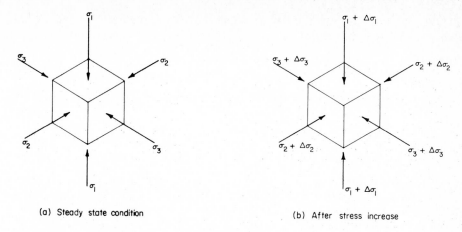

Figure 10-2 A soil element subjected to incremental principal stress increases.

opposite in direction. The element strains can be shown in a similar manner, as also shown in Fig. 10-1.

Figure 10-2 illustrates an orientation of the element of Fig. 10-1 such that no shear stresses exist on the sides of the element. This orientation produces principal axes, and the normal stresses on the element faces are *principal stresses*; by analogy, there exists a set of principal axes for *principal strains*. The principal axes for principal stresses may not be coincident with the axes for principal strains. In either case, however, there is always an axis orientation which produces principal stresses or principal strains.

Shown in Fig. 10-2a is the general compressive stress state existing in the soil prior to application of some soil load resulting in an incremental increase in stress of $\Delta\sigma_i$. Figure 10-2b is the stress state after the application of the stress increment $\Delta\sigma_i$. We are generally interested in the deformations resulting from application of these stress increments. Deformation is related to strain by

$$\text{Deformation } \delta = \int_0^M \varepsilon \, dM$$

where M may be either a length or a volume with ε the corresponding linear or volumetric strain.

10-3 THEORY OF ELASTICITY CONCEPTS USED IN SOIL MECHANICS PROBLEMS

A *tensor* is a vector quantity describing a physical state (such as a stress) or a physical phenomenon (such as strain), and requires three or more components for a complete description. In general, any stress or strain requires three direction cosines for a complete description of magnitude and direction. In plane stress (or strain), one of the direction cosines is 0. Referring to Fig. 10-1, the stress and strain tensors (or matrices) can be developed by interchanging τ_{ij} with σ_{ij} (i.e., $\tau_{xy} = \sigma_{xy}$; $\tau_{zy} = \sigma_{zy}$, etc.) for stresses and similarly interchanging γ_{ij} with ε_{ij} for strains ($\gamma_{xy} = \varepsilon_{xy}$; $\gamma_{zx} = \varepsilon_{zx}$, etc.). The resulting stress and strain tensors are:

Stress			Strain		
σ_{xx}	σ_{xy}	σ_{xz}	ε_{xx}	ε_{xy}	ε_{xz}
σ_{yx}	σ_{yy}	σ_{yz}	ε_{yx}	ε_{yy}	ε_{yz}
σ_{zx}	σ_{zy}	σ_{zz}	ε_{zx}	ε_{zy}	ε_{zz}

The reader can readily observe that a tensor is shorthand notation for summing the stress or strain vectors along the three cartesian axes orienting the element of interest. The subscript convention is standard (refer again to Fig. 10-1) in that

1. The first subscript is the axis normal to the plane of interest.
2. The second subscript is an orthogonal axis parallel to the vector.

This subscripting produces *ii* subscripts for normal stresses or strains; *ij* subscripts are parallel to the plane and are either shear stresses or shear strains.

From moment equilibrium, and neglecting second- and higher-order differentials, we obtain three of the shear and strain quantities as complementary values:

$$\sigma_{xy} = \sigma_{yx} \qquad \sigma_{xz} = \sigma_{zx} \qquad \sigma_{yz} = \sigma_{zy}$$

$$\varepsilon_{xy} = \varepsilon_{yx} \qquad \varepsilon_{xz} = \varepsilon_{zx} \qquad \varepsilon_{yz} = \varepsilon_{zy}$$

which gives a symmetrical tensor, and only six quantities are needed for a complete description.

10-4 OCTAHEDRAL STRESSES

In most soil mechanics work the existing stresses are taken as a *steady state* condition. Geologically, or well aged, deposits tend to satisfy this criterion. Recent delta (and similar) deposits, recently filled in areas, and sites where the water table has changed datum are in an unsteady state as settlements are ongoing.

In the steady state condition a soil element surrounded by soil is subject to a vertical σ_1 and lateral pressure σ_3. No shear stresses exist on the surface of the element since it is no longer undergoing strain in a steady state condition. When the sides of the element are oriented parallel to certain axes (x, y, z) and there are no shear stresses on the sides the stresses are termed *principal stresses*. The axis orientation produces *principal axes*. This stress state is commonly referred to as a *triaxial stress state*, where "tri" refers to stresses along the three axes. A similar state is approximately developed in a laboratory triaxial test.[†]

[†] See a laboratory manual for test details, e.g., Bowles (1978).

Strictly a cube shaped element of sides dx, dy, dz can have a different principal stress on each of the sides, i.e., $\sigma_1 \neq \sigma_2 \neq \sigma_3$. The largest stress is the major principal stress defined as σ_1 and may be the vertical (but not always) stress. The smallest stress is the minor principal stress defined as σ_3; the intermediate value is σ_2. When the soil is preconsolidated, one or more of the horizontal stresses may be larger than the vertical one which is nominally computed as $\sigma_v = \gamma h$. Effective stresses are produced in a soil mass at steady state conditions.

It is common in soil mechanics work to either ignore σ_2 or assume that it is equal to σ_3. The computational effort is greatly simplified (without too much error in most cases) by using the resulting two-dimensional stress state.

Strictly, the stress state is three-dimensional as stated earlier. If we idealize a point with a cube centered with respect to the principal axes and (1) look at one-fourth of the cube and (2) pass a section to produce plane ABC in Fig. 10-3a, the result is one-eighth of the surface of an eight-sided (octahedron) figure. The plane ABC is an octahedral plane and the stresses of interest on this plane are the

1. Normal or octahedral stress (as vector OR in Fig. 10-3b), σ_{oct}
2. Tangent or octahedral shear stress, τ_{oct}

The stresses from the conventional triaxial test fall in plane $OCFE$ in Fig. 10-3a since the confining stress $\sigma_c = \sigma_3$ is equal all around and acts on planes AOC and BOC. This results in a vector $OE = \sqrt{2}\sigma_3$. The vertical stress σ_1 ($EF = OC$) produces the octahedral stress ($\sigma_{\text{oct}} = OF$) vector *normal* to the octahedral plane.

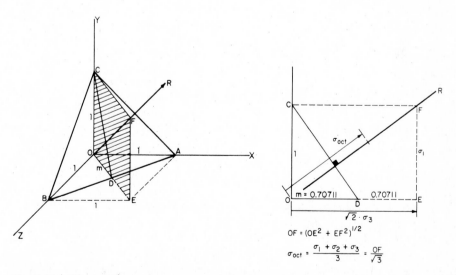

Figure 10-3 The octahedral stress state from three-dimensional stresses. Note that stresses in plane $OCEF$ are computed as shown in part b. Line OF is a stress vector perpendicular to the octahedral plane but is larger than the octahedral (normal) stress as shown.

The resultant stress vector σ_{oct} on the octahedral plane ABC can be obtained for the general axis orientation of Fig. 10-3a by summing forces in the three orthogonal directions and using the appropriate direction cosines m_i. Since force equilibrium is necessary because of the partial areas, one cannot obtain the stresses on the octahedral plane as a direct vector summation of the normal stress components.

From Fig. 10-3a, using the equilateral tetrahedron shown, we have $OA = OB = OC = 1$; thus areas $AOC = BOC = AOB = (1 \times 1)/2 = 0.5$ unit. The area ABC is computed from the relationships $m = 0.707111$, $AB = \sqrt{2}$, and $OC = 1$ to obtain $ABC = 0.86602$ units.

The direction cosines are established as the ratio of areas as follows:

$$m_1 = \frac{AOB}{ABC} = \frac{0.5}{0.866} = \frac{1}{\sqrt{3}} \qquad m_2 = \frac{BOC}{ABC} = \frac{1}{\sqrt{3}} \qquad m_3 = \frac{AOC}{ABC} = \frac{1}{\sqrt{3}}$$

These will always be the direction cosines for an equilateral tetrahedron.

From the relationship $\sum F = 0$ along any axis, and defining the *normal stress* on plane ABC as $R = \sigma_{oct}$, we have, for the area A_{ABC}

$$R(A_{ABC})m_i - \sigma_i A_i = 0$$

For the actual area values of this example, this produces

$$R(0.866)\left(\frac{1}{\sqrt{3}}\right) - \sigma_1(0.5) = 0$$

$$R(0.866)\left(\frac{1}{\sqrt{3}}\right) - \sigma_2(0.5) = 0$$

$$R(0.866)\left(\frac{1}{\sqrt{3}}\right) - \sigma_3(0.5) = 0$$

and adding, $\qquad 3R(0.866)\left(\frac{1}{\sqrt{3}}\right) - 0.5(\sigma_1 + \sigma_2 + \sigma_3) = 0$

The octahedral normal stress is then obtained as

$$\sigma_{oct} = R = \frac{\sigma_1 + \sigma_2 + \sigma_3}{3} \qquad (10\text{-}1)$$

Note that this equation for the octahedral normal stress does not depend on equal values for the principal stresses. In a triaxial test under initial conditions of isotropic consolidation (consolidation at a constant cell pressure σ_3) the octahedral stress σ_{oct} = cell pressure. Most in situ initial stress states are not isotropic.

With an excess pore pressure Δu from some loading the effective stress according to Eq. (2-20) is

$$\sigma' = \sigma_i - \Delta u$$

Substitution of σ'_i for σ_i in Eq. (10-1) gives

$$\sigma'_{oct} = \frac{\sigma_1 + \sigma_2 + \sigma_3 - 3\Delta u}{3} \qquad (10\text{-}1a)$$

which is identical to using Eq. (2-20) directly and using σ_{oct} for σ_t to obtain the effective stress.

The octahedral shear stress can be obtained (see a textbook on theory of elasticity) as

$$\tau_{oct} = \tfrac{1}{3}\sqrt{(\sigma_1 - \sigma_2)^2 + (\sigma_2 - \sigma_3)^2 + (\sigma_3 - \sigma_1)^2} \qquad (10\text{-}2)$$

Equation (10-2) shows that:

1. No shear stresses are produced with $\sigma_1 = \sigma_2 = \sigma_3$—commonly called a *hydrostatic stress condition*. This condition can be produced in a soil test as an *isotropic* (equal all-round stresses) *consolidation*, considered in more detail in Chap. 13.
2. Shear stresses are produced when $\sigma_1 > \sigma_2$ or $\sigma_1 > \sigma_3$ as in a triaxial test. The stress difference $\sigma_1 - \sigma_3$ is the *deviator* stress.
3. In situ soil has initial shear and normal stresses on the octahedral plane in most cases since any stress inequality between the three principal stresses produces a shear stress.
4. Shear stresses are independent of any pore water pressure Δu.

If we use matrix notation for the octahedral normal stress we obtain

$$|\sigma_i|\{m_i\} = \{\sigma_{oct}\} \qquad (10\text{-}3)$$

Rearranging, dividing through by the direction cosine matrix $\{m_i\}$, and expanding, we obtain

$$\begin{bmatrix} \sigma_1 - \sigma_{oct} & 0 & 0 \\ 0 & \sigma_2 - \sigma_{oct} & 0 \\ 0 & 0 & \sigma_3 - \sigma_{oct} \end{bmatrix} = 0$$

Expanding the determinant (the matrix) to 0, we obtain

$$\sigma^3_{oct} - J_1\sigma^2_{oct} + J_2\sigma_{oct} - J_3 = 0$$

where

$$J_1 = \sigma_1 + \sigma_2 + \sigma_3 = 3\sigma_{oct}$$

$$J_2 = \sigma_1\sigma_2 + \sigma_2\sigma_3 + \sigma_3\sigma_1 \qquad (10\text{-}4)$$

$$J_3 = \sigma_1\sigma_2\sigma_3$$

The J factors are termed *invariants* since the principal stresses are independent (nonvarying) of arbitrary axis orientation.

The resultant stress vector is

$$\sigma_r = (\sigma_{oct} + \tau_{oct})^{1/2} \qquad (10\text{-}5)$$

and makes an angle α with the octahedral normal stress of

$$\alpha = \tan^{-1} \frac{\tau_{oct}}{\sigma_{oct}}$$

Triaxial tests produce data which plot in plane $OCFE$ in Fig. 10-3a, but the stress resultant of Eq. (10-5) may be either above (most common) or below the octahedral normal stress vector.

Example 10-1 What are the octahedral normal and shear stresses when $\sigma_1 = 50$ kPa, $\sigma_2 = \sigma_3 = 25$ kPa, and the pore pressure $u = 10$ kPa? The sample was isotropically consolidated at a cell pressure of $\sigma_1 = \sigma_3 = \sigma_c = 25$ kPa.

SOLUTION Using Eq. (10-1), obtain

$$\sigma_{oct} = \frac{50 + 25 + 25}{3} = 33.3 \text{ kPa}$$

Using Eq. (10-2), obtain

$$\tau_{oct} = \tfrac{1}{3}\sqrt{(50 - 25)^2 + (25 - 25)^2 + (25 - 50)^2} = 11.8 \text{ kPa}$$

These values as well as several intermediate values of σ_1 in the following table have been plotted in Fig. 10-4.

σ_1	$\Delta\sigma_1$	σ_{oct}	τ_{oct} (Total stresses)
25	0	25.0	0.00
30	5	26.7	2.40
35	10	28.3	4.70
40	15	30.0	7.07
45	20	31.7	9.42
50	25	33.3	11.8 (failure)

The radial vector through point O in plane $OECF$ of Fig. 10-3 and with side view of Fig. 10-3b can be computed as

$$\sigma_r = \frac{1}{\sqrt{3}} [(\sqrt{2} \cdot 25)^2 + \sigma_1^2]^{1/2}$$

For initial conditions of all $\sigma_i = 25$, this gives

$$\sigma_r = \sigma_{oct} = 25 \qquad \text{(vector 01'' in Fig. 10-4} \times 1/\sqrt{3})$$

The *effective octahedral stresses* with $u = 10$ kPa are computed as follows:

$$\sigma'_{oct} = \frac{50 + 25 + 25 - 3(10)}{3}$$

$$= 23.3 \text{ kPa} \qquad \text{(same as } 33.3 - 10)$$

$$\tau_{\text{oct}} = \tfrac{1}{3}\sqrt{(40-15)^2 + 0^2 + (15-40)^2}$$

$$= 11.8 \text{ kPa} \qquad \text{(same as before)}$$

Since the pore water can carry only hydrostatic stress and cannot support any shear stress, we see that τ_{oct} is unaffected by the excess pore pressure of 10 kPa developed during the test.

The effective stress resultant σ_r on the octahedral plane is

$$\sigma'_r = (\sigma'^2_{\text{oct}} + \tau^2_{\text{oct}})^{1/2} = (23.3^2 + 11.8^2)^{1/2} = 26.11 \text{ kPa}$$

The resultant acts above the normal vector in the octahedral plane since the vertical stress σ_1 is increasing. The angle between the stress resultant and the octahedral stress vector σ_{oct} is

$$\alpha = \tan^{-1}\frac{11.8}{33.3} = 19.5°$$

For effective stresses the angle α is

$$\alpha = \tan^{-1}\frac{11.8}{23.3} = 26.85°$$

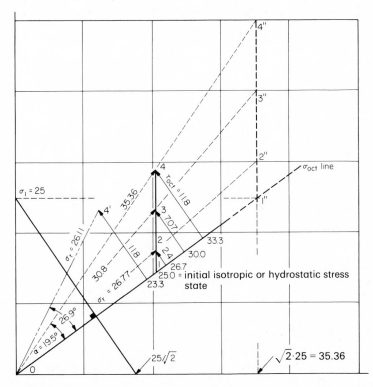

Figure 10-4 Stress path plot of octahedral normal and shear stresses. Part of the data shown are computed in Example 10-1.

The locus of points 1 through 4 traces an octahedral (total) stress path as shown in Fig. 10-4. The locus of double-primed points (1″ through 4″) traces a total stress path. An effective stress path lies between point 1 and 4′, but because it has only the two end points, the shape of the path is not known but is generally curved and not straight as is the octahedral total stress path. Note carefully that path 1″–4″ can be obtained geometrically—paths 1 through 4 must be computed.

////

10-5 STRESS-STRAIN MODULUS AND HOOKE'S LAW

The relationship between deformation and stress in a solid is of considerable interest. To put deformation on a common basis, it is usually reported as a ratio by dividing the deformation by the length over which it occurs so that the result is for a unit length. Values of m/m, in/in, ft/ft, etc., result which are, in fact, dimensionless since the resulting numerical ratio would be the same for all three of the given "units." The deformation ratio is called *strain* and usually uses the symbol ε. For example, if the total deformation is 0.5 mm in a length of 0.5 m (obviously an average value of deformation), we have

$$\varepsilon = \frac{\delta}{L} = \frac{0.5}{500} = 0.001 \text{ mm/mm}$$

$$= \frac{0.0005}{0.5} = 0.001 \text{ m/m}$$

$$= \frac{0.0005(0.3048)}{0.5 \times 0.3048} = 0.001 \text{ ft/ft}$$

This strain is equivalent to saying the elongation is 0.10 percent of the length.

Strain is produced by a stress which is defined as

$$\sigma = \frac{\text{Force}}{\text{Area}}$$

If we plot a curve of stress versus strain, it is straight over a range for steel and several other materials. It is curved over nearly the entire range for concrete, soil, and a large number of other materials as qualitatively shown in Fig. 10-5a. One of several elastic parameters used in deformation analyses of solids is given by the slope of the straight line portion of the stress-strain curve. This parameter is the modulus of elasticity E given as

$$E = \frac{\text{Change in stress}}{\text{Change in strain}} = \frac{\Delta\sigma}{\Delta\varepsilon} \tag{10-6}$$

When the stress-strain plot is curved some interpretation is required and we obtain either a tangent or secant modulus as illustrated in Fig. 10-6a. The

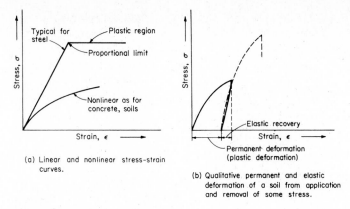

(a) Linear and nonlinear stress-strain curves.

(b) Qualitative permanent and elastic deformation of a soil from application and removal of some stress.

Figure 10-5 Stress-strain characteristics.

modulus of elasticity is a measure of both deformation and stiffness of the material.

When a solid is loaded as in Fig. 10-6b it shortens and becomes thicker or elongates and becomes thinner. This can be described by lateral strains. Poisson (ca. 1811) showed that the ratio of the lateral to vertical strain is a constant for materials within the proportional limit (straight section of stress-strain curve). This constant is termed Poisson's ratio μ and defined as

$$\mu = \frac{-\varepsilon_h}{-\varepsilon_v} \qquad (a)$$

Signs are used and ε_v is $(-)$ for elongation strains.

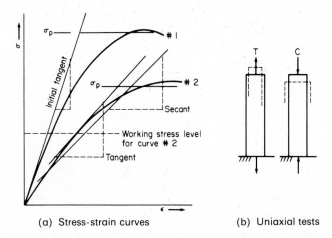

(a) Stress-strain curves

(b) Uniaxial tests

Figure 10-6 (a) Methods of computing the initial tangent, tangent, and secant stress-strain modulus. The "yield" stress σ_p corresponding to the "plastic" region for steel as in Fig. 10-5a is not clearly defined for soil and must be estimated as shown. (b) Geometry changes in sample in a uniaxial test to define lateral strains for Poisson's ratio.

The stress-strain curves for soils, as previously noted, are nonlinear or curved, over the entire range of interest. There is some evidence that at extremely small strains, say, 10^{-4} and less, there *may* be a linear region. Deformations necessary to plot strain this small are seldom obtained, however; so for all practical purposes a nonlinear curve is obtained from laboratory compression tests. Coupling this with the particulate nature of soil so that recoverable (elastic) strains are negligible (Fig. 10-5b), the modulus of elasticity and Poisson's ratio are not very optimistic parameters for use in soil deformation studies. Nevertheless, both parameters are widely used, although it is more conventional to use the term "stress-strain modulus" rather than the modulus of elasticity. Usually the stress-strain modulus is subscripted to obtain

$$E_s = \frac{\Delta\sigma}{\Delta\varepsilon}$$

for soil. Commonly, the initial tangent modulus is used to compute E_s which somewhat indirectly allows for the possibility of a linear range. Both the tangent and secant modulus in the region of the working stresses are also used—particularly in finite-element studies. Since the stress-strain modulus is dependent on the method, reported values should indicate how it was determined.

We should note that foundations, pavements, dams, fills, and similar, are placed on a soil producing compressive stresses. Soil retained by a wall is compressed in the load carrying zone. Soil is never relied on to carry tension and the few laboratory tests to evaluate tensile strength of clay find the tensile stress to be negligible. Thus only compression loads (producing compressive stresses and strains) are of consequence in soil stress-deformation studies. This produces a positive value of μ in Eq. (a) since ε_v is ($+$) for compression strain.

Rearranging Eq. (a), we have for an x, y, z coordinate system (y axis vertical)

$$\varepsilon_x = \mu\varepsilon_y \qquad \varepsilon_z = \mu\varepsilon_y \qquad\qquad (b)$$

Since $\varepsilon_i = \sigma_i/E_s$ we have for a soil element beneath a loaded area as in Fig. 10-7 a restoring effect on the axial strain. Using Eqs. (10-6) and (b), we obtain

$$\varepsilon_y = \frac{\Delta\sigma_y}{E_s} - \frac{\mu\,\Delta\sigma_x}{E_s} - \frac{\mu\,\Delta\sigma_z}{E_s} \qquad\qquad (c)$$

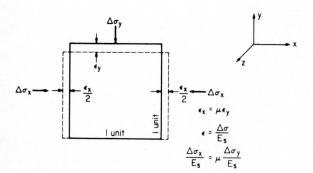

Figure 10-7 The confining effect producing modifications in strains for a sample with a stress increment $\Delta\sigma_y$ shown. The surrounding soil "pushes" against the soil element as shown with a stress intensity increase (over the existing value) based on Poisson's ratio and computed as shown.

Simplifying Eq. (c) and interchanging subscripts, we see that the three coordinate strains are:

$$\varepsilon_y = \frac{1}{E_s} [\Delta\sigma_y - \mu(\Delta\sigma_z + \Delta\sigma_x)]$$

$$\varepsilon_x = \frac{1}{E_s} [\Delta\sigma_x - \mu(\Delta\sigma_z + \Delta\sigma_y)] \tag{10-7}$$

$$\varepsilon_z = \frac{1}{E_s} [\Delta\sigma_z - \mu(\Delta\sigma_x + \Delta\sigma_y)]$$

These three strain equations are called Hooke's generalized stress-strain law.[†] In mechanics of materials courses, the Poisson's ratio effects are commonly neglected, but in soil mechanics work the Poisson's ratio effects are generally too large to neglect (although it has been commonly done).

The *shear modulus* G is widely used in geotechnical work—particularly in soil dynamics studies. The value G is related to E_s as

$$G = \frac{E_s}{2(1 + \mu)} \tag{10-8}$$

The volumetric strain $\varepsilon_v = \varepsilon_x + \varepsilon_y + \varepsilon_z$ produces the *bulk modulus* E_b as

$$E_b = \frac{\sigma_x + \sigma_y + \sigma_z}{3\varepsilon_v} = \frac{\sigma_{oct}}{\varepsilon_v} \tag{10-9}$$

With some manipulation of Hooke's law, we can also obtain

$$E_b = \frac{E_s}{3(1 - 2\mu)} \tag{10-9a}$$

Since neither G nor E_b can be $(-)$, we obtain the limits of Poisson's ratio for any "elastic" material from Eqs. (10-8) and (10-9a) as

$$-1 < \mu < 0.5$$

It appears the range of μ for soils is between 0 and 0.5 (note that we often use 0.5 although a discontinuity theoretically exists at 0.5). If a soil is considered "elastic"—questionable in any case for a particulate medium—the range of μ is as just given. If a computed value of μ is out of this range the concept of an elastic medium for the mathematical model is no longer valid and the use of "elasticity" concepts increases the risk factor.

Uniaxial compression and tension tests are widely used (Fig. 10-6b) for many engineering materials (steel, concrete, etc.). Since these materials form structural members which are isolated in space as columns and beams, slabs, and similar

---

[†] When ε_y, ε_x, and ε_z are not principal strains, there are also three shearing-strain equations as part of this stress-strain law.

with atmospheric (0 gage) confining pressure, these tests produce very satisfactory design parameters without including Poisson's ratio. On the other hand soil is always surrounded by other soil which produces a confining effect. Rarely, exceptions such as rock or soil pillars (columns) for support of mine roofs and similar are used. Even here, the major load capacity may be on the interior portion of the pillar which is surrounded (confined) by the exterior and is a major reason for these types of members being rather large.

From Eq. (10-7) it is evident that uniaxial compression tests without confinement will produce larger strains and thus a smaller stress-strain modulus. This is a major reason for preferring triaxial rather than unconfined compression tests. Actual laboratory versus field comparisons have found unconfined compression tests to be as small as one-fourth the in situ value. Even triaxial compression tests may produce E_s values of only 50 to 60 percent of in situ. Obviously a major factor is soil sample disturbance.

If a fully saturated, consolidated, sample is tested in compression the application of a load increment $\Delta\sigma_1$ will immediately increase the pore pressure from 0 (consolidated) to $u = \Delta\sigma_1$ until some drainage occurs. If no drainage is allowed, there will be no volume change unless $\Delta\sigma_1$ is very large since water (and the soil grains) are relatively incompressible at usual loads. Any vertical movement (strain) is directly related to the lateral strain since the volume = constant and Poisson's ratio computes 0.5. As soon as some drainage occurs, however, Poisson's ratio will be something different since further vertical strain is not directly related to lateral strain but rather to "elastic" properties of the soil skeleton. Strains (and resulting settlements) involving compression stresses on saturated soils will be considered in some detail in Chaps. 11 and 12.

Example 10-2 Given: stress-curves in Fig. E10-2; the curve for $\sigma_3 = 0$ is for an unconfined compression test (special case of a triaxial test). The other curve is for a triaxial test using the confining pressure $\sigma_3 = 30$ kPa as shown. The reader should note that the justification for plotting the two curves on the same graph for comparison requires that the samples be from the same source and at the same (as close as practical) water content. It is estimated that Poisson's ratio $\mu = 0.3$ and the soil is not saturated.

REQUIRED Estimate the initial tangent modulus E_s for both tests.

SOLUTION Draw initial tangents to both curves as shown and extend to obtain a convenient set of values for $\Delta\sigma$ and $\Delta\varepsilon$. Here the curves pass through the origin so the intercept values can be directly read.

For unconfined compression: $E_s = \dfrac{78}{0.05} = 1560$ kPa

For the triaxial test: $E_s = \dfrac{82}{0.025} = 3280$ kPa

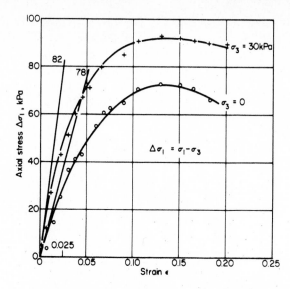

Figure E10-2

This latter is the common procedure for obtaining E_s from a triaxial test. We can also use Eq. (10-7) rearranged and interpreted to obtain

$$\varepsilon_v = \varepsilon_y = \frac{1}{E_s}\left[\Delta\sigma_1 - 2\mu(\sigma_3)\right]$$

rearranging and solving for E_s and using given value of Poisson's ratio

$$E_s = \frac{1}{0.025}\left[82 - 2(0.3)(30)\right] = 2560 \text{ kPa}$$

Either of these latter values is substantially larger than the unconfined compression value. This latter computation also suggests that we might obtain both E_s and μ by using tangents at several points along the stress-strain curve of a triaxial test with $\sigma_3 > 0$. For each tangent, the two unknowns are E_s and μ. Using two points with the corresponding $\Delta\sigma$ and $\Delta\varepsilon$ values gives two equations with the two required unknowns. We also might note at this point that E_s depends on μ and both seem to be dependent on the stress level at which they are evaluated.

////

10-6 ANISOTROPIC SOIL

Hooke's law as given by Eqs. (10-7) is based on an isotropic, elastic solid (or soil mass). An isotropic material has uniform engineering properties such as E_s, μ, and G_s in any direction and is produced from a homogeneous soil, or one having uniform physical properties (composition, density, etc.) throughout. No real soil meets this criterion.

Stratification from normal geologic and deposition processes and overburden pressures produce at the very minimum, a soil with unit weight increasing with depth. Confining (lateral) pressure also increases with depth since it is a fraction of the overburden pressure. These factors produce an in situ soil with anisotropic properties.

An anisotropic soil mass has vertical parameters E_s, G_s, μ, which are different from the lateral values, and requires definition:

$$E_i = \frac{\text{Stress in } i \text{ direction}}{\text{Strain in } i \text{ direction}}$$

$$\mu_i = \frac{\text{Strain normal to applied stress}}{\text{Strain parallel to applied stress}}$$

$$G_i = \frac{E_i}{2(1 + \mu_i)}$$

In general, there will be three values from using $i = x, y, z$ in the above equations. In most soils, however, it is reasonable that the x and z values will be the same but different from the vertical or y value. A soil with the x and z parameters equal is said to be *cross-anisotropic*. The modified Hooke's law for a cross-anisotropic soil is given in Bowles (1982).

In many cases the assumption of an isotropic soil mass provides reasonably accurate answers—at least as good as the values of E_s, μ, and G_s used. It is not a trivial task to determine reliable engineering parameters of any soil. It is even more difficult if the additional parameters defining a cross-anisotropic soil are required. This is because soil samples are tested from vertical borings so that only the vertical parameters are directly determined. In situ testing tends to obtain horizontal parameters.

In passing, we might note that unless the engineering parameters are accurately quantified, highly sophisticated analysis techniques founded in the theory of elasticity are not justified.

10-7 TWO-DIMENSIONAL STRESSES AT A POINT

It has been a computational convenience to take one of the horizontal stresses (or strains) in Fig. 10-1 as zero. The resulting two-dimensional state is called either (1) plane stress (but the strain > 0) or (2) plane strain (but the stress > 0). Generally, the strain is ignored in plane stress or the stress is ignored in plane strain. In terms of principal stresses (or strains), the σ_2 (or ε_2) is removed, leaving σ_1 and σ_3 or the corresponding strains as the parameters of interest.

Actually, in many soil problems a two-dimensional stress or strain state is a reasonable model. For example, long walls retaining earth tend to lean forward under the earth pressure producing two-dimensional strains in the soil along the length except at the ends. Wall footings tend to punch into the ground producing

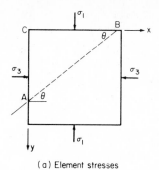

(a) Element stresses

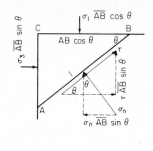

(b) Shear and normal stresses on AB.
Note AB is same as EF of Fig. 10-9

Figure 10-8 Plane stress with principal stresses shown. A stress plane through the element at θ to horizontal produces the shear and normal stresses of interest as shown in part b. Take thickness perpendicular to page as one unit. Also take AB as one unit to simplify computations although a length AB can be used as shown.

strains vertical and normal to the length except at the ends. A square footing and a triaxial compression test both produce three-dimensional strains. Using these illustrations as a base, the reader should be able to deduce other situations of either two- or three-dimensional stress and strain.

Figure 10-8 illustrates the conditions for a two-dimensional stress condition with the principal stresses σ_1 and σ_3 acting on the principal planes shown (element faces). A similar analysis could be made for any two of the three principal stresses, but the major and minor values are generally chosen as here.

Two of the principal planes (AC and CB of Fig. 10-8) are selected for detailed inspection, as in Fig. 10-8. We will assume plane AB is 1 unit × 1 unit perpendicular to the plane of the page, giving an area of 1 unit. From this assumption it follows that

$$BC = AB \cos \theta$$

$$AC = AB \sin \theta$$

Summing forces parallel to the X axis, we obtain ($\sum F_x = 0$)

$$\sigma_3(AB \sin \theta) + \tau(AB \cos \theta) - \sigma_n(AB \sin \theta) = 0 \qquad (d)$$

Summing forces parallel to the Y axis, we obtain

$$\sigma_1(AB \cos \theta) - \tau(AB \sin \theta) - \sigma_n(AB \cos \theta) = 0 \qquad (e)$$

Eliminating AB in both Eqs. (d) and (e), we obtain

$$\sigma_3 \sin \theta + \tau \cos \theta - \sigma_n \sin \theta = 0$$

$$\sigma_1 \cos \theta - \tau \sin \theta - \sigma_n \cos \theta = 0 \qquad (f)$$

This gives two equations in σ_1 and σ_3 and two unknown values, shear stress τ and normal stress σ_n. By the process of elimination and making use of the trigonometric relationships

$$\cos^2 \theta = 1 - \sin^2 \theta \qquad \sin^2 \theta = \tfrac{1}{2}(1 - \cos 2\theta) \qquad \sin \theta \cos \theta = \tfrac{1}{2} \sin 2\theta$$

we obtain the following equations (found in any mechanics of materials textbook but presented here for completeness and review):

$$\sigma_n = \frac{\sigma_1 + \sigma_3}{2} + \frac{\sigma_1 - \sigma_3}{2} \cos 2\theta \qquad (10\text{-}10)$$

$$\tau = \frac{\sigma_1 - \sigma_3}{2} \sin 2\theta \qquad (10\text{-}11)$$

Observe that these two equations are based on principles of mechanics and have nothing to do with material properties. Theory of elasticity deals with stresses and material properties of E_s and μ. It should be noted in passing that a similar set of equations could have been derived for the general case where planes AC and CB of Fig. 10-8 are not principal planes. The principal difference would be that shearing stresses on planes AC and CB would also have to be included. Since soil tests where Eqs. (10-10) and (10-11) are used involve starting with known principal stresses, the presentation here is preferred.

10-8 MOHR'S STRESS CIRCLE

Equations (10-10) and (10-11) are the parametric equations of a circle of stress in the XY plane known as a *Mohr diagram* (Fig. 10-9). Otto Mohr (ca. 1882) is credited with devising this circle to obtain the stresses at a point by graphic means. We will be concerned here with two-dimensional stresses based on the major (σ_1) and minor (σ_3) principal stresses; however, Mohr's circle can also be drawn to include the intermediate principal stress (σ_2) as shown in Fig. 10-9b. This latter shows that the critical shear stress can always be obtained using a two-dimensional Mohr circle—if the minor principal stress (σ_3) is properly identified.

In using Eqs. (10-10) and (10-11) to plot a Mohr's circle, we:

1. Set $2\theta = 0°$ and: From Eq. (10-10), find $\sigma_n = \sigma_1$.
   From Eq. (10-11), find $\tau = 0$.
   Using σ_n and $\tau = 0$, plot point B in Fig. 10-9a.
2. Set $2\theta = 180°$ and: From Eq. (10-10), find $\sigma_n = \sigma_3$.
   From Eq. (10-11), find $\tau = 0$.
   Using these values, plot point A in Fig. 10-9a.
3. Set $2\theta = 90°$ and: From Eq. (10-10), find $\sigma_n = (\sigma_1 + \sigma_3)/2$.
   From Eq. (10-11), find $\tau = (\sigma_1 - \sigma_3)/2$.

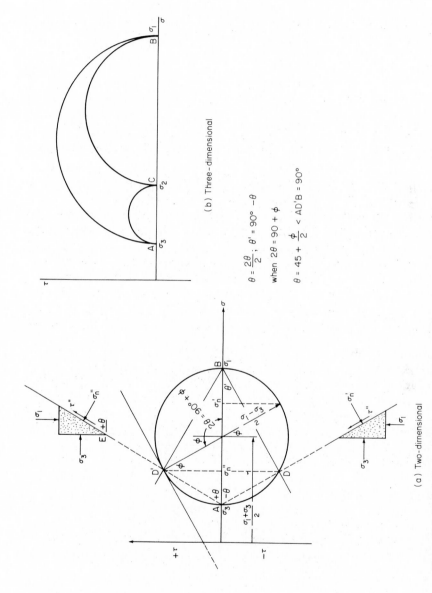

(a) Two-dimensional

(b) Three-dimensional

$$\theta = \frac{2\theta}{2}; \ \theta' = 90° - \theta$$

when $2\theta = 90 + \phi$

$$\theta = 45 + \frac{\phi}{2} \quad < AD'B = 90°$$

Figure 10-9 Mohr's circles. (a) Ray for $+\theta$ produces plane EF with same orientation as plane AB in Fig. 10-8. Ray for $-\theta$ seldom used but shown here to orient τ opposite to direction on plane EF. (b) Note that (half) circle AB includes all critical shear stresses produced by any possible combination of principal stresses, including the two Mohr circles using σ_2.

329

From this step and knowing (or plotting a number of additional points at values of 2θ using computed coordinates of σ_n, τ to produce a circle) we have a circle it is readily deduced that

$$\sigma_n = \frac{\sigma_1 + \sigma_3}{2}$$

is midway between A and B and is thus the origin of the circle. Also from step 3

$$\tau = \frac{\sigma_1 - \sigma_3}{2}$$

which is one-half the distance AB and is thus the radius of the Mohr circle.

In mechanics of materials courses where tensile stresses are taken as $(+)$, it is conventional to plot σ_1 along the horizontal axis to the right of the shear axis. In nearly all cases Mohr's circle fell in the first quadrant. It is also customary to plot only the top half of the circle since all needed information can be obtained from the resulting semicircle.

Since the principal stresses in soil masses are nearly always compression, it has become convention to plot them in the first quadrant as well—and using the largest compression stress as σ_1. Actually using the standard mechanics of materials convention of $(-)$ for compression would produce a problem since, for example, $\sigma_1 = 50$; $\sigma_3 = 10$ kPa with a strict sign convention of -50 and -10 produces a situation where -50 is smaller than -10. We avoid this by taking both values as $(+)$ and plotting in the first quadrant.

Referring to Fig. 10-9a and the geometry of a circle which obtains θ as shown, we can make the following observations:

1. Angle θ represents the orientation of a plane as in Fig. 10-8 through the element (or point). That is, by taking the origin of the ray at A (or B), any circle intercept is related to 2θ.
2. When $\theta = 45°$ $(2\theta = 90°)$, the shear stress τ is a maximum.
3. When $\theta = 90°$ $(2\theta = 180)$, the plane is coincident with the principal plane on which σ_3 acts.
4. When $\theta = 0$ the stress plane EF is coincident with the plane on which σ_1 acts.
5. Angle θ' produces planes 90° from θ planes.
6. For compressive principal stresses the shear and normal stresses on the plane defined by θ from point A are *always as shown on Fig. 10-9a*.
7. To determine the orientation of τ and σ_n, it is necessary to use a force summation as used in the derivation of Eqs. (10-10) and (10-11) and as shown on Fig. 10-8 and as used in Example 10-3 (following).

Chapter 13 will apply Mohr's circle more specifically to geotechnical stress analysis and include the particular significance of the angle ϕ shown in Fig. 10-9a.

Example 10-3 Given principal stresses $\sigma_1 = 100\,$kPa and $\sigma_3 = 20\,$kPa.

REQUIRED What are the normal and shear stresses on a plane at an angle θ produced at $\tau = 35$ and $\sigma_n = 40\,$kPa and their orientation?

SOLUTION We will plot a Mohr's circle, obtain the shear and normal stresses, and compute θ. Using these data, we will refer to Fig. 10-8 and sum forces on the stress block to verify the stress orientation.

Step 1 Plot $A = 20$ and $B = 100$ and compute the center of the circle as $\sigma_n = (100 + 20)/2 = 60\,$kPa. The radius is, by inspection, 40 kPa, and with these data the circle shown in Fig. E10-3 is drawn.

Step 2 Locate point C at $\tau = 35\,$kPa and $\sigma_n = 40\,$kPa within plot accuracy.

Step 3 Compute θ and θ':

$$\theta = \tan^{-1}\frac{35}{20} = 60.25° \qquad \theta' = \tan^{-1}\frac{35}{60} = 30.25°$$

By inspection, $\theta + \theta' = 90°$: here $60.25 + 30.25 = 90.5°$ (close enough).

Step 4 Verify the block orientation (assume τ and σ_n as on plane EF in Fig. E10-3). Compute forces on block faces along reference axes for a statics force summation.

Along σ_3: $F_3 = \sigma_3 \cos 30° = 20(0.866) = 17.3$
Along σ_1: $F_1 = \sigma_1 \sin 30 = 100(0.5) = 50$

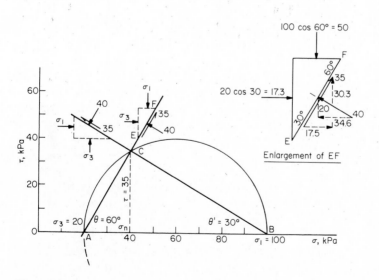

Figure E10-3

On the shear plane defined by θ:

$$F'_1 = 35 \cos 30 + 40 \sin 30 = 50.3$$

$$\sum F_1 = 50 - 50.3 = -0.3 \qquad \text{(should be zero but close enough)}$$

$$F'_3 = -35 \sin 30 + 40 \cos 30 = 17.1$$

$$\sum F_3 = 17.3 - 17.1 = 0.2 \qquad \text{(also close enough to zero)}$$

This verifies the stress magnitude and orientation as shown by plane EF in Fig. E10-3 are correct.

////

10-9 MOHR'S CIRCLE FOR INCLINED PRINCIPAL AXES

On occasion the principal axes are inclined, as when the stress planes coincide with the coordinate axes (and contain both normal and shear stresses). Mohr's circle can be used to obtain the normal and shear stresses for this situation by a somewhat indirect means. The procedure, illustrated in Example 10-4, is as follows:

1. Draw Mohr's circle as usual with the given major and minor principal stresses.
2. Draw a line from either principal stress point at the angle of inclination α which is parallel to the plane of interest on which the corresponding stress acts. For example, if σ_1 acts on a plane 15° counterclockwise to the horizontal, draw a ray BA' from B that makes an included angle of 15° between ABA' (refer to Fig. 10-9) and with A' above A.
3. From the pole point A' defined by the ray BA' we have the new origin of planes. Obtain stresses on any plane of interest by drawing a line parallel to that plane (in the stressed element) through A' and reading the stresses at the intercept of the new ray and Mohr's circle.

Example 10-4 Given the same principal stresses as in Example 10-3. Also, the stress block is oriented at $\alpha = 15°$ as in Fig. E10-4 (inset).

REQUIRED Stresses on planes $C1$ and $C2$.

SOLUTION

Step 1 Draw Mohr's circle as in Example 10-3.

Step 2 Draw ray BC at 15° to horizontal as shown.

Step 3 From C draw rays $C1$ and $C2$ parallel to $C1$ and $C2$ of inset figure.

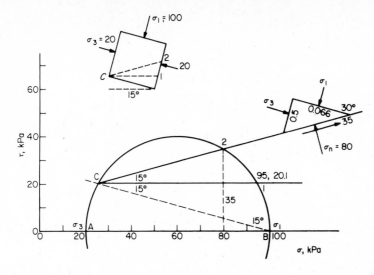

Figure E10-4

Step 4 Read values of normal and shear stress as shown on Mohr's circle. Since ray $C1$ or $C2$ is identical to the ray defining plane EF in Fig. 10-9a, the stress orientation is correctly obtained as shown. Alternatively, using the same procedures of making force summations along the X and Y axes as in Example 10-3, we can also obtain the directions shown. Note, however, with the inclined orientation of the principal axes, simultaneous equations are required to compute the shear and normal stresses which can then be compared to the graphic values.

////

10-10 BOUSSINESQ STRESSES IN AN ELASTIC HALF SPACE

Boussinesq (ca. 1885) applied some complicated mathematical concepts to some of the equations of elasticity together with the following boundary conditions (refer to Fig. 10-10):

1. Both stresses and strains vanish at $r \rightarrow \infty$.
2. Shear stresses τ are zero at the ground surface ($y = 0$).
3. Normal stresses are zero at the ground surface except at the point of load application.
4. Statics is satisfied, i.e., $\sum F_y = 0$.

In real soils the region of soil disturbance due to the load P as zone ABC in Fig. 10-10 is substantially less than infinity.

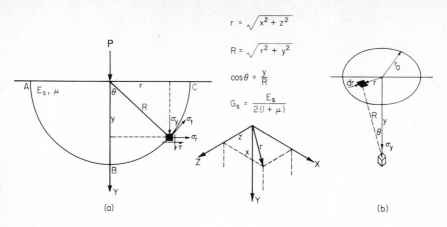

Figure 10-10 (*a*) Point load on the surface of a semi-infinite, elastic, half space for Boussinesq equations. (*b*) Application of the Boussinesq equation in reverse to obtain subsurface stresses due to application of an area load.

For the point load at the surface of a semi-infinite, homogeneous, isotropic half space, and using the symbols shown in Fig. 10-10, Boussinesq obtained the following equations:

$$\Delta y = \frac{P}{4\pi R}\frac{1}{G_s}[2(1-\mu)+\cos^2\theta] \qquad \text{(vertical settlement of point)} \quad (10\text{-}12)$$

where G_s = shear modulus of elasticity as defined earlier.

Thus

$$\sigma_y = -\frac{3P}{2\pi R^2}\cos^3\theta \qquad \text{(vertical stress)} \qquad (10\text{-}13)$$

$$\sigma_r = \frac{P}{2\pi R^2}\left(-3\cos\theta\sin^2\theta+\frac{1-2\mu}{1+\cos\theta}\right) \qquad \text{(radial stress)} \quad (10\text{-}14)$$

$$\sigma_t = \frac{P}{2\pi R^2}(1-2\mu)\left(\cos\theta-\frac{1}{1+\cos\theta}\right) \qquad \text{(tangent stress)} \quad (10\text{-}15)$$

$$\tau = -\frac{3P}{2\pi R^2}(\cos^2\theta\sin\theta) \qquad \text{(shear stress)} \qquad (10\text{-}16)$$

The Boussinesq equations are widely used to obtain the vertical and lateral stresses in a soil mass from a surface (or near surface) loading. The equations for Δy, σ_t, and τ are not used much. Most soil loads are distributed over a finite area rather than being a point load so that the equations require adjustments for practical use.

10-11 BOUSSINESQ SOLUTION FOR ROUND PLATES

Equation (10-13) can be used to obtain an exact numerical calculation of stress from a round plate (or footing) with a load producing a uniform contact pressure q_o. The same procedure can be used for square or rectangular plates which are converted to equivalent round plates of radius r_o as

$$r_o = \sqrt{\frac{A}{\pi}}$$

where $A = B^2$ for a square and BL for a rectangular plate (dimensions $B \cdot L$).

The resulting solution for a square plate is very nearly "exact" but deteriorates as L/B increases from 1. All solutions improve as the depth of interest increases since the loaded surface area converges to a point as $\theta \to 0$.

The solution for the vertical pressure beneath a round plate requires rewriting Eq. (10-13) with θ and R in terms of r and y as shown on Fig. 10-10b to obtain

$$\sigma_y = \frac{3P}{2\pi y^2 [1 + (r/y)^2]^{2.5}} \tag{10-17}$$

Now take the product of the contact pressure σ_o on area dA as $P = \sigma_o dA$ and we have

$$d\sigma_y = \frac{3\sigma_o}{2\pi y^2} \frac{1}{[1 + (r/y)^2]^{2.5}} dA \tag{10-17a}$$

With dA taken as $2\pi r \cdot dr$ and integrating along the radius from 0 to r_o we obtain

$$\sigma_y = \sigma_o \left\{ 1.0 - \left[1 + \left(\frac{r_o}{y} \right)^2 \right]^{-1.5} \right\} \tag{10-18}$$

where σ_y = stress at depth of interest in units of σ_o
$\quad r_o$ = radius (or equivalent of a rectangle) of plate
$\quad \sigma_o$ = plate contact stress = P/A

Example 10-5 What is the increase in stress at 4 m beneath the center of a circular footing of diameter = 3 m for a load $P = 1200$ kN?

Solution $\sigma_o = 1200/(0.7854 \times 3^2) = 170$ kPa. From Eq. (10-18) with $y = 4$ and $r_o = 1.5$,

$$\sigma_y = 170\{1.0 - [1.0 + (1.5/4)^2]^{-1.5}\} = 30.4 \text{ kPa}$$

////

Example 10-6 What is the increase in stress at 1.5 and 3 m beneath the center of a square footing of $B = 3$ m? The footing load is 1800 kN.

SOLUTION $\sigma_o = 1800/3^2 = 200$ kPa; $r_o = \sqrt{A/\pi} = \sqrt{9/\pi} = 1.69$ m.

For $y = 1.5$ m: $\sigma_y = 200\{1.0 - [1.0 + (1.69/1.5)^2]^{-1.5}\} = 141.5$ kPa
For $y = 3$ m: $\sigma_y = 200\{1.0 - [1.0 + (1.69/3)^2]^{-1.5}\} = 67.7$ kPa

////

10-12 NUMERICAL INTEGRATION OF THE BOUSSINESQ EQUATIONS

Since integration is the summation of all of the differential areas dA over the region of area A, the method used to obtain Eq. (10-18) suggests that the stresses at a depth y in a stratum from a rectangular surface load can be obtained by:

1. Computing $\sigma_o = P/A$.
2. Dividing area $A = B \cdot L$ into a number of "unit" areas A_i each loaded with $P_i = \sigma_o A_i$. A unit area 1×1 ft or 0.3×0.3 m is the smallest subdivision of area necessary.
3. Using Eq. (10-17) with

$$r = (x^2 + z^2)^{1/2}$$

where x and z are the surface coordinates with respect to the projected vertical position of the point of interest, $y = $ constant depth for summation of all A_i, and $P_i = $ constant for any depth y.
4. Obtaining the σ_{yi} contribution for each "unit" area with coordinates x, z.
5. Summing the σ_{yi} contribution for all of the unit areas to obtain σ_y at depth y.
6. Obtaining a vertical stress profile by incrementing y.

This procedure is readily adapted to programming on a digital computer. The location of the stressed element may be beneath the center, at midside, corner, or any other point. It is only necessary to set up the reference axes through the point of interest. This method is illustrated in Example 10-7.

Lateral stresses can be obtained similar to the above using Eq. (10-14) with

$$P = \sigma_o A_i \text{ (as for the vertical stress method)}$$

$$R = (x^2 + z^2 + y^2)^{1/2}$$

with $y = $ constant for an A_i summation then incrementing y to obtain a lateral pressure profile as against a wall.

Example 10-7 What is the stress at 2, 4, and 6 m beneath a rectangular footing 3×4 m in plan carrying 2000 kN?

SOLUTION Grid the footing into 48 units of 0.5 m, as shown in Fig. E10-7.

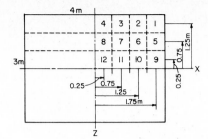

Figure E10-7

Find the stresses at 2, 4, and 6 m. Due to symmetry only $\frac{1}{4}$ footing needs to be used. Coordinates are (partial list):

Element	x	z	Element	x	z
1	1.75	1.25	5	1.75	0.75
2	1.25	1.25	6	1.25	0.75
3	0.75	1.25	7	0.75	0.75
4	0.25	1.25	8	0.25	0.75

The y coordinate is successively 2, 4, then 6 m. A computer program is used.

After inputting the 12 values, summing, recalling, and multiplying by 4, we obtain

$$\sigma_{y(0\ m)} = 166.7\ \text{kPa}\left(= \frac{2000}{12}\ \text{as contact pressure } \sigma_o;\ \text{not computed,}\atop \text{as the equation is discontinuous at this point}\right)$$

$$\sigma_{y(2\ m)} = 103.8\ \text{kPa}$$
$$\sigma_{y(4\ m)} = 45.2\ \text{kPa}$$
$$\sigma_{y(6\ m)} = 23.2\ \text{kPa}$$

////

10-13 OTHER SOLUTIONS FOR THE BOUSSINESQ EQUATIONS

Graphical solutions for the vertical stress in the form of influence charts (Newmark, 1942), or curves (Fadum, 1948) are sometimes used. Newmark (1935) integrated Eq. (10-13) to obtain the pressure beneath the corner of a rectangular plate of dimensions $B \cdot L$ at a depth y to obtain:

$$\sigma_y = \frac{\sigma_o}{4\pi}\left[\frac{2BLy\sqrt{V_0}}{y^2 V_0 + (BL)^2} \cdot \frac{B^2 + L^2 + 2y^2}{V_0} + \tan^{-1}\left(\frac{2BLy\sqrt{V_0}}{y^2 V_0 - (BL)^2}\right)\right] \quad (10\text{-}19)$$

where $V_0 = B^2 + L^2 + 1$

This equation can be conveniently solved to produce a table of influence coefficients to obtain

$$\sigma_y = \sigma_o I_\sigma \tag{10-19a}$$

on a digital computer. This is done with ratios M and N and introducing variables V and V_1 as follows:

$$M = \frac{B}{y} \qquad N = \frac{L}{y} \qquad V = M^2 + N^2 + 1 \qquad V_1 = (MN)^2$$

Substituting these ratios into Eq. (10-19), we obtain

$$\sigma_y = \sigma_o \frac{1}{4\pi} \left[\frac{2MN\sqrt{V}}{V + V_1} \cdot \frac{V + 1}{V} + \tan^{-1} \left(\frac{2MN\sqrt{V}}{V - V_1} \right) \right] \tag{10-19b}$$

When $V_1 > V$ the arctan term is $(-)$ and it is necessary to add π. In passing, note that Newmark allowed use of either $\tan^{-1}$ or $\sin^{-1}$ (but with appropriate changes in the terms and application of π when $V_1 > V$). This equation can be programmed to produce a table of influence coefficients as in Table 10-1. Inspection of the ratios used in this table indicates that it is rather complete and that little interpolation is required. The vertical soil stress at a depth y can be obtained for almost any surface load configuration by using both real and fictitious "rectangles" as illustrated for Example 10-9.

Example 10-8 Redo Example 10-6, using Table 10-1.

SOLUTION Make sketch E10-8, dividing the footing into four rectangles with their common corners directly above the point of interest. Obtain $y = 1.5$ and 3 m and $\sigma_o = 200$ kPa.
For $y = 1.5$ m

$$M = \frac{1.5}{1.5} = 1.0 = N$$

Obtain $I_\sigma = 0.175$ from Table 10-1. For the four contributing corners, obtain

$$\sigma_y = 4\sigma_o I_\sigma = 4(200)(0.175) = 140 \text{ kPa}$$

This value compares to the approximate value of 141.5 using an equivalent round footing.

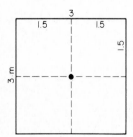

Figure E10-8

Table 10-1 Values of I_σ for depth ratios given for use in Eq. (10-19a)
Note that $M = A/y$ and $N = B/y$ are interchangeable.

$N=B/y$ \ $M=L/y$	0.100	0.200	0.300	0.400	0.500	0.600	0.700	0.800	0.900	1.000	1.100
0.1	0.005	0.009	0.013	0.017	0.020	0.022	0.024	0.026	0.027	0.028	0.029
0.2	0.009	0.018	0.026	0.033	0.039	0.043	0.047	0.050	0.053	0.055	0.056
0.3	0.013	0.026	0.037	0.047	0.056	0.063	0.069	0.073	0.077	0.079	0.082
0.4	0.017	0.033	0.047	0.060	0.071	0.080	0.087	0.093	0.098	0.101	0.104
0.5	0.020	0.039	0.056	0.071	0.084	0.095	0.103	0.110	0.116	0.120	0.124
0.6	0.022	0.043	0.063	0.080	0.095	0.107	0.117	0.125	0.131	0.136	0.140
0.7	0.024	0.047	0.069	0.087	0.103	0.117	0.128	0.137	0.144	0.149	0.154
0.8	0.026	0.050	0.073	0.093	0.110	0.125	0.137	0.146	0.154	0.160	0.165
0.9	0.027	0.053	0.077	0.098	0.116	0.131	0.144	0.154	0.162	0.168	0.174
1.0	0.028	0.055	0.079	0.101	0.120	0.136	0.149	0.160	0.168	0.175	0.181
1.1	0.029	0.056	0.082	0.104	0.124	0.140	0.154	0.165	0.174	0.181	0.186
1.2	0.029	0.057	0.083	0.106	0.126	0.143	0.157	0.168	0.178	0.185	0.191
1.3	0.030	0.058	0.085	0.108	0.128	0.146	0.160	0.171	0.181	0.189	0.195
1.4	0.030	0.059	0.086	0.109	0.130	0.147	0.162	0.174	0.184	0.191	0.198
1.5	0.030	0.059	0.086	0.110	0.131	0.149	0.164	0.176	0.186	0.194	0.200
1.8	0.031	0.061	0.088	0.113	0.134	0.152	0.167	0.180	0.190	0.198	0.205
2.0	0.031	0.061	0.089	0.113	0.135	0.153	0.169	0.181	0.192	0.200	0.207
2.5	0.031	0.062	0.089	0.114	0.136	0.155	0.170	0.183	0.194	0.202	0.209
3.0	0.031	0.062	0.090	0.115	0.137	0.155	0.171	0.184	0.195	0.203	0.211
5.0	0.032	0.062	0.090	0.115	0.137	0.156	0.172	0.185	0.196	0.204	0.212
10.0	0.032	0.062	0.090	0.115	0.137	0.156	0.172	0.185	0.196	0.205	0.212

$N=B/y$ \ $M=L/y$	1.200	1.300	1.400	1.500	1.800	2.000	2.500	3.000	5.000	10.000
0.1	0.029	0.030	0.030	0.030	0.031	0.031	0.031	0.031	0.032	0.032
0.2	0.057	0.058	0.059	0.059	0.061	0.061	0.062	0.062	0.062	0.062
0.3	0.083	0.085	0.086	0.086	0.088	0.089	0.089	0.090	0.090	0.090
0.4	0.106	0.108	0.109	0.110	0.113	0.113	0.114	0.115	0.115	0.115
0.5	0.126	0.128	0.130	0.131	0.134	0.135	0.136	0.137	0.137	0.137
0.6	0.143	0.146	0.147	0.149	0.152	0.153	0.155	0.155	0.156	0.156
0.7	0.157	0.160	0.162	0.164	0.167	0.169	0.170	0.171	0.172	0.172
0.8	0.168	0.171	0.174	0.176	0.180	0.181	0.183	0.184	0.185	0.185
0.9	0.178	0.181	0.184	0.186	0.190	0.192	0.194	0.195	0.196	0.196
1.0	0.185	0.189	0.191	0.194	0.198	0.200	0.202	0.203	0.204	0.205
1.1	0.191	0.195	0.198	0.200	0.205	0.207	0.209	0.211	0.212	0.212
1.2	0.196	0.200	0.203	0.205	0.210	0.212	0.215	0.216	0.217	0.218
1.3	0.200	0.204	0.207	0.209	0.215	0.217	0.220	0.221	0.222	0.223
1.4	0.203	0.207	0.210	0.213	0.218	0.221	0.224	0.225	0.226	0.227
1.5	0.205	0.209	0.213	0.216	0.221	0.224	0.227	0.228	0.230	0.230
1.8	0.210	0.215	0.218	0.221	0.227	0.230	0.233	0.235	0.237	0.237
2.0	0.212	0.217	0.221	0.224	0.230	0.232	0.236	0.238	0.240	0.240
2.5	0.215	0.220	0.224	0.227	0.233	0.236	0.240	0.242	0.244	0.244
3.0	0.216	0.221	0.225	0.228	0.235	0.238	0.242	0.244	0.246	0.247
5.0	0.217	0.222	0.226	0.230	0.237	0.240	0.244	0.246	0.249	0.249
10.0	0.218	0.223	0.227	0.230	0.237	0.240	0.244	0.247	0.249	0.250

For $y = 3$ m

$$M = \frac{1.5}{3} = 0.5 = N \quad \text{and} \quad I_\sigma = 0.084$$

$$\sigma_y = 4(200)(0.084) = 67.2 \text{ kPa (vs. 67.7)}$$

////

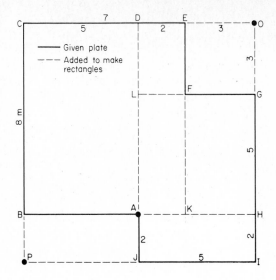

Figure E10-9

Example 10-9 Find the stress at $y = 6$ m below point A in Fig. E10-9 for some contact stress σ_o.

SOLUTION Add dotted lines shown and number corners as shown so the contributing rectangles can be identified. Note that point P is for a home problem. By inspection, the stress σ_y will be due to corner contributions as follows:

$$ABCD + AHIJ + ALGH + ADEK - ALFK$$

Note that we subtract area $ALFK$ since it is included twice (in $ALGH$ and in $ADEK$). Since M and N are interchangeable, it is not necessary to formally identify the length and width of any rectangle as long as we compute both ratios M and N.

For $ABCD$: $M = 5/6 = 0.8$ $\qquad N = 8/6 = 1.3$; $\qquad I_\sigma = 0.171$
For $AHIJ$: $M = 2/6 = 0.3$ $\qquad N = 5/6 = 0.8$; $\qquad I_\sigma = 0.073$
For $ALGH$: $M = N = 5/6 = 0.8$; $\qquad\qquad\qquad I_\sigma = 0.146$
For $ADEK$: $M = 2/6 = 0.3$ $\qquad N = 8/6 = 1.3$; $\qquad I_\sigma = 0.085$
For $ALFK$: $M = 2/6 = 0.3$ $\qquad N = 5/6 = 0.8$; $\qquad I_\sigma = 0.073$

$$\sigma_y = \sigma_o \sum I_\sigma = \sigma_o(0.171 + 0.073 + 0.146 + 0.085 - 0.073) = \sigma_o(0.402)$$

The reader should verify that the stress at point 0 at depth y is

$$OIJD + OHBC - OHAD - OGFE$$

We see that both $OIJD$ and $OHBC$ include $OHAD$, so we have to take it away once. This still leaves one inclusion too many for $OGFE$, which we also remove.

////

10-14 PRESSURE BULBS AND STRESS CONCENTRATION FACTORS

The pressure from Eq. (10-13) produces a bell-shaped pressure distribution on a horizontal plane at depth y as illustrated in Fig. 10-11. Some early experiments indicated that the shape of the bell depends on the soil and that Eq. (10-13) should be written (Cummings, 1936) as

$$\sigma_y = -\frac{nP}{2\pi R^2} \cos^n \theta \qquad (10\text{-}20)$$

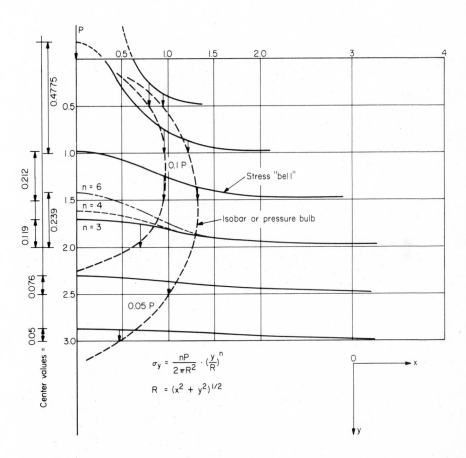

Figure 10-11 Pressure profile along X axis at depths shown from Eq. (10-13) using $n = 3$ except as shown. From scale plot one can obtain pressure bulbs from a trace of the locus of points of equal pressure. Note the pressure concentration when $n = 4$ and 6. Bell is symmetrical so only one side shown. This diagram illustrates that a point away from the loaded area may be subjected to a superposition of stresses from several surface loads.

where n is a concentration factor or index that is soil-dependent. If we define the ratio of the lateral σ_3 to vertical σ_1 stress ratio as

$$K = \frac{\sigma_3}{\sigma_1}$$

we can use Hooke's law [Eqs. (10-7)] to obtain a relationship between K and μ as

$$K = \frac{\mu}{1 - \mu}$$

Krynine in the discussion by Cummings (1936) shows that $n = 2 + 1/K$, so that for an incompressible material such as water, $\mu = 0.5$ giving $K = 1$ and $n = 3$. It would appear that the range of n is $3 < n \le 6$. Near the ground surface some experimental evidence indicates that $n > 3$ but may well approach 3 at depths in the ground where the soil is more dense, stiffer, and substantially "confined."

It is common to use $n = 3$ and all the material here, unless specifically noted, is based on using $n = 3$. Newmark's discussion of Cummings (1936) showed that the factors M and N of Eq. (10-19b) can be multiplied by $\sqrt{n/3}$ to produce the equivalent of using a concentration factor other than 3.

If we draw contours of equal pressure through several horizontal planes beneath a loaded area, the pressure isobars form a bulb of pressure as for 0.1 and $0.05P$ in Fig. 10-11. This is more commonly done for surface loaded plates than for a concentrated load which was used here for illustration.

10-15 THE VERTICAL PRESSURE PROFILE

The vertical pressure profile is of considerable use in obtaining the stress increase at some depth beneath a loaded area or the average stress increase in a stratum of some thickness H. This latter is used in Chap. 11 for settlement computations.

Vertical pressure profiles are widely available in the form of pressure bulbs for round, square and strip (infinitely long) plates. The methodology is as shown for a point load in Fig. 10-11, but using Eq. (10-19b). A rectangle is divided into a series of four smaller rectangles with a common corner, and the vertical pressure profile is obtained. Points from the footing center to about $0.75B$ from the center, using $B/8$ increments (the last two points are off the footing), gives seven vertical profiles which are adequate to extrapolate a series of pressure bulbs.

Generally only the pressure profile is required at the center of the footing or at a corner. For these cases it may be more convenient to use a series of vertical pressure profiles as Fig. 10-12 which are for the center of a footing of type labeled. The corner pressure is one-fourth of the center value. These pressure profiles are readily obtained by using Eq. (10-19b), where y is expressed in terms of the least lateral plate dimension B.

Figures 10-11 and 10-12 illustrate that the zone of significant stress influence is on the order of $2B$ and is the basis for the depths of borings being on this

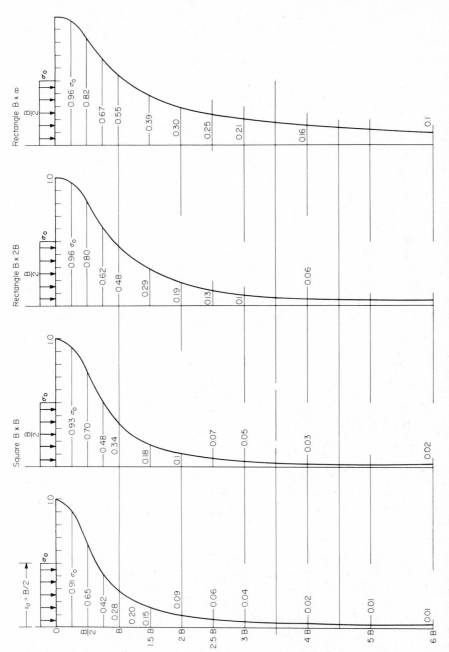

Figure 10-12 Vertical profile in terms of B for several footings shown. Pressures are at center of footing as shown. For footing corner, use one-fourth of pressure coefficient given on profile.

order. Whether $2B$ is strictly valid (and obviously is not for rectangular footings) depends on the soil profile and the E_s and μ values of the several strata in the zone of interest. In many cases, such as sand and silt layers over clay, a homogeneous mass is taken. There is some evidence that the computed results are not significantly in error when this is done.

Where allowance for the several soil strata is deemed necessary, one may find published solutions—usually in terms of a modular ratio E_1/E_2 (and $\mu = 0$ or a constant) as in Poulos and Davis (1974) or make recourse to a finite-element solution. This latter requires sufficient soil tests to reliably determine E_s and μ—including anisotropy—or the resulting computer output will only be an expensive computational exercise.

Example 10-10 Redo Example 10-6 using Fig. 10-12.

SOLUTION Example 10-6 required the stress increase at 1.5 and 3 m beneath a square footing of $B = 3$ m loaded with 1800 kN.

$$\sigma_o = \frac{1800}{3^2} = 200 \text{ kPa}$$

At $y = 1.5$ m, the depth ratio $= 1.5/3 = 0.5B$ and from Fig. 10-12

$$\sigma_y = 0.7\sigma_o = 0.7(200) = 140 \text{ kPa} \qquad \text{(vs. 141.5 previously)}$$

At $y = 3$ m, the depth ratio $= B$ and

$$\sigma_y = 0.34\sigma_o = 0.34(200) = 68 \text{ kPa} \qquad \text{(vs. 67.7 previously)}$$

////

10-16 THE AVERAGE PRESSURE

The average stress increase in a stratum of thickness H can be found by numerically integrating over the depth of interest to obtain the area A_s of the stress profile so that

$$\sigma_{y(av)}H = A_s$$

The trapezoidal formula can be used to numerically integrate an area bounded by curved lines (as in Fig. 10-12) if the vertical increment Δy between points is held constant. Using the trapezoidal formula to compute A_s, we have

$$\sigma_{y(av)}H = A_s = \Delta y \left[\frac{\sigma_{y1} + \sigma_{yk}}{2} + \sigma_{y2} + \sigma_{y3} + \cdots + \sigma_{y(k-1)} \right] \qquad (10\text{-}21)$$

Sufficient points k should be taken that the variation in stress σ_{yk} between any two points (if plotted) appears linear. It should be evident that at depths greater than B and for small values of H (thin layers) the average stress may be taken as the first term of Eq. (10-21) with $\Delta y = H$.

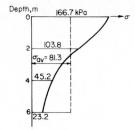

Figure E10-11

Example 10-11 What is the average stress increase in the 6-m depth beneath the footing in Example 10-7?

SOLUTION We must also use the pressure at the ground surface of $\sigma_o = 166.7$ computed in the example.

We will arbitrarily plot the pressure profile to illustrate the concept of average pressure as in Fig. E10-11, which shows the pressures computed in the example. From the profile and using Eq. (10-21) with $H = 6$ and $\Delta y = 2$ m,

$$\sigma_{y(av)}H = 2\left(\frac{166.7 + 23.2}{2} + 103.8 + 45.2\right) = 487.9 \text{ kN/m}$$

$$\sigma_{y(av)} = \frac{487.9}{6} = 81.3 \text{ kPa}$$

Some slight improvement in the average pressure might be obtained by obtaining 3 additional pressures so that $\Delta y = 1$ m.

////

10-17 THE NEWMARK INFLUENCE CHART

Direct numerical solutions of the Boussinesq equations are convenient especially when using a computer program (or programmable calculator) as illustrated by Example 10-7. Direct solution using Table 10-1 or Fig. 10-12 are also very convenient. Fadum (1948) plotted Table 10-1 in the form of curves (typically I_σ vs. M with N = constant), but there is no particular advantage unless the curve scale is large.

Newmark (1942) reinterpreted Eq. (10-18) by rearranging it to read

$$\frac{r_o}{y} = \left[\left(1 - \frac{\sigma_y}{\sigma_o}\right)^{-2/3} - 1.0\right]^{1/2}$$

The (+) root is used and r_o/y is interpreted as the relative size of a circular bearing area which gives a unique pressure ratio σ_y/σ_o at depth y in the stratum.

We may take values of σ_y/σ_o from 0 to 1.0 in 0.1 (or other constant series) increments to obtain corresponding r_o/y values. These r_o/y values can be used to draw a "Newmark chart" as in Fig. 10-13. Drawing rays from point 0 to make approximate squares and counting the "squares" Z thus formed (or that would be formed if the full range of σ_y/σ_o is used), we obtain the influence factor for the chart as $I_N = 1/Z$.

The scale used to draw the concentric circles is the distance AB shown on the chart. In use we draw the footing (or footing group) to scale $AB = y$ and place the point where the pressure at depth y is desired over "0." We then count the

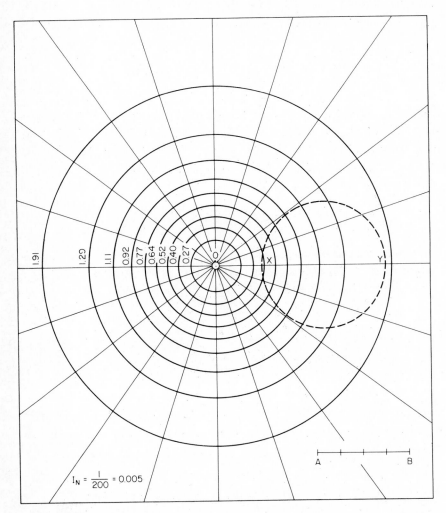

Figure 10-13 Newmark's chart with I_N is shown. Numbers such as 1.91 are units of AB used to draw concentric rings based on σ_y/σ_o increment $= 0.1$. Dotted circle XY is Example 10-12 superimposed. Use numbers shown to make a larger Newmark chart for general use.

"squares" Z_f falling within the footing area (estimating partial squares) and obtain the required pressure as

$$\sigma_y = \sigma_o Z_f I_N \qquad (10\text{-}22)$$

For a vertical pressure profile, we redraw the footing plan at successive scales of $AB = y_i$ and repeat the above procedure. The Newmark chart has major use for very irregular shaped and round footings and when the point of interest is not directly beneath the footing.

Example 10-12. Determine the stress at $y = 6$ m beneath point A, which is 3 m from the footing of $D = 8$ m. Take $\sigma_o = 300$ kPa.

SOLUTION We can do one of two things:

(a) Use Newmark's chart.
(b) Convert the footing to an equivalent square and use Table 10-1.

We will do both of these and compare results. Note that normally a transparent footing to scale is used to lay over a Newmark chart so the "squares" can be seen for counting. Here we will superimpose the footing to scale directly in Fig. 10-13 along points O-X-Y.

The depth y used with AB gives distance

$$OX = 3\,\frac{(AB)}{6}$$

$$XY = 8\,\frac{(AB)}{6}$$

We locate OX as shown, although any orientation could have been used as long as point A is coincident with "0." Using the given orientation allows use of symmetry, so only one-half of the footing plan needs to be counted. Counting "squares," it is estimated $Z_f = 10.5$ and doubling since only one-half was counted, we obtain $Z_f = 21$ (total) and using Eq. (10-22), obtain

$$\sigma_y = 300(21)(0.005) = 31.5 \text{ kPa}$$

Let us now compute the stress using Table 10-1. For this refer to Fig. E10-12

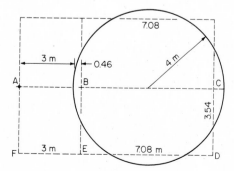

Figure E10-12

where the footing is converted to an equivalent square (same area), so that the side B is

$$B^2 = \pi R^2 \qquad B = (\pi \times 4^2)^{1/2} = 7.08 \text{ m}$$

We will adjust the distance from the edge of the footing to point A as shown so that it is still 3 m, although it appears that it should be larger. With this we next develop the contributory rectangles (two) at point A and note that we will have

$$2(ABCDEF - ABEF)$$

We also note that M and N are interchangeable and

$$ABCDEF: \quad \text{For } y = 6 \text{ m} \qquad M = \frac{3.54}{6} = 0.59 \quad \text{and} \quad N = \frac{10.08}{6} = 1.68$$

$$ABEF: \qquad M = \frac{3}{6} = 0.5 \qquad N = \frac{3.54}{6} = 0.59$$

Rounding: For $M = 0.6$ and $N = 1.7$ $\qquad I_\sigma = 0.151$

For $M = 0.5$ and $N = 0.6$ $\qquad I_\sigma = 0.095$

$$\sigma_y = 2(0.151 - 0.095)(300) = 33.6 \text{ kPa}$$

(vs. 31.5 from Newmark chart). We might note that this latter is considerably simpler than using the Newmark chart. The Newmark chart would be more useful if there were several footings grouped about point A (unless symmetry could be used).

////

10-18 WESTERGAARD'S STRESS

Westergaard (1938) proposed that the vertical stress in an elastic solid consisting in alternating thin layers of rigid reinforcements (such as sand seams in a clay stratum) could be computed as

$$\sigma_y = \frac{Pa}{2\pi^2 y^2} \left(\frac{1}{a^2 + (r/y)^2} \right)^{1.5} \tag{10-23}$$

where

$$a = \sqrt{\frac{1 - 2\mu}{2 - 2\mu}}$$

P = point surface load
μ = Poisson's ratio for the soil between the rigid reinforcements

We may treat Eq. (10-23) like the Boussinesq equation and obtain the vertical stress at the center of a round footing as

$$\sigma_y = \sigma_o \left(1.0 - \sqrt{\frac{a}{a + (r_o/y)^2}} \right) \tag{10-24}$$

Solving Eq. (10-24) for r_o/y and using the $(+)$ root, we obtain

$$\frac{r_o}{y} = \sqrt{\frac{a}{(1.0 - (\sigma_y/\sigma_o)^2} - a} \tag{10-25}$$

This latter equation can be used to obtain r_o/y values for arbitrary values of σ_y/σ_o (0.1, 0.2, etc.) as with Eq. (10-18). An influence chart similar to Newmark's chart in Fig. 10-13 can be drawn, but now dependent on Poisson's ratio μ.

Westergaard's solution tends to produce larger stresses in the immediate vicinity of the footing base between 0 and 2B. This solution is not much used, however, since few soils truly meet the alternating rigid layer criterion. It may be used with the Boussinesq solution to produce a range of possible stresses.

Another reason for the lack of popularity of this equation is the need for a realistic value of Poisson's ratio. Values of 0 and 0.5 are commonly used for computational convenience but unless realistic values are used the Boussinesq equations provide adequate answers without the additional computational effort.

10-19 SOIL DEFORMATION OR SETTLEMENT

When a load is applied to any elastic body, strains are produced. The summation of strain over the stressed length is the deformation or settlement. With soil, the strain (deformation/unit length) is produced primarily by a combination of particle rolling and slippage and sliding displacements and minimally by elastic distortions of the particles. The deformation ΔH is the statistical summation of the strains in the direction of interest over the stressed depth. We may formally evaluate the settlement ΔH as

$$\Delta H = \int^L \varepsilon \, dL \tag{10-26}$$

where $\varepsilon = \sigma/E_s$ = strain and depends on Hooke's law [Eq. (10-7)] with μ
    L = length (or depth) over which the stress σ is significant

With a concrete or steel column axially loaded, ΔH is a relatively simple computation since the strain (load, area, and E) and length are rather well defined. For soil, however, the problem is formidable. In general, the Boussinesq equations indicate stress decreases with depth at a nonlinear rate. We have already noted that the soil "stiffens" with depth from confinement and increased packing; thus E_s increases (nonlinearly) at the same time that σ decreases. We also have

Poisson's ratio, which appears to be stress-dependent, to contend with. Anisotropy in the soil mass means that we do not know the depth L over which the significant stress σ acts. It is evident that the effective stress produces the collective soil skeleton changes termed "strain." In summary, then, ΔH for geotechnical work is at best an estimate.

Since effective stresses produce the strain (and settlements), pore pressure, and the coefficient of permeability k which controls pore drainage are parameters relating how suddenly the settlement ΔH occurs. It becomes evident that settlements which depend on k are time-dependent. Those in which k is not a factor occur rapidly. In geotechnical work we distinguish between these types of settlements as:

1. Consolidation—time-dependent settlements which occur in saturated or partially saturated fine-grained soils that have relatively low coefficients of permeability. The estimated time for these settlements may be from months to several hundred years.
2. Immediate—settlements which occur in hours to less than a month after application of load. Typical soils are sands, sandy gravels, silty sands, and fine-grained soils with a low degree of saturation S. The degree of S is generally not important for sands and sandy soils. In all these soils pore drainage is very rapid, so the settlement occurs "immediately" after the application of load. These settlements, illustrated in Fig. 10-14 as ΔH, will be considered in Chap. 15.
3. Creep—long-term settlements which tend to occur at the end of consolidation settlements but may also occur after "immediate" settlements. These settlements represent the final positioning of the soil grain matrix under load. Soils typically are fine-grained and/or organic. Both this and consolidation settlement are considered in Chaps. 11 and 12.

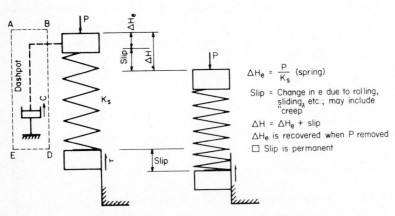

Figure 10-14 A rheological soil model commonly used (excluding *ABDE*) for immediate soil settlements. Slip above is a combination of creep and other permanent soil skeleton changes resulting in a permanent reduction in void ratio. In general, ΔH_e is a very small value.

Equation (10-26) is used in all three types of settlement computation, as will be shown in the following chapters. Semiempirical equations are often used for immediate settlements, but if a strain and depth of influence estimate can be made, we can always use Eq. (10-26)—or a numerical equivalent—to obtain an estimate of ΔH.

10-20 RHEOLOGICAL MODELS

Rheology is a study of the behavior of a material in a fluid state. The statistical "flow" of the soil skeleton from particle rolling, sliding, and slipping in the direction of interest and producing a permanently reduced void ratio can be considered a problem in rheology. The change in void ratio is time-dependent if water must be squeezed from the soil voids in order for the change in e to occur. The length of time is dependent on the coefficient of permeability k and will be much larger for fine-grained soils, where k is small, compared to sands and gravels, in which k is often several hundred times larger.

The simplest rheological model is the spring element or *hookean* model (the spring shown in Fig. 10-14). A dashpot element (also shown in Fig. 10-14) resists immediate displacements similar to the shock absorber of an automobile. In soil the dashpot or *newtonian* element constant C is directly related to the coefficient of permeability. The permanent reduction in void ratio is related to a *yield stress* model (shown producing the "slip" in Fig. 10-14). The yield or slip occurs at some minimum stress level f and in soil may halt after some amount of strain either from strain hardening or other factors (sufficient remolding and reduction in e produces a new material).

In most real soils we have some combination of these three models called *model coupling*. For example, a soil might have the hookean and yield stress models coupled in series as shown in Fig. 10-14. If pore water is present, the newtonian model may be included in parallel with one of the other basic elements. Figure 10-14 displays the dashpot in parallel with the hookean element.

The spring and dashpot in parallel can also be illustrated as in Fig. 10-15 where the pore water pressure and subsequent drainage through the valve produce the dashpot effect. The spring in parallel with a dashpot is termed a *Kevin model*.

The rheological model of Fig. 10-14 can be idealized as a Kevin model in series with a yield stress model as in Fig. 10-16 and using unit stress σ and strain ε for a convenience. We can obtain an expression for the instantaneous strain as follows:

Passing a plane at A-A and summing stress, we have

$$\sigma = K\varepsilon + C\frac{d\varepsilon}{dt} \tag{10-27}$$

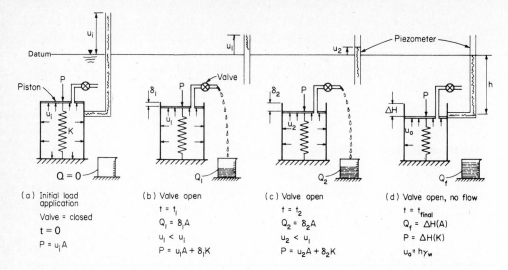

Figure 10-15 The Kevin model part of Fig. 10-14 as a piston-spring combination. The valve opening is related to the coefficient of permeability and dashpot C. Note that a steady state condition for the model has been obtained in part d when the pore pressure stabilizes and the spring carries the applied load. In real soils some additional creep settlement will occur so that some type of yield stress model needs to be added in series.

where $d\varepsilon/dt$ = rate of change in strain (velocity of the deformation)

ε = instantaneous unit strain

K = spring constant (force/length)

σ = unit stress (force/area)

C = dashpot coefficient (force $\cdot$ time/area)

Since Eq. (10-27) is of the standard form of $dx/dy + Mx = N$ for differential equation solution, it is readily solved, and taking a steady state value of σ and with $\varepsilon = 0$ at time $t = 0$, we obtain

$$\varepsilon = \frac{\sigma}{K} (1 - e^{-Kt/C}) \qquad (10\text{-}28)$$

Here $e = 2.71828$ the base of natural logarithms. With a yield stress model in series with the above (and assuming superposition is valid), the strain is the sum of the above and that from the yield stress model based on f = limiting stress to initiate "yield." Adding this strain, we obtain

$$\varepsilon = \frac{f}{E_s} + \frac{\sigma}{K} (1 - e^{-Kt/C}) \qquad (10\text{-}29)$$

A qualitative plot of strain versus time for this equation is shown in Fig. 10-16b.

The crude model used to obtain Eq. (10-29) illustrates the complexity of the problem. Here we are attempting to use superposition in coupling the elemental

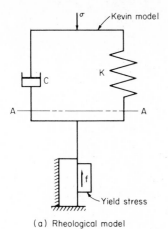

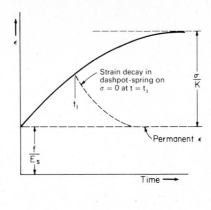

(a) Rheological model   (b) Time-displacement diagram

Figure 10-16 Coupling of a spring and dashpot to produce a "Kevin" model which is in series with a "yield stress" model. The resulting time-displacement diagram is as shown. Note the permanent "set" from the yield stress model.

models, but how can superposition be valid? It is very likely a part of the creep or "yield" occurs in advance of the coupled spring-dashpot response, some must occur in parallel (as pore water is squeezed out and e changes) and some after; does this mean that we should use three yield stress models? We use a spring to model the soil skeleton elasticity, but available evidence indicates that soil grain elastic deformations are only a very small fraction of the total strain. Finally, what numerical values do we assign for $K, f,$ and the dashpot constant C?

Because of these several problems, rheological models are not used at present to make settlement analyses. Their principal use is in illustrating the idealized interacting components of a soil mass under a stress which produces time-dependent strains.

10-21 SUMMARY

This chapter has presented an introduction to the following concepts:

1. Theory of elasticity—particularly as applicable to geotechnical work.
2. Octahedral stresses—since the octahedral plane contains the stresses obtained from the widely used triaxial test.
3. Principal stresses—since many soil stresses begin with a reference set of principal stresses from an initial steady state condition. Note in particular that the principal stress difference $\sigma_1 - \sigma_3$ is termed the *deviator stress*. This stress term is widely used.
4. Mohr's circle for plane stress conditions—since Mohr's circle is widely used (as in Chap. 13) to obtain soil strength parameters. Note that the diameter of Mohr's circle is the deviator stress.

5. The Boussinesq equations for stress analysis—since these equations are widely used to obtain stresses within a soil mass from some surface loading. We used the equations both directly and also presented
   a. Numerical methods for stress evaluation
   b. Newmark's chart for stress evaluation
   c. The concept of these stresses producing a bulb of pressure
   d. The concept of a stress profile and numerical integration to obtain an "average" stress over this stress profile
   e. Some of the limitations on using the Boussinesq method and noted the Westergaard alternative procedure
6. Elements of rheology and the three basic models—we also coupled the models into a "soil" model and discussed the several limitations that are involved in attempting to model the soil in this manner.

The conceptual relationship depicted in Fig. 10-15 should be carefully understood as it is immediately used in Chap. 11 along with a vertical pressure profile from which an "average" stratum pressure is obtained.

PROBLEMS

10-1 Replot Fig. 10-4, using the data from Example 10-1. Also plot the effective stress path if the pore water pressure for $\sigma_1 = 40$ is $u = 5$.

10-2 For an initial triaxial stress condition of $\sigma_1 = \sigma_2 = \sigma_3 = 40$ kPa, what are the octahedral stresses? When the deviator stress is 70 kPa, what are the new octahedral stresses?
  Partial ans.: $\tau_{oct} = 33$ kPa

10-3 For the stress-strain plot in Fig. E10-2, what is the tangent modulus for a tangent at $\varepsilon = 0.05$ for curve $\sigma_3 = 0$? What is the secant modulus between $\varepsilon = 0.05$ and $\varepsilon = 0.10$ for both curves?
  Partial ans.: $E_t = 540$ kPa; $E_{sec} = 410$ kPa ($\sigma_3 = 0$)

10-4 A two-dimensional stress condition has $\sigma_1 = 65$ kPa and a deviator stress of 40 kPa. The angle ϕ as shown on Fig. 10-9 is 35°. Find the shear stress at failure and display the orientation of the failure surface.

10-5 The deviator stress is 60 kPa, and $\tau = 10$ kPa when $\sigma_n = 30$ kPa. Draw a Mohr's circle and obtain both σ_1 and σ_3.
  Partial ans.: $\sigma_1 = 83$ kPa (scaled)

10-6 Draw the bell of pressure to a suitable scale for a point load using Eq. (10-20) with $n = 3$, 4, and 6. Take $y = 4$ m (depth) and $P = 10$ kN. Draw only one-half of the bell since it is symmetrical.
  Ans.: σ_y at center = 4.79, 6.37, and 9.55 kPa

10-7 Compute the vertical pressure profile in terms of contact pressure q_0 beneath a round footing of diameter = 6 m from a depth of 0.75B to 2.5B. Find the average pressure in this depth using Eq. (10-21).
  Partial ans.: $\sigma_{y(av)} = 0.16q_0$ (programmed on a calculator)

10-8 What is the average increase in pressure in terms of the contact pressure q_0 beneath a square footing of $B = 6$ m in the depth from 3 to 7 m beneath the center of the footing (stratum 4 m thick)?
  Ans.: $0.53q_0$

10-9 Do Prob. 10-8, assuming that the footing is 6×9 m (rectangle).

    Ans.: $\sigma_{y(av)} = 0.55q_0$

10-10 Redo Example 10-9 for point P.

10-11 Redo Example 10-12 (using Newmark's chart) if $y = 5$ m instead of 6 m below point A.

10-12 Redo Example 10-12, using Table 10-1 if $y = 5$ m instead of 6 m below point A.

    Ans.: 31.2 kPa (it is less than at 6 m; why?)

ELEVEN

CONSOLIDATION AND CONSOLIDATION SETTLEMENTS

11-1 SOIL CONSOLIDATION AND SETTLEMENT PROBLEMS

All soils subjected to stress undergo strain within the soil skeleton. This strain is caused by rolling, slipping, sliding, and to some extent by crushing at the particle contact points, and elastic distortions. The statistical accumulation of these deformations in the direction of interest is the strain. The integration of strain (deformation per unit length) over the depth of influence (the total length) is the "settlement." This method of producing settlement is mostly nonrecoverable when the stress is removed since a permanent reduction in void ratio has been produced. Noting that we are involved with a change in the soil skeleton it follows that the effective stress defined by either Eq. (2-19) or (2-20) produces the strain. Strains in all coarse grained, and dry or partially saturated fine-grained, soils occur very soon (termed *immediate*) after application of stress. Application of stress to saturated (and nearly saturated) fine-grained soils produces strains that are time-dependent. The resulting settlements are time-dependent and are termed *consolidation* settlements. The length of time involved is based on the rate of consolidation.

The length of time for consolidation settlements to take place depends on how fast the excess pore pressure caused by the applied load can dissipate (see Fig. 10-15). The coefficient of permeability is thus a significant factor as well as how great a distance the pore fluid being expelled from the reduced-in-size pores must travel to dissipate the excess pressure. Since the clay minerals tend to have adsorbed layers of water the effective pressure defined by Eq. (2-20) seems to be most applicable—both theoretically and from laboratory measurements.

In the past many problems (buildings, levees, dams, road fills, etc.) were caused by failure to recognize that settlements may be rate processes, and could continue for years with large final total settlements. The tilt in the Leaning Tower of Pisa is at least partly due to differential consolidation settlements that have been ongoing for around 700 years. Large settlements in the Mexico City area of up to 5+ m occurred over spans exceeding 50 years. Areal settlements around Houston, Texas of 1 to 5 m in places are consolidation type settlements. Both these latter settlements are believed to be a combination of stress and from pumping water (and oil in Texas) from compressible strata.

On a more local basis, the observant reader may observe bumps in roads at transitions from cuts to fills and at bridge abutments. This is not a universal occurrence and is sometimes due to subsidence in poorly compacted fills rather than consolidation settlement within the underlying soil. In other cases, however, these settlements are due to consolidation in one or more strata underlying the fill. These settlements are obviously time-dependent; otherwise, when the fill was constructed, settlements would occur. The roadbed would then be filled and graded to final elevation and the pavement would be poured. The result with no further settlement would be a smooth pavement surface. Where it is practical, road engineers prefer to place the larger fills during one construction season and lay the pavement the next (or later) to allow both fill subsidence and consolidation settlement to occur. This effectively produces a preload as briefly discussed in Chap. 7.

In this chapter we will find that the potential for consolidation settlements can be reliably determined by soil exploration. Methods of predicting the magnitude of the consolidation settlement and the length of time (in Chap. 12) for this to occur are substantially less reliable. We will find that the best method for predicting will involve making laboratory consolidation tests on "undisturbed" samples. Empirical equations using index properties or the void ratio may provide approximations for preliminary studies. Since settlement predictions (or estimates) are not overly reliable, it is a chance occurrence if the computed and measured settlement values for a project will be equal. Of course, if we predict 20 mm and the measured value is 15 or 25 mm, our results will be "successful" for most projects. What we hope to avoid is predicting 20 mm when the measured value will be 40, 60 mm, or more.

Natural soil heterogeneity, test, and theory limitations produce some of the discrepancies noted above. Another source of error is that after consolidation settlements take place, *secondary consolidation*, or *creep* may continue under the applied stress for some additional time. Soil creep is a major portion of the total settlement for many organic and peaty soils. Creep settlement estimates will also be considered in this chapter. It appears, further, that the poorer the soil (soft, stratified, thick layers) the less the likelihood for a reliable settlement estimate; success more likely results from experience and some luck.

11-2 SOIL CONSOLIDATION

When the compression of a soil stratum is time-dependent, the effect is termed a *consolidation settlement* or, more commonly, *consolidation*. The general theory including the concept of pore pressure and effective stress was one of the few original† developments of Terzaghi and occurred during 1920–1924.

The Terzaghi consolidation theory makes the following assumptions:

1. The soil is, and remains, saturated ($S = 100$ percent). Consolidation settlements can be obtained for nonsaturated soil, but the predicted time for settlement to occur is extremely unreliable.
2. Water and soil grains are incompressible.
3. There is a linear relationship between applied pressure and volume change [$a_v = \Delta e/\Delta p$, as defined in Eq. (11-7)].
4. The coefficient of permeability k is a constant. This is essentially true in situ, but in the laboratory there may be large errors associated with this assumption which will tend to produce error in the time for settlement to occur.
5. Darcy's law is valid ($v = ki$).
6. There is a constant temperature. A change in temperature from about 10 to 20°C (typical field and laboratory temperature, respectively) results in about a 30 percent change in the viscosity of water. It is important that the laboratory test be performed at a known temperature, or preferably at the in situ temperature.
7. Consolidation is one-dimensional (vertical), that is, there is no lateral flow of water or soil movements. This is exactly true in the laboratory test and is generally nearly so in situ.
8. Samples are undisturbed. This is a major problem in that no matter how carefully the sample is taken, it is unloaded of the in situ overburden. Additionally, the static water table pore pressure is usually lost. In sensitive soils serious errors may result; in other soils the effects may be much less. Careful interpretation of data can reduce the effect of sampling errors somewhat.

The consolidation characteristics (or parameters) of a soil are the *compression index* C_c and the *coefficient of consolidation* c_v. The compression index relates to how much consolidation or settlement will take place. The coefficient of consolidation relates to how long it will take for an amount of consolidation to take place.

The consolidation parameters can be obtained (or at least estimated) from a laboratory consolidation test, schematically shown in Fig. 11-1. The carefully trimmed soil specimen [usual diameter from 6.3 to 11.3 cm (2.5 to $4\frac{7}{16}$ in)] is placed inside a metal confining ring. Uniform soil pressure is applied through the

---

† In the author's opinion Terzaghi's major contribution to the geotechnical profession was in keeping up with published developments in soil mechanics and synthesizing the work of others into practical solutions.

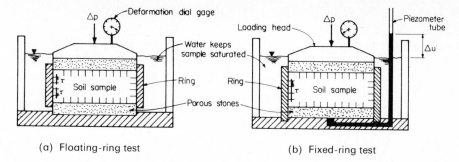

Figure 11-1 Schematics of the consolidation test. The load is applied to a sample through the load head onto porous stones to the sample. In the *floating-ring test*, the base porous stone provides lateral drainage from beneath the confining ring and compression takes place from both faces of the soil sample. Friction effects for both tests are qualitatively shown. In the *fixed-ring test*, the sample can drain only through the top porous stone, as the ring fits tightly to the base, so that water at the sample base (qualitatively) rises in the piezometer tube immediately on application of Δp and then slowly drains back through the base and up through the soil sample. The piezometer tube (say 100-ml burette) can be used to perform a falling-head permeability test to obtain the coefficient of permeability at the end of each load increment to obtain k versus e.

loading block, and the porous stones allow the excess pore pressure due to the load increment to freely escape as the soil voids are compressed. A dial gage or LVDT† is used to measure the amount of compression at varying time intervals; thus, volume changes can be computed.

A new increment of load is periodically applied to the soil. Research (Leonards, 1962) has found that best results are obtained when the load is doubled, producing a ratio of $\Delta p/p = 1$; thus, a typical sequence would be 25, 50, 100, 200, 400, 800, 1600, 3200 kPa, ‡ or 0.25, 0.50, 1, 2, 4, 8, 16 tons/ft² in the fps system. There is also evidence (Leonards, 1962) that if the initial load increment is too low, the excess pore pressure gradient may not be sufficient to initiate pore water flow in some clay soils. This should only affect the initial part of the curve of void ratio versus pressure, as the later load increments will be large enough to avoid this problem; also, initial loads on the order of 25 kPa appear adequate to avoid it.

The test loads are changed on the sample when consolidation under the current load increment is complete. This may be taken as the time when the dial reading has remained relatively unchanged for three successive readings, where the elapsed time of each reading is approximately double that of the previous reading. One may arbitrarily change loads every 24 h (commonly done in commercial testing laboratories), which is generally satisfactory for samples of the usual 2- to 3-cm thickness and using the floating-ring (Fig. 11-1a) test equipment.

† Linear voltage displacement transducer.
‡ The reader may consult Bowles (1978) for a method of converting existing fps consolidation test equipment to use this load sequence directly.

11-3 INTERPRETING THE CONSOLIDATION TEST

The immediate data from a consolidation test are presented in the form of settlement (or dial reading) versus time as shown in Figs. 12-4 and 12-5. These curves are used to obtain the rate of consolidation taken up in more detail in Chap. 12. The settlement dial readings are converted by computations to either void ratio or strain and using the sample area and the load increments to compute stress (p = load/area), plots of either

$$\varepsilon \quad \text{vs.} \quad p \text{ or } \log p$$

$$e \quad \text{vs.} \quad p \text{ or } \log p$$

are made.

The load increments initially produce a total stress state with the pore water carrying most (or all) of the applied load. After some elapsed time the excess pore pressure dissipates via drainage and the load is carried by the soil skeleton—an effective stress state. As a consequence of this sequence of stress states the stress p is taken as an *effective* stress.

Arithmetic plots of ε, or e, versus p are nonlinear over nearly the entire range of stress increments used. Semilog plots allow an estimation of sample preconsolidation, which is a very important factor in making settlement computations/estimates. The end branch of semilog plots is often linear, or nearly so, for most soils making for easier interpretation. For these two reasons the semilog plot is most often used; however, the arithmetic plot will be introduced in Sec. 11-7.

The e versus Log p Plot

The void ratio e versus log p plot has been used from at least the early 1930s to obtain the compression index C_c. The void ratio e_i at the end of any load increment is

$$e_i = e_o - \Delta e$$

where e_o = void ratio at the beginning of the load increment (or the reference void ratio)

Δe = change in void ratio from load application

The change in void ratio is readily computed from the change in sample height ΔH and sample area A using Eq. (2-2) as

$$\Delta e = \frac{V_v}{V_s} = \frac{\Delta H(A)}{H_s(A)} = \frac{\Delta H}{H_s} \qquad (a)$$

The height of soil solids H_s can be computed for a saturated soil using the total sample height H, the natural water content w_N, and the dry weight W_s of the sample as follows:

$$W_w = w_N W_s$$

and

$$V_w = V_v = \frac{W_w}{\gamma_w} = H_v A \qquad \text{(when saturated)}$$

From which the initial voids height H_v is

$$H_v = \frac{W_w}{\gamma_w A}$$

Since the total height H is the sum of the voids + solids height (see Fig. 2-2),

$$H_s = H - H_v = H - \frac{W_w}{\gamma_w A} \qquad\qquad (b)$$

The test procedure provides excess water around the sample (see Fig. 11-1) to maintain saturation. This coupled with load increments reducing the voids will nearly always produce a saturated state ($S = 100$ percent) at the end of the test, and regardless of the initial degree of saturation (often close but not 100 percent). For this reason, void ratio computations should be made by using the H_s computed from the measured change in sample height ΔH and the final water content as

$$H_s = H - \frac{W_w}{\gamma_w A} - \Delta H \qquad\qquad (11\text{-}1)$$

Here W_w is based on the change in weight of the soil cake at the end of the consolidation test from oven drying. This method is preferable to using the specific gravity and either the initial water content w_N, or the weight of the oven dry soil cake W_s.

With H_s (solids), the initial sample height H and the several values of ΔH from the load applications the initial, several intermediate, and final void ratios can be readily computed. The principal disadvantage of this procedure is that the data are all computed at the end of the test which is usually several days duration.

The ε versus Log p Plot

The strain ε versus log p plot is a relatively recent method of presenting the compression data and results in a curve of exactly the same shape as the e versus log p curve (see Fig. 11-2). The ε versus log p plot, however, has several advantages recommending its use, including:

1. The plot can progress from points obtained at the end of each load increment.
2. The plot requires less computations and thus reduces the chance for errors. Strain is computed simply as $\varepsilon = \Delta H/H$.
3. It is easier to use electronic data acquisition equipment to directly make (and plot) the computations.
4. The settlement computation simply becomes $H\varepsilon$.

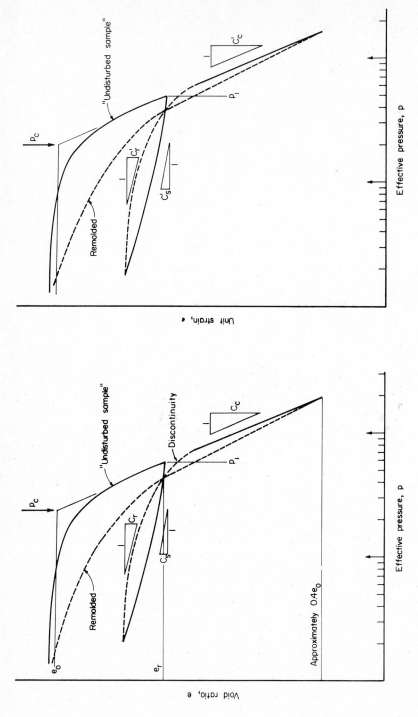

Figure 11-2 Compression plots illustrating the qualitative effect of a load-reload cycle. Note that either void ratio versus log p or strain versus log p gives a curve of similar shape.

Semilog Plot Characteristics

The semilog plot of either ε (or e) versus log p for an *undisturbed* cohesive soil has the following laboratory determined characteristics:

1. The initial branch of the curve has a relatively flat slope. If the sample is loaded with a series of incremented loads and then unloaded, the soil allowed to swell, and then reloaded with the same load sequence, this branch would be reasonably reproduced but at a smaller void ratio since swell only recovers a small part of the total reduction in void ratio.
2. At some pressure the plot curvature sharply increases. Extensive research has discovered this point to be close to the maximum past (effective) apparent pressure to which the soil has been subjected, i.e., loading and unloading one or more cycles produces curve "breaks" at the previous maximum loading(s). Curve sets produced by successive load-unload cycles are called *hysteresis loops*.
3. If the soil is loaded with a series of increasing load increments the curve beyond the initial branch tends to become somewhat linear for most inorganic soils (Figs. 11-2 and 11-3). For soils which are heavily organic, very sensitive to disturbance, or very silty, the end branch may have varying degrees of concavity as qualitatively shown in Fig. 11-4.
4. Load-unload cycles tend to produce discontinuities in the end branch as in Fig. 11-2. This discontinuity (return to "virgin" curve below the exit) is partly

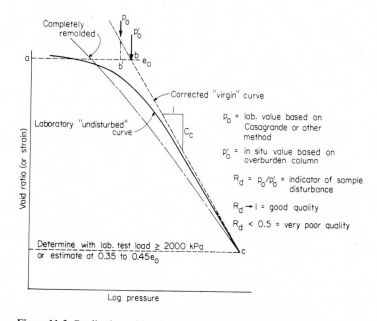

Figure 11-3 Qualitative void ratio (or strain) versus log p curve for a normally consolidated clay with a sensitivity $S_t < 4$.

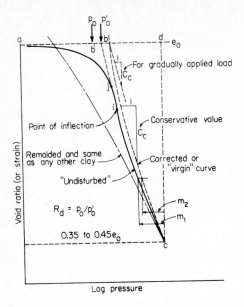

Figure 11-4 Qualitative void ratio (or strain) versus log p curve for a sensitive soil and the method of correcting C_c. *(After Terzaghi and Peck, 1967.)*

due to the disturbance from the mechanics of load/unload but also seems to be partly from creep effects.

When a fully remolded sample is tested in parallel with an "undisturbed" sample the qualitative result is as shown on Figs. 11-2 through 11-5. The following comments can be made:

1. There is no distinctive initial branch for remolded soils—unless we cycle the loading to produce one or more hysteresis loops.

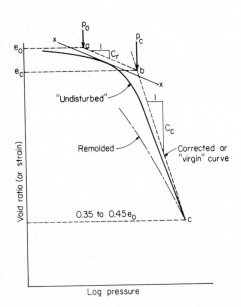

Figure 11-5 Correcting the laboratory void ratio (or strain) versus log p curve for pre-consolidated soils. *(After Schmertmann, 1955.)*

2. The remolded curve always falls to the left and below the "undisturbed" curve. This may be due in part to the more dispersed clay structure obtained from remolding as well as a loss of geologic effects (aging, cementation, etc.).
3. The end branch tends to be linear and the slope is less than for the "undisturbed" sample.

From the comments on the "undisturbed" versus remolded curves, the following conclusions may be drawn:

1. The undisturbed curve must be either below or at best coincident with the field "virgin" curve.
2. The distance between the remolded and undisturbed curves may be indicative of the "undisturbed" sample quality.
3. The initial branch of the "undisturbed" curve must be a recompression branch of the field (or virgin) curve.
4. The slope of the "undisturbed" curve is likely to be too low because of sample disturbance. The true slope may be somewhat larger and corrections may be required.

11-4 CORRECTION OF THE COMPRESSION CURVE FOR DISTURBANCE

Suggested correction procedures are shown on Figs. 11-4 and 11-5. Note that laboratory testing to very high load increments indicates that regardless of sample quality, the remolded curve merges with the virgin compression curve at a void ratio between 0.35 and 0.45 for normally consolidated and low to moderately preconsolidated soils.

In making the corrections for disturbance, three values are needed: the in situ effective pressure p_o, the apparent preconsolidation pressure p_c, and the in situ void ratio e_o. The in situ void ratio e_o is generally not the sample value at the start of the consolidation test since some swell has occurred from loss of in situ overburden pressure. The actual value is somewhere between this and the value (slightly low) where the sample is recompressed to p_o (or p_c) at the curve discontinuity. Unless a fairly reliable value of e_o can be obtained, one probably should not attempt to correct the end branch slope.

11-5 PRECONSOLIDATION AND ESTIMATION OF THE PRECONSOLIDATION PRESSURE

When the curvature sharply increases ("breaks") close to the in situ effective overburden pressure p_o, we can assume the soil is *normally consolidated*. That is the soil structure developed under pressure accumulations as the existing deposit increased in depth.

If the "break" occurs at a pressure greater than p_o, we say that the soil is *preconsolidated*. This preconsolidation may be due to

1. A greater depth of past overburden which has since eroded away (an occasional cause).
2. Cycles of wetting and drying (shrinkage, which is a very common cause).
3. Cycles of wetting and drying in the presence of certain sodium, calcium, and magnesium salts as in uplifted marine deposits (also a very common cause).
4. Effective pressure changes from water table fluctuations.

The relative amount of preconsolidation is usually reported as the *overconsolidation ratio* (OCR), defined as

$$OCR = \frac{p_c}{p_o} \tag{11-2}$$

The OCR will not be a unique value for most soils since $p_o = \gamma z$ increases with depth (refer to Fig. 11-6). Very large values of OCR may be obtained near the surface of many clay deposits from a combination of small p_o and p_c. Regardless of the cause producing the OCR, it is important to detect preconsolidation in a deposit where settlements are under consideration and to use a value appropriate for the depth of interest.

The preconsolidation pressure can be estimated with reasonable precision by using judgment and extending the straight line portions of the ε (or e) versus log p curve to a point approximately midway between the two branches. Note from Fig. 11-7 that point A represents approximately the lower limit of p_c and point B is at the upper limit. The true value is somewhere between and since the curve has a discontinuity is likely to be closer to point B than to A.

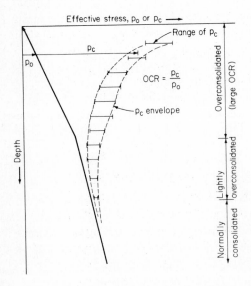

Figure 11-6 Qualitative overburden (p_o) and preconsolidation pressure profile. Note that consolidation tests at any given depth will tend to give a range of p_c rather than a single value. The results would then produce a p_c envelope as shown. Near the top a combination of large p_c and small p_o gives a very large *OCR*.

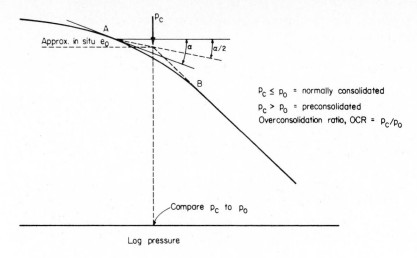

$$p_c \le p_0 = \text{normally consolidated}$$
$$p_c > p_0 = \text{preconsolidated}$$
$$\text{Overconsolidation ratio, OCR} = p_c/p_0$$

Figure 11-7 Casagrande method for obtaining preconsolidation pressure. In situ void ratio e_o as shown may be slightly low.

Alternatively, one may use a method proposed by Casagrande (1936) to obtain the approximate p_c as follows:

1. By eye estimate the sharpest point of curvature and draw a tangent line.
2. Through the point of tangency draw a horizontal line to form angle α in Fig. 11-7.
3. Bisect the α angle (by eye is sufficiently accurate).
4. Project the straight line part of the end branch to intersect the α-angle bisector of step 3.
5. Project the intersection of step 4 to the pressure axis and obtain p_c.

If $p_o \simeq p_c$, the soil is probably normally consolidated with sampling disturbance, causing the small discrepancies. If $p_c < p_o$, the soil is probably normally consolidated—but any large discrepancy should be investigated. If $p_c > p_o$, the soil may be taken as preconsolidated. Note the horizontal projection of step 4 above may give a reasonable value of e_o.

11-6 THE COMPRESSION PARAMETERS

The ε or e versus log p curves describe the settlement of the soil sample in the laboratory under loading. Thus, when the sample has been reloaded back to the in situ state of p_c, any new increment of pressure Δp produces an approximately linear (semilog) response of $\Delta\varepsilon$ or Δe along the end branch of the laboratory curve.

The in situ stress increase Δp produces an effective stress state of

$$p_2 = p_1 + \Delta p$$

where p_1 is the reference stress of p_o or p_c. We must now check the position of p_2 along the corresponding laboratory curve. For a normally consolidated soil, $p_1 = p_o$ and the pressure increment produces settlement along the end (or virgin) branch. When $p_c > p_o$, we have $p_1 = p_c$ and must first transition a portion of Δp along the recompression (initial) branch from p_o to p_c and then the remainder $\Delta p'$ along the end branch. The increment $\Delta p'$ along the end branch is

$$\Delta p' = \Delta p - (p_c - p_o) \qquad (c)$$

The Compression Coefficients

The slope of the end branch (Figs. 11-3 and 11-4) of the e versus log p curve is denoted as the *compression index* C_c and computed as

$$C_c = \frac{e_2 - e_1}{\log p_2 - \log p_1} = \frac{\Delta e}{\log p_2/p_1} \qquad (11\text{-}3)$$

It is convenient to take the $p_2 - p_1 = 1$ log cycle so that log $p_2/p_1 = 1$ (and extending the curve slope as necessary). When no clearly identified linear region exists, the slope defining C_c should be taken in the general region applicable to the stress increase Δp. Strictly, C_c is negative $(-)$, but conventional practice ignores the sign and the value is usually reported as $(+)$.

Solving Eq. (11-3) for Δe, we obtain

$$\Delta e = C_c \log \frac{p_2}{p_1} \qquad (11\text{-}4)$$

For the ε versus log p plot we can obtain in a similar manner

$$\Delta \varepsilon = C_c' \log \frac{p_2}{p_1} \qquad (11\text{-}5)$$

The parameter C_ε' is generally called the *compression ratio*, although other terms are sometimes used.

By analogy, we can obtain the following table for other indices (refer to Fig. 11-2 for identification of terms):

e versus log p	ε versus log p
For swell:	
$\Delta e = C_s \log \dfrac{p_2}{p_i}$	$\Delta \varepsilon = C_s' \log \dfrac{p_2}{p_i}$
For recompression:	
$\Delta e = C_r \log \dfrac{p_2}{p_i}$	$\Delta \varepsilon = C_r' \log \dfrac{p_2}{p_1}$

where Δe, $\Delta \varepsilon$ = change in void ratio or strain between p_i and p_2
p_i = any pressure along the appropriate curve

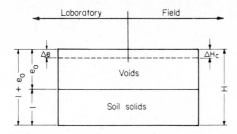

Figure 11-8 Settlement of a soil sample or layer of soil of thickness H in situ.

The swell relationships above are useful to estimate heave in the bottom of excavations (soil is unloaded). The recompression indices describe settlements for added stresses (in situ) falling between p_o and p_c.

A relationship exists between C_c and C_c' (and similar to values in the table above) as follows. Referring to Fig. 11-8 and since

$$\Delta e = \frac{\Delta H}{H_s} = \Delta H \qquad (\text{for } H_s = 1 \text{ as shown})$$

We have from Eq. (11-4) another expression for Δe so that

$$\Delta e = \Delta H = C_c \log \frac{p_2}{p_1} \qquad (d)$$

The definition of $\Delta\varepsilon$ is

$$\Delta\varepsilon = \frac{\Delta H}{H} = \frac{\Delta H}{1 + e_o} = C_c' \log \frac{p_2}{p_1} \qquad (e)$$

Now equating Eqs. (d) and (e) for ΔH, we obtain

$$C_c' = \frac{C_c}{1 + e_o} \qquad (11\text{-}6)$$

11-7 COMPRESSION PARAMETERS FOR ARITHMETIC PLOTS

Occasionally ε or e versus p plots are made as in Fig. 11-9. From a tangent to the curve (preferable in the region p_1-p_2 defined in the preceding section), we obtain the coefficient of compressibility a_v as

$$a_v = \frac{\Delta e}{\Delta p} \qquad (11\text{-}7)$$

or a compressibility ratio a_v' as

$$a_v' = \frac{\Delta\varepsilon}{\Delta p} \qquad (11\text{-}8)$$

As with C_c, the negative sign should be ignored.

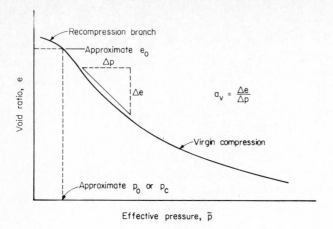

Figure 11-9 Arithmetic plot of e versus pressure (effective) to obtain the coefficient of compressibility a_v. Note that a_v is heavily dependent on tangent location and is not a constant.

The coefficient of volume compressibility m_v is defined as

$$m_v = \frac{a_v}{1 + e_o} = a_v'$$

(11-9)

The reciprocal of m_v has the units of modulus of elasticity and is often referred to as the *constrained modulus*. This term is used to identify the stress-strain relationship in a laterally confined compression test. This is an alternative description of the standard consolidation test with the sample confined against lateral strain by the metal ring.

A relationship between the stress-strain modulus E_s and $1/m_v$ can be obtained by using Eqs. (10-7) and setting $\varepsilon_x = \varepsilon_y = 0$ to obtain

$$\sigma_x = \sigma_z = \sigma_y \frac{\mu}{1 - \mu}$$

Substituting into the equation for ε_y, we obtain

$$\frac{1}{m_v} = \frac{E_s(1 - \mu)}{1 - 2\mu^2}$$

(11-10)

11-8 EMPIRICAL APPROXIMATIONS FOR C_c AND C_c'

Because of the time and expense involved in performing consolidation tests to obtain the compression indices, it is often desirable to obtain approximate values by using other soil indices which are more easily determined. Approximate values are of particular value in preliminary studies and to provide some indication of the order of magnitude (or gross errors) for laboratory data since the number of consolidation tests for a given project are usually limited. Azzouz et al. (1976)

lists several equations, some of which are given in Table 11-1. In general, it is suggested that overly simple equations using only the void ratio or unit weight (Herrero, 1983) should be used cautiously and only for inorganic clays and silts of low plasticity. The overwhelming evidence is that clay structure, geology history, and other factors strongly influence the compression index so that any correlation equation should be used cautiously. Correlation equations using more than one independent index property should be preferred and as shown in Table 11-1 tend to provide the best reliability.

Table 11-1 Empirical and regression equations for C_c and C_c' as indicated†

(a) Equations from several literature sources for C_c and C_c'

Equation	Regions of applicability
$C_c = 0.007(w_L - 7)$	Remolded clays
$C_c = 0.01w_N$	Chicago clays
$C_c = 1.15(e_o - 0.35)$	All clays
$C_c = 0.30(e_o - 0.27)$	Inorganic cohesive soil; silt, silty clay, clay
$C_c = 0.0115w_N$	Organic soils, peats, organic silt and clay
$C_c = 0.0046(w_L - 9)$	Brazilian clays
$C_c = 1.21 + 1.055(e_o - 1.87)$	Motley clays from São Paulo city
$C_c = 0.009(w_L - 10)$	Normally consolidated clays
$C_c = 0.75(e_o - 0.50)$	Soils with low plasticity
$C_c' = 0.208e_o + 0.0083$	Chicago clays
$C_c' = 0.156e_o + 0.0107$	All clays
$C_c = 0.141\ G_s^{1.2}\left(\dfrac{1 + e_o}{G_s}\right)^{2.38}$	All clays (Herrero, 1983)
$C_c = 0.5 I_p G_s$	All remolded nc clays (Wroth and Wood, 1978)

(b) Regression equations used, with reliability R and coefficient of variation S indicated

R	$\bar{C}$‡	Regression equation
0.85	0.077	$C_c = 0.40(e_o - 0.25)$
0.86	0.074	$C_c = 0.37(e_o + 0.003w_L - 0.34)$
0.85	0.007	$C_c = 0.40(e_o + 0.001w_N - 0.25)$
0.81	0.085	$C_c = 0.009w_N = 0.002w_L - 0.10$
0.86	0.074	$C_c = 0.37(e_o + 0.003w_L + 0.0004w_N - 0.34)$
0.74	0.038	$C_c' = 0.14(e_o + 0.007)$
0.76	0.039	$C_c' = 0.126(e_o + 0.003w_L - 0.06)$
0.74	0.038	$C_c' = 0.142(e_o - 0.0009w_N - 0.006)$
0.71	0.040	$C_c' = 0.003w_N + 0.0006w_L - 0.004$
0.76	0.037	$C_c' = 0.135(e_o + 0.01w_L - 0.002w_N - 0.06)$

† From Azzouz et al. (1976) and edited except as noted.
‡ A small value of $\bar{C}$ is preferred, indicating not too much interdependence between variables.
Symbols: e_o = in situ void ratio; w_N = in situ water content; w_L = liquid limit.
I_p = plasticity index; G_s = specific gravity.

Terzaghi and Peck (1967), on the basis of research on undisturbed clays of low to medium sensitivity and on an earlier equation proposed by Skempton, proposed

$$C_c = 0.009(w_L - 10) \tag{11-11}$$

This equation seems to have a reliability range of about ± 30 percent. This equation has been widely used in spite of the poor reliability to make initial consolidation estimates. This equation should not be used where the sensitivity (defined in Chap. 13) is greater than 4.

The use of empirical equations such as in Table 11-1 have an additional very serious shortcoming—namely, they give no information on the previous stress history (preconsolidation) of the soil mass. The natural water content w_N might be of some use on a qualitative basis as indicated in Sec. 11-9.

11-9 SOIL STRUCTURE AND CONSOLIDATION

Consolidation is the gradual modification of the soil skeleton. The shape of the e versus log p curve is an indicator of the changes taking place within the skeleton. Considering Fig. 11-3, one may note that little structure change takes place under increasing pressure until the preconsolidation pressure (p_o or p_c) has been reached. The reason is that the soil has been accustomed to a pressure up to the maximum past pressure and has reached equilibrium under this stress. Only when the applied pressures exceed this equilibrium (historical) condition does the soil skeleton undergo significant changes displayed by the much steeper slope of the end branch of the compression curve. Since the skeleton formed under previous stresses and the contact points have aged, we see that the soil mass history is a significant parameter in response to stress.

When the soil is sensitive to disturbances which can cause a structure collapse, a compression curve such as that in Fig. 11-4 can be obtained. Again, up to the previous preconsolidation pressure, the skeleton has little response to pressure. When applied pressure exceeds some critical value, the structure essentially fails (or collapses) and a large amount of settlement occurs. Settlement computations may be too large if the applied additional stresses barely extend into the region where the e versus log p curve has a steep slope. This may be very dangerous as well if a future addition produces a sudden collapse. On the other hand, the additional stress may produce a skeleton collapse and the settlements may be substantially underestimated—particularly if care is not taken to obtain the compression index in this region.

A completely remolded soil loses all its previous stress history and a new skeleton is produced. When it is loaded in a consolidation test, the particles are more likely to be in a dispersed state and are more easily forced together under stress so that the resulting curve is relatively smooth. Actually, it is probably a virgin curve for some soil—but with loadings far more rapid than occurring naturally. Since a remolded soil has no previous stress history, it is self-evident that it is not possible to obtain a preconsolidation pressure p_c unless load-unload cycles are made.

Since a remolded soil has no preconsolidation pressure, it follows that the accuracy of determining the preconsolidation pressure of an undisturbed sample is very dependent on sample quality.

Particle packing should be more complete in a preconsolidated ($p_c > p_o$) soil so that the void ratio is reduced from that of a normally consolidated soil. If the soil is saturated, it naturally follows that the in situ water content w_N would be less (unless the apparent preconsolidation is from other factors such as cementation). On a water content scale it would appear that if the w_N is closer to the plastic limit w_P than to the w_L, the soil may be preconsolidated. If the w_N is closer to the w_L (or larger), it may be normally consolidated. This concept should be used with other evidence, however, to decide whether the soil is preconsolidated. In any case, the quantitative evaluation of p_c requires a consolidation test.

11-10 SETTLEMENT COMPUTATIONS

The settlement ΔH of any soil mass under an applied stress Δp is composed of an "immediate," a "consolidation," and a "secondary compression" or "creep" component. In equation form the settlement is

$$\Delta H = \Delta H_i + \Delta H_c + \Delta H_s \qquad (11\text{-}12)$$

In some soils ΔH_c, $\Delta H_s \to 0$ and in others $\Delta H_i \to 0$ or is so small that it may be built out of the project as it occurs.

In this section we will focus on evaluation of ΔH_c and in the next, on ΔH_s. The immediate settlement component ΔH_i will be considered in Chap. 15.

From Fig. 11-8, by proportion we may write

$$\frac{\Delta H_c}{H} = \frac{\Delta e}{1 + e_o} \qquad \left(\text{note } \frac{\Delta H}{H} = \Delta \varepsilon, \text{ strain} \right)$$

From which the in situ consolidation settlement numerically integrated over depth H is

$$\Delta H_c = H(\Delta \varepsilon) = H \frac{\Delta e}{1 + e_o}$$

Now we substitute Δe from Eq. (d) given earlier and obtain

$$\Delta H_c = \frac{H C_c}{1 + e_o} \log \frac{p_2}{p_1} \qquad (11\text{-}13)$$

Using the definition for $\Delta \varepsilon$ in Eq. (e), we obtain

$$\Delta H_c = H C_c' \log \frac{p_2}{p_1} \qquad (11\text{-}14)$$

Using the arithmetic plot data and noting $m_v = 1/E$ and from mechanics of materials $\Delta\varepsilon = \Delta p/E$, we directly obtain

$$\Delta H_c = m_v \,\Delta pH = a'_v \,\Delta pH \qquad (11\text{-}15)$$

In these computations

$\quad p_2 = p_1 + $ a pressure increment referenced to p_1

$\quad p_1 = $ reference pressure, may be either p_o or p_c depending on problem

$\quad e_o = $ reference void ratio corresponding to reference pressure p_1

In practice p_o, e_o, and the pressure increment are referenced to the midheight (at $H/2$) of the stratum which will consolidate. The variation of p_o and e_o with depth is usually nearly linear so that an "average" value at the middle of H is sufficiently accurate. The stress increase due to project loading is approximately parabolic (decreasing) with depth so that special procedures are required to obtain the "average" since a large error results in using a linear average unless H is very small (say, under 2 m).

The consolidation settlement ΔH_c will consist in two components when $p_c > p_o$. One component is from p_o to p_c and using C_r or C'_r and the other is from p_c to $p_c + \Delta p'$ [where $\Delta p'$ is defined by Eq. (c) given earlier] and using C_c or C'_c.

11-11 SECONDARY COMPRESSION (CREEP) SETTLEMENTS

Secondary compression (or creep) occurs after the consolidation settlement develops. We define consolidation settlement as that compression occurring during the time excess pore pressure exists in the stratum of interest (or in the laboratory sample). When we talk about excess pore pressure, we are implying that it is a measurable quantity—strictly, in a saturated soil it would not be possible to have any void ratio reduction without some excess pore pressure developing. During creep settlements the strain rate is generally so low that the excess pore pressure is taken as not measurable.

Secondary compression is arbitrarily defined, then, as that soil skeleton adjustment continuing for some time after the excess pore pressure has dissipated. It is thus time-dependent and may continue for a very long time—perhaps hundreds of years. We will be concerned with that which occurs within a lesser time span. Past evidence indicates that creep eventually stops or becomes so small that it is not measurable.

Secondary compression is difficult to evaluate. It may be relatively insignificant for many inorganic soils. It may be a significant—or the dominant—portion of many highly organic soils (Weber, 1969). Problems of evaluating the creep parameter in the laboratory include factors such as equipment corrosion, temperature control, sand particles causing hang-ups (and producing abrupt discontinuities in the data as the particle breaks loose), and similar.

From observations of settlement versus log time curves as in Fig. 12-4 and taken over times sufficient to plot well into the secondary compression range, it appears that the slope for remolded soils is relatively constant in the creep region. This may not hold true, however, for "undisturbed" soils, and further it appears that for these latter, the slope is stress dependent.

From the secondary compression region of settlement versus log time curves as in Fig. 12-4, we can obtain a coefficient of secondary compression C'_α as

$$C'_\alpha = \frac{\Delta H/H}{\log (t_p + \Delta t)/t_p} = \frac{\Delta \varepsilon}{\log t_2/t_p} \tag{11-16}$$

where terms not previously identified are

H = total sample height in units of ΔH
t_p = time when primary consolidation is complete
Δt = time increment producing ΔH

The secondary compression or creep in an in situ time increment Δt for a stratum of thickness H is

$$\Delta H_s = H(\Delta \varepsilon) = HC'_\alpha \log \frac{t_p + \Delta t}{t_p} \tag{11-17}$$

We may have some difficulty in determining a reliable value of t_p in situ. We are free, however, to use any value of Δt which appears reasonable.

If we were to plot a curve of e versus $\log t$ as in Fig. 11-10, the slope of the secondary compression branch would be

$$C_\alpha = \frac{\Delta e}{\log t_2/t_p} \tag{11-18}$$

Similarly, the secondary compression [using Eq. (11-13)] is

$$\Delta H_s = \frac{HC_\alpha}{1 + e_o} \log \frac{t_2}{t_p} \tag{11-19}$$

Equating Eq. (11-17) and Eq. (11-19), we find that

$$C'_\alpha = \frac{C_\alpha}{1 + e_o} \tag{11-20}$$

The value of void ratio should be that at the end of primary consolidation (as is t_p). This would be extremely critical for laboratory samples but much less so for in situ unless the soil undergoes enormous settlement (and void ratio reduction). In most cases the initial in situ void ratio is adequate for e_o in Eq. (11-19).

Several studies (e.g., Mesri and Godlewski, 1977) have indicated a relationship between C_c and C_α. It appears that the general range is

$$0.025 \leq \frac{C_\alpha}{C_c} \leq 0.10$$

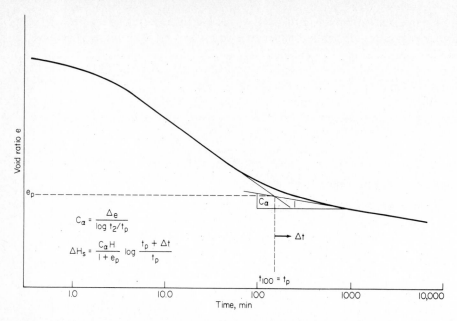

Figure 11-10 Plot of any load increment to obtain a secondary compression index C_α as shown. Use load increment closest to field conditions with new load Δp.

As an estimate one might use

$$\frac{C_\alpha}{C_c} \leq 0.05$$

for inorganic soils and 0.07 to 0.10 for organic soils with larger values for the more organic (and peat) deposits. Note, however, that we must still estimate a value of Δt during which most of the "creep" will occur. The settlement-time curve for that sample loading closest to anticipated field loading should be used for the secondary compression index.

11-12 ILLUSTRATIVE EXAMPLES

The following examples illustrate the several concepts presented in this chapter and limitations on the methodology. It is suggested that the examples be carefully studied as a text supplement. It is also suggested that the literature be scanned for illustrative consolidation curves for different soil types. The following references provide a number of e versus log p curves covering a wide range of soils:

Rutledge (1944) (see also the discussion)
Casagrande and Fadum (1944) (see also the discussion)
Schmertmann (1955)
Clevenger (1958)

Kaufman and Sherman (1964)

Taylor (1948)

McDonald and Sauer (1970)

Example 11-1 Given the following data from a consolidation test for the load increment of 800 kPa:

Dial reading	$t_0 = 2246$	(units when 800 kPa is applied)
Dial reading	$t_{100} = 2035$	(from plot of DR vs. log time)
Dial reading	$t_f = 1995$	(when new load is added)
Dial constant	$= 0.01$ mm/div	

The sample dimensions are: diameter $= 63$ mm; initial height 20 mm; total change in height for all loads $= 4.37$ mm; final wet weight of soil cake $= 110.3$ g; oven dry weight of soil cake $= 91.1$ g.

REQUIRED What is the change in void ratio, change in strain (for this load increment), and the final void ratio e_f?

SOLUTION Note that, strictly, the ε or e versus (log) p plots should be based on the consolidation settlements (not including "creep") and thus based on the dial reading at t_{100} and not t_f. We would thus have to plot time-settlement curves to obtain this (see Fig. 12-4). Since there are additional computations with this method, it is common to base the settlement-pressure plots on the total settlement (or at some constant time interval) for the load increments.

We will compute both void ratio changes and indicate how they would be used in obtaining curve points.

Step 1 Find H_s and the initial and final void ratio using the water content at the end of the test from oven drying the saturated soil cake.

$$A = \left(\frac{63}{10}\right)^2 (0.7854) = 31.17 \text{ cm}^2$$

$$H_v = \frac{W_w}{\gamma_w A}$$

$$H_v = \frac{110.3 - 91.1}{1 \times 31.17} = 0.615 \text{ cm}$$

$$H_s = H - H_v - \Delta H = 2.000 - 0.615 - \frac{4.37}{10} = 0.948 \text{ cm}$$

$$e_f = \frac{H_v}{H_s} = \frac{0.615}{0.948} = 0.649$$

$$e_i = \frac{2.000 - 0.948}{0.948} = 1.11$$

Step 2 Find Δe for the 800-kPa load increment. For consolidation:

$$\Delta H = \frac{(2246 - 2035)0.01}{10} = 0.211 \text{ cm}$$

$$\Delta e_c = \frac{H}{H_s} = \frac{0.211}{0.948} = 0.222$$

For total change in sample height:

$$\Delta H = \frac{(2246 - 1995)0.01}{10} = 0.251 \text{ cm}$$

$$\Delta e_t = \frac{0.251}{0.948} = 0.265$$

Step 3 Find $\Delta \varepsilon$ for the load increment. Use ΔH from step 2 and $H = 20$ mm. For consolidation:

$$\Delta \varepsilon_c = \Delta H / H = 0.211/2.00 = 0.106.$$

For total change in sample height:

$$\Delta \varepsilon_t = 0.251/2.00 = 0.126.$$

One may use either incremental void ratio changes (as here) or total changes to obtain the plot points. Using the incremental void ratio changes, we obtain for consolidation (or constant time increment) only:

First load: $\quad e_1 = e_o - \Delta e_{c(1)}$

Second load: $e_2 = e_o - \Delta e_{t(1)} - \Delta e_{c(2)}$

Third load: $\quad e_3 = e_o - \Delta e_{t(1)} - \Delta e_{t(2)} - \Delta e_{c(3)} \cdots$

For total settlement we have simply

First load: $\quad e_1 = e_o - \Delta e_{t(1)}$

Any load i: $e_i = e_{i-1} - \Delta e_{t(i)}$

which is much simpler and less subject to computation errors.

////

Example 11-2 Given the e versus log p curve shown in Fig. E11-2. The sample was reloaded to the approximate in situ p_o using smaller than usual increments to 25 kPa. The sample was then unloaded, rebounded, and reloaded as shown. Note that p_c was somewhat arbitrarily taken using a visual inspection of the curve.

REQUIRED Compute C_c, C_r, C_s, and the primed equivalents and the overconsolidation ratio (OCR). Take $e_o = 1.00$.

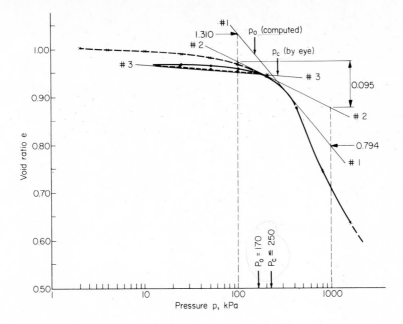

Figure E11-2

SOLUTION Draw curve slopes (lines 1, 2, and 3) as shown. Note that these are somewhat arbitrary but that "2" is in the region between p_o and p_c, where it would be applicable. C_c is taken near p_c since any foundation load producing Δp will traverse along slope 2 and then the remainder (if any) along curve "1." All slopes have been extended over 1 log cycle, so the log of the pressure ratio is 1.

Step 1 Finding C_c, C_r, and C_s:

From curve 1: $C_c = \dfrac{\Delta e}{\log p_2/p_1} = \dfrac{1.310 - 0.794}{\log 1000/100} = \dfrac{0.526}{1} = 0.526$

From curve 2: $C_r = \dfrac{\Delta e}{1} = 0.974 - 0.879 = 0.095$

From curve 3: $C_s = 0.952 - 0.968 = -0.016$ [use $(-)$ since others are $(+)$]

The OCR $= p_c/p_o = 250/170 = 1.5$ (data barely justify this precision).

These values are obviously very sensitive to the curve slope and the accuracy of reading the void ratio intercepts.

Signs are not usually required; however, if we take C_c and C_r as $(+)$, we might report C_s as $(-)$ for some consistency. Actually, C_c and C_r are $(-)$ and C_s is $(+)$ with strict attention to signs.

Step 2 Find C_c', C_r', and C_s'

$$C_c' = \frac{C_c}{1 + e_o} = \frac{0.526}{1 + 1.0} = 0.263$$

$$C_r' = \frac{C_r}{1 + e_o} = \frac{0.095}{1 + 1.0} = 0.0475$$

$$C_s' = \frac{C_s}{1 + e_o} = \frac{-0.016}{1 + 1.0} = -0.008$$

It is only coincidental that the compression ratios are exactly one-half the compression indices.

////

Example 11-3 Given the ε versus log p plot and soil profile shown in Fig. E11-3.

REQUIRED Compute the expected settlement for the load causing the *net pressure* profile shown in the clay stratum.

SOLUTION Settlement computations are based on *average* values of p_o, C_c', and e_o in some stratum thickness H. Assume that C_c' is constant in the entire thickness of 9 m of clay.

Step 1 Compute average p_o. Since p_o varies linearly, the middepth value is also the averge value in the stratum.

$$3(17.3) = 51.9 \text{ kPa}$$

$$+4.5(19.8 - 9.807) = 44.97 \text{ kPa}$$

$$\overline{p_o = 96.87} \text{ kPa} \qquad \text{(use 97)}$$

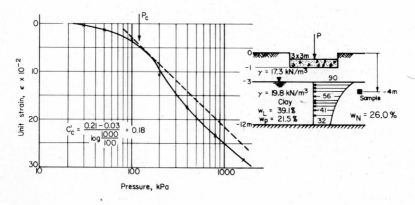

Figure E11-3

Since the estimated p_c from the ε versus log p curve shows 120 kPa [and also, w_N (for any G_s) is slightly closer to w_P than to w_L], assume that the stratum is moderately preconsolidated.

Step 2 Compute average Δp using the pressure profile shown (which could be obtained using the Boussinesq method of Sec. 10-10). Use numerical integration to find the area of the diagram, and note that the area is $H_c\,\Delta p_{average}$. Use the trapezoidal rule [Eq. (10-21)] to obtain the area.

$$A = H\,\Delta p = 3\left(\frac{90 + 32}{2} + 56 + 41\right) = 474$$

$$\Delta p = \frac{474}{9} = 53 \text{ kPa} \qquad \text{(simply averaging top and bottom gives 61 kPa)}$$

Step 3 Compute the expected settlement using C_c', shown on Fig. E11-3. Note that the value shown did not use any virgin curve correction.

$$\Delta H = C_c'\,H \log \frac{p_o + \Delta p}{p_o}$$

$$= 0.18(9) \log \frac{97 + 53}{97}$$

$$= 0.18(9)(0.217) = 0.352 \text{ m}$$

This is a very large settlement, which few structures could take. Piles might be required (but the underlying soil is not given). We would first "refine" the computations as in the next example.

$/\!/\!/\!/$

Example 11-4 Redo Example 11-3, allowing for p_o and p_c and using both C_r' and C_c'.

SOLUTION The settlement will be produced by an increment of Δp from p_o to p_c of $120 - 97 = 23$ kPa and the remainder as $\Delta p' = 53 - 23 = 30$ kPa.

Step 1 Find C_r' and the first part of settlement:

$$C_r' \cong \frac{0.01 - 0.04}{\log 100/30} = 0.06 \text{ (not using 1 log cycle)}$$

$$\Delta H_1 = C_r'\,H \log \frac{p_2}{p_1} = 0.06(9) \log \frac{97 + 23}{97} = 0.05 \text{ m}$$

Step 2 Find remaining amount of settlement using C_c' with p_c:

$$\Delta H_2 = C_c'\,H \log \frac{p_c + p'}{p_c} = 0.18(9) \log \frac{120 + 30}{120} = 0.157 \text{ m}$$

The total settlement $\Delta H = \Delta H_1 + \Delta H_2 = 0.05 + 0.157 = 0.207$ m, and the percent $= 0.207/0.352 = 0.59$ or 59 percent.

This example illustrates the importance of identification of any precon-solidation pressure and its considerable effect on reducing computed settlements.

////

11-13 CONTROLLING CONSOLIDATION SETTLEMENTS

The material, and equations, presented in this chapter indicate that several methods of controlling settlements might be possible.

A method which is suggested by noting that $p_2 = p_1 + \Delta p$ is to remove soil by excavation so that

$$p_o - p_{\text{exc}} + \Delta p \rightarrow p_o \qquad \text{(or at least not much more than } p_c\text{)}$$

If this is done, $\log p_2/p_1 \rightarrow 0$ and consolidation settlements (and creep) will generally be negligible. This is a common construction practice and produces what is called a "floating" foundation.

We might note that C_r is much less than C_c (or the primed equivalents), so that if sufficient time is available, preloading the site so that the working pressure

$$\Delta p + p_o \leq p_c \qquad \text{(from the preload)}$$

and settlements will be greatly reduced. This is also a very common construction practice. Note that this procedure does not eliminate settlements but rather controls when they occur (during the preloading period). Several methods of producing the preload were discussed in Chap. 7—the most common being that of placing excess fill which is later partially (or totally) removed.

Where it is not practical to preload or to excavate to "float" the foundation, it may be necessary to carry the project loads through the consolidating stratum into more competent underlying soil and/or rock. This is a major reason requiring the use of piles or caissons to support structures (dams, buildings, etc.).

11-14 RELIABILITY OF CONSOLIDATION SETTLEMENT COMPUTATIONS

In spite of the problems of sample disturbance, if (1) the e or ε versus $\log p$ curve is interpreted reasonably with some correction for sample disturbance, (2) the sample disturbance is not great (as determined by a parallel test on a remolded sample), (3) the field exploration correctly located all the consolidating strata, and (4) the deposit is not very soft, computed settlements will usually be reliable to within about 10 to 15 percent of measured values.

When the soil mass is stratified and soft, where large settlements take place, the reliability of the settlement estimate falls off rapidly. The computed settlement may be either quite low or quite high. Wu et al. (1978) presented work from a very large number of consolidation tests for an embankment over a stratified clay. Computed to measured settlement ratios ranged from 0.57 to 3.1. If the extremes (three values) are discarded and the remaining ratios averaged, a value of 0.99 is obtained. Averages of this kind are highly misleading, however, since the specific site settlement is of interest and determines the success (or failure) of a project. This particular data set was cited since clients are seldom willing to pay for any follow-up measurements on their projects or to allow the geotechnical engineer to make any—even at no client expense.

Finite-element and finite-difference methods may be used to model settlements and time rates. Olsen and Ladd (1979) present some comparative analyses using finite-differences versus classical theory. These presentations did not show sufficient improvement over the classical method to be useful in most field situations. These more sophisticated methods are not widely used for this and several other reasons, including:

1. Lack of general familiarity with the computer methodology.
2. Necessity for computer access (and resulting additional costs).
3. Problems are as sensitive to the several soil parameters (C_c, etc.) as the classical methods, i.e., output is no better than the input.

This latter factor, in particular, coupled with possibly only having one or two C_c (or C_c') values for a stratified soil, where more values are needed to produce a statistically reliable "parameter," can produce a substantial error.

In passing, a minimum of two independently determined values of the compression index and parameters should be determined which are in agreement before a "value" is selected for a stratum. Soil variability and sample disturbance are such that a single value—which may actually be correct—is introducing unnecessary risk both to the owner and the geotechnical engineer.

11-15 SUMMARY

Several methods of using consolidation test data to obtain settlement parameters have been presented and are summarized as follows:

e versus log p	ε versus log p	e or ε versus p
Compression index C_c	Compression ratio C_c'	Coefficient of compressibility a_v, a_v'
Recompression index C_r	Recompression ratio C_r'	Coefficient of volume compressibility m_v
Swell index C_s	Swell ratio C_s'	

Several empirical relationships (Table 11-1) for C_c and C_c' have also been presented.

The use of the settlement parameters has been illustrated in computation of consolidation settlements. Concepts of previous stress history and the effect of preconsolidation p_c have been examined. From observations of the load-unload-reload curves of e versus log p for remolded samples, a method of obtaining p_c has been developed. The reader should carefully note that all the consolidation settlement equations are of the general form

$$\Delta H = \int_{H_1}^{H_2} \varepsilon \, dh \rightarrow \sum_{i=1}^{n} H_i \varepsilon$$

which is identical to using

$$\Delta H = \frac{C_c H}{1 + e_o} \log \frac{p_i + \Delta p}{p_i}$$

In this latter equation particularly note:

1. Δp is the average increase in pressure in the increment of stratum H_i or of the total stratum thickness, which is preferred since its use reduces computations.
2. e_o is the average value of the *in situ* void ratio and usually not the sample value. Equation (11-3) shows that e_o varies with depth logarithmically. For, say, $H_i = 1$ to 2 m one may use the average of top and bottom or simply the midheight value. For thicker strata several values should be computed and numerical integration using Eq. (10-15) performed to obtain the "best" value of e_o.
3. p_i is the average *effective overburden pressure* existing in a stratum of thickness H_i. The value at midheight is commonly used, since the variation is linear with depth if we assume a constant unit weight. This will produce a small error, since there is some increase in unit weight with depth along with the corresponding decrease in e_o. Note: p_i may be either p_o or p_c depending on Δp.

Secondary compression, or creep, has also been considered. Careful analysis of secondary settlement computations shows that this is simply computed as

$$\Delta H_s = H \varepsilon$$

since the C_α term is simply a time-dependent strain.

It was noted that strictly ε (or e) versus (log) p curves should be based on "consolidation" settlements—not including any secondary compression or "creep" effects.

Several practical considerations and limitations were noted to put the reliability of consolidation computations into perspective. Since consolidation tests are time-consuming (and expensive) to run, the number made for most projects are too few to utilize statistical methods for reliability predictions; thus experience is a primary factor in making a successful settlement estimate.

PROBLEMS

11-1 In Example 11-2 it is known that $e_0 = 1.00$ and $G_s = 2.72$. What is (a) the approximate corrected value of C_c and (b) the in situ value of w_N?

Ans.: $C_c \cong 0.37$; $w_N = 36.8$ percent

11-2 Referring to Example 11-3, if $G_s = 2.70$, what is e_o? What is the OCR?

Ans.: $e_o = 0.67$; OCR $= 1.2$

11-3 Given are the following void ratio versus pressure data from laboratory tests. The initial sample height is 20.00 mm and diameter $= 62.3$ mm.

p, kPa	Test 1 void ratio e	Test 2 void ratio e
0	1.02	1.11
25	0.98	1.09
50	0.975	1.085
100	0.954	1.079
200	0.0880	0.942
400	0.781	0.831
800	0.688	0.762
1600	0.575	0.705
3200	—	0.638

Required: plot e versus p and e versus log p of the assigned test and obtain:

(a) Coefficient of compressibility a_v
(b) Coefficient of volume compressibility m_v
(c) Compression index C_c
Partial ans.: Test 1 $C_c = 0.35$.

11-4 For the assigned soil test of Prob. 11-3, back compute the void ratio data using the given height and diameter to obtain corresponding strains, and plot ε versus log p and obtain C_c'.

Partial ans.: Test 1 $\varepsilon_{25} = 0.0198$; $\varepsilon_{1600} = 0.220$; $C_c' = 0.16$, and using $e_0 = 1.02$

11-5 If the sample used in Prob. 11-3 to plot and obtain C_c' or C_c came from the strata as shown in Fig. P11-5,

(a) Obtain the preconsolidation pressure p_c and determine if the soil is preconsolidated.

Ans.: Test 1 $R_d \cong 1.06$.

(b) Obtain the expected settlement ΔH if the soft clay stratum is 3 m thick and the average increase in pressure is $\Delta p = 30$ kPa.

Ans.: Test 1 $\Delta H \cong 60$ mm.

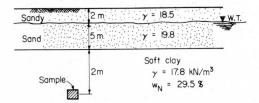

Figure P11-5

11-6 Given the following consolidation test data, plot e versus log p, compute C_c and C_r, and find p_c.

	Void ratio e		
p, kN/m²	First load	Unload	Reload
0	1.20	1.04	—
5	1.18	1.00	1.01
12.5	1.17	0.981	0.990
25	1.15	0.949	0.968
50	1.12	0.923	0.945
100	1.07	0.880	0.917
200	0.99	0.865	0.888
400	0.855		0.843
800			0.734
1600			0.589

Partial ans.: $p_c \cong 110$ kPa.

11-7 Back compute the void ratio data of Prob. 11-6 to obtain the strain ε if the sample was 20×63.5-mm diameter, and obtain C'_c, C'_r, and p_c.

11-8 If the sample used in Prob. 11-6 (or 11-7) came from in situ as shown in Fig. P11-8, is the soil preconsolidated? If the pressure profile due to the surface load is as shown, what is the expected consolidation settlement?

Partial ans.: Yes; $p_c \cong 110$ versus 51.3 kPa, $\Delta H \cong 33$ mm.

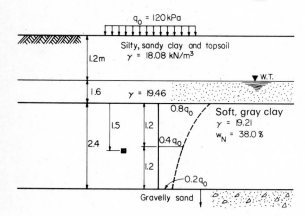

Figure P11-8

RATE OF CONSOLIDATION

12-1 THE COEFFICIENT OF CONSOLIDATION

Consolidation settlement studies involve two factors:

Total estimated settlement (including creep) using C_c and C_α (or the primed equivalents) parameters given in Chap. 11
Estimated settlement rate (to be considered here)

The settlement rate can be estimated using a *coefficient of consolidation* c_v. It is important to be able to estimate the rate of settlement of a structure so that if the total estimated settlement ΔH is large, the owner can anticipate when a significant portion will have taken place. When the total settlement is small, its rate is usually of little importance—it would make little difference that the settlement is spread over 5, 10, or even 50 years.

The consolidation rate parameter c_v is obtained from the theory presented in the following text material and using a time value obtained from a plot of deformation (settlement, strain, or void ratio) versus time. The time scale may be either logarithmic or $\sqrt{\text{time}}$. The deformation and time data are obtained from the standard laboratory consolidation test described in Chap. 11 (see also Fig. 11-1).

The equation for c_v is developed based on *one-dimensional flow* and *saturated soil conditions* (assumptions 1 and 7 of Sec. 11-2). Since consolidation under these conditions is directly dependent on the extrusion of pore water from the soil

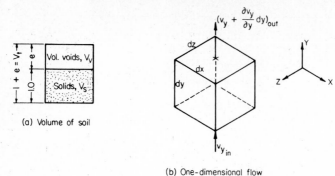

(a) Volume of soil

(b) One-dimensional flow

Figure 12-1 Idealization of soil and fluid flow for development of the one-dimensional theory of consolidation.

voids, one may develop the needed equations by considering continuity of flow (referring to Fig. 12-1) as

$$Q_{in} = v_y \, dx \, dz \, dt \qquad Q_{out} = \left(v_y + \frac{\partial v_y}{\partial y} \, dy \right) dx \, dz \, dt$$

and the volume change as a rate process is

$$\text{Volume change} = \frac{\partial V}{\partial t} \, dt = Q_{out} - Q_{in}$$

Equating as indicated, one obtains

$$\frac{\partial v_y}{\partial y} \, dy \, dx \, dz \, dt = \frac{\partial V}{\partial t} \, dt \tag{a}$$

Since the soil grains and water are incompressible for all practical purposes and for the pressures involved, the change in soil volume must be due to a change in the volume of soil voids, or

$$\frac{\partial V}{\partial t} = \frac{\partial V_v}{\partial t} \tag{b}$$

The definition of void ratio gives

$$V_v = eV_s$$

and differentiating with respect to time and using the product rule, obtain

$$\frac{\partial V_v}{\partial t} = e \frac{\partial V_s}{\partial t} + V_s \frac{\partial e}{\partial t}$$

Observing that $e(\partial V_s/\partial t) = 0$, since there is no change in the volume of soil grains, we have

$$\frac{\partial V_v}{\partial t} = V_s \frac{\partial e}{\partial t} \tag{c}$$

From Fig. 12-1 note that $V_s/V_t = 1/(1 + e)$, and solving for V_s, obtain

$$V_s = \frac{V_t}{1 + e} = \frac{dx\ dy\ dz}{1 + e} \tag{d}$$

Substituting Eq. (d) into Eq. (c) and multiplying by dt,

$$\frac{\partial V}{\partial t}\ dt = \frac{dx\ dy\ dz}{1 + e}\ \frac{\partial e}{\partial t}\ dt \tag{e}$$

Equating Eq. (e) and Eq. (a) and canceling the product of the four differentials, obtain

$$\frac{\partial v_y}{\partial y} = \frac{1}{1 + e}\ \frac{\partial e}{\partial t} \tag{f}$$

From assumption 3 of Sec. 11-2, that there is a linear relationship between applied pressure and volume change, and noting that volume change depends on pore water extrusion from the soil voids when the soil is saturated, obtain

$$\frac{\partial e}{\partial t} = \frac{de}{dp}\ \frac{\partial u}{\partial t} = a_v\ \frac{\partial u}{\partial t} \tag{g}$$

where $a_v = de/dp$ (or $\Delta e/\Delta p$), the coefficient of compressibility
  u = pore water pressure

From Darcy's law, the velocity of water is $v = ki$, and with the total head $h = u/\gamma_w$, we have

$$\frac{\partial v_y}{\partial y} = \frac{k}{\gamma_w}\ \frac{\partial^2 u}{\partial y^2} \tag{h}$$

Substituting Eqs. (h) and (g) into Eq. (f), obtain

$$\frac{a_v}{1 + e}\ \frac{\partial u}{\partial t} = \frac{k}{\gamma_w}\ \frac{\partial^2 u}{\partial y^2}$$

Rearranging, and substituting Eq. (11-9)

$$\frac{\partial u}{\partial t} = \frac{k}{m_v \gamma_w}\ \frac{\partial^2 u}{\partial y^2}$$

which can be written as

$$\frac{\partial u}{\partial t} = c_v\ \frac{\partial^2 u}{\partial y^2} \tag{i}$$

where

$$c_v = \frac{k}{m_v \gamma_w} = \frac{k(1 + e)}{a_v \gamma_w} \tag{12-1}$$

Use dimensional consistency in units of k, m, v, and unit weight of water γ_w.

The solution of Eq. (*i*) takes the form of a series solution (Taylor, 1948) to give the instantaneous value of the excess pore water pressure u at a specified point in the soil mass as

$$u = \sum_{n=1}^{\infty} \left(\frac{1}{H} \int_0^H u_i \sin \frac{n\pi y}{H} \, dy \right) \left(\sin \frac{n\pi y}{H} \right) \exp \left(-\frac{1}{4} n^2 \pi^2 T \right) \qquad (12\text{-}2)$$

where n = any integer (generally 0, 2, 3, and 4 are sufficient)
$\quad y$ = depth into a stratum along drainage path H
$\quad H$ = length of longest drainage path in soil sample or mass
$\quad T$ = dimensionless number termed a *time factor*, or

$$T = \frac{c_v t_i}{H^2} \qquad (12\text{-}3)$$

$\quad t_i$ = time of interest
$\quad u_i$ = initial pore pressure distribution; use constant, linear variation, sine wave, or other shape

For the case of constant (actually any *linear* variation) initial hydrostatic pressure, Eq. (12-2) simplifies to

$$u = \sum_{m=0}^{\infty} \frac{2u_i}{M} \sin \frac{My}{H} \exp \left(-M^2 T \right) \qquad (12\text{-}4)$$

where u_i = constant or $u_i = u_o + u_1(H - y/H)$ as in Fig. 12-2
$\quad M = \frac{1}{2}\pi(2m + 1)$, where m is any integer from 0 to ∞

Referring to Fig. 12-2, if we apply a pressure increment Δp to a fully saturated soil which has fully consolidated ($u = 0$) under the existing pressure p_1, the new total pressure is $p_2 = p_1 + \Delta p$. The pore water carries the load increment Δp at $t = 0$ (Fig. 12-2a), since drainage is not instantaneous, and $u_i \cong \Delta p$ for $S = 100$

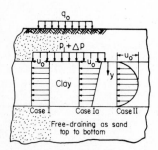
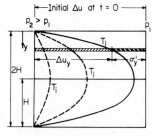
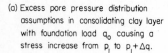

(a) Excess pore pressure distribution assumptions in consolidating clay layer with foundation load q_o causing a stress increase from p_1 to $p_1 + \Delta q$.

(b) Pore pressure distribution with times T_i (noting T = f(time)) and depth in stratum.

Figure 12-2 Excess pore pressure distribution assumptions for an increased effective stress within a stratum and qualitative pore pressure distribution as a function of elapsed time.

percent. We also assume a thin stratum such that Δp is essentially constant with depth. At $t = 0$ the consolidation has just begun, or the percent consolidation $U = 0$ percent. At some times t_i (as in Fig. 12-2b), the pore pressure patterns (isochrones) due to drainage being more rapid at or near the free surfaces are at, or approaching, zero as shown. At any point y from the free surface one may compute the status of the pore fluid using the percent consolidation concept of the following section.

12-2 PERCENT CONSOLIDATION

At the free surfaces, $u = 0$ and consolidation is complete ($U = 100$ percent). At interior points the consolidation U_y would, by inspection of Fig. 12-2b, be

$$U_y = \frac{u_o - u_i}{u_o} = 1 - \frac{u_i}{u_o} \tag{12-5}$$

To obtain U_y in percent, multiply the ratio by 100. One can now use Eq. (12-4) to plot a graphic display of the variation of pore pressure (or percent consolidation) with depth in a stratum of thickness $2H$ (half thickness $= H$) as

$$U_y = 1 - \sum_{m=0}^{\infty} \frac{2}{M} \sin \frac{My}{H} \exp - M^2 T \tag{12-6}$$

Equation (12-6) is obtained by dividing Eq. (12-4) by u_o and subtracting 1, as indicated in Eq. (12-5). A solution can be easily obtained by programming Eq. (12-6) on a computer for the following variables:

$M = \frac{1}{2}\pi(2m + 1)$, which depends only on incrementing the integer m

$T = $ constant, say, $0.05, 0.1, 0.15, \ldots, 0.90$

$y/H = 0, 0.1, 0.2, \ldots, 1.0$.

Values larger than 1 are not needed, since the resulting curve is symmetrical about the half depth of $y = H$.

Sufficiently precise results are obtained by varying m from 0 to about 4 or 5 for computing M. A plot of Eq. (12-6) for selected values of T is shown in Fig. 12-3.

While plots of the type shown in Fig. 12-3 give an indication of pore pressure variations within the stratum, it is often of more immediate interest to obtain the *average* consolidation U for the entire stratum, for example, when an amount of consolidation of, say, 10, 50, or 80 percent is complete. For this estimate it is necessary to integrate U_y for the entire stratum thickness as

$$U = 1 - \frac{\displaystyle\int_0^{2H} u_i \, dy}{\displaystyle\int_0^{2H} u_o \, dy} \tag{12-7}$$

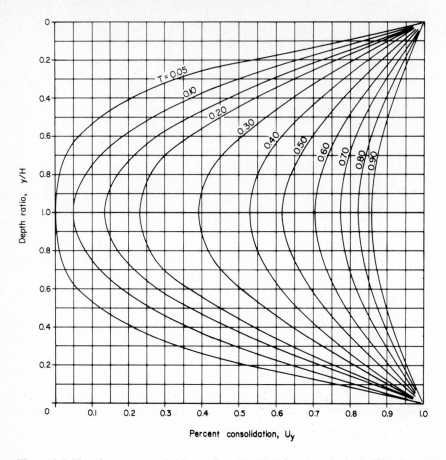

Figure 12-3 Plot of pore pressure isochrones from Eq. (12-6) for selected values of time factor T.

Substituting Eq. (12-3) in the numerator and u_o = constant in the denominator of Eq. (12-7) gives

$$U = 1 - \sum_{m=0}^{\infty} \frac{y}{M^2} \exp(-M^2 T) \tag{12-8}$$

Often the first term (with $m = 0$) provides a solution of sufficient precision for practical purposes, and tabular values can be made once and for all in terms of percent consolidation U versus T by rearranging and solving Eq. (12-8) for T, to obtain

$$T \cong \frac{\ln(2/M^2) - \ln(1 - U)}{M^2} \tag{12-9}$$

Note that the time factor T is not defined when $U = 1.00$ (100 percent consolidation, which theoretically never occurs), since the logarithm of 0 is ∞. Selected solutions of Eq. (12-8) are given in Table 12-1 (with $m > 0$) for several assumed

Table 12-1 Time factors for percent consolidation and indicated pore pressure distributions shown

U, percent	Cases I and Ia	Case II
0	0.000	0.000
10	0.008	0.048
20	0.031	0.090
30	0.071	0.115
40	0.126	0.207
50	0.197	0.281
60	0.287	0.371
70	0.403	0.488
80	0.567	0.652
90	0.848	0.933
100	∞	∞

Case I   Case Ia Case II

initial pore pressure distributions caused by a stress increase Δp as shown. Superposition of Cases I and II or Cases Ia and II can provide solutions for other pore pressure distributions.

Of some interest, the basis of the $\sqrt{\text{time}}$ method used in Sec. 12-4, the percent consolidation U for one-way drainage of a layer of soil extending to an infinite depth $(2H \to \infty)$ is

$$U = \sqrt{\frac{T}{\pi}} \tag{12-10}$$

12-3 METHODS OF OBTAINING THE TIME OF INTEREST FOR COMPUTING c_v

The coefficient of consolidation is seldom obtained from Eq. (12-1); rather, it is computed from Eq. (12-3) rearranged to

$$c_v = \frac{TH^2}{t_i}$$

and using t_i obtained from deformation versus time curves. The value of $t_i = t_{50}$ (the "laboratory time for 50 percent of the primary deformation to occur") is usually used. The time may be obtained from either a semilog or $\sqrt{\text{time}}$ plot.

Semilog Plot

Usually one plots the deformation against logarithm time as illustrated in Figs. 12-4 and 11-10 for the purpose of obtaining the time at a given percent consolidation for a given stress increment. The same shaped curve will be obtained whether one plots the deformation as:

$$\left.\begin{array}{l} \text{Dial readings} \\ \text{Strain } \varepsilon \\ \text{Void ratio } e \end{array}\right\} \quad \text{vs.} \quad \text{log time}$$

The difference between dial readings is directly related to the settlement. The strain is a settlement percentage.

Since the time for a given percent consolidation will be different from the several deformation versus log time plots from a laboratory consolidation test, it is evident that there are some limitations in theory—or in making the several assumptions previously given in Sec. 11-2. At least two of the factors of apparently major importance are the stress level and the fact that small deformations produce significant changes in the sample void ratio and, hence, in k as used in Eq. (12-1).

The use of deformation versus log time curves requires finding the apparent initial value D_0 at $t = 0$ which cannot be plotted since log $0 = \infty$. This initial deformation value of D_0 may be arbitrarily taken as the actual value at $t = 0$ for that load increment. Most often, however, the initial branch of the deformation versus log time curve is parabolic with an apparent origin different from the actual D_0. This apparent discrepancy may be due to dial gage slack, sample seating, soil grain interference with consolidation ring and/or porous stone, and other factors.

When the initial branch of the curve is parabolic or nearly so—as observed by eye—one may obtain the apparent initial deformation reading D_0 as follows:

1. Find some time t_i of, say, 0.1 or, 0.2 min (illustrated in Fig. 12-4).
2. Take a time $4t_i$ (i.e., $4 \times 0.1 = 0.4$ or $4 \times 0.2 = 0.8$ min).
3. Obtain the vertical offset from t_i and $4t_i$ as Δy.
4. Lay off the Δy distance above t_i to represent D_0. This may be repeated one or more times, and the average of the several points is taken as D_0.

To find the deformation at $U = 100$ percent consolidation (D_{100}), draw a tangent to the middle branch of the settlement curve by eye and another tangent to the end branch of the curve. The intersection of these two tangents is arbitrarily taken as D_{100}, and a projection to the curve† and then down to the time axis is taken as t_{100}. The deformation for D_{50} is usually used to obtain t_{50},

---

† There is not universal agreement on this—some obtain t_{100} by directly projecting to the time scale from the intersection point.

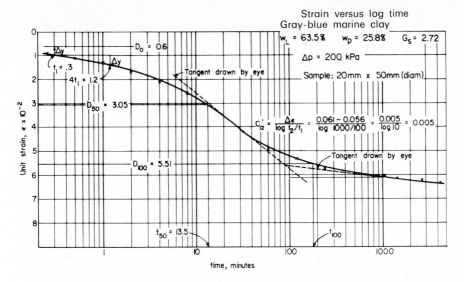

Figure 12-4 Plot of strain versus log time and data obtained from plot. Note that a plot of dial reading or void ratio would give a curve of identical shape.

however, and is obtained as

$$D_{50} = \frac{D_0 + D_{100}}{2}$$

and t_{50} can be found by entering at D_{50} and projecting to the curve and down to the time axis as shown in Fig. 12-4.

$\sqrt{\text{Time}}$ Plot

The plot of $U = 2\sqrt{T/\pi}$ is essentially a straight line in the initial stage of consolidation, before it curves and becomes asymptotic at $U = 1$. Taylor (1948) proposed that one could use this observation as an alternative method of presenting time-settlement curves to obtain the time at various percents of consolidation. That is, we can plot deformation versus $\sqrt{t}$ as in Fig. 12-5 and *draw as a best fit* a straight line to the deformation ordinate of the first several data points, locating point A and continuing this straight line to the time axis to locate a point b. Since this neglects the constant $2/\sqrt{\pi} \simeq 1.15$ (actually 1.13), we draw a second line $\overline{Ac}$ from A which is 15 percent larger than $\overline{Ob}$. The point d where the experimental curve intercepts Ac is at $U \cong 0.9$ (approximately 90 percent consolidation). Now that we know the deformation for 90 percent consolidation (the distance Ae of Fig. 12-5), we can obtain the time at 90 percent consolidation by projecting to the time axis, and $t_{90} = N_{90}^2$. We can find the deformation for 50 percent consolidation by assuming that D_0 occurs at point A (this is simpler than the semilog

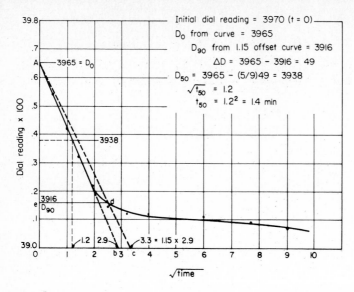

Figure 12-5 Plot of dial readings versus $\sqrt{\text{time}}$ to obtain D_i and corresponding t_i. Values for D_{50} and t_{50} are shown.

plot) and that $\frac{5}{9}Ae$ is the deformation corresponding to t_{50}; then rearranging Eq. (12-3) and using T factors from Table 12-1, we obtain

$$c_v = \frac{0.848H^2}{t_{90}} \quad \text{or} \quad c_v = \frac{0.197H^2}{t_{50}}$$

Both values computed above should, of course, be equal—in actual practice they will be close. These values of c_v should compare reasonably well with those from the semilog plot.

With c_v computed, one can estimate the field consolidation time as

$$t_{i(\text{field})} = \frac{T_i H_{\text{field}}^2}{c_v} \tag{12-11}$$

12-4 RATE OF CONSOLIDATION BASED ON STRAIN

Experimental evidence from both laboratory and field observations indicates that estimated rates of consolidation are rather poor. Some opinion exists that there is no constant of proportionality between Δp and the resulting pore pressure u. In fact, it is regularly observed with consolidometers using pore pressure measuring equipment that there is a lag between applying a new load increment and a change in pore pressure. One would expect this effect to be even more pronounced in a thick soil mass where the application of a load increment produces a pressure profile which decreases exponentially with depth (as in Fig. E11-3 and used in Example 12-6).

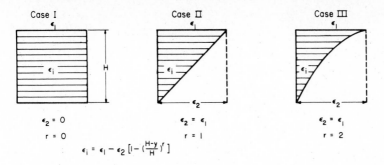

Figure 12-6 Strain distribution assumption in any increment of stratum thickness H as defined by the integer r.

From these observations, some persons are of the opinion that Eq. (*i*) would be more correctly written as

$$\frac{\partial \varepsilon}{\partial t} = \frac{k}{m_v \gamma_w} \frac{\partial^2 \varepsilon}{\partial y^2}$$

i.e., consolidation as a strain rate process.

Since strain (and Δp) are not constant with depth, except possibly for very thin laboratory samples (20 to 30 mm thick), a procedure should be used which incorporates the stress (and strain) into the solution. Janbu (1965) proposed using the soil strain at various depths to obtain a strain profile, as in Fig. 12-6. The percent consolidation U could be computed using

$$U = \frac{U_0 - f_s F_{r(T)}}{1 - f_s} \qquad (a)$$

where $f_s = \dfrac{r\varepsilon_2}{(1 + r)\varepsilon_1}$ $\qquad (b)$

r = integer = 0, 1, and 2
ε_1 = strain at top of layer
ε_2 = strain at base or midheight of layer, depending on drainage
U_0 = value of percent classical consolidation, Eq. (12-5); also values of $F_{r(T)}$ for $r = 0$

$$F_{r(T)} = 1 - 2(r + 1) \sum \frac{\sin^{2+r} N}{N^{2+r}} \exp - N^2 T \qquad (c)$$

$N = k\pi/2; k = 1, 3, 5, 7, \ldots$
T = input values of time factor, say, 0.0, 0.001, 0.002, 0.005, 0.01, 0.02, 0.05, 0.1, 0.2, 0.5, 1.0, 2.0, 5.0, 9999999

Several values of $F_{r(T)}$ are as follows:

T	$r = 0$	$r = 1$	$r = 2$
0.01	0.1128	0.0199	0.0276
0.05	0.2523	0.0999	0.1247
0.10	0.3568	0.1977	0.2285
0.20	0.5040	0.3703	0.3981
1.00	0.9313	0.9125	0.9164

Figure 12-6 gives the significance of r, ε_1, and ε_2. From the figure the values of f_s are 0.0, 0.5, and 0.667 for Case I, II, and III, respectively. A U versus log T plot can be made, say, for a Case II strain distribution as

$$\text{For } T = 0.01: \quad U = \frac{0.3568 - 0.5(0.1977)}{0.5} = 0.514$$

$$T = 1.00: \quad U = \frac{0.9313 - 0.5(0.9125)}{0.5} = 0.9501$$

Other points are obtained similarly. This illustrates that curve 2 in Fig. 12-7 is a direct plot of the data for $r = 0$ (typical values given in the table above).

Using superposition of Cases I through III, the six general cases of strain distribution shown on Fig. 12-9 can be obtained (note "2" is exactly Case I). From extended tables of $F_{r(T)}$ for the three values of r, the six curves in Figs. 12-7 and 12-8 can be drawn. For example, curve 3 is based on using Eq. (a) with U_0 values equal to $F_{r(T)}$ at $r = 0$ and using f_s based on $r = 1$ and $e_d/e_s = 0.5$ and the $F_{r(T)}$ values for $r = 1$. Similarly curve 4 uses $e_d/e_s = 0.75$ and $r = 2$.

The most convenient method of computing the rate of consolidation based on strain is that used in the table shown as Fig. 12-9. The steps are as follows:

1. Prepare a separate table of y values. These values are to the center of a subdivision of the total stratum thickness H (refer to Example 12-6).
2. Compute the effective overburden pressures p_o at depth y, and from any consolidation curves obtain p_c values.
3. Compute the average strain in the several elements of thickness y making up the total thickness H. Use the compression ratio C_c' given by Eq. (11-5). Make adjustments as necessary in the ratio log p_2/p_1 as outlined in Sec. 11-6 depending on whether the stratum is normally or preconsolidated.
4. Compute the strain at the top of the consolidating stratum similar to step 3.
5. Make a table similar to that shown in Fig. 12-9 (or a tracing) which includes both the strain profile and the columnar data. From the data in steps 1 through 4, fill in the left side of the table and plot y/H against $\varepsilon_2/\varepsilon_1$ onto the strain profile.

6. From the curves shown in Fig. 12-9 and the curve given in step 5, obtain the curve number which best fits. This curve number identifies the appropriate curve from Fig. 12-7 or 12-8 (depending on single or double drainage).
7. Compute time factors T, using

$$T = \frac{c_v t_i}{H^2} \quad \text{[Eq. (12-3) rearranged]}$$

where c_c = coefficient of consolidation from laboratory test (may use an average value if you have access to more than one)

t_i = time for desired percent consolidation (say, $t_i = 0.1, 0.5, 1, 5, 10$ years, etc.)

H = *total* thickness of stratum (note that the length of the drainage path has already been taken into account by use of either Fig. 12-7 or Fig. 12-8)

8. Using T from step 7, enter either Fig. 12-7 or 12-8 to the curve number from step 6 and obtain the percent consolidation U.
9. Fill in the right part of Fig. 12-9 with time t_i, computed T, and curve interpolated values of U and compute the portion of the total settlement occurring to this time t (using U as a decimal) as

$$\Delta H_i = \Delta H_t(U)$$

where ΔH_t = total estimated settlement based on $\sum \varepsilon y$ of the several layers of thickness y making up H from step 3

10. Plot ΔH_i against time (either log or arithmetic). A comparison of this plot with a superimposed field settlement plot will display any discrepancies between computations and reality.

This method of obtaining the rate of settlement is recommended for thicker consolidating strata, say, greater than 5 m. For thinner layers the average method with values obtained from Table 12-1 or Fig. 12-3 will be sufficiently correct. Several reasons for this include:

1. The difficulty of obtaining a reliable value (or values) of c_v. In the laboratory, when thin samples are consolidated an amount ΔH of 2 mm in a H of 20 mm, this is a 10 percent change in height and a large change in void ratio and the resulting coefficient of permeability k. In the field a 20-mm change in a thickness of, say, 2 or 3 m is a negligible change in void ratio and in k; thus c_v may be considerably in error for the field rate of consolidation.
2. The coefficient of consolidation c_v is dependent on the viscosity of the pore fluid, which, in turn, depends on the temperature of the test. Laboratory tests are commonly run at the temperature of the laboratory (20 to 25°C). In the field the groundwater temperature is much less—generally on the order of 8 to 12°C.

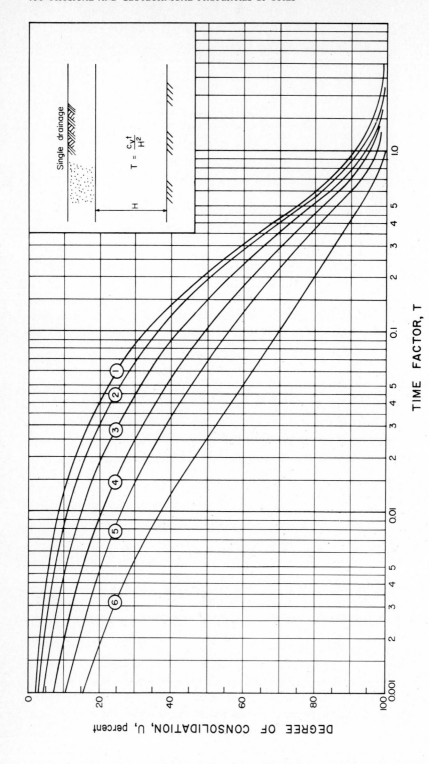

Figure 12-7 Time factor versus degree of consolidation curves for single drainage. Note that curve 2 is a plot of Case I in Table 12-1. Refer to Fig. 12-6 for pore pressure distributions for curve numbers. (*After Lade and Lee, 1976.*)

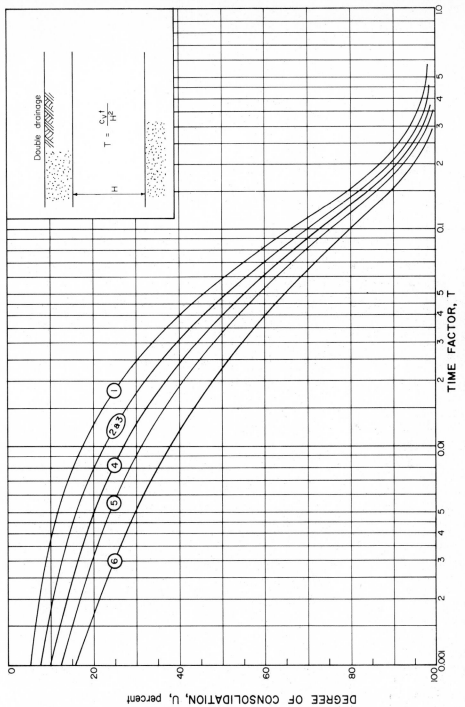

Figure 12-8 Time factor versus degree of consolidation curves for double drainage. Refer to Fig. 12-6 for pore pressure distributions for curve numbers. *(After Lade and Lee, 1976.)*

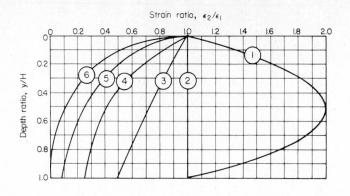

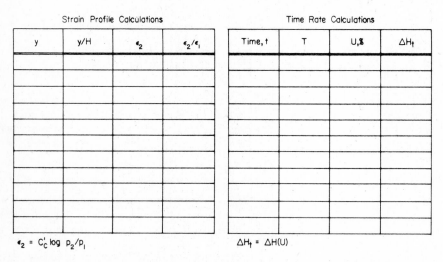

Strain Profile Calculations				Time Rate Calculations			
y	y/H	ϵ_2	ϵ_2/ϵ_1	Time, t	T	U,%	ΔH_t

$\epsilon_2 = C'_C \log p_2/p_1$ 　　　　　　$\Delta H_t = \Delta H(U)$

Figure 12-9 Table for calculating rate of settlement. Plot strain ratio on the figure to find the correct T versus U curve. Use c_v and arbitrary selected times t to compute T. Enter the T versus U curve and find U. Using the total settlement, compute ΔH_t.

3. The difficulty of obtaining the increase in effective stress Δp at the various depths in the strata. The Boussinesq method (as in Chap. 11) is commonly used for obtaining the stresses. The Boussinesq method, however, idealizes the soil to a homogeneous, isotropic, elastic, semi-infinite half space. A layered soil mass does not satisfy these assumptions, and the amount of deviation from the ideal soil mass may be substantial—or at least enough to not justify elaborate or refined computations.

12-5 ILLUSTRATIVE EXAMPLES

The following examples will be presented to illustrate the material presented in this chapter.

Example 12-1 Data from a consolidation test are as follows:

$t_{50} = 12.2$ min (see Fig. 12-4 for method of obtaining).
Load increment = 1600 kPa (approx. 8 tons/ft^2).
Conditions are such that Case I, constant pore pressure through H, exists at
    $t = 0$ and drainage is on *both* faces of the sample.
The average one-half sample height is:

$$(D_f + D_0)/2 = 0.7913 \text{ cm} = 7.913 \times 10^{-3} \text{ m}$$

REQUIRED Compute c_v for this load increment.

SOLUTION

$$c_v = \frac{TH^2}{t_{50}}$$

From Table 12-1, at $U = 50$ percent, which corresponds to t_{50}, obtain
$T = 0.197$.

$$c_v = \frac{0.197(7.913 \times 10^{-3})^2}{12.2} = 1.011 \times 10^{-6} \text{ m}^2/\text{min}$$

////

Example 12-2 Data from a consolidation test are as in Example 12-1 except
that the sample drains from *one* side.

REQUIRED Compute the coefficient of consolidation.

SOLUTION With one-side drainage, $H = $ full sample height, but T remains
0.197 as obtained from Table 12-1 at $U = 50$ percent.

$$H = 2(0.7913) = 1.5826 \text{ cm} = 1.583 \times 10^{-2} \text{ m}$$

$$c_v = \frac{0.197(1.583 \times 10^{-2})^2}{12.2} = 4.04 \times 10^{-6} \text{ m}^2/\text{min}$$

From these two computations (Examples 12-1 and 12-2), one may conclude
that doubling the length of the drainage path will increase the time of consolidation four times.

////

Example 12-3 In this example we will compute time for field consolidation.
Given laboratory data of Example 12-1: $H = 0.7913$ cm (one-half height),
$t_{50} = 12.2$ minutes, $c_v = 1.011 \times 10^{-6}$ m^2/min computed.

REQUIRED Length of time it will take for consolidation settlement to occur
for the field conditions shown in Fig. E12-3.

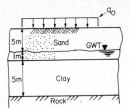

Figure E12-3

SOLUTION Inspection of the figure indicates drainage on one face; therefore, $H = 5$ m. For full consolidation, $U \to 100$ percent, obtain $T = 0.848$ from Table 12-1 (Case I or Ia).

$$t = \frac{TH^2}{c_v}$$

$$= \frac{0.848(5)^2}{1.011 \times 10^{-6}} = 2.097 \times 10^7 \text{ min} = 14562 \text{ days} = 39.9 \text{ years}$$

With a 5-m-thick layer involved, the time for consolidation might be better computed by using the "strain" method (illustrated in Example 12-6). The method illustrated here is widely used because of its greater simplicity.

////

Example 12-4 In this example we will find the instantaneous pore pressure at a point in a soil mass. Given the fact that a load is applied to the soil in Example 12-3 (see Fig. E12-3). At the instant of load application, a piezometer located 2 m below the top of the clay layer indicates an excess pore pressure Δu of 2 m, as shown in Fig. E12-4. Assume Case I conditions (Table 12-1) for the initial pore pressure distribution in the clay layer.

REQUIRED What is the pore pressure at this point when settlements corresponding to $T = 0.10$ and $T = 0.50$ have taken place?

SOLUTION Referring to Fig. 12-3 at $y/H = \frac{2}{5} = 0.4$, obtain the approximate values of U_y as

T	U_y
0.10	0.375
0.50	0.780

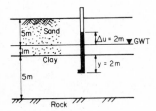

Figure E12-4

The corresponding pore pressures in meters of water are computed as follows:

$$u_{10} = 2 - 2(0.375) = 1.25 \text{ m}$$

$$u_{50} = 2 - 2(0.78) = 0.44 \text{ m}$$

Note that the values of U_y of 0.375 and 0.78 are the percent consolidations at the point $y = 2$ m below the top of the clay stratum.

////

Example 12-5 In this example we will compute the average consolidation at selected values of T. Given time factors of $T = 0.10$ and 0.50, as in Example 12-4.

REQUIRED The average percent consolidation U of the soil stratum when the pore pressures of $u = 1.25$ and $u = 0.44$ m of water remain as computed in Example 12-4.

SOLUTION Draw a curve of U versus T using the data of Table 12-1 and Case I as in Fig. E12-5.

From Fig. E12-5, obtain $U = 0.35$ for $T = 0.10$ and $U = 0.76$ when $T = 0.50$, or

$$T = 0.10 \quad \text{Percent consolidation is 35 percent}$$

$$T = 0.50 \quad \text{Percent consolidation is 76 percent}$$

These average values compare to 37.5 and 78 percent, respectively, in Example 12-4. One would expect the average values to be smaller than point values near free surfaces, larger than point values at the maximum depth of the drainage path, and approximately the same near the middle of the drainage path.

////

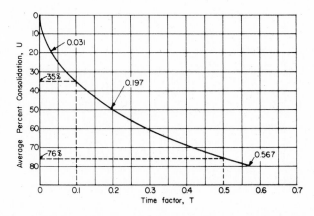

Figure E12-5

Example 12-6 Given the soil profile of Fig. E12-6a and the e versus log p curve of Fig. E12-6b. Assume that the several plots of settlement versus log time give

$$c_v = 0.032 \text{ cm}^2/\text{min}$$

Unit weights on the soil profile for the cohesive soil can be obtained using displacement methods of placing a sample of known weight W_t in a container of known volume V_c and filling it with water V_w to obtain the sample volume as $V = V_c - V_w$. From these data the unit weight is readily computed as $\gamma_{\text{wet}} = W_t/V$.

REQUIRED Compute the expected total settlement for the footing load shown and make a plot of time versus settlement so that settlement rates can be predicted.

SOLUTION From the e versus log p curve of Fig. E12-6b, the value of p_c can be plotted on the effective stress profile of Fig. E12-6a, and the soil appears to be *normally* consolidated.

Spot-check the data for reasonableness:

$$C_c \cong 0.009(w_L - 10) = 0.009(52.3 - 10) = 0.38 \text{ versus } 0.36 \qquad \text{(O.K.)}$$

$$e_o \cong 1.13 \text{ from } e \text{ versus } \log p \text{ curve};$$

$$e_o \cong w_N G_s$$

from which, solving for G_s, $G_s \cong 1.13/0.409 = 2.76$ which is not unreasonable for clay. With $G_s = 2.76$ and a block diagram as shown in Fig. E12-6c, and

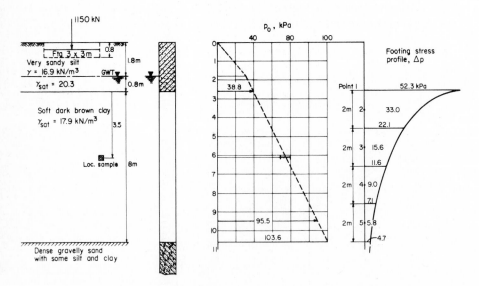

Figure E12-6a

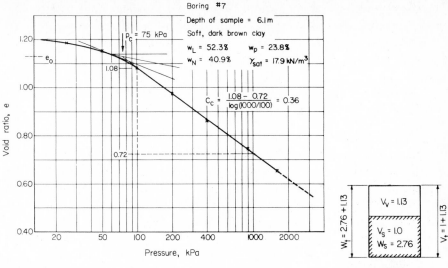

Figure E12-6b

Figure E12-6c

taking $V_s = 1.0$, we have

$$\gamma_{\text{sat}} = \frac{1.13 + 2.76}{1 + 1.13} (9.807) = 17.9 \text{ kN/m}^3 \text{ versus } 17.9 \qquad \text{(O.K.)}$$

Therefore, the data appear satisfactory and we shall continue. With 8 m of soil involved in consolidation, use the method of Sec. 12-4 instead of a single layer. Divide the soil into four sublayers of 2 m each as shown in Fig. E12-6a.

Step 1 Calculate the effective overburden pressure p_o at the top of the stratum and at the midheight of each sublayer:

Point	y, m		p_o
1 (top)	2.6	$1.8(16.9) + 0.8(20.3 - 9.8)$	$= 38.8$ kPa
2	3.6	$38.8 + 1(17.9 - 9.8)$	$= 46.9$
3	5.6	$46.9 + 2(8.1)$	$= 63.1$
4	7.6	$63.1 + 2(8.1)$	$= 79.3$
5	9.6	$79.3 + 2(8.1)$	$= 95.5$
Bottom	10.6	$95.5 + 1(8.1)$	$= 103.6$

Some of these points are shown on the soil stress profile of Fig. E12-6a.

Step 2 Obtain e_o from Fig. E12-6b as $e_o \cong 1.13$, obtain $C_c = 0.36$ from the computation shown on Fig. E12-6b, and compute C_c' as

$$C_c' = \frac{C_c}{1 + e_o} = \frac{0.36}{1 + 1.13} = 0.17$$

One might adjust e_o for each H_i, but this is generally too much refinement.

Step 3 Obtain Δp due to the footing load of 1150 kN on a footing of 3×3 m using the Boussinesq method of Sec. 10-10, assuming a homogeneous soil and using a computer program to treat the footing as a series of 36 point loads (unit area of 0.5×0.5 m) with a point load value on each unit area of $1150/36 = 31.94$ kN. Summing the contribution from the 36 unit areas gives the Δp values shown on the stress profile of Fig. E12-6a. One could average Δp at the midpoint of each layer but the values shown are from a scaled plot.

Step 4 Develop data to fill in the left side of the table in Fig. E12-6d. In the table, $y = $ distance below top of consolidating cohesive layer; $H = $ total thickness of consolidating layer ($=8$ m); ε_1 and $\varepsilon_2 = $ strain values using C'_c from step 3, and computations are as follows:

| Point | H_i | At midheight of any sublayer | | | ΔH_i |
		p_o, kPa	Δp, kPa	ε_2	
1	—	38.8	52.3	$0.063 = \varepsilon_1$	—
2	2 m	46.9	33.0	0.039	0.078
3	2	63.1	15.6	0.016	0.032
4	2	79.3	9.0	0.008	0.016
5	2	95.5	5.8	0.004	0.008
Bottom	—	103.6	4.7	0.003	—
				$\sum \Delta H_i = \Delta H = 0.134$ m	

Typical computations:

$$\varepsilon_1 = C'_c \log \frac{p_o + \Delta p}{p_o} = 0.17 \log \frac{38.8 + 52.3}{38.8} = 0.063 \text{ m/m}$$

$$\varepsilon_2 = 0.17 \log \frac{46.9 + 33.0}{46.9} = 0.039 \text{ (second line of above table)}$$

$$\Delta H_1 = H_1(\varepsilon_2) = 0.078 \text{ m (also second line of above table)}$$

The remaining table entries are computed similarly. Note that the point number identifies the point plotted on the stress profile in Fig. E12-6a.

These data can be used to fill in the left side of the table as shown. Plot y/H versus $\varepsilon_2/\varepsilon_1$ as shown on the graph above the table of Fig. E12-6d to obtain the curve number. It appears curve 6 is a best fit.

Step 5 Fill in the right side of Fig. E12-6d as follows:

(a) Rearrange Eq. (12-11) to give

$$T = \frac{c_v t_i}{H^2}$$

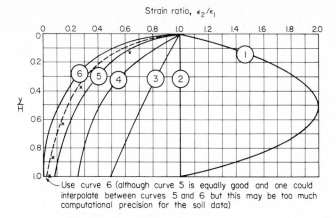

Strain ratio, ϵ_2/ϵ_1

Use curve 6 (although curve 5 is equally good and one could interpolate between curves 5 and 6 but this may be too much computational precision for the soil data)

Strain Profile Calculations

y, m	y/H	ϵ_2	ϵ_2/ϵ_1
1	0.125	0.039	0.619
3	0.375	0.016	0.254
5	0.625	0.008	0.127
7	0.875	0.004	0.063
8	1.000	0.003	0.048
$\epsilon_1 = 0.063$			

Time Rate Calculations

Time, t	T	U,%	ΔH_i
0.1 yr.	0.003	25	0.033
0.5	0.013	40	0.054
1.0	0.026	52	0.070
2.0	0.053	66	0.088
4.0	0.105	85	0.114
6.0	0.157	94	0.126

$\epsilon_2 = C_c' \log p_2/p_1$

$y/H = 1/8 = 0.125$

$\epsilon_2/\epsilon_1 = 0.039/0.063 = 0.619$

$\Delta H_i = \Delta H(U)$ $\Delta H_i = 0.134(.25) = 0.034\,m$

$T = 0.0263t = 0.0263(.1) = 0.003$

Figure E12-6d

where t = time in field (assume several values sufficient to draw curve)

H = full thickness of stratum in field (8 m in this example)

$c_v = 0.032$ cm^2/min $= 0.032(1440 \times 365)/(100)^2 = 1.682$ m^2/year

from which

$$T = \frac{1.682t_i}{8^2} = 0.0263t_i$$

Use the values of t_i of 0.1, 0.5, 1, etc., as shown in the table.

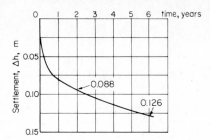

Figure E12-6e

(b) For the computed values of T using t_i, enter Fig. 12-8 (double drainage, from Fig. E12-6a) and obtain the corresponding U values in percent. For $t_i = 0.1$,

$$T = 0.0263(0.1) = 0.003$$

Entering Fig. 12-8 at $T = 0.003$, project to curve 6 (from step 4) and obtain

$$U \cong 25 \text{ percent}$$

(c) Compute the amount of settlement at the various times t_i using the equation shown in Fig. E12-6d,

$$\Delta H_i = \Delta H(U)$$

In the example from step (b), with the value of ΔH obtained from Step 4 of 0.126 m,

$$\Delta H_i = 0.134(0.25) = 0.033 \text{ m}$$

Step 6 Plot t_i versus ΔH_i as shown in Fig. E12-6e.

$////$

12-6 CONSOLIDATION RATES FOR LAYERED MEDIA

When the stratum consists of several layers of soil subject to consolidation theory, the total settlement can be computed by obtaining C_c or C'_c from consolidation tests on samples from the several strata, computing the ΔH_i values for the individual strata, and summing to obtain

$$\Delta H = \sum_{1}^{n} \Delta H_i$$

The rate of consolidation will be quite complicated, however, as the settlements of the lower or interior strata depend on c_v of the particular stratum as well as on c_v of the exterior strata. It is evident that drainage (see Fig. 12-10) must travel through both the stratum under consideration and the exterior strata to the free surface.

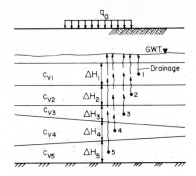

Figure 12-10 Layered soil with one-way drainage. Drainage from stratum 5 must travel through the other four strata at a rate dependent on $c_{v(i)}$ of the other strata.

Drainage, then, depends on the coefficient of permeability, void ratio, stress response, and applied stress in the strata, and from Eq. (12-1),

$$c_v = f(k, e, a_v, \gamma_w)$$

From this it is evident that to even approximate a solution, consolidation tests are required for each stratum so that values of c_v and C_c for that stratum can be obtained. Theoretical solutions for this case are beyond the scope of this text, and the reader should refer to Rowe (1964), De Leeuw (1965), Martins (1965), or Poskitt (1969) for analytical solutions. Finite-difference solutions are also available and require the use of a computer. Finite-difference solution references include Abbot (1960) and Gibson and Lumb (1952).

12-7 THREE-DIMENSIONAL CONSOLIDATION

Three-dimensional consolidation will occur in situations involving drainage to a central source, such as sand drains used beneath fills to accelerate drainage by reducing the drainage path and therefore accelerate consolidation. This type of analysis is also beyond the scope of this text, and the reader is referred to references such as Berry and Wilkinson (1969), Rowe (1964), and Aboshi and Monden (1961).

12-8 SUMMARY

This chapter has presented a means of estimating the rate of one-dimensional consolidation settlement using the coefficient of consolidation c_v. Note that the consolidation settlement is only a portion of the total settlement—the other two components being "immediate" and the secondary compression (or "creep"). This latter settlement occurs after the consolidation settlement has ended.

The coefficient of consolidation is obtained by using a time factor T and the time for some percent of consolidation (usually t_{50}) to occur from a settlement versus time plot. The time scale may be logarithmic (common) or $\sqrt{\text{time}}$. A semilog plot is used if the secondary compression parameter (C_α or C'_α) is to be obtained.

The theory of consolidation has been used to obtain dimensionless time factors T_i which are found to depend on the pore (or strain) distribution within the soil layer under some pressure increment.

We found that the rate of consolidation (as defined by c_v) depends heavily on the coefficient of permeability k and the length of the drainage path H. In the laboratory with the use of a floating-ring consolidometer, H is one-half the sample thickness; for a fixed ring, H is the total sample thickness. The time for primary consolidation for a constant sample thickness is approximately four times as great for the fixed-ring device. In the field the drainage path is determined by the location of the water table, and the arrangement of free-draining (sand or silt-sand seams) and consolidating soil layers.

With t_i from a settlement-time plot, and corresponding values of T and H, the coefficient of consolidation is computed as

$$c_v = \frac{TH^2}{t_i} = \frac{k}{m_v \gamma_w}$$

This interdependency between k and c_v suggests that if we obtain c_v by using T, H, and t_i, the coefficient of permeability can be obtained using c_v, m_v, and γ_w. This would be done by plotting ε against time and computing $a'_v = \Delta\varepsilon/\Delta p = 1/m_v$ (as in Chap. 11). One would use either an averaged k as computed for the several load increments in the consolidation test or for that load closest to in situ.

We note that Eq. (12-3), which is generally used for c_v, indicates that t_i is critical in setting its magnitude. We must look at Eq. (12-1) to see that the load increment Δp (or the corresponding strain) is also a very critical factor. From a practical standpoint, it is well recognized that a large fill (or other load) settles faster than a small one. It is common to use preloads which are larger than service loads to accelerate (and later control) consolidation and to effectively produce a "preconsolidated" soil for the structure.

PROBLEMS

12-1 Redo Example 12-1 if $t_{50} = 9.1$ minutes.

12-2 Redo Example 12-3 using the value of c_v from Prob. 12-1 above ($t_{50} = 9.1$ minutes).

12-3 Redo Example 12-4 for a piezometer at a 2.5-m depth instead of 2 m.

12-4 Redo Example 12-4 for a piezometer at a 5-m depth. When will U be 10 percent at the 5-m depth using $c_v = 1.011 \times 10^{-6}$ m²/min?
   Ans.: 137 days

12-5 Redo Example 12-5 for the piezometer at 5 m into the soil instead of 2 m.
   Ans.: $U_{av} = 35$ percent for $T = 0.10$ when $U_i = 5$ percent.

For Probs. 12-6, 12-7, and 12-8, use the data in Table P12-1 following as assigned.

12-6 Using data from Table P12-1 as required,
   (a) Draw ε versus log t curves and compute c_v. Average the values from the several curves for a final value.
   (b) Draw a curve of ε versus log p and compute C'_c.
   Ans.: $C'_c = 0.202$

12-7 For data of Table P12-1,
   (a) Draw settlement versus log t curves and compute c_v.

Table P12-1 Time–dial reading data for a consolidation test*

Time	25 kPa	Time	50 kPa	Time	100 kPa	Time	200 kPa	Time	400 kPa
0	0000	0	305	0	570	0	975	0	1490
0.25	102	0.25	386	0.25	663	0.25	1051	0.25	1544
0.50	121	0.50	397	0.50	681	0.50	1065	0.50	1566
1.0	133	1.0	410	1.0	701	1.0	1079	1.0	1595
2.0	154	2.0	428	2.0	726	2.0	1107	2.0	1626
4.0	184	4.0	451	4.0	757	4.0	1146	4.0	1674
8.0	208	8.0	468	8.0	803	8.0	1190	8.0	1727
16	229	16	493	16	841	32	1321	16	1800
32	244	32	510	32	879	60	1370	60	1976
60	254	231	543	60	902	120	1408	137	2045
126	269	406	553	120	925	240	1434	256	2075
250	274	600	559	285	944	1150	1465	1276	2115
883	285	1585	570	415	960	1440	1471	2510	2130
1043	296	1746	570	1440	972	2756	1481	5761	2154
1440	302			4272	975	4320	1490		
2320	305								

Time	800 kPa	Time	1600 kPa	Time	3200 kPa
0	2154	0	2652	0	3300
0.25	2164	0.25	2703	0.25	3317
0.50	2195	0.50	2721	0.50	3373
1.0	2205	1.0	2736	1.0	3395
2.0	2253	2.0	2771	2.0	3431
4.0	2296	4.0	2813	4.0	3475
8.0	2362	8.0	2876	8.0	3537
32	2487	16	2954	16	3603
60	2542	32	3039	32	3684
224	2602	60	3128	60	3755
1164	2633	105	3183	147	3828
1539	2642	360	3245	211	3838
2664	2652	1395	3278	475	3871
		1640	3281	1465	3892
		2886	3300	4725	3912
				5670	3913

* Soil sample data: $G_s = 2.73$; $w_L = 62.1$ percent; $w_p = 27.1$ percent;
$H = 25$ mm; diameter $= 50$ mm;
$H_f = 25 - 3913(0.0025) = 15.218$ mm;
dial gage: 1 div $= 0.0025$ mm
$H_s = 10.774$ mm; $e_o = 1.361$

    (b) Draw e versus log p curves and compute C_c and C_c'.
    Ans.: $C_c = 0.47$

12-8 For the data of Table P12-1,
    (a) Draw settlement versus $\sqrt{\text{time}}$ curves and compute c_v.
    (b) Draw a curve of e versus p and obtain a_v and m_v.
    Ans.: $m_v \cong 8.62 \times 10^{-5}$ m^2/kN

12-9 Redo Example 12-6 if the footing load is 792 kN.

THIRTEEN

SHEAR STRENGTH OF SOILS

13-1 INTRODUCTION

A load placed on a soil mass will always produce stresses of varying intensity in a bulb-shaped zone beneath the load. We used material in Chap. 10 to estimate their magnitude at the several depths of interest. These stresses were used with the "consolidation theory" described in Chaps. 11 and 12 to estimate long-term, or consolidation, settlements. We will consider "immediate" settlements together with bearing capacity (allowable soil loading) in Chap. 15, but it will be necessary to first consider soil strength. This is because the load placed on a soil mass requires two considerations:

1. The magnitude of the total settlement. For many soils this is the consolidation settlement discussed in Chap. 11—including any "creep."
2. The possibility of a soil failure. This may be as a rotational movement of soil from beneath the loaded area as illustrated in Fig. 13-1a or, sometimes, as a punching failure (Fig. 13-1c). This latter is usually a limited movement; however, the magnitude may be large enough to produce structural distress in the superstructure.

All the failures shown in Fig. 13-1 are "shear" failures since the movement is a slip between two surfaces. Actually this slip plane is idealized and is made up of substantial particle rolling, sliding, and slipping in the zone shown as a "surface." The direction of the largest statistical accumulation of movement defines the plane and the slip. Since soil failures are in a shearing mode, the soil strength of interest is the shear strength.

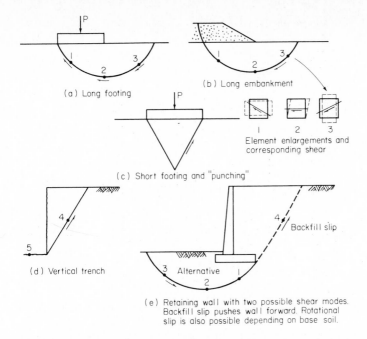

Figure 13-1 Typical failure modes requiring shear strength evaluation. Note that the shear is different for points 1, 2, and 3 in parts *a* and *b* (as shown with element enlargements) as well as *d*4, *e*4, and *d*5.

We will find that the shear strength (often determined in a compression test) is not a unique value, but is heavily influenced in situ by factors such as:

1. Soil state—void ratio, grain size, and shape.
2. Soil type—sand, sandy, gravel, clay, etc., and/or the relative amounts of these materials present.
3. Water content—particularly for clay (often ranging from very soft to stiff depending on the instant value of w).
4. Load type and rate. From consolidation theory we might note that a rapid load may produce excess pore pressures.
5. Anisotropy. Strength normal to bedding plane is different from that parallel to plane.

In the laboratory the shear strength is heavily influenced by:

1. Test method—development of excess pore pressures
2. Sample disturbance—decreases strength
3. Water content
4. Strain rate—usually increases strength

13-2 SHEAR STRENGTH PARAMETERS c AND ϕ

The first hypothesis on the shear strength of a soil was presented by Coulomb (ca. 1773) as

$$s = c + v\sigma \tag{13-1}$$

where s = shear strength (we will use this instead of τ for the remainder of the text) on the plane of interest

c = cohesion, or interparticle attraction effect, nearly independent of the normal stress on the plane

σ = normal stress on the plane of interest

v = coefficient of friction between the materials in contact

Equation (13-1) is a superposition of the cohesion and friction resistance, and referring to Fig. 5-1 we see that $v = \tan \phi$. When the effective stress σ' is used for σ, we obtain

$$s = c' + \sigma' \tan \phi' \tag{13-2}$$

From the computations involving τ_{oct} in Chap. 10 we found the shear strength was independent of pore pressure, so we do not prime s. It is evident then that ϕ and c will depend on whether we use Eq. (13-1) or (13-2). The angle ϕ is termed the *angle of internal friction* and is not a constant as we have just shown. Since $\phi \neq \phi'$ care should be exercised to note whether total [Eq. (13-1)] or effective [Eq. (13-2)] strength parameters are being used in the analysis.

Mohr (of Mohr's circle) presented a theory of failure at a limiting stress state as

$$s_f = f(\sigma') \tag{a}$$

which simply states that on a failure plane the shear stress at failure (s_f) is some function of the effective normal stress on that plane. One such expression for Eq. (*a*) is obviously

$$s_f = \sigma'(v) = \sigma' \tan \phi'$$

and suggests that Eq. (*a*) is a special case of Coulomb's Eq. (13-2). Generalizing, we obtain what is now commonly called the *Mohr-Coulomb failure criterion* as

$$s_f = c'_f + \sigma'_f \tan \phi' \tag{b}$$

It is customary to drop the f subscripts as we are always only interested in the most critical (or failure) case. Thus we will use the Mohr-Coulomb equation as given by *either* Eq. (13-1) or (13-2).

Analysis of soils at this failure state is called by some *critical states* and by others *limit states*; also *limiting equilibrium* analysis.

Early researchers in mechanics of soils noted that a triaxial compression test using a cell pressure $\sigma_c = \sigma_3$ would produce a maximum (at failure) pressure σ_1 which was sufficient to plot a Mohr's circle. A second test at a larger cell pressure σ_3 would produce a larger σ_1 and a new Mohr's circle to the right of the first and similarly for other values of cell pressure. Of course, a large number of

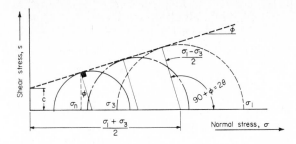

Figure 13-2 Mohr-Coulomb failure envelope for obtaining the limiting soil parameters c and ϕ based on σ_1, σ_3 at failure.

Mohr's circles could be drawn at intermediate σ_1 stress levels ranging from $\sigma_3 < \sigma_1 \leq \sigma_{1(f)}$, but if we used only the ones at "failure," we would obtain the limiting case.

Since Eq. (13-2) or (13-1) is an equation of a straight line of limiting shear stress and all the points on each Mohr's (failure) circle also constitute a limiting stress, it appeared reasonable to examine the envelope line produced by drawing a tangent to a series of Mohr's circles. When this was done, the relationships shown on Figs. 13-2 and 13-3 are obtained. That is, the failure shear strength of a soil could be reasonably well predicted from the line (or envelope) obtained from drawing a tangent line to a series of Mohr's circles. The equation of best fit for this line was either Eq. (13-1) or (13-2) depending on whether total or effective stresses were used for σ_1 and σ_3.

Let us examine Fig. 13-3, which gives qualitative (without numerical values) Mohr's circles for several cases:

1. *Cohesionless soil.* Note that the cohesion intercept is zero. Tests on damp sands may give a small cohesion intercept from surface tension accumulations.

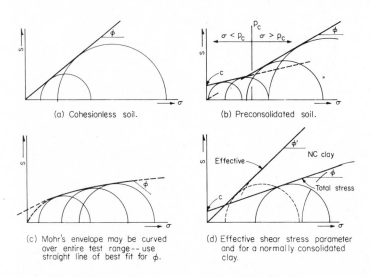

(a) Cohesionless soil.

(b) Preconsolidated soil.

(c) Mohr's envelope may be curved over entire test range -- use straight line of best fit for ϕ.

(d) Effective shear stress parameter and for a normally consolidated clay.

Figure 13-3 Qualitative failure envelopes for several soils as indicated.

The rubber membrane used to confine the sample may also produce a small apparent cohesion. The failure envelope for most cohesionless soils is straight within testing precision and over the likely working range of σ.

2. *Preconsolidated soil.* Here for stress combinations up to the preconsolidation pressure p_c a cohesion parameter and reduced ϕ is produced (Fig. 13-3b); beyond p_c the soil behaves as a normally consolidated material as Fig. 13-3d.

3. *A curved Mohr envelope.* For nearly all soils, Mohr's envelope is curved and requires interpretation. A secant or tangent line in the region of $\sigma = \sigma_{\text{working}}$ can be used to obtain ϕ and a projection to the shear axis can be used to obtain c.

4. *Effective stress parameters.* When pore pressures are measured, the Mohr's circles can be corrected as shown. For a normally consolidated (NC) clay, the cohesion intercept is nearly zero for ϕ'. Refer to item 2 and observe that if all of a series of triaxial tests were run so that $\sigma_n > p_c$ the soil would plot as normally consolidated. Correcting tests at $\sigma_n < p_c$ for pore pressure does not necessarily produce $c = 0$ but does produce a value of ϕ' which is not the same as ϕ' in the normally consolidated region.

It is not necessary to draw Mohr's circles to obtain the failure envelope of Eq. (13-1) or (13-2). We might simply apply σ_n to a soil and measure the maximum s. Two or more data sets allow the equations to be solved simultaneously for c and ϕ, or, preferably, the data are plotted and an "averaged" graphic solution obtained. The use of stress paths as in Sec. 13-5 may also be used to obtain the shear strength parameters c and ϕ.

The maximum shear stress s is not defined by Eqs. (13-1) and (13-2) for most soils. The Mohr-Coulomb equations define the critical shear stress producing failure and are a combination of shear and normal stresses on the critical plane.

13-3 SOIL FAILURE, CRITICAL VOID RATIO, AND RESIDUAL STRENGTH

Soil failure in situ is often a settlement which is larger than wanted, a mass movement, or similar. In all of these cases the failure zone enlarges until the shear strength is sufficient to halt movement (although the mass may end up resting in the bottom of a valley as in the case of a landslide). In all cases the movement halts with the soil at the perimeter of the movement in a highly remolded state. This corresponds to the state resulting in a compression test after a considerable amount of strain. We may term the strength remaining after this large strain (where the sample load never drops to zero) as the *residual strength*. Since remolding very nearly destroys the cohesion parameter, the residual strength parameter of interest is the residual angle of internal friction ϕ_r.

Soil stress-strain curves may exhibit a *brittle*, or relatively sudden strength decrease from some peak value or may gradually build to a nearly constant value which is held over a large value of strain. Brittle behavior can be obtained when:

1. The soil is a dry or wet granular material in a dense state
2. The soil is cohesive but is either heavily overconsolidated or has a low degree of saturation S (i.e., is fairly dry)
3. The soil has a substantial amount of natural cementation
4. The soil has been compacted and tested at a water content on the dry side of optimum
5. The confining pressure is much larger than about 70 kPa in triaxial tests

Loose sands and most remolded cohesive soils, and soils compacted on the wet side of optimum tend to a more progressive failure, particularly in q_u tests or in triaxial tests using low cell pressures. These conditions are qualitatively illustrated in Figs. 13-4a and 13-4b. Brittle failure occurs at low strains—often on the order of 1 to 3 percent (0.01 to 0.03 m/m). Progressive, or residual, failure may not be clearly defined and some arbitrary strain—often 20 percent (0.2 m/m) is used (Fig. 13-4b).

Our discussion so far has not clearly defined "dense" and "loose," and we might, from comments on cell pressure, deduce that a "dense" material at a low cell pressure might behave similarly to a "loose" material at a high cell pressure.

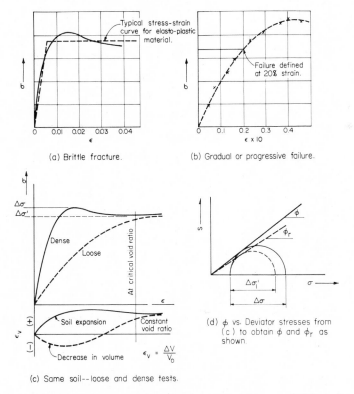

(a) Brittle fracture.

(b) Gradual or progressive failure.

(c) Same soil--loose and dense tests.

(d) ϕ vs. Deviator stresses from (c) to obtain ϕ and ϕ_r as shown.

Figure 13-4 Stress-strain relationships for dense and loose soil. "Loose" and "dense" as used here are relative terms and depend on the confining pressure.

Thus we find that "dense" and "loose" become relative terms which must be used together with cell (or consolidation) pressure.

The gradual development of the ultimate stress can be termed a "progressive action" (Fig. 13-4b) compared to the more brittle failure illustrated in Fig. 13-4a. In progressive failure, which is the likely mode of failure for most in situ cases, weak portions of the soil matrix "fail" first with a stress transfer to adjacent zones. The stress transfer results in further "failure" and stress transfers, etc. In situ, it is very probable that soil failure is progressive over finite areas rather than having a singular stress (point) value.

For cohesionless samples, experimental evidence indicates that the brittle or dense curve in Fig. 13-4a and the "loose" curve in Fig. 13-4b will converge as in Fig. 13-4c for the same soil and using the same confining pressure. The dense soil will expand under shear to produce a new void ratio which further modifies to the *critical void ratio* at some strain. Similarly, the loose soil will decrease in void ratio to some constant value at some amount of strain. This constant or critical void ratio is generally between the two void ratios used for the test samples.

The increase in void ratio from e_{min} to e_{crit} for a dense soil is from the *dilation* or increase in soil volume during shear as the interlocked soil grains roll about (over, between, around) with a statistical accumulation of increase in volume. The decrease in void ratio from e_{max} to e_{crit} of a loose sample is the collapse of the structure and grains "falling into" the voids so that the statistical accumulation is a decrease in volume.

The critical void ratio may be approximately determined by measuring volumetric changes for a "dense" and "loose" soil sample in the direct shear test to obtain the type of data illustrated in Fig. 13-4c. The critical void ratio is less easily determined in a triaxial test since it is more difficult to measure volume changes. While volume change under shear is common to all soils, the concept of critical void ratio is primarily applicable to cohesionless soils.

The residual strength from a triaxial test may be used instead of the maximum strength to plot a Mohr's circle. The residual shear strength from a direct shear test may also be used similarly. Where the residual strength is not clearly defined, the strength at some arbitrary strain value may be used (but should be noted). The resulting Mohr-Coulomb failure envelope gives ϕ_r as in Fig. 13-4d. In general, the cohesion intercept is negligible and ϕ_r is several degrees less than ϕ since the residual Mohr's circle is smaller.

13-4 SOIL TESTS TO DETERMINE SHEAR STRENGTH PARAMETERS

The soil tests commonly employed to obtain the strength parameters include (in order of increasing costs) the following:

1. Unconfined compression or q_u tests. The compressive strength obtained from this test is always identified as q_u. This test is also called an *unconsolidated-undrained* or *U test*. The undrained shear strength is usually identified as s_u.

Dilation → increase in soil volume.

2. Direct shear and direct simple shear (DSS) tests.
3. Confined compression or triaxial tests

These tests are schematically illustrated in Fig. 13-5. Note that the unconfined compression test is a triaxial test with $\sigma_3 = \sigma_c = 0$. Figure 13-6a is an unconfined compression test in progress, and Figs. 13-6b–d illustrate several steps to assemble a triaxial test for testing using isotropic consolidation. Most testing laboratories use equipment of a configuration similar to that shown in Fig. 13-6.

Historically, the direct shear test was the first attempt to quantify the Mohr-Coulomb strength parameters c and ϕ. Around 1930, nearly simultaneously European and American researchers developed the triaxial test for soils. A number of purported improvements to include plane strain, allow a controlled triaxial stress state ($\sigma_1 \neq \sigma_2 \neq \sigma_3$), measure pore pressure, and to produce a state of pure shear have been proposed. Generally these "improvements" have been so

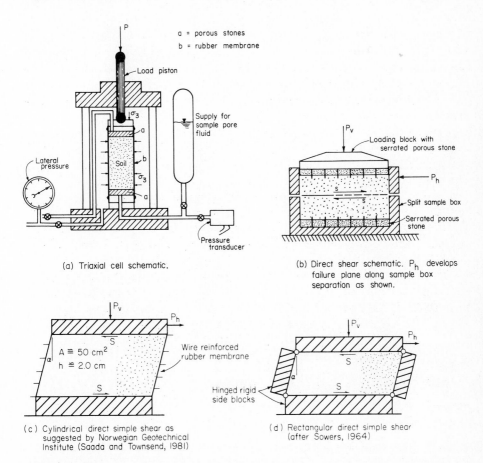

(a) Triaxial cell schematic.

(b) Direct shear schematic. P_h develops failure plane along sample box separation as shown.

(c) Cylindrical direct simple shear as suggested by Norwegian Geotechnical Institute (Saada and Townsend, 1981)

(d) Rectangular direct simple shear (after Sowers, 1964)

Figure 13-5 Schematics of triaxial and direct shear tests. (See Fig. 13-7a for schematic of unconfined compression test.)

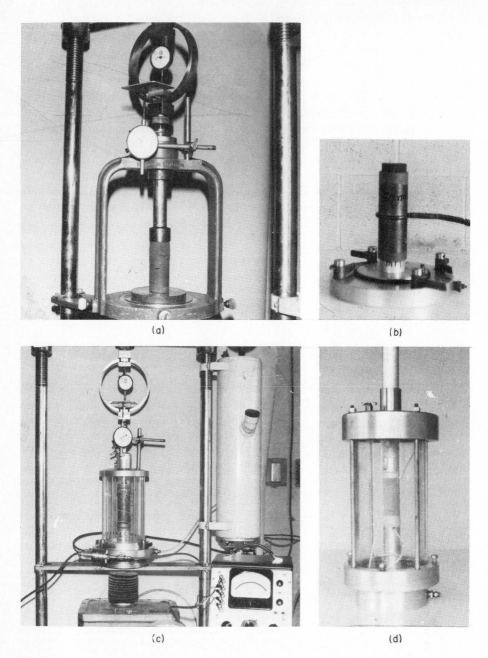

Figure 13-6 Laboratory shear tests: (*a*) unconfined compression test using CBR loading piston; (*b*) cohesive sample partially inserted into sample membrane using oversized membrane stretcher; (*c*) triaxial test with pore pressure transducer and electronic pressure indicator; (*d*) triaxial cell modified so that the piston is the same size as the sample for anisotropic consolidation.

complicated (or in fact do not fulfill the criteria) that few laboratories use them. Pore pressure measurements are commonly made in most modern soils laboratories. The other relatively recent introduction, the direct simple shear test, claiming to develop a close state of nearly pure shear, has also survived. The direct simple shear test may be done by two methods as schematically illustrated in Fig. 13-5c and 13-5d. It appears that a major factor in its popularity is that it is very amenable to direct attachment to electronic data acquisition equipment and can be used for cyclic shear testing. In cyclic shear testing the shear force P_h is periodically reversed, and usually the shear strain (defining the angle α in Fig. 13-5c) is kept small.

There is a substantial body of literature on shear strength testing and laboratory equipment limitations. The ASTM has published several Special Technical Publications, including STP 131 (1953), STP 361 (1964), and STP 740 (1981). The state-of-the-art reports in these latter two provide an in-depth critique of the several procedures and limitations.

Shear tests are commonly considered as:

1. *Unconsolidated-undrained* (U or UU). Here the sample is tested as is without any control of pore water drainage. Drainage control may be from using either a closed system or a high strain rate so that failure occurs faster than drainage.
2. *Consolidated-undrained* (CU). Here the sample is stressed with all around stresses with drainage allowed. When the volumetric change is complete (consolidated) as determined by volume measurements or measuring pore drainage (or both), the drainage outlets are closed and the sample loaded to failure. Sometimes pore pressures are measured. Note also that an undrained test is a *constant-volume test* since drainage is not allowed.
3. *Consolidated-drained* (CD). This test is similar to the CU test, except that after consolidation, drainage is allowed during loading and the load rate is sufficiently slow that large pore pressures do not develop. It is impossible to have a truly "drained" case since any strain can produce an excess pore pressure u (depending on the degree of saturation S), but a very slow strain rate tends to keep u small. It can take a week or more to properly perform a CD test on a saturated clay. With drainage, we might note that the test is not at constant volume.

We should observe that:

1. Any of the above test procedures will give the same results for a dry soil.
2. Drainage can occur fairly rapidly for granular materials with a large coefficient of permeability k so that a CD test may not take very long compared to a cohesive sample.
3. A CU test on clay can produce a wide range of strength parameters depending on S since volume change during shear cannot be controlled unless the soil is saturated.

4. It is preferable to make a CU test and measure the pore pressure at failure than to perform a CD test.
5. It is realistic to test cohesive soils in a saturated state since there is seldom any control over the saturation state in situ and the saturated sample produces a "worst case."

The Unconfined Compression Test

The unconfined compression test is very widely used worldwide. It is a simple test where atmospheric pressure surrounds the soil sample. The corresponding Mohr's circle is illustrated in Fig. 13-7c. From the single circle averaged for a series of tests we can only extrapolate the slope of the failure envelope as $\phi = 0$ and the undrained shear strength s_u is

$$s_u = c = \frac{q_u}{2}$$

Some hold the opinion that if water is present, an effective "confining" pressure exists internally in an unconfined compression test sample due to capillary effects and that these effects may effectively confine the sample just as if it were in situ. If this be correct, then it is also true that degree of saturation, grain size, stress

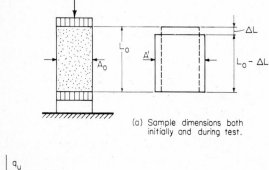

(a) Sample dimensions both initially and during test.

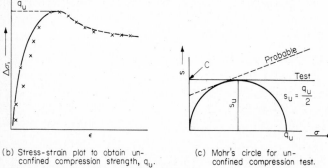

(b) Stress-strain plot to obtain un-confined compression strength, q_u.

(c) Mohr's circle for un-confined compression test.

Figure 13-7 The unconfined compression test.

cracks or fissuring, and laboratory humidity would be very important considerations. In any case, few (if any) laboratories attempt to allow for capillary confinement; q_u is taken as the maximum compressive stress and plotted on Mohr's
circle as in Fig. 13-7c. Since each test would be plotted with $\sigma_3 = 0$, the strength
parameters c (with $\phi = 0$) generally provide a conservative solution as long as the
working stresses are greater than $q_u/2$.

The unconfined compression (and triaxial) test is made by obtaining a thin-
walled tube or other sample as nearly "undisturbed" as possible. Sometimes q_u is
gotten from the sample obtained from the standard penetration test of Sec. 6-9.
The sample is divided, where the length is sufficient,† into several samples with a
length/diameter ratio

$$2 < \frac{L}{d} < 3$$

to ensure that a failure plane does not intersect the loading heads. The sample is
made as square as practical (no simple matter when small pieces of gravel are
present) and placed in a compression machine adjusted to a deformation rate on
the order of under 1.5 mm/min, and deformations vs. corresponding sample loads
are obtained. These data are used to plot a stress-strain curve of σ versus ε to
obtain the maximum value of compressive stress, which is q_u for the unconfined
compression test. It is evident that an unconfined compression test can be made
only on cohesive soils.

The compressive stress σ (or $\Delta\sigma_1$ for a triaxial test) is computed differently
than for steel, concrete, or other materials. For these materials the compressive
(deviator) stress is computed as

$$\Delta\sigma_1 = \frac{P}{A}$$

with A being the original cross-sectional area, which is a conservative computation. As the soil strains vertically, it produces lateral strains which increase the
effective cross-sectional area which resists the stress. Since this always occurs with
the laboratory sample, and on a similarly enlarged area in the field due to the
Poisson effect, it is considered more correct to base the compressive stress on this
enlarged cross-sectional area A'. This area can be computed using the assumption
that the volume of the sample remains constant (Fig. 13-7a); thus

$$A_o L_o = A'(L_o - \Delta L)$$

from which we obtain, since strain $\varepsilon = \Delta L/L_o$,

$$A' = \frac{A_o}{1 - \varepsilon} \qquad (13\text{-}3)$$

---

† Some organizations obtain "undisturbed" tube samples on the order of 120-mm diameter so
that three cylindrical samples of a diameter of approximately 5 cm can be trimmed from any vertical
location to obtain 3 tests with the depth variable minimized.

with all terms identified in Fig. 13-7. The instantaneous deviator stress $\Delta\sigma_1$ is

$$\Delta\sigma_1 = \frac{P}{A'} \tag{13-4}$$

where $\Delta\sigma_1$ is the change in major principal stress, as it is assumed that there is no shear stress on the ends of the soil sample. Since friction is developed as the sample attempts to expand, a small error is introduced; however, research indicates the error is essentially negligible (Barden and McDermott, 1965). The error may not be negligible in tests where the sample is in contact with the end platens for a very long time, as may happen in research projects using cohesive soils, and it is necessary to fully consolidate the sample prior to testing. Research by the author indicates that Eq. (13-3) does describe the bulging reasonably well (i.e., measure the sample diameter and compare change to corresponding strain). The use of smooth metal end platens with silicone grease may reduce end restraint effects.

The unconfined compression test is classified as an unconsolidated-undrained (U or UU) test because of the test procedures. In the literature this is often the terminology used for the unconfined compression test since it tends to give the procedure more dignity. The context of the material will generally indicate how the test was made. This distinction is necessary since some triaxial compression tests are run by quickly applying some cell pressure and then immediately compressing the sample to produce U-test soil data.

The Direct Shear and Direct Simple Shear Tests

The *direct shear* test is a simple, straightforward test to perform and is next in order of increasing cost. The test is made by placing a soil sample into the shear box illustrated in Fig. 13-5b. The box is split as shown, with the bottom half fixed and the top half free to float and translate. The box is available in several sizes but commonly is 6.4 cm in diameter or 5.0×5.0 cm square. The sample is carefully placed in the box; a loading block, which includes a serrated porous stone for rapid drainage, is placed on the sample. Next a normal load P_v is applied. The two halves are separated slightly and the loading block and top half are clamped together and either

1. The test is begun immediately and horizontal displacements obtained for the corresponding horizontal load P_h, so that a plot of P_h versus δ_h can be made to find $P_{h,\,max}$. This is called an *undrained* (U) test.
2. The test is begun after consolidation of the soil under P_v is complete as determined by observation of a dial gage in contact with the load head. If the test proceeds:
   (a) Rapidly, pore pressures will develop in wet or saturated cohesive soils due to the low coefficient of permeability and the test is called a *consolidated-undrained* (CU) test.

(b) Very slowly, so that no pore pressures develop, the test is called a *consolidated-drained* (CD) test. This test usually gives for normally consolidated soils a value of $c' \cong 0$ and

$$s = \sigma'_n \tan \phi' \tag{13-5}$$

Two or more additional tests at larger values of P_v are performed to make a scaled plot of

$$s = \frac{P_v}{A_o} \quad \text{versus} \quad \sigma_n = \frac{P_h}{A_o}$$

as in Fig. 13-8 so that a graphical solution of Eq. (13-1) can be obtained. This is necessary, as there are two unknowns (c and ϕ) in Eq. (13-5) or (13-6) and a minimum of two tests is required to obtain two values of s so that a simultaneous equation solution can be made. Qualitative results of the U, CU, and CD direct shear tests are shown in Fig. 13-8b, from which it is apparent that the shear strength parameters are not unique values but depend heavily on the test procedure.

The thickness of the direct shear test sample is on the order of 20 to 30 mm; thus, drained conditions are quite probable for cohesionless soils (and are commonly assumed). In the undrained test with cohesive soils, some drainage is likely to take place; consolidation usually does not take a very long time for CU tests, which represents a considerable economic advantage. Pore water conditions cannot be controlled or measured in this test, however. Some researchers have attempted to correlate the water content along the failure surface (say the zone 4 to 6 mm along the shear box separation line) to shear strength. The water content in this zone will be different from the average due to particle displacement, resulting in a soil void ratio decrease.

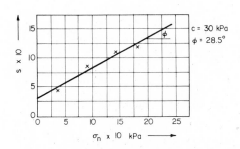

(a) Actual plot of direct shear test on cohesive soil. Seldom does data plot a straight line—some interpolation is necessary as shown.

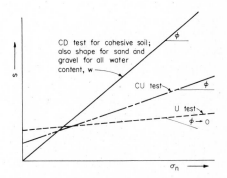

(b) Qualitative results of the several types of direct shear tests on normally consolidated soils.

Figure 13-8 Direct shear test.

The direct shear test forces the direction and location of the failure plane, i.e., at the location of the box split and parallel to the horizontal load. Practically, this condition may not be obtained, but the results are nevertheless considered to be satisfactory. This test is considered by some persons as meeting the requirements of plane strain sufficiently to term this a plane strain test. Considering that the sample is confined so that only lateral and vertical movements (strains) can take place, which is by definition plane strain, the reader can decide whether to use the data as "plane strain" data. A second test deficiency is that the shear area A_o decreases as the test proceeds. A correction for area reduction can easily be made with square shear boxes, but it is not very practical to make this correction for round boxes due to the considerable mathematics involved in computing the instantaneous area as a function of δ_h.

It appears that the best application of the direct shear test is in measuring the residual shear strength parameter ϕ_r. The cohesion parameter is (or should be) nearly zero resulting from the large strains at the residual strength state.

The direct simple shear (DSS) test devices are attempts to produce a state of pure (or simple) shear in the soil in plane strain without the limitations of the direct shear apparatus. With a closed system it is possible to both consolidate the sample and to measure pore pressures during shear. It is highly doubtful that these devices produce a state of pure shear. There is also some disagreement that the DSS is an improvement over the direct shear test—particularly since the equipment is substantially more complicated.

The direct shear test would provide the soil parameters for the zone "2" in Fig. 13-1 if a zoned strength analysis is used in this class of problems.

The Triaxial Test

The *triaxial test* is considered to provide the best soil parameters and stress-strain data (for E_s, μ, and G_s). This is true only if "undistorted" soil samples are obtained and great care is exercised in trimming them to size and inserting them into the rubber membrane. The general sample dimensions range from 35 to 75 mm in diameter with an L/d ratio of 2.2 to 3.0 as with the unconfined compression test. Samples 100 mm and larger in diameter may be tested in special cells, but recovery of samples this large from borings becomes expensive.

It is necessary to test a sample starting from in situ stress and pore pressure conditions if reliable parameters are to be obtained. This is the particular advantage of the triaxial test since

1. It is possible to both control and measure the pore pressure. If the pore pressure is given an initial (as in situ) value at the start of the test, this is termed applying "back pressure."†

---

† Back pressure is used to reproduce the in situ pore pressure or to ensure sample saturation. When back pressure is used, the cell pressure must be increased by the same amount to produce the same effective stress within the sample.

2. It is possible to apply a range of confining pressures—both isotropic (or constant all round) or anisotropic (lateral different from vertical)—to initially consolidate the sample to some predetermined state.
3. Special strain tests may be run such as increasing (compression) or decreasing (extension) the axial load.

It is common in triaxial tests to isotropically consolidate the sample at some cell pressure $\sigma_c = \sigma_3$. Considerable evidence indicates that anisotropic consolidation produces better strength and stress-strain data. In particular, the consolidation of a sample back to its in situ state is desirable. From Chap. 10 the ratio of the lateral to vertical stress at steady state (zero strain) was

$$K_o = \frac{\sigma_h}{\sigma_v} = \frac{\sigma_h}{\gamma y}$$

where $\sigma_v = \gamma y$ is the in situ vertical stress from the soil at depth y. This state can be obtained in a consolidation test by applying a vertical load to produce σ_v since the metal ring prevents any lateral strain. In this case when the vertical settlement is completed the sample is K_o-consolidated (but may include some sampling disturbance). Special triaxial equipment is required (e.g., Bishop and Wesley 1975; Campanella and Vaid, 1972) to obtain this state of anisotropic consolidation. These special triaxial cells allow both extension as well as the more common compression tests.

We may identify the following "consolidated" triaxial tests for future reference as:

CAU   Anisotropically consolidated–undrained test.
$CK_o U$ K_o-consolidated–undrained test.
CIU   Isotropically consolidated–undrained test.
$CK_o E$ K_o-consolidated–extension (and undrained test). In this test the lateral stress is held constant and the vertical stress decreased or the vertical stress held constant and the lateral stress increased.

It is generally understood that U tests are undrained compression tests. It is also generally understood that if U is omitted, the test is "undrained" unless it is denoted specifically as "drained." It is most common to obtain the effective stress parameters c', and σ' from consolidated undrained CU-type tests with pore pressure measurements.

The pore water pressure is commonly taken from the base of the sample using a pore pressure measuring device (a transducer if electronic equipment is used), as schematically illustrated in Fig. 13-5a, but can be taken from midheight if a pore pressure needle is used (which requires a large sample and a fluid-tight entry through the rubber sample membrane). Unless the pore pressure is measured exactly in the failure plane, the value may not be correct, since it is on the failure plane that maximum soil structure reorientation and resulting maximum developed excess pore pressure are occurring.

A triaxial stress-strain plot is made exactly the same way as that for an unconfined compression test, computing the deviator stress as

$$\Delta\sigma_1 = \frac{P}{A'}$$

with the stress-strain plot as in Fig. 13-9a. Figure 13-9b shows a plot of σ_1 versus ε. This plot is not used, since the stress $\Delta\sigma_1 = \sigma_3$ at $\varepsilon = 0$. For the triaxial test the major principal stress is computed as

$$\sigma_1 = \sigma_3 + \Delta\sigma_1$$

and the instantaneous values could be plotted with Mohr's circles as illustrated in Fig. 13-9c. Usually the maximum, or failure, circle which produces a point on the Mohr failure envelope is the only one of interest for a particular confining pres-

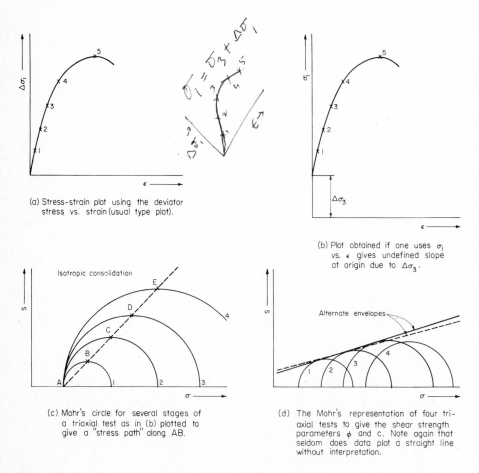

(a) Stress-strain plot using the deviator stress vs. strain (usual type plot).

(b) Plot obtained if one uses σ_1 vs. ε gives undefined slope at origin due to $\Delta\sigma_3$.

(c) Mohr's circle for several stages of a triaxial test as in (b) plotted to give a "stress path" along AB.

(d) The Mohr's representation of four triaxial tests to give the shear strength parameters ϕ and c. Note again that seldom does data plot a straight line without interpretation.

Figure 13-9 Presentation of triaxial stress-strain and Mohr's circle data.

sure. The results of several tests are plotted, as in Fig. 13-9d, and the failure envelope draws as a best fit to obtain c and ϕ.

In general, three strength tests should be performed as an absolute minimum, as averaging the results of two tests may give a poor value if one is high and the other low. With three tests, an unreasonably high or low value can be discarded.

Note carefully that the Mohr-Coulomb failure envelope is seldom straight except over a short region at low confining pressures. This is true for both clay and sand so that the curve will require some interpretation. Part of the curve for clay may be from preconsolidation. For sand, the curvature seems to result from particle contact crushing (Banks and MacIver, 1969), particularly at higher pressures.

Example 13-1 Given: a cohesive soil was tested in a direct shear test with the following values: Square shear box 5.5×5.5 cm; height $= 2.1$ cm.

			Computed stresses, kPa	
Test	P_v, kg	P_h, kg	σ_n	s
1	4	2.9	13.0	9.4
2	8	4.3	26.0	13.9
3	12	5.1	39.0	16.5

REQUIRED

(a) Obtain c and ϕ (c in kilopascals).
(b) Find the orientation of principal planes for test 2.

SOLUTION

Step 1 Compute σ_n and s as shown in the table above.

$$\sigma_n = \frac{4(9.807) \times 10^4}{(5.5)^2 \times 10^3} = 13.0 \text{ kPa}$$

The remaining stresses are computed similarly.

Step 2 Plot values and "fair" the failure envelope as shown in Fig. E13-1a, and measure

$$\phi = 17.5°$$

$$c = 5.0 \text{ kPa}$$

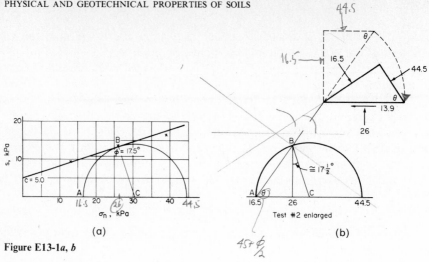

Figure E13-1a, b

Step 3 To obtain the orientation of principal planes, construct a Mohr's circle for test 2 and lay off AB at θ as shown in Fig. E-13-1b, noting that principal planes are oriented with θ as shown and 90° apart and that with the actual failure plane horizontal, the principal planes are as shown.

////

Example 13-2 Given: a direct simple shear test K_o-consolidated (with measured) values with $\sigma_v = 50\ \text{kPa}$ and $\sigma_h = K_o \sigma_v = 25\ \text{kPa}$.

REQUIRED For a known cohesion of $c = 10\ \text{kPa}$, what is the angle ϕ produced when the measured shear stress at failure is 20 kPa?

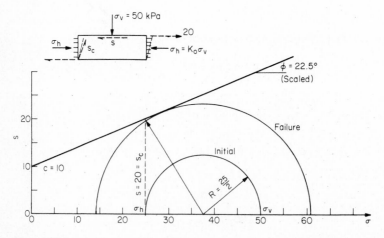

Figure E13-2

cases and reduces the graphic congestion. The trace of the stress path of Fig. 13-10a is termed the K_f line (f for failure). Since we earlier defined K as $K = \sigma_3/\sigma_1$ and used K_o to define the in situ stress ratio, we see slope α of the K_f line defines the failure ratio

$$K_f = \frac{\sigma_3}{\sigma_{1(f)}}$$

by substitution into Eq. (13-6) for $\sigma_3 = K_f \sigma_1$ to obtain

$$\tan \alpha = \frac{q}{p} = \frac{1 - K_f}{1 + K_f} \tag{13-7}$$

This method of plotting stress paths seems to have been first used by Simons (1960).

The effective stress path can be plotted by using

$$p' = p - u_f \quad \text{and} \quad q' = q$$

obtained by using effective stresses $\sigma_i' = \sigma_i - u_f$ in Eqs. (13-6). Note that because q is a shear stress, it is not affected by the pore pressure. The pore pressure moves p' to the left of p along the normal stress axis by the pore pressure u_f.

The four general cases which can be made in a triaxial test can be plotted to produce a stress path as illustrated in Fig. 13-10b. The slopes of the two K_f lines shown do not have to be equal.

The relationship between the K_f line (of a pq plot) and the Mohr's failure envelope (called the ϕ line) is as shown in Fig. 13-11. Since the origin is common for both lines (Fig. 13-11b), we have

$$\frac{q}{m + x} = \tan \alpha = \sin \phi = \frac{a}{m}$$

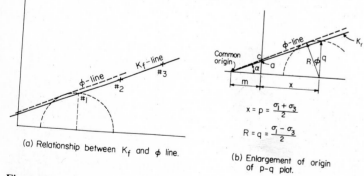

(a) Relationship between K_f and ϕ line.

$$x = p = \frac{\sigma_1 + \sigma_3}{2}$$

$$R = q = \frac{\sigma_1 - \sigma_3}{2}$$

(b) Enlargement of origin of p-q plot.

Figure 13-11 Relationship between shear strength parameters and the K_f line from the stress path plot.

SOLUTION What you have here is a part of the data from a series of DSS tests. We will draw a Mohr's circle to obtain ϕ graphically since it will be simpler than trying to compute it:

1. Draw an initial Mohr's circle for the sample at the end of consolidation.
2. Next, noting that with shear, both σ_v and σ_h are "normal" stresses on a shear plane on which we measure the shear stress of 20 kPa. This means that a second Mohr's circle with the same center but through coordinates σ_v, $-s$ and σ_h, $+s_c$ can be drawn. Note that the shear stress in the insert is reversed to the normal ($+$) direction, but this does not affect the values or the circle. We would have to consider direction (signs) if we want the correct orientation of the principal planes. Scale from Fig. E13-2 a value of $\phi = 22.5°$.

////

Example 13-3 Given a CU triaxial test on an unsaturated soil with the following data:

Test	σ_3, kPa	$\Delta\sigma_1$ = deviator stress	σ_1 (computed)
1	40	63.0	103.0
2	80	86.0	166.0
3	120	102.0	222.0

$\sigma_1 = \sigma_3 + \Delta\sigma_1$

REQUIRED Find the apparent soil parameters ϕ and c and orient the failure plane for test 1.

SOLUTION

Step 1 Compute

$$\sigma_1 = \sigma_3 + \Delta\sigma_1$$

$$= 40 + 63.0 = 103.0 \text{ for test 1, etc.}$$

Step 2 Plot Mohr's circle for the three tests as in Fig. E13-3, fair a failure envelope, and scale

$$\phi = 12°$$

$$c = 17.5 \text{ kPa}$$

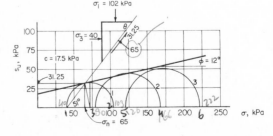

Figure E13-3

Step 3 From σ_3 of test 1, draw line AB at $\boxed{\theta = 45 + \phi/2} = 51°$ as shown, and place on the element the stresses shown. Values of σ_n and s can be computed from Eqs. (10-10) and (10-11) but can be scaled with satisfactory precision.

////

$\theta = 45 + \boxed{\frac{\square}{2}} = \boxed{\square}$

Example 13-4 Given triaxial tests on a cohesionless soil.

Test	σ_1	σ_3, kPa
1	78.5	20
2	186.0	50

handwritten:
$\sigma_3 = 96.6 \text{ kN/m}^2$
$\Delta\sigma_f = 67.7$ "
$\Delta\sigma_f = \sigma_3 + \Delta\sigma_f$
$\sigma_1 = \sigma_3 + \Delta\sigma_f$
$= 163.6 \text{ kN/m}^2$

REQUIRED The angle of internal friction ϕ.

SOLUTION Compute the angle from the relationships shown in Fig. E13-4. From Fig. E13-4:

$$\sin \phi = \frac{R}{X} = \frac{\sigma_1 - \sigma_3}{\sigma_1 + \sigma_3}$$

For test 1,

$$\phi = \sin^{-1} \frac{78.5 - 20}{78.5 + 20} = 36.4°$$

For test 2,

$$\phi = \sin^{-1} \frac{136}{236} = 35.2°$$

Take the average as the best value to obtain $\phi = 35.8°$.

With only two tests, it is probably better to compute ϕ, but with three or more tests, a graphic plot would be best so that a better average could be obtained—or a test rejected if the plot were too far out of line.

////

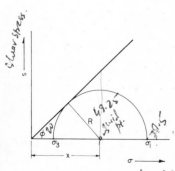

Figure E13-4

13-5 STRESS PATHS

A trace of the locus of points describing the instant stress on a soil element stress path. The octahedral stress plot given in Fig. 10-4 produced a stress p Since a general three-dimensional stress state using octahedral stresses was u we might call that stress trace an *octahedral stress path*. When total stresses used, we have a total stress path; effective stresses plot an effective stress path.

In geotechnical work where a two-dimensional stress state is commonly u it would initially appear that we could simply use the Mohr-Coulomb equati to obtain a stress path. This has two major drawbacks, however, since we do know the values of c and ϕ until we plot the curve, and we cannot plot the cu until a series of tests are performed. A more convenient means must be use stress paths are to be practical. One method which allows plotting results im diately is to use stress coordinates of

$$p = \frac{\sigma_1 + \sigma_3}{2} \qquad q = \frac{\sigma_1 - \sigma_3}{2} \qquad (13$$

to make the stress path trace. The p-q points are directly related to Mohr's circ as they are the center (p) along the normal stress axis with radius $q = $ maximu shear stress. A stress path may be drawn based on a single compression test path $ABCDE$ in Fig. 13-9c using values of σ_1 from any initial value to $\sigma_{1(\text{failu}}$ More commonly a stress path is made for a series of tests as in Fig. 13-10a. Figure 13-10a shows both the p-q values from each test as in Fig. 13-10a. Figure 13-10a shows both the p-q values from each test and just the p-q points. This latter is preferable in mo

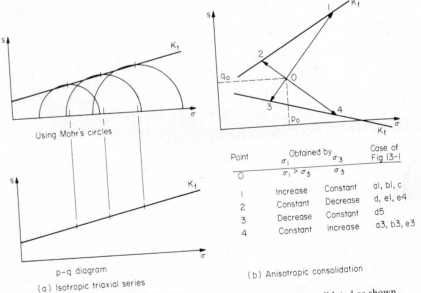

Point	Obtained by		Case of
	σ_1	σ_3	Fig 13-1
O	$\sigma_1 > \sigma_3$	σ_3	
1	Increase	Constant	a1, b1, c
2	Constant	Decrease	d, e1, e4
3	Decrease	Constant	d5
4	Constant	Increase	a3, b3, e3

p-q diagram

(a) Isotropic triaxial series

(b) Anisotropic consolidation

Figure 13-10 Stress paths using p,q coordinates for triaxial tests consolidated as shown.

SOLUTION What you have here is a part of the data from a series of DSS tests. We will draw a Mohr's circle to obtain ϕ graphically since it will be simpler than trying to compute it:

1. Draw an initial Mohr's circle for the sample at the end of consolidation.
2. Next, noting that with shear, both σ_v and σ_h are "normal" stresses on a shear plane on which we measure the shear stress of 20 kPa. This means that a second Mohr's circle with the same center but through coordinates σ_v, -s and σ_h, $+s_c$ can be drawn. Note that the shear stress in the insert is reversed to the normal ($+$) direction, but this does not affect the values or the circle. We would have to consider direction (signs) if we want the correct orientation of the principal planes. Scale from Fig. E13-2 a value of $\phi = 22.5°$.

////

Example 13-3 Given a CU triaxial test on an unsaturated soil with the following data:

Test	σ_3, kPa	$\Delta\sigma_1 =$ deviator stress	σ_1 (computed)
1	40	63.0	103.0 ✓
2	80	86.0	166.0
3	120	102.0	222.0 ✓

$$\sigma_1 = \sigma_3 + \Delta\sigma_1$$

REQUIRED Find the apparent soil parameters ϕ and c and orient the failure plane for test 1.

SOLUTION

Step 1 Compute

$$\sigma_1 = \sigma_3 + \Delta\sigma_1$$
$$= 40 + 63.0 = 103.0 \text{ for test 1, etc.}$$

Step 2 Plot Mohr's circle for the three tests as in Fig. E13-3, fair a failure envelope, and scale

$$\phi = 12°$$
$$c = 17.5 \text{ kPa}$$

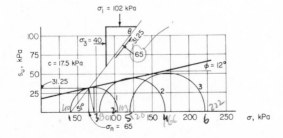

Figure E13-3

Step 3 From σ_3 of test 1, draw line AB at $\boxed{\theta = 45 + \phi/2} = 51°$ as shown, and place on the element the stresses shown. Values of σ_n and s can be computed from Eqs. (10-10) and (10-11) but can be scaled with satisfactory precision.

$\theta = 45 + |\boxed{\phi/2}| = \boxed{53.75}$

Example 13-4 Given triaxial tests on a cohesionless soil.

Test	σ_1	σ_3, kPa
1	78.5	20
2	186.0	50

REQUIRED The angle of internal friction ϕ.

SOLUTION Compute the angle from the relationships shown in Fig. E13-4. From Fig. E13-4:

$$\sin \phi = \frac{R}{X} = \frac{\sigma_1 - \sigma_3}{\sigma_1 + \sigma_3}$$

For test 1,

$$\phi = \sin^{-1} \frac{78.5 - 20}{78.5 + 20} = 36.4°$$

For test 2,

$$\phi = \sin^{-1} \frac{136}{236} = 35.2°$$

Take the average as the best value to obtain $\phi = 35.8°$.

With only two tests, it is probably better to compute ϕ, but with three or more tests, a graphic plot would be best so that a better average could be obtained—or a test rejected if the plot were too far out of line.

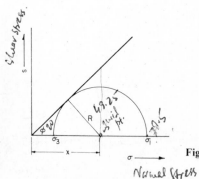

Figure E13-4

13-5 STRESS PATHS

A trace of the locus of points describing the instant stress on a soil element
stress path. The octahedral stress plot given in Fig. 10-4 produced a stress p
Since a general three-dimensional stress state using octahedral stresses was u
we might call that stress trace an *octahedral stress path*. When total stresses
used, we have a total stress path; effective stresses plot an effective stress path.

In geotechnical work where a two-dimensional stress state is commonly u
it would initially appear that we could simply use the Mohr-Coulomb equati
to obtain a stress path. This has two major drawbacks, however, since we do
know the values of c and ϕ until we plot the curve, and we cannot plot the cu
until a series of tests are performed. A more convenient means must be use
stress paths are to be practical. One method which allows plotting results imi
diately is to use stress coordinates of

$$p = \frac{\sigma_1 + \sigma_3}{2} \qquad q = \frac{\sigma_1 - \sigma_3}{2} \qquad (13$$

to make the stress path trace. The p-q points are directly related to Mohr's cir
as they are the center (p) along the normal stress axis with radius $q = $ maximu
shear stress. A stress path may be drawn based on a single compression test
path $ABCDE$ in Fig. 13-9c using values of σ_1 from any initial value to $\sigma_{1(failu}$
More commonly a stress path is made for a series of tests using only the "failur
p-q values from each test as in Fig. 13-10a. Figure 13-10a shows both the p
points on Mohr's circles and just the p-q points. This latter is preferable in mo

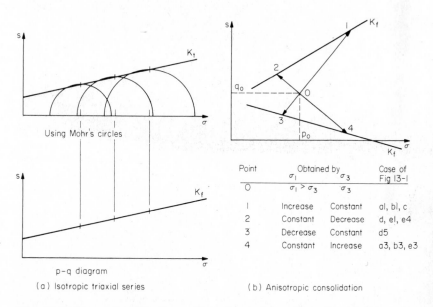

Point	Obtained by		Case of
	σ_1	σ_3	Fig 13-1
O	$\sigma_1 > \sigma_3$	σ_3	
1	Increase	Constant	a1, b1, c
2	Constant	Decrease	d, e1, e4
3	Decrease	Constant	d5
4	Constant	Increase	a3, b3, e3

(a) Isotropic triaxial series

(b) Anisotropic consolidation

Figure 13-10 Stress paths using p,q coordinates for triaxial tests consolidated as shown.

cases and reduces the graphic congestion. The trace of the stress path of Fig. 13-10a is termed the K_f line (f for failure). Since we earlier defined K as $K = \sigma_3/\sigma_1$ and used K_o to define the in situ stress ratio, we see slope α of the K_f line defines the failure ratio

$$K_f = \frac{\sigma_3}{\sigma_{1(f)}}$$

by substitution into Eq. (13-6) for $\sigma_3 = K_f \sigma_1$ to obtain

$$\tan \alpha = \frac{q}{p} = \frac{1 - K_f}{1 + K_f} \tag{13-7}$$

This method of plotting stress paths seems to have been first used by Simons (1960).

The effective stress path can be plotted by using

$$p' = p - u_f \qquad \text{and} \qquad q' = q$$

obtained by using effective stresses $\sigma_i' = \sigma_i - u_f$ in Eqs. (13-6). Note that because q is a shear stress, it is not affected by the pore pressure. The pore pressure moves p' to the left of p along the normal stress axis by the pore pressure u_f.

The four general cases which can be made in a triaxial test can be plotted to produce a stress path as illustrated in Fig. 13-10b. The slopes of the two K_f lines shown do not have to be equal.

The relationship between the K_f line (of a pq plot) and the Mohr's failure envelope (called the ϕ line) is as shown in Fig. 13-11. Since the origin is common for both lines (Fig. 13-11b), we have

$$\frac{q}{m + x} = \tan \alpha = \sin \phi = \frac{a}{m}$$

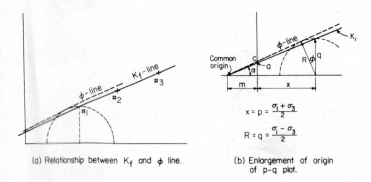

(a) Relationship between K_f and ϕ line.

(b) Enlargement of origin of p-q plot.

Figure 13-11 Relationship between shear strength parameters and the K_f line from the stress path plot.

Also

$$\frac{c}{\tan \phi} = \frac{a}{\tan \alpha}$$

But with $\tan \alpha = \sin \phi$ we have for the cohesion intercept

$$c = \frac{a}{\cos \phi} \qquad (13\text{-}8)$$

From this, the soil parameters c and ϕ may be obtained from a pq diagram by (1) scaling α, a, and m, (2) computing $\alpha = \tan^{-1}(a/m)$ (or scaling it), and (3) computing $\cos \phi$ and then computing the cohesion c.

When anisotropy is a factor (and using anisotropic consolidation), it may be preferable to use the octahedral stress path which may be in terms of either total or effective stresses. As noted in Chap. 10, the plane $OCBA$ in Fig. 13-12a contains the range of triaxial stresses. Figure 13-12b illustrates plotting the four triaxial test cases from an initially isotropic consolidation represented by point D on the hydrostatic stress line. Note that the drained shear strength is the vertical distance from the initial conditions to the intersection of the failure line (as line $D1$). The undrained shear strength is the vertical distance from the initial conditions to the intersection of the effective stress path with the failure line (as point $D5$ for s_u).

In Fig. 13-12b isotropic consolidation stresses fall on the hydrostatic (equal all around as in water) stress line. Anisotropic consolidation with $\sigma_1 > \sigma_3$ as for normally and lightly overconsolidated soils will plot a point D above the hydrostatic pressure line. For $\sigma_1 < \sigma_3$, as for heavily overconsolidated soils, a point D will plot below the hydrostatic pressure line.

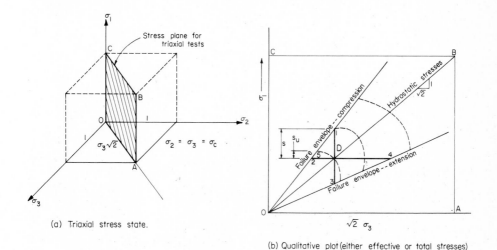

(a) Triaxial stress state.

(b) Qualitative plot (either effective or total stresses)
for stresses on plane of interest shown in (a) and
for both compression and extension type tests.

Figure 13-12 Plotting of triaxial stress data in stress plane $OABC$. The plot in part b is on the stress plane after rotation, so that the plot is two-dimensional. (*After Henkel, 1960.*)

Example 13-5 Given: triaxial data from Example 13-3.

REQUIRED Plot a p-q diagram and obtain c and ϕ and compare with Example 13-3 and also compute K_f. If $K_o = 0.5$, does the K_o line lie above or below the K_f line?

SOLUTION

Step 1 Compute p and q for each test (note that $q = \Delta\sigma_1/2$):

Test	σ_3	σ_1	p	q, kPa
1	40	103.0	71.5	31.5
2	80	166.0	123.0	43.0
3	120	222.0	171.0	51.0

Step 2 Plot p versus q as in Fig. E13-5, fair the K_f line and scale $a = 18$ kPa, $m = 94$ kPa, and $\alpha = 11°$.

Step 3 Compute

$$\sin \phi = \tan 11° = \frac{a}{m} = \frac{18}{94} = 0.19145$$

$$\phi = \sin^{-1}(0.19145) = 11.04°$$

From Eq. (13-8)

$$c = \frac{a}{\cos \phi} = \frac{18}{\cos 11.04} = 18.3 \text{ kPa}$$

These values compare to $12°$ and 17.5 kPa in Example 13-3. The differences are due to scale effects and "fairing" in the K_f line but are well within the overall precision of the test method.

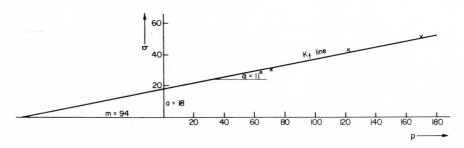

Figure E13-5

Step 4 Find K_f. From Eq. (13-7) we have

$$\tan 11° = 0.19148 = \frac{1 - K_f}{1 + K_f}$$

Solving $K_f = 0.80852/1.19148 = 0.679$.

Step 5 Where is the K_o line with respect to the K_f line? From Eq. (13-7), using K_o, we have

$$\tan \alpha' = \frac{1 - K_o}{1 + K_o} = \frac{1 - 0.5}{1 + 0.5} = \frac{0.5}{1.5} = 0.3333$$

$$\alpha' = \tan^{-1} 0.3333 = 18.4°$$

Since $18.4 > 11°$, the K_o line is above the K_f line. This is not a good comparison, however, since the K_f line here is for total stresses and the K_o line is an effective stress trace.

////

Example 13-6 Given: a sample was recovered from a saturated clay stratum. A consolidation test indicated the stratum is normally consolidated. The sample was recovered with negligible disturbance, returned to the laboratory, and prepared for triaxial tests. The triaxial test is a $CK_o U$ type with pore pressure measurements.

REQUIRED Sketch the qualitative stress paths for these conditions.

SOLUTION Refer to Fig. E13-6:

OA = the K_o consolidation in situ (effective stress) and passes through the origin since a normally consolidated soil is compressed from a gradual buildup of overburden.

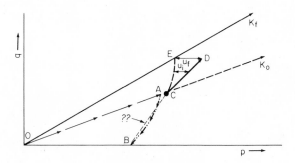

Figure E13-6

AB = stress path as the sample is unloaded in situ and is a combination of locked in stresses from point A and negative pore pressures (as the sample expands from unloading).

C = stress path point from B as a result of K_o consolidation. As pore pressures are not measured, the trace from CD is not known. Point C represents effective stresses.

CD = total stress path produced during the triaxial test from the deviator stress application.

CE = qualitative effective stress path based on pore pressure measurements.

OE = effective stress K_f line. Note this will pass through the origin as well as the K_o line since the soil is normally consolidated ($c' \cong 0$).

AC = disturbance. A perfect sample should reconsolidate from B back to A.

////

13-6 SHEAR STRENGTH OF COHESIONLESS SOIL

The shear strength of cohesionless soil is simply

$$s = \sigma_n \tan \phi \tag{13-9}$$

In situ there may be a cohesion intercept from natural cementation but is lost with disturbed laboratory samples. It is not possible to perform unconfined compression tests on cohesionless soils, but U, CU, and CD tests using either direct shear or triaxial equipment can be made. The relatively large coefficient of permeability allows sample drainage to occur so that it is quite practical to obtain the drained CD parameter ϕ' in Eq. (13-9).

Direct simple shear equipment can provide a closed system to that U, CU, and CD tests can be performed with corresponding values of ϕ being obtained ranging from 0 to ϕ' depending on drainage, degree of saturation, and strain rate.

The triaxial test also being a system with controlled drainage allows U, CU, and CD testing with a range of ϕ from 0 to ϕ' depending on drainage, degree of saturation, and strain rate. Since this test uses larger samples it is necessary to use a somewhat slower strain rate to allow adequate drainage to dissipate pore pressures to produce a drained angle of internal friction ϕ'.

It is nearly impossible to test undisturbed cohesionless samples in either the direct shear or in a triaxial test. As a consequence, the direct shear test being simpler and more rapid is most commonly used. Also there is no advantage to using undrained tests on cohesionless soils since they tend to drain in situ in most cases since loads are applied relatively slowly (as far as drainage is concerned). The principal exception is earthquake or blast loadings which occur in fractions of seconds.

Triaxial tests tend to give values of ϕ that are 2 to 8° smaller than comparable plane strain tests. The DSS is considered to be a plane strain test and many consider the direct shear test to also be a plane strain test. In any case both these tests give larger angles of internal friction than the triaxial test. An extensive study by Lee (1970) and later published results by Lade and Lee (1976) indicate that the plane strain ϕ_{ps} might be taken as

$$\phi_{ps} = 1.5\phi_{tr} - 17° \qquad (\phi_{tr} > 34°) \qquad (13\text{-}10)$$

$$\phi_{ps} = \phi_{tr} \qquad (\phi_{tr} \leq 34°) \qquad (13\text{-}10a)$$

where ϕ_{tr} = triaxial value of the angle of internal friction.

Factors affecting ϕ in cohesionless soils include:

1. Density (or void ratio)—decreasing e increases ϕ.
2. Angularity of grains—crushed sand and gravel has a larger ϕ than bank run (as found in river beds and glacial deposits).
3. Water—negligible unless S is large and/or free drainage is limited. Damp sand may give an apparent cohesion intercept from capillary tension.
4. Grain size distribution—well-graded material has a larger ϕ than poorly graded soil.
5. Confining (or consolidation) pressure—increasing σ_3 decreases ϕ even though the shear strength increases (nonlinearly). An increase in confining pressure from 100 to 1000 kPa may decrease ϕ by 6 or 8°.

Typical values of ϕ are given in Table 13-1. From the range of values given, and noting that most natural deposits consist in rounded grains and that undisturbed samples are nearly impossible to obtain, it is common to "estimate" values as follows:

Sand, gravelly sands	30–36°
Gravel, sandy gravels	35–45°

The *angle of repose* obtained as the slope of a pile of granular material (e.g., stockpiles of sand and gravel, etc.) is the minimum angle of internal friction produced from a combination of "loose" and minimal to no confinement.

Table 13-1 Typical range of values of true angle of internal friction ϕ' for several soils

Soil	ϕ' Loose	ϕ' Dense
Sand, crushed (angular)	32–36°	35–45
Sand, bank run (subangular)	30–34	34–40
Sand, beach (well rounded)	28–32	32–38
Gravel, crushed	36–40	40–50
Gravel, bank run	34–38	38–42
Silty sand	25–35	30–36
Silt, inorganic	25–35	30–35
Clay	See Figs. 13-21 and 13-22	

The in situ shear strength of cohesionless deposits will generally be larger than laboratory values using Eq. (13-9), and even if the test is at the same void ratio. This is from the inability to reproduce the structure, loss of natural cementation, and the very difficult problem of consolidating the sample to K_o conditions, as well as other factors.

13-7 SHEAR STRENGTH OF COHESIVE SOILS

The shear strength of cohesive soils depends on type and state of the soil, and the test procedure. Type and state refer to whether the sample is undisturbed, or remolded (as in fills), and if undisturbed whether it is normally or over-consolidated, i.e., the stress history. State also includes the degree of saturation.

Since the geotechnical engineer has no control over the environment, it is common to treat cohesive soils in a saturated state—although for many projects, the "as is" strength is also used. The saturated state is certainly a "worst case" analysis which should be conservative since the strength is usually a minimum at $S = 100$ percent. In many cases it may be desirable to saturate the sample using back pressure to both ensure $S = 100$ percent and to speed up the saturation process.

A total stress strength analysis is often used but from Eq. (2-20) for effective stress

$$\sigma' = \sigma_i - u$$

we may write

$$\sigma'_1 = \sigma_1 - u$$
$$\sigma'_3 = \sigma_3 - u$$

Subtracting, we obtain

$$\sigma'_1 - \sigma'_3 = \sigma_1 - \sigma_3 \qquad (a)$$

which says the stress difference is a constant regardless of the type of analysis.

Now let us consider the several soils and test methods.

Cohesive Soil—Remolded

Figure 13-13a illustrates a remolded, saturated sample in a rubber membrane as for a triaxial test. A cell pressure σ_3 is applied with no volume change allowed (no drainage). With no drainage we have an induced pore pressure of $u = \sigma_3$ as shown in Fig. 13-13b. This produces effective stresses (neglecting any body weight) of $\sigma'_1 = 0$, $\sigma'_3 = 0$. With an isotropic stress of σ_3 acting, we can use Eq. (10-10) for σ_n to find the effective pressure on any plane is

$$\sigma'_n = \sigma_n - u = 0$$

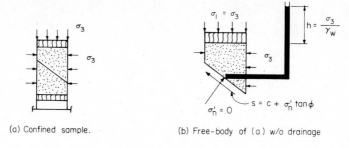

(a) Confined sample.   (b) Free-body of (a) w/o drainage

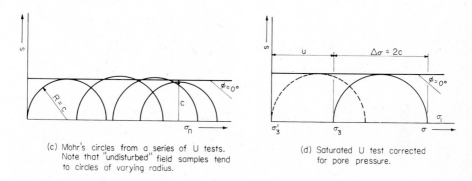

(c) Mohr's circles from a series of U tests. Note that "undisturbed" field samples tend to circles of varying radius.

(d) Saturated U test corrected for pore pressure.

Figure 13-13 General state of unconsolidated-undrained (U) testing and resulting Mohr's circles.

Now we increase the vertical stress by $\Delta\sigma_1$, which induces a pore pressure of

$$\Delta u = \Delta\sigma_1 \qquad (b)$$

and the effective stresses are

$$\Delta\sigma_1' = \Delta\sigma_1 - \Delta u \qquad \Delta\sigma_3 = -\Delta u \qquad (c)$$

Also, we can write the total and effective stresses as

$$\sigma_1 = \sigma_3 + \Delta\sigma_1 \qquad \sigma_1' = \Delta\sigma_1 - \Delta u + \sigma_1'$$

$$\sigma_3 = \sigma_3 \qquad \sigma_3' = -\Delta u + \sigma_3'$$

and subtracting

$$\sigma_1 - \sigma_3 = \Delta\sigma_1 \qquad \sigma_1' - \sigma_3' = \Delta\sigma_1 \qquad (d)$$

This verifies the effective pressures produced by pore pressure Δu as in Eq. (c) since the constant value of deviator stress produced in Eq. (d) is as in Eq. (a). From Eqs. (b) and (c) it can be shown that the value of σ_n' on any potential failure plane remains zero. In the shear stress equation

$$s = c + \sigma_n' \tan \phi$$

we now have $\sigma'_n \tan \phi = 0$ from σ'_n being zero; thus we apparently have $s = c$. We call this case a

$\phi = 0$ test   (and analysis)

$s_u = c$ (identify the "undrained" shear stress with the symbol s_u)

From Eq. (10-11) and with the apparent $\phi = 0$, we have for the maximum stress a θ-angle $= 45°$ and obtain the shear (using s for τ) as

$$s_u = \frac{\sigma_1 - \sigma_3}{2} = c$$

This gives the shear strength as the radius of a Mohr's circle and the value of c obtained when a tangent at zero slope is drawn to the corresponding circle. This is about the only situation where the "maximum" shear stress is used.

If a series of consolidated undrained (CU) tests are performed it is possible to obtain a small angle of internal friction as in Fig. 13-14. A single CU test would, of course, give $\phi = 0$ as our best estimate of the failure envelope which would be a horizontal line.

Strictly, the unconfined compression test may not produce a state of $\sigma'_n = 0$ on the possible shear planes since the pore pressure state at the start of—and during—the test is indeterminate. With the confining pressure $\sigma_3 = 0$ (actually atmospheric taken as zero), we can only plot a Mohr's circle as shown in Fig. 13-7c. Where several tests are made on the same soil it would be better to average the several values of q_u $(= \sigma_1)$ and plot a single circle with the cohesion thus obtained being the average value.

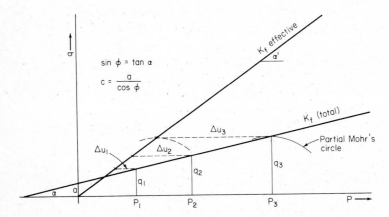

Figure 13-14 Consolidated-undrained test with pore pressure measurements. Use of a p-q plot greatly simplifies data presentation by avoiding overlapping Mohr's circles.

Cohesive Soils—from In Situ

If we take a sample from in situ, the loss of overburden pressure produces some sample expansion. This is resisted by capillary attraction from the water films surrounding the particles and if no water is available to be drawn into the sample negative pore pressures develop. The negative pore pressure which develops may approach the overburden pressure σ'_o depending on S and care in sample recovery. The recovered sample may be tested in either an unconfined compression test or in a triaxial test. Unless we CK_o-consolidate the sample in the triaxial test, the data from a q_u test will generally be satisfactory.

In the q_u test we have (1) sample disturbance during recovery and (2) loss of σ_o and $K_o \sigma_o$—the confining pressure. These factors will reduce the measured shear strength. Compensating, however, is any negative pore pressure and the testing strain rate which is larger than that produced in situ from any proposed loading. Both of these factors increase the measured shear strength so that the strength obtained (for the cost) may be adequate. Unfortunately, there is no way to quantify either the loss or gain in strength so we have only observations which tell us that most of the time q_u tests work satisfactorily. In any case, the q_u test is the most widely used and gives $s_u = c = q_u/2$ and $\phi = 0$ (since the resulting Mohr's circle with $\sigma_3 = 0$ and $\sigma_1 = q_u$ is as shown in Fig. 13-7c).

Since the unconfined compression test gives measured strengths ranging from as low as 25 to about 60 to 70 percent of in situ values, it may not be appropriate to use q_u tests for important projects or in very soft clay deposits. In these cases the economics of conservative design may make the project infeasible unless strengths are determined more closely. In most projects of a routine nature the low strength from a q_u test may be quite adequate and produce only a minimally conservative design, i.e., will not be grossly overdesigned to compensate.

Where triaxial tests are used they should be of the consolidated-undrained or CD type. There is little point in performing unconsolidated-undrained tests since the application of a cell pressure and immediate testing may damage and further degrade the sample quality. In the consolidated tests the sample is carefully placed in the cell and either isotropically or anisotropically consolidated in increments, e.g.,

$$\text{Isotropically:} \qquad \sigma_c = \sigma'_o \left(\frac{1 + 2K_o}{3} \right)$$

$$\text{Anisotropically:} \qquad \sigma_1 = \sigma'_o \qquad \sigma_3 = K_o \sigma'_o$$

where σ'_o = existing overburden pressure and $K_o = \sigma'_h/\sigma'_o$. It is, of course, no simple task to obtain the value of K_o. Values used in anisotropic consolidation that are larger than in situ will overconsolidate the sample. There is some opinion that we should test a sample set at in situ consolidation and perhaps two additional sets at higher OCR—perhaps 2 and 4 or 4 and 8. Since all three results should plot on the K_f line (of a p-q plot) an indication of sample quality and strength reliability might be obtained.

In any case a CU test will develop pore pressures as the deviator stress is applied. This is usually (+) but may be (−) if the soil skeleton collapses. The results of a series of CU tests are shown in Fig. 13-14 and produce an apparent c and ϕ as shown. If pore pressures are measured and the Mohr's circles drawn (dashed) for effective stresses, it is found that the cohesion intercept is very nearly zero and the shear strength in terms of the effective stress parameters may be taken as

$$s = \sigma_n' \tan \phi' \qquad (13\text{-}11)$$

for *normally* consolidated soils.

Generally pore pressure measurements for effective stress parameters will not be of much value unless $S = 100$ percent since:

1. Water content tends to decrease on the shear plane as the soil void ratio decreases in the shear zone.
2. The location of the shear plane is not known in advance unless the sample has a visible defect.
3. Pore pressures will not be uniformly distributed through the sample.
4. Pore pressures are measured only at the location of the pick-up unit—usually at one end of the sample.

Normally Consolidated Clay

Figure 13-15a illustrates the field consolidation path for a cohesive soil. Under normal field loadings of overburden (Fig. 13-16), water table, etc., the soil consolidates along path ABC. The present status of the sample is point B, corresponding to the *effective* overburden pressure $\sigma_o' = p_o$. No excess pore pressure exists in the sample by definition, since it is consolidated. When the sample is removed from the ground, the overburden pressure is lost and hydrostatic pressure conditions are considerably changed. This is qualitatively shown as dashed line BD, which represents sample expansion due to the loss of confining stress. If drainage is prevented, i.e., the sample is not allowed to adsorb water from any source including laboratory humidity, negative pore pressures will develop as capillary effects resist sample expansion.

If the sample is placed in the triaxial membrane and a confining cell pressure is applied in increments (*and without drainage*), as in the consolidation test, to a level which reproduces the in situ value of p_o, we obtain the qualitative curve branch DB'. At this point the pore pressure is again zero. The sample is also some to considerably disturbed, since B' is below B. The relative positions of B and B' would, if it were possible to reproduce curve ABC, indicate the degree of disturbance (refer also to Fig. 13-18). We can only qualitatively state that:

1. If the clay is sensitive, the disturbance will be large.
2. If we use isotropic recompression stresses instead of the in situ anisotropic stress state of $\sigma_3' = K_o \sigma_1'$, we will also introduce a disturbance.

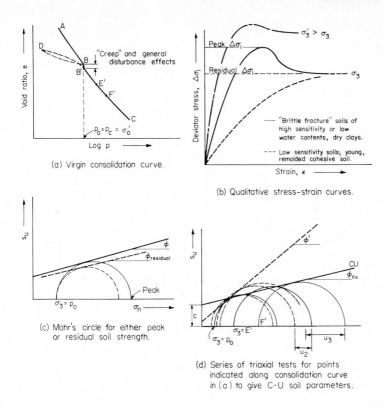

(a) Virgin consolidation curve.

(b) Qualitative stress-strain curves.

(c) Mohr's circle for either peak or residual soil strength.

(d) Series of triaxial tests for points indicated along consolidation curve in (a) to give C-U soil parameters.

Figure 13-15 Consolidated-undrained tests on normally consolidated cohesive soil.

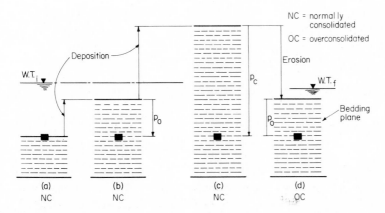

Figure 13-16 Natural formation of clay deposit.

3. If the recompression stress $\sigma_c < p_o$, the pore pressure will remain slightly negative; if $\sigma_c > p_o$, there will be a positive pore pressure, or in general,

$$u = \sigma_c - p_o$$

If we apply a deviator stress $\Delta\sigma_1$, with no drainage allowed, we obtain the consolidated-undrained stress-strain curve qualitatively shown in Fig. 13-15b, and using the peak or residual value of strength, we may plot the solid or dashed Mohr's circles of Fig. 13-15c.

We may now take a second sample from the same in situ elevation and as close as practical, put it in the membrane and recompress it to p_c (point B'), then further consolidate *with drainage* along the compression curve $BB'C$ to point E'. When the volume change is complete, the pore pressure is again zero in this sample. Closing the drainage system and applying the deviator stress, we obtain a second Mohr's circle, and with at least one additional sample at $\sigma_c = F'$, we obtain the series of circles shown in Fig. 13-15d. The failure envelope obtained as a "best fit" gives the CU shear strength parameters, which may be:

Peak shear strength values	Residual strength values
ϕ_{cu}	ϕ_r
c_{cu}	$c_r \to 0$

If we measure the pore pressure which develops during application of the deviator stress, we will obtain the dashed circles shown in Fig. 13-15d, which give Mohr's rupture line as approximately

$$s = \sigma' \tan \phi'$$

where ϕ' is the drained angle of internal friction. Any cohesion intercept will be small and is usually neglected. The reader might observe the greater ease of interpreting the p-q plot in Fig. 13-14 compared to that in Fig. 13-15d even though the angle ϕ must be computed.

If a plane strain problem is to be zone analyzed (refer to Fig. 13-1) CK_0-triaxial tests should be appropriate. We note that certain zones require extension tests, others compression tests and that the zone labeled "2" requires direct shear tests. The triaxial test labels are shown in Fig. 13-10.

We should also note that if the plane of anisotropy is to be considered that samples should be obtained and tested with proper orientation. For example, referring to Fig. 13-16 and taking the short horizontal lines to define the bedding or horizontal plane, the strength obtained from a vertical sample as obtained from a vertical boring will give a larger strength than for a sample tested at an angle of 30, 45 or 90° from the vertical. A 90° orientation would be parallel to the plane of anisotropy (or bedding plane). The strength from this orientation is generally a minimum for normally consolidated soils. Similarly, the ratio R of the strength of triaxial compression tests s_{tc} to extension tests

$$R = \frac{s_{te}}{s_{tc}}$$

is about 0.5.

Normally consolidated clays and remolded cohesive soils generally do not contain cracks and fissures from shrinkage effects. Many overconsolidated deposits, however, are fissured or contain defined discontinuities which are visible to the eye. These deposits are characterized by being overconsolidated from factors other than loss of overburden pressure. Because of their nature, it is necessary to consider them separately as either intact or fissured materials.

Intact Overconsolidated Clay

An intact overconsolidated clay may be produced as in Fig. 13-16. Figure 13-16a represents a clay deposit, as in the bed of a lake or bay. Subsequent deposition produces the overburden of Fig. 13-16b and c. At this time the deposit may be several hundred to 1000 or more meters in depth. Subsequent erosion produces the overconsolidated condition of Fig. 13-16d. The removal of pressure is accompanied by a slight increase in water content, but far less than the decrease occurring during the consolidation process. Thus, although the clay in b and d is at the same present overburden pressure p_o, the density and shear strength are larger and the water content is smaller for d. Other factors such as desiccation, changes in salt content, and creep may produce the effect of overconsolidation. Figure 13-17 illustrates how creep may produce the effect of preconsolidation.

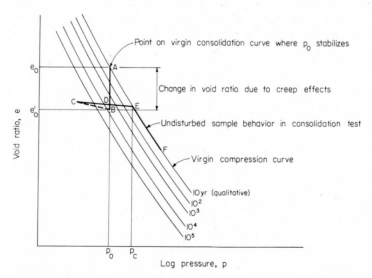

Figure 13-17 Creep (or secondary compression) producing apparent preconsolidation. (*After Bjerrum, 1972.*)

The curve shown is interpreted as follows:

1. Soil is deposited to an in situ overburden pressure of p_o, which produces e_o at the end of primary consolidation at point A.
2. Subsequent creep reduces the in situ void ratio e_o to e'_o at point B.
3. A sample removed would expand to C, then in a consolidation test recompress along $CDEF$, giving an apparent preconsolidation of p_c at E.

Note that this concept is highly idealized and that at least three factors may act to produce a small p_c:

1. Many clays are postglacial; thus the creep duration is not more than 15 000 years.
2. Creep is not large in inorganic clays.
3. Creep is dependent on p_o, which may not be very large for shallow deposits.

During glacial periods the ice depth produced overconsolidation of some deposits, although clays deposited as a result of ice melt or as glacial till are generally normally consolidated unless desiccation effects have since taken place.

From Fig. 13-16, an overconsolidated soil is defined as having an over-consolidation ratio (OCR) computed as

$$\text{OCR} = \frac{p_c}{p_o} > 1 \qquad [\text{See also Eq. (11-2)}]$$

An OCR = 1 is a normally consolidated clay. Figure 13-18a illustrates the shear strength envelope ($fdB'A$) of an overconsolidated clay which was preconsolidated to p_c over a period of time and is currently in situ under a present overburden pressure of p_o. The curve ABC of Fig. 13-18b represents the virgin compression curve during preconsolidation (without creep effects) and later loss of effective overburden pressure along curve branch BD. Sample recovery produces the unload branch DE and CK_o consolidation produces the branch EFD. A similar branch would be obtained if the sample had preconsolidated to point C and unloaded of overburden to the current status of p_o.

At each stage along the unload-reload branch we can estimate the shear strength and produce the overconsolidation branch $fdB'A$ of the s versus p curve as follows (for the two points D and F):

1. At points D and F in Fig. 13-18b, erect vertical lines as shown and locate D' and F' on the normally consolidated s versus p curve in Fig. 13-18a.
2. Project from D and F horizontally and locate points 1 and 2 on the virgin compression curve.
3. From points 1 and 2 extend vertical lines to the s versus p curve and locate points 1 and 2. These points are the theoretical shear strengths of this clay at the void ratio in situ corresponding to these points on the e versus p curve.
4. Extend horizontal lines from point 1 and 2 to intersect with vertical extensions from D' and F' to locate points d and f.

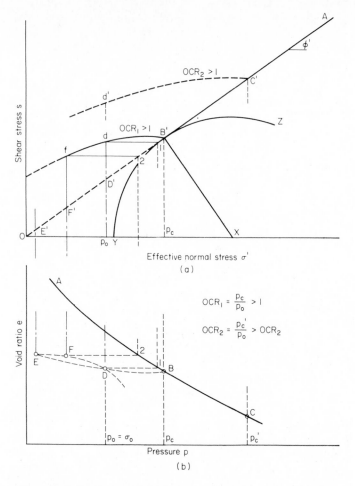

Figure 13-18 Qualitative effects of overconsolidation on shear strength.

5. Construct the overconsolidation branch of the s versus p curve using points d, f, and B'. One would have to interpret the slope to obtain c and ϕ since this plot (and one using Mohr's circles) is usually curved.

We may observe that unloading the sample produces negative pore pressures. They may or may not be zero at the current p_o. The negative pore pressures will either develop or increase from D to E and be partially to fully reduced during recompression from E back to D.

Application of the deviator stress $\Delta\sigma_1$ in the CU test will produce one of the following:

1. Positive pore pressures if $p_c/p_o < 4^+$; this will produce a reduced shear strength since $\phi < \phi'$ (unless we measure the pore pressure).

2. Negative pore pressures if $p_c/p_o > 4^+$ as the sample tends to expand at constant volume. This volumetric expansion is similar to that occurring in a dense sand under shear.

For case 2 above, the apparent shear strength will be higher than the undrained shear strength (since the pore pressure adds and moves the "drained" Mohr's circle to the right). This higher shear strength should not be used for design since the negative pore pressure in the field will attract water (since the volume cannot be held constant), which will both soften the soil by reducing the cohesion and dissipate the negative pore pressures as the attracted water breaks the capillary films producing the suction effect.

If we draw a perpendicular $B'X$ as in Fig. 13-18a, we can draw a Mohr's circle (partially shown as $YB'Z$). If a perfect sample is CK_o-consolidated with $\sigma_3 = K_o p_c = OY$ and $\sigma_1 = p_c$, we should obtain this Mohr's circle. We see that in this case the initial stress state is anisotropic with a Mohr's circle (not shown) of diameter Yp_c. Up to this point CK_o consolidation would require the triaxial cell be able to provide $\sigma_h > \sigma_v$.

The qualitative effect of increased density (particle packing) from consolidation is shown by the location of the second OC curve using points $C'd'$ (not to scale). The difference between points d and d' is the beneficial effect of consolidation. This is the increase in the vertical shear strength; the lateral increase will only be a small fraction of this. This is an important factor in vane shear testing (Sec. 13-11) where the lateral shear strength is measured.

The cohesion intercept shown in Fig. 13-18 may be close to the true cohesion of the clay. It probably results from a combination of interparticle forces from particle packing, reduction in water content from the reduced void ratio and cementation caused by mass contamination from external sources together with by-products from crushing and grinding at the particle contacts.

Shear strength behavior for overconsolidated clays at pressures greater than p_c (branch $B'A$ in Fig. 13-18a) will be the same as for normally consolidated clays. This is self-evident as we are back on the "virgin compression" curve idealized by Fig. 13-18b.

Since overconsolidated clays exhibit brittle fracture with a peak deviator stress at a relatively low strain, the residual (also termed ultimate) shear stress parameters may be preferred for certain analyses. The residual shear strength is very nearly

$$s_r = \sigma' \tan \phi_r \qquad (13\text{-}12)$$

where ϕ_r is obtained as shown on Fig. 13-15c. It is expected that at large strains the soil is sufficiently remolded that no cohesion intercept remains. Laboratory tests indicate this to be very nearly so.

The residual strength should be investigated in situations where a progressive failure may be possible. The residual shear strength may be difficult to obtain in the standard triaxial test because of the influence of end platen restraint and the confining effect of the rubber membrane as the sample cross section substantially enlarges at the large strains. A direct shear test may be useful to obtain the residual angle of internal friction (Saada and Townsend, 1981).

Fissured Overconsolidated Clays

In many areas, particularly if the clay is heavily overconsolidated, the mass will contain a network of discontinuities. If the discontinuities consist of fine hairline cracks, they may be termed *fissures*. The term *joint* is also used to describe discontinuities in the soil mass. There is no unanimity of agreement at present as to what constitutes a fissure or joint. In any case, these discontinuities may be from a few centimeters to several meters in length. *Slickensides* are the somewhat polished interfaces developed when adjacent blocks of clay slip with respect to each other. Slickensides may also be formed at the interface of a landslide in clay. Figure 13-19 illustrates fissuring, joints, and slickensides using the author's definitions. These discontinuities are caused by cyclic wetting and drying, loss of overburden pressure causing expansion or tension cracks, earthquakes and other tectonic movements, and laminations occurring during sedimentation processes. Crack contamination with dust or organic materials when the fissure is open or exposed tends to preserve the discontinuity even though later saturation (and swell) or subsequent overburden pressure cause crack closure.

The shear strength of fissured clays will vary widely, depending on whether the test is being made on a small, intact sample or on one that is larger and/or contains one or more discontinuities. The triaxial test produces reasonably reliable results, since the cell pressure will tend to close the fissure as in situ. The

Figure 13-19 Slickensides, fissures, and joints. (*a*) Slickenside. Clay has been exposed to the atmosphere and slightly dried. (*b*) Fissures and joints. Fissures are large cracks.

major problem with these soils is not in the testing, however, but in the prediction of the in situ shear strength after disturbance. This is particularly important in slopes constructed in this material, since the loss of overburden pressure coupled with crack opening and resultant softening from entering water can result in a failure either shortly after construction or as much as 10 to 20 years later. This phenomenon, with a number of references, is considered in detail by Morgenstern (1967).

Example 13-7 Given two consolidated-undrained triaxial tests on an unsaturated cohesive soil as follows:

Test	σ_1	σ_3	u, kPa
1	190	65	35
2	340	130	60

REQUIRED

(a) Apparent shear strength parameters c and ϕ.
(b) Effective shear strength parameters c' and ϕ'.

SOLUTION

Step 1 Plot Mohr's circles using total stresses as in Fig. E13-7 (solid lines). Obtain the cohesion intercept as $c = 14$ kPa and $\phi = 23.1°$.

Step 2 Subtract u from σ_3 to obtain $\sigma_3' = 30$ for test 1 and 70 for test 2. Using the same radius for each test as in Step 1 and the σ_3' starting points, draw the dashed circles of effective stresses as shown and scale $c' = 18.0$ kPa and $\phi' = 31.5°$.

////

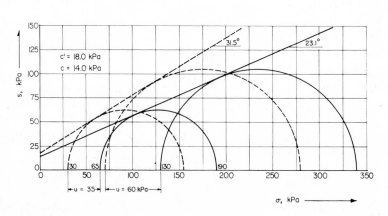

Figure E13-7

13-8 PORE PRESSURE EFFECTS IN CONSOLIDATED-DRAINED TESTS

Consolidated-drained tests are performed by consolidating the sample as for CU tests. The test is then performed so slowly that "no pore pressures develop." Practically, this is impossible for soils with $S > 80$ to 90 percent, since drainage is necessary for the soil structure to reorient in shear and drainage can only occur with an excess pore pressure. This requires the test to be performed so slowly that small excess pore pressures develop which do not seriously invalidate the test.

Consolidated-drained tests are seldom performed for three major reasons:

1. CU tests with pore pressure measurements will give about the same results at a considerable saving in time.
2. CD tests take too long to be practical.
3. For most in situ loading conditions, U or CU tests adequately describe the shear strength requirements.

The Coulomb shear strength equation for CD tests on normally consolidated clays is (based on observations) given by Eq. (13-11).

The shear strength for overconsolidated clays for $\sigma'_n < p_c$ is given by Eq. (13-2). When the effective normal stresses are greater than p_c, for these clays the shear strength is defined by Eq. (13-11), as illustrated in Fig. 13-18.

13-9 SENSITIVITY OF COHESIVE SOILS

The sensitivity of cohesive soils is defined as

$$S_t = \frac{\text{undisturbed strength}}{\text{remolded strength}}$$

where the strength may be the undrained, consolidated-undrained, or consolidated-drained value. The undrained value is most commonly used. The remolded strength is obtained by remolding the specimen used for the undisturbed strength so that the water content is as nearly the same as practical. The sensitivity description is as follows:

Insensitive	$S_t < 2$
Medium sensitivity	$2 < S_t < 4$
Sensitive	$4 < S_t < 8$
Very sensitive	$8 < S_t < 16$
Quick	$S_t > 16$

Highly overconsolidated clays tend to be insensitive. This is at least partially due to the low natural water contents in these deposits. Most cohesive soils, such as glacial till clays and those found in the *B* horizon of residual deposits, are of medium sensitivity. A few glacial clays and most fresh-water deposits are very sensitive. A few of the fresh-water and marine deposits are quick. The sensitivity of the large majority of cohesive deposits will range from 2 to 8. Sensitivities greater or less than this are much less commonly encountered. Most quick clays seem to be found (or at least reported) in Canada and Scandinavia.

13-10 EMPIRICAL METHODS FOR SHEAR STRENGTH

Numerous correlations for shear strength or shear strength parameters have been proposed in the literature. Several will be presented here to illustrate some of those available.

One of the earliest correlations is that between the SPT (Sec. 6-9) and the unconfined compression strength, as was illustrated in Table 6-1.

Correlations between ϕ and plasticity index I_P are shown in Fig. 13-20. A relationship between ϕ and percent clay fraction (Skempton, 1964) is shown in Fig. 13-21. Both of these curves should be used cautiously, as there are several major exceptions which can be found in the literature as well as substantial scatter in the data points used to establish the curves. For routine soil work, however, particularly in regions where w_L is on the order of 20 to 45 and I_P on the order of 15 to 30, these curves will be reasonably reliable.

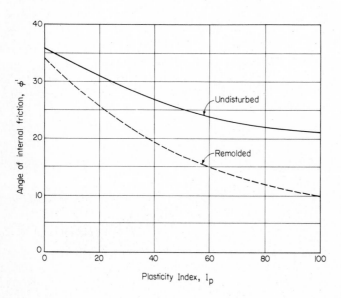

Figure 13-20 Correlation between angle of internal friction ϕ' (true) and plasticity index for both undisturbed and remolded soil. *(After Bjerrum and Simons, 1960.)*

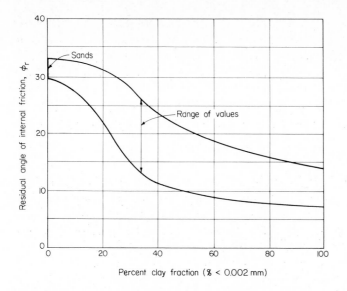

Figure 13-21 Correlation between residual angle of internal friction and percent clay. *(After Skempton, 1964.)*

Figure 13-22 illustrates the torvane, which is commonly used to obtain the shear strength of soft to very soft clay. Because of its small size, several tests can be made for statistical determination of s_u from SPT samples or in the sides of test pits.

Figure 13-23 (also Fig. 6-3) illustrates the pocket penetrometer device, which can be used in test pits or in the base of borings 75 cm or larger in diameter where a person can be lowered into the hole, as in caisson work. This device works well in any fine-grained cohesive soil. The operator simply selects a gravel-free location, pushes the piston rod into the soil to the calibration mark, and

Figure 13-22 The torvane.

Figure 13-23 Pocket penetrometer.

simultaneously reads q_u on the graduated scale. Again due to its small size, several tests can be made for a statistical determination of q_u.

The "Rimac" device, shown in Fig. 13-24, was developed for testing automobile valve springs and has been modified for field testing. This device is widely used in field boring operations to determine q_u as the borehole advances. The operator obtains a reasonably intact sample from the SPT test and measures the diameter after carefully trimming a length (determined directly from the scale on machine). The sample is placed between the platens of the Rimac tester and compressed to the maximum load. The load and final sample length are simultaneously read, and the unconfined-compression strength is computed as

$$q_u = \frac{P_{max}(L_o - \Delta L)}{A_o L_o}$$

which is, of course, the same as Eq. (13-4).

Figure 13-24 Rimac tester used for unconfined compression test. Note the small pointer indicating sample length. The large center dial indicates the sample load in pounds and kilograms.

13-11 IN-SITU DIRECT MEASUREMENT OF SHEAR STRENGTH

A borehole shear device has been recently developed, as schematically shown in Fig. 13-25. This device is inserted in the borehole, expanded into the soil, then pulled. The force used to expand the device into the soil is P_v, and the pull to shear is P_h in the direct shear test as used in Fig. 13-5d. This device essentially performs the direct shear test in situ. The test should approximate the CU test if a reasonable time is allowed for drainage, as the shear teeth or serrations are not very deep.

A vane shear test, schematically shown in Fig. 13-26, is widely used in soft cohesive deposits where sample disturbance is critical. This test consists of inserting a vane of small volume into the soil and applying a torque. From a free-body diagram as in Fig. 13-26b, the consolidated-undrained vane shear strength s_{uv} can be computed as

$$s_{uv} = \frac{4T}{\pi(2d^2h + ad^3)} \qquad (13\text{-}13)$$

where T = applied torque
  d = diameter of vane (commonly 50 to 150 mm)
  h = length of vane (commonly 100 to 225 mm)
  $a = \frac{2}{3}$ for uniform end shear (for vane shown)
  $\frac{3}{5}$ for parabolic end shear
  $\frac{1}{2}$ for triangular end shear (some ends are triangle-shaped)

In practice, a borehole is advanced to less than 1 m from the point of the test. The hole is cleaned and the vane inserted and carefully pushed to the desired elevation and a torque applied until the cylinder of soil defined by the vane perimeter shears.

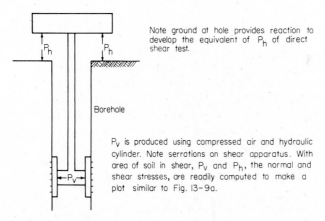

Note ground at hole provides reaction to develop the equivalent of P_h of direct shear test.

Borehole

P_v is produced using compressed air and hydraulic cylinder. Note serrations on shear apparatus. With area of soil in shear, P_v and P_h, the normal and shear stresses, are readily computed to make a plot similar to Fig. 13-9a.

Figure 13-25 Borehole shear device. [For additional details, see Wineland (1975).]

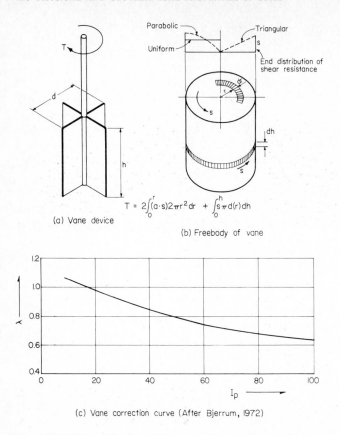

(a) Vane device

(b) Freebody of vane

$$T = 2\int_0^r (a \cdot s)2\pi r^2 dr \; + \; \int_0^h s\pi d(r)dh$$

(c) Vane correction curve (After Bjerrum, 1972)

Figure 13-26 Schematic of vane shear test and strength corrections.

The CU shear strength is obtained since the sample is consolidated in situ by the existing overburden. One of the two lateral values of s_{uv} is obtained—the other would not be practical since the vane would have to be horizontal.

The vane test is widely used in Europe but much less in the United States. Its application is primarily in sensitive, gravel-free clay and silt deposits, where quality "undisturbed" samples are hard to recover. Using s_{uv} in design has had mixed success. It appears that s_{uv} is too large for a conservative design and the shear strength should be corrected (at least for plane strain problems) according to Bjerrum (1972) as

$$s_u = \lambda s_{uv} \tag{13-14}$$

where λ is a reduction factor which has been related to the plastic limit I_p. Plotting of "failure" points in Fig. 13-26c shows a very large scatter so that λ is at best a crude estimate.

Several researchers (Arman et al., 1975; Eden, 1965) have reported that s_{uv} is approximately two times the value of s_u from unconfined compression tests but somewhat less than s_u from CU triaxial compression tests. On the other hand,

Table 13-2 Strength anisotropy in several soils

Numbers ( ) refer to numbers in inset of Fig. 13-1. Data from Bjerrum (1972) except as noted.

Soil	w_L	I_p	TC(1)	s_u/p_o‡ TE(3)	DSS(2)	Vane
Bangkok clay	150	85	0.70	0.40	0.41	0.59
Matagami clay	85	47	0.61	0.45	0.39	0.46
San Francisco Bay mud*	88	45	0.35	—	0.25	—
AGS clay*	71	40	0.32	0.20	0.25	—
Dramman plastic clay	61	29	0.40	0.15	0.30	0.36
Haney sensitive clay*	44	18	0.27	0.17	—	—
Boston blue clay*	41	21	0.33	0.16	0.20	—
Vaterland clay	42	16	0.32	0.09	0.26	0.22
Studenterlunden clay	43	18	0.31	0.10	0.19	0.18
Drammen lean clay	33	11	0.34	0.09	0.22	0.24†
Connecticut valley varved clay*	Variable		0.25	0.21	0.17	—

* Ladd et al. (1977).

† Simons (1960) reported a vane value of 0.16 for this type clay.

‡ TC = triaxial compression; TE = triaxial extension; DSS = direct simple shear.

Simons (1960) found s_{uv} approximately equal to the unconfined compression s_u. In passing, this type of contradiction is common in geotechnical literature. This is also somewhat illustrated in comparing the data in Table 13-2.

The large actual and s_{uv} strength discrepancies (would not be a problem except that several large embankment failures have occurred from apparently overestimating the vane shear strength) may be due to

1. Strain rate (usually 2 to 3 minutes for a test) with the vane may be too high. This may only account for about 10 to 20 percent strength increase, however.
2. Anisotropy—wrong orientation of bedding plane to the shear surface on the vane perimeter.
3. Progressive failure—sensitive soils with a peak ultimate strength and lower residual strength may attract load initially to a smaller load zone. When a local failure occurs the strength reduces to residual with a load transfer to adjacent zones which then fail, and so on to a general failure.
4. Use of a single s_{uv} for the entire potential shear surface when it is in fact zoned as in Fig. 13-1.

This latter factor, together with "1," may be particularly serious; for example, if $s_{uv} = 100$ kPa and, for example,

Strain rate accounts for 20 percent   (leaves 80 kPa)

Zone 3 is 50 percent of 1   (extension-type test)

Zone 2 is 67 percent of 1   (direct simple shear test)

Zone 1 is 100 percent   (usual compression test)

and we have equal lengths of the slip surface (Fig. 13-1a) at one-third for each strength to obtain

$$s'_{uv} L = \frac{80 \times 1 \times L}{3} + \frac{80 \times 0.67 \times L}{3} + \frac{80 \times 0.5 \times L}{3} = 57.8L$$

$$s'_{uv} = 57.8, \quad \text{say}, \quad 58 \text{ kPa}$$

and an equivalent $\lambda = 58/100 = 0.58$. Comparing this with Fig. 13-27c, we see the minimum λ is approximately 0.65. These strength percentages are not unrealistic according to Table 13-2.

13-12 FACTORS AFFECTING SHEAR STRENGTH

It should be evident at this point that the shear strength is not a unique value but, rather, is both state- and method-dependent. Some of these factors are:

1. Effective or intergranular pressure.
2. Interlocking of particles; thus angular particles give greater interlocking and higher shear strength (larger ϕ) than do rounded particles found in bank run gravel and glacial deposits.
3. Particle packing or density.
4. Cementation of particles, naturally occurring or otherwise.
5. Particle attraction or cohesion. Be aware that the apparent cohesion from damp grains may be a transient state.
6. Water content for cohesive soils. This is illustrated by the "strength" of a dry versus wet lump of clay.
7. Sample quality (relating to disturbance, cracks, fissures, and similar).
8. Test method as U, CU, or CD and including whether triaxial, direct shear, in situ, etc. For the laboratory the consolidation procedure as isotropic, aniso- tropic, and K_o or some other consolidation ratio being used.
9. Other effects such as humidity, temperature, testing skill, motivation of the laboratory personnel, care in transporting samples to the laboratory, condi- tion of the laboratory equipment, and similar.

For cohesive soils the shear strength is also very considerably influenced by:

1. The strain rate—higher shear strengths are obtained at higher strain rates.
2. Anisotropy of the mass—vertical strength not same as lateral strength.
3. Effects of progressive failure.

13-13 NORMALIZED SOIL PARAMETERS

It is a convenience in many cases to normalize the parameter of interest. For example, the strain

$$\varepsilon = \frac{\text{Elongation, } \delta}{\text{Length, } L}$$

is a normalized parameter which is unitized or independent of either the length or the absolute value of δ once it is reported as ε. We also note that steel strains plotted versus stress produce a linear curve of great value over the straight-line portion of the plot. The corresponding curve for a soil is of some less value since it is curved. We might further normalize a soil stress-strain plot so that instead of plotting $\Delta\sigma_1$ versus ε we plotted a stress ratio

$$\frac{\Delta\sigma_1}{\sigma_3} = \frac{\sigma_1 - \sigma_3}{\sigma_3}$$

This is illustrated in Fig. 13-27, and we see that we can sometimes replace a series of stress-strain curves with a single normalized curve which contains all needed

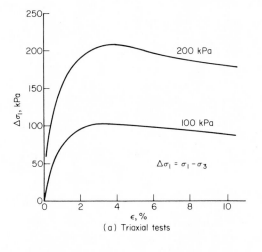

(a) Triaxial tests

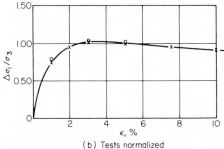

(b) Tests normalized

Figure 13-27 Results from normalizing triaxial stress-strain curves.

data. The data from several normalized stress-strain curves will seldom plot a single curve, rather, one draws a best fit curve. For most soil work, and considering the normal sample variation, soil heterogeneity, and other factors, an averaged curve (as in Fig. 13-27b) is adequate.

The use of normalized data is to obtain relationships independent of the specific sample or that become apparent from less cluttered graphics in some form of

$$y = a + kx^n \qquad \text{or} \qquad y = n \ln x$$

There is little advantage in normalizing parameters if the resulting equations are more complicated than this (and generally for $n \le 2$).

Normalizing soil data is basically a trial process that has been on going for some time with the δ/L, $\Delta\sigma_1/\sigma_3$, and s_u/p_o (of the next section) most commonly used.

13-14 THE s_u/p_o RATIO

One normalized soil parameter of much use is the ratio obtained by dividing the undrained shear strength s_u by the *current effective in situ overburden pressure p_o*. In the laboratory the effective consolidation pressure may be substituted for p_o. This ratio has been used since at least the mid-1930s (Hvorslev, 1937).

In order to establish some kind of relationship there must be a second variable of interest—preferably more easily obtained than s_u. Figure 13-28 illustrates plotting the s_u/p_o ratio against I_p and w_L for normally consolidated soils. These plots illustrate the ease of obtaining an estimated value of s_u for a normally consolidated soil from the Atterberg limits and the present in situ overburden pressure p_o.

From Fig. 13-29 we can obtain approximately the following:

$$\frac{s_u}{p_o} = 0.11 + 0.0037 I_p \qquad \text{(for } I_p > 10 \text{ and scatter of } \pm 0.05\text{)} \qquad (13\text{-}15)$$

$$\frac{s_u}{p_o} = 0.45 w_L \qquad \text{(for } w_L > 0.40 \text{ and scatter of } \pm 0.10\text{)} \qquad (13\text{-}16)$$

When the soil is overconsolidated, we might use Fig. 13-29 to obtain a s_u/p_o ratio. This diagram is based on a limited number of soils, but with widely diverse Atterberg limits. The properties at large OCR were laboratory consolidated. The ordinate axis is based on normalizing a normalized parameter. The method of application is as shown in Figure 13-29. Note that Fig. 13-29 is based on $CK_o U$ direct simple shear and not triaxial tests. This figure might be used with Table 13-2 to extrapolate DSS to triaxial s_u data.

Correlations as shown on both Figs. 13-28 and 13-29 should be used cautiously and, preferably, supplemented with laboratory data to verify the applicability of the graphs to the area or to the soil. The major value of correlations is in preliminary studies where soil data are often very limited. The use of correlations for final design is taking an unnecessary and excessively high risk.

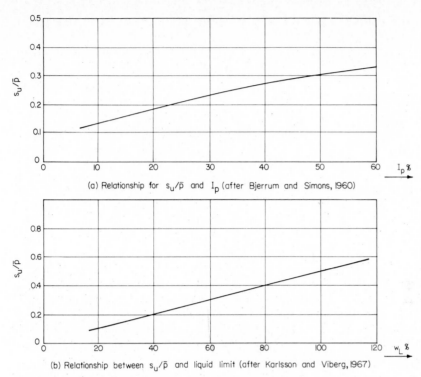

(a) Relationship for $s_u/\bar{p}$ and I_p (after Bjerrum and Simons, 1960)

(b) Relationship between $s_u/\bar{p}$ and liquid limit (after Karlsson and Viberg, 1967)

Figure 13-28 Relationships between $s_u/\bar{p}$ and soil index properties. Use only for normally consolidated soils. Interpret $\bar{p}$ as the effective consolidation pressure.

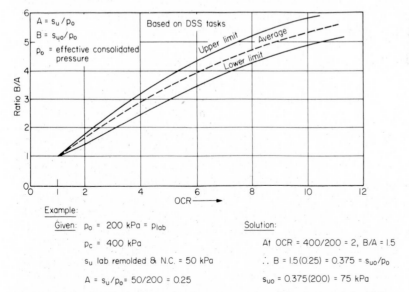

Figure 13-29 Normalized relationship between overconsolidated s_{uo} and normally consolidated s_u shear strengths for CK_oU direct simple shear (DSS) tests. (*After Lade and Lee, 1976.*)

13-15 PORE PRESSURE PARAMETERS

Since pore pressure u† developed during shear has a significant effect on the shear strength of soil some effort has been expended to predict it. One obvious method is to directly measure it, but this has definite limitations since pore pressure is strain-dependent as well as being heavily influenced by the degree of saturation S. These two factors make it very difficult to obtain a reliable u since it varies from point to point in a sample (or in the mass) and the value that is measured is that at the detection unit (piezometer, pressure transducer, probe, etc.). With these limitations we will briefly discuss the pore pressure problem.

One method (Lo, 1969) of expressing the pore water pressure is:

$$u = aJ_1 + bJ_2^{1/2} + cJ_3^{1/3} \tag{13-17}$$

where J_i are the principal stress invariants defined in Eq. (10-4). In an isotropic stress state, $\sigma_1 = \sigma_2 = \sigma_3$ and $a = 1$, so that we obtain

$$u = J_1 + bJ_2^{1/2} + cJ_3^{1/3} \tag{13-18}$$

The difficulty is that the coefficient b and c are dependent on the mechanical properties of the soil mass and are not easily determined.

If the mass were perfectly elastic, no pore water pressures would be caused by J_2 and J_3 and

$$u = \tfrac{1}{3}(\sigma_1 + \sigma_2 + \sigma_3)$$

which, for isotropic consolidation and $S = 100$ percent, gives $u = \sigma_3$ in triaxial tests.

For real soils the coefficients b and c depend on the stress-strain state of the soil, and no unique values are obtained. If we neglect the effect of stress-invariant J_3, and with triaxial conditions of $\sigma_2 = \sigma_3$, for saturated soils Eq. (13-18) may with some algebraic manipulation be transformed to

$$\Delta u = \Delta\sigma_3 + A(\Delta\sigma_1 + \Delta\sigma_3) \tag{13-19}$$

where

$$A = \frac{1}{3} + \frac{b}{3}$$

To give some additional generality to the problem, we may rewrite Eq. (13-19) as

$$u = B[\sigma_3 + A(\sigma_1 - \sigma_3)] \tag{13-20}$$

which is the form suggested by Skempton (1954). Table 13-3 gives some ranges of values for the pore pressure parameter A.

---

† Some persons use u while others use Δu for pore pressure. In any case, the pore pressure of interest is that in excess of hydrostatic conditions caused by changes in applied stresses.

Table 13-3 Values of the pore pressure parameter A†

Soil	A
Loose, fine sand	2–3
Sensitive clay	1.5–2.5
Normally consolidated clay	0.7–1.3
Lightly overconsolidated clay	0.3–0.7
Heavily overconsolidated clay	−0.5–0.0
Compacted sandy clay	0.25–0.75

† After Skempton (1954).

Particular limitations on Eq. (13-20) include:

1. When a stress difference $(\sigma_1 - \sigma_3 = \Delta\sigma_1)$ is applied, an induced pore pressure u is set up. When $\Delta\sigma_1$ is removed, a large part of u remains for some time. If $\Delta\sigma_1$ is repeatedly applied and removed, the magnitude of the residual pore pressure will increase.
2. Under sustained application of $\Delta\sigma_1 = $ constant, the pore pressure continues to increase to a maximum. In consolidation theory we often assume that the change is instant.
3. For sensitive clays the pore pressure continues to increase after failure even though $\Delta\sigma_1$ remains constant or somewhat decreases.
4. Difficulty of reliable prediction of u when S is less than 100 percent.

These several observed phenomena indicate that the pore pressure is strain-dependent rather than stress-dependent. Lo (1969) recommended separating u into parts as

$$u = u_i + u_s$$

where $u_i = $ pore pressure from applying a stress [Eq. (13-18)] with $\varepsilon_v = \varepsilon_1 = \varepsilon_2 = \varepsilon_3 = 0$, i.e., no volume change as obtained for $S = 100$ percent and isotropic stress conditions. This gives $u_i = \sigma_3$ for an isotropic consolidated triaxial test.

Application of a stress difference $\Delta\sigma_1$ will produce shearing strains (even for constant volume conditions) and additional pore pressures u_s. The pore pressures due to strain effects can be expressed as

$$\frac{u_s}{p_o} = b'\varepsilon_1$$

for triaxial stress conditions of $\varepsilon_2 = \varepsilon_3$. For plane strain conditions of $\varepsilon_2 = 0$, the pore pressure is

$$\frac{u_s}{p_o} = b''\varepsilon_1$$

where the coefficients b' and b'' would have to be determined somehow. The pore pressure is normalized using p_o = effective consolidation pressure in both equations for dimensional analysis. In practice, the values of u_i are combined (Lo, 1969), resulting in the following:

For anisotropic consolidation at p_o,

$$\frac{u}{p_o} = 1 + \varepsilon_1 - K \tag{13-21}$$

For isotropic consolidation ($K = 1$),

$$\frac{u}{p_o} = 1 + \varepsilon_1 \tag{13-22}$$

In both of these equations the principal strain ε_1 is a percent. It is, of course, desirable to use stress differences instead of strain, since stresses are more likely to be known in a given pore pressure problem.

Equations (13-20), (13-21), and (13-22) can be used to estimate and/or compute the excess pore pressure u to obtain the effective stress parameters c' and ϕ'. The best estimate is from direct measurements. Unless $S = 100$ percent, the B parameter in Eq. (13-20) is not 1.0. (For saturated soils we can apply an isotropic cell pressure in a triaxial test and measure u to make a direct computation since $\sigma_1 - \sigma_3$ is zero.)

If we somehow obtain the pore pressure parameter A (preferably at $\Delta\sigma_1 =$ maximum) and the pressure ratio K and the drained angle of internal friction ϕ', the s_u/p_o ratio can be computed (Skempton and Bishop, 1954) as

$$\frac{s_u}{p_o} = \frac{[K + (1 - K)A] \sin \phi'}{1 + (2A - 1) \sin \phi'} \tag{13-23}$$

Example 13-8 Given two samples from the same stratum initially subjected to consolidation stresses as follows:

Test 1	Test 2
Isotropic: $\sigma_3 = 70$ kPa	$\sigma_2 = \sigma_3 = 55$ kPa
$u = 70$ kPa (measured)	$\sigma_1 = 100$ kPa during test

REQUIRED

(a) Under what circumstances can u be the same for both tests at the stresses indicated?

(b) If the measured u for soil 2 is 110 kPa for the σ_1 given, how much axial strain has occurred?

(c) What are the A and B parameters of Eq. (13-20)?

SOLUTION

(a) For test 1:

$$\sigma_{oct} = \frac{\sigma_1 + \sigma_2 + \sigma_3}{3} = \Delta\sigma = 70 \text{ kPa}$$

$$\sigma_1 - \sigma_3 = 0$$

$$u = \sigma_3 + A(\sigma_1 - \sigma_3) = 70 \text{ kPa} \qquad \text{using Eq. (13-19) or Eq. (13-20)}$$

For test 2:

$$\sigma_{oct} = \frac{100 + 55 + 55}{3} = 70 \text{ kPa}$$

$$u = \sigma_3 + A(\sigma_1 - \sigma_3) = 55 + A(100 - 55) = 55 + 45A$$

For u to be the same, we have

$$45A + 55 = 70$$

$$A = \frac{70 - 55}{45} = \frac{1}{3}$$

(b) Assume

$$K = 1 - \sin \phi = 0.50$$

With anisotropic conditions and $p_o = 55 \text{ kPa}$,

$$u = p_o \varepsilon_1 + p_o(1 - K)$$

$$110 = 55\varepsilon_1 + 55(1 - 0.5)$$

Solving, we obtain $\varepsilon_1 = 1.5$ percent (0.015 m/m). Thus at $\sigma_1 = 100 \text{ kPa}$, the deviator stress is $\Delta\sigma_1 = 45 \text{ kPa}$ and the strain is approximately 0.015 m/m.

(c) The B parameter is 1 for the soil, since from test 1 we have

$$u = B[\sigma_3 + A(\sigma_1 - \sigma_3)] = B\sigma_3 = 70$$

therefore,

$$B = \frac{70}{70} = 1.0$$

For test 2 we have

$$110 = 1[55 + A(100 - 55)]$$

from which

$$A = \frac{55}{45} = 1.22$$

////

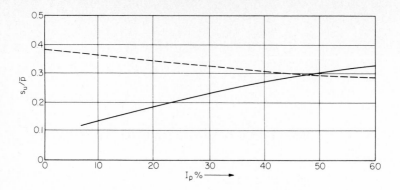

Figure E13-9

Example 13-9 Using Eq. 13-23, taking $K_o = 1 - \sin \phi'$, the pore pressure parameter $A = 1.00$ (average from Table 13-3), and several values of ϕ' from Fig. 13-20 for I_p (undisturbed clay), compute the several s_u/p_o ratios and plot on a tracing of Fig. 13-28a for comparison.

SOLUTION Obtaining the several values of ϕ' shown and corresponding I_p and programming Eq. 13-23 on a programmable calculator for s_u/p_o, we obtain:

$I_p =$	0	20	40	60	80	100
$\phi' =$ 36		31	27	24	22.5	21°
$\dfrac{s_u}{p_o} =$ 0.37		0.34	0.31	0.29	0.28	0.26

The appropriate values of I_p versus s_u/p_o are plotted as shown on Fig. E13-9 and show little agreement.

////

Example 13-10 Given: overconsolidated lean clay with $w_L = 38$; $I_p = 20$. Consolidation test gives $p_c = 150$ kPa. In situ $p_o = 50$ kPa; $s_{uo} = 90$ kPa (measured in a triaxial test).

REQUIRED Estimate s_u for a normally consolidated sample of this soil.

SOLUTION We probably should remold a sample and either isotropically or anisotropically consolidate it to $p_o = 50$ kPa and directly measure the undrained shear strength. Here we will use correlations to make the estimate.

From Table 13-2 it appears that

$$\frac{s_u}{p_o} \cong 0.32 \qquad \text{(triaxial compression)}$$

$$\frac{s_u}{p_o} \cong 0.25 \qquad \text{(DSS)} \qquad \text{for "lean" clays with } w_L < 50$$

Adjusting s_{uo} to DSS, we obtain $s_{uo} = 90(0.25/0.32) = 70$ kPa (rounding).
For using Fig. 13-29:

$$B = \frac{70}{50} = 1.4$$

$$\text{OCR} = \frac{p_c}{p_o} = \frac{150}{50} = 3$$

We obtain at OCR = 3 the ratio $B/A = 2.2$ (average):

$$A = \frac{B}{2.2} = \frac{1.4}{2.2} = 0.64$$

$$\frac{s_u}{p_o} = 0.64 = A$$

$$s_u = 0.64(50) = 32 \text{ kPa} \qquad \text{(DSS)}$$

$$s_u = 32\left(\frac{0.32}{0.25}\right) = 41 \text{ kPa} \qquad \text{(triaxial)}$$

////

13-16 SUMMARY

The concept of shear strength has been considered in some detail. We find that it is dependent on both soil state and test method.

For cohesionless soils the shear strength is

$$s = \sigma_n \tan \phi$$

where σ_n is usually the effective stress since the equipment is such that the soil can usually drain (if very wet or saturated) if a reasonable strain rate is used. We should not use damp sand since an apparent cohesion from capillary tension will be obtained. Direct shear tests are preferable, but plane strain values of ϕ tend to be larger than triaxial values. Principal test deficiencies are in duplication of the in situ structure and any natural cementation. Under usual conditions large variations in water content do not affect the shear strength of sands and gravels.

Cohesive soils are the most troublesome because small variations in water content can greatly alter the shear strength. Fortunately, the worst case is at full saturation so much testing is in that state. The shear strength is heavily dependent on state and method.

For undrained (unconsolidated-undrained and unconfined compression) tests

$$s_u = c \qquad (\phi = 0)$$

and this is the most common strength value obtained for these soils. This strength is usually adequate for most routine projects. Most problems seem to occur with embankments over very soft soils.

For consolidated-undrained tests without pore pressure measurements or CD tests where $\sigma_n < p_c$ the strength may be given by the Mohr-Coulomb strength equation as

$$s = c + \sigma_n \tan \phi$$

For CD tests and CU tests with pore pressure measurements on normally consolidated soils and CD tests on overconsolidated soils with $\sigma_n > p_c$ the drained shear strength is

$$s = c' + \sigma'_n \tan \phi'$$

The values of c' and ϕ' are effective strength parameters. The cohesion c' is usually very small and is neglected; however, it is shown to make the equation general since, as noted in the text, the geotechnical literature is replete with "exceptions to the rule."

The concepts of brittle failure, progressive failure, and critical void ratio have been introduced. We find that all "dense" soils expand (or dilate) during shear and "loose" soils decrease in volume. We also find that "dense" and "loose" are relative and depend on the confining pressure.

The concept of residual strength and the use of ϕ_r are appropriate for problems where large shear strains are possible. In this case the cohesion intercept is negligible and is usually ignored.

Mohr's circle and p-q diagrams can be used to obtain the Mohr-Coulomb shear strength parameters. Both p-q diagrams and octahedral stress plots can be used to produce stress paths. Stress paths are traces of point stresses during shear and are used to give insight into soil response to loading.

The effect of overconsolidation to increase shear strengths was examined. Also, we find that unloading soil can produce negative pore pressures. Since these may be quite large if the soil is overconsolidated, care must be taken to identify or eliminate them during testing so that a realistic shear strength is used.

We have found that the effective stress has major influence on shear strength and the effective stress is

$$\sigma' = \sigma - u$$

When a soil is unloaded, the pore pressure u may be $(-)$, which temporarily increases σ'. Reliable pore pressures can be obtained only by direct measurements although means to compute estimates were given.

We examined several common laboratory and in situ methods for measuring shear strength. We found that all of these current methods have some deficiencies.

The concept of normalizing select soil parameters was given along with several correlations. It was suggested that correlations might be used in preliminary site studies but should be verified with tests on local soil before they are used in the final design.

PROBLEMS

13-1 Describe the cohesion and friction components for a cohesive soil.

13-2 Why is there not cohesion intercept for a cohesive soil in a CD test?

13-3 Why is the cohesion intercept negligible for the residual strength?

13-4 Why can no data be obtained for an undrained test on a sand at $S = 100$ percent?

13-5 Explain why the pore pressure can never be zero in a drained test as long as $S > 0$.

13-6 Compute s_u at $\sigma_n = 65$ with $c = 17.5$ and $\phi = 12°$. Compare to the scaled value of 31.25 shown in Example 13-3.

13-7 Referring to Example 13-2, what is ϕ if the measured shear strength is 25 instead of 20?
   Ans.: Approximately $31°$

13-8 Draw stress paths on a p-q diagram for the following stress conditions. State the type of consolidation for each test:

Test	Initial stresses, kPa		Final stresses, kPa	
	σ_1	σ_3	σ_1	σ_3
1	200	200	600	200
2	200	200	200	700
3	200	200	100	100
4	100	200	400	250
5	150	300	600	150

13-9 A triaxial test (CU) was performed and the following data obtained:

Test	σ_3	σ_1, failure, kPa
1	60	309
2	120	640
3	180	900

Plot Mohr's circles and find c and ϕ. What type of soil do you think this is?
   Ans.: $c \cong 0$; $\phi = 42°$

13-10 An unconfined compression test yielded $q_u = 175$ kPa (as an average for three tests). What is the percent error, and is it conservative or unconservative to use $s_u = q_u/2$ when it is known that the soil has a CU angle of $10°$?

13-11 The following data were obtained from a series of drained triaxial tests on a damp sand. Plot the assigned data and determine both ϕ' and $\phi_{residual}$. For residual, use $\Delta\sigma_1$ at $\varepsilon \geq 0.25$. Comment on the effect of density both on the slope of the curves and on ϕ. Assume that the moisture had a negligible effect on $\Delta\sigma_1$. The data has been plotted by the author and points taken from the curves to construct the following table:

	Test					
	1	2	3	4	5	6
σ_c, kPa	70	140	210	70	140	210
Void ratio e	0.79	0.78	0.80	0.46	0.47	0.46
Strain ε	Deviator stress $\Delta\sigma_1$, kPa					
0	0	0	0	0	0	0
0.01	50	155	245	140	330	600
0.02	90	220	410	180	420	713
0.04	125	300	460	230	480	710
0.06	140	320	470	236	460	680
0.08	152	316	455	230	440	650
0.10	150	305	450	220	425	640
0.12	145	300	440	215	418	630
0.15	140	290	435	210	408	615
0.20	140	285	430	205	398	600
0.25	140	280	425	200	395	590

13-12 A vane shear test was performed and the following data obtained:

$$T = 61 \text{ N} \cdot \text{m} \qquad w_L = 68.4 \text{ percent}$$

$$d = 65 \text{ mm} \qquad w_P = 34.1$$

$$h = 110 \text{ mm} \qquad w_N = 71.3$$

Assuming uniform end resistance, compute s_{uv} and indicate the value of s_u for use in design for a long embankment foundation. Do you think that the soil is overconsolidated or sensitive? What laboratory test will give the desired value of s_u?

13-13 Sketch a qualitative shear envelope for an overconsolidated clay. Show values for OCR = 1, 2, and 5. Would a q_u test give a satisfactory value of shear strength for design? Refer to Fig. 13-18.

13-14 In a direct shear test on a clay with $\sigma_n = 180$ kPa, $s = q_o$ kPa and from other data c is known to be 40 kPa. Find the angle of internal friction and the principal stresses σ_1 and σ_3.
    Partial ans.: $\phi = 15°$; $\sigma_3 = 113$ (scaled)

13-15 Plot the several s_u/p_o values from Table 13-2 which are in the tabular range on a copy of Fig. 13-28a and 13-28b, using I_p or w_L versus s_u/p_o. Also plot the I_p versus s_u/p_o data computed in Example 13-9 using a different symbol. Comment on the results.

13-16 Referring to Example 13-10, could the same result ($s_u = 41$ kPa) have been obtained without the adjustment ratios using 0.32 and 0.25?

13-17 Three CU triaxial tests were performed as follows:

Test	σ_3	σ_1	u, kPa
1	0	138.5	? (Unconfined-compression test)
2	70	351.8	56.0
3	140	515.0	105.5

What are the apparent and effective soil parameters c and ϕ? Is the soil saturated?

13-18 In a CU test on a normally consolidated clay with pore pressure measurements, the following data were obtained at failure:

$$\sigma_c = 60 \text{ kPa} \qquad \Delta\sigma_1 = 68.5 \text{ kPa} \qquad u = 21.4 \text{ kPa}$$

What is the apparent and effective angle of internal friction?

Ans.: $\phi = 0°$; $\phi' = 27°$

STATIC AND DYNAMIC STRESS-STRAIN CHARACTERISTICS

14-1 STRESS-STRAIN DATA

Stress-strain data are obtained from the several shear strength tests described in Sec. 13-4. Where compression tests are made it is conventional practice to correct the area using Eq. (13-3) and use the corrected area to compute the deviator stress as in Eq. (13-4). The resulting values of $\Delta\sigma_1$ versus strain ε are plotted as qualitatively illustrated in Figs. 13-4, 13-7, 13-9, and 14-1b. Shear stress can also be plotted against strain, but it is more common to use compression tests.

From the stress-strain curve one may obtain an elastic modulus. This is the modulus of elasticity E_s if the data are from a compression test and the shear modulus G using shear stress-strain plots. One may also obtain Poisson's ratio μ from a compression test.

The stress-strain modulus and Poisson's ratio are used in computing elastic settlements in Chap. 15. These values are necessary material properties for finite-element analyses (computer) which model the soil as an elastic continuum.

14-2 THE STRESS-STRAIN MODULUS

The modulus of elasticity, a property of elastic materials, is defined as a constant of proportionality between stress and strain as

$$E = \frac{\Delta\sigma}{\Delta\varepsilon} \tag{14-1}$$

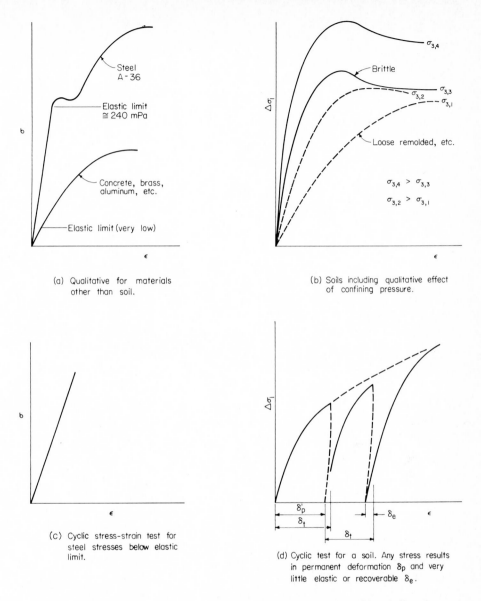

Figure 14-1 Typical qualitative stress-strain curves for the materials and conditions indicated.

Most engineering materials exhibit this linear behavior over some region of stress (and the corresponding strain). Steel (Fig. 14-1*a*) exhibits linear behavior over the largest stress range (on the order of 250 MPa); concrete, aluminum, cast iron, wood, etc., are linear over a very limited stress range. Further, even though the stress-strain curve is nonlinear for these materials, if the sample is not stressed beyond the elastic limit, it will return to its original dimensions on removal of the stress. This is a part of the definition of a homogeneous, isotropic, elastic material.

Soil exhibits linear stress-strain characteristics only at extremely low strain amplitudes—strains generally on an order of magnitude of 10^{-4} and smaller. In conventional triaxial tests no data are obtained at these strain levels, and as a consequence the stress-strain curve has no linear region. Also, unless the test is halted when the sample has strained the low amplitude indicated, the sample will not recover its original shape but will remain permanently deformed, as illustrated in Fig. 14-1d. This is because the principal soil deformations are state changes caused by relative particle motion. Only the very small amount of soil deformation occurring from particle distortion is elastically recoverable. Because soil deformation is due primarily to relative particle motion, modulus of elasticity is not a correct term to use; rather, one should use the *stress-strain modulus*. This becomes a value describing the relationship between applied stresses and resulting strains. Conventional terminology, however, uses modulus of elasticity, and the two descriptive terms will be used interchangeably in this text. The user should be aware that modulus of elasticity is a term of convenience rather than a true elastic property for soil materials. Some persons use "elastoplastic" or "theory of plasticity" as descriptive terms for soil distortion. Both of these terms are based on an elastic continuum, and thus for soil are also convenience terms, since soil is a particulate mass and not a continuum.

There are two commonly used methods for computing the stress-strain modulus from nonlinear soil stress-strain curves:

Tangent modulus. Modulus based on the slope of a line which is just tangent to the stress-strain curve at a point. The *initial* tangent modulus is commonly used (a tangent at the origin), since the slope at the origin is not highly subject to environmental factors such as type of test and confining pressure.

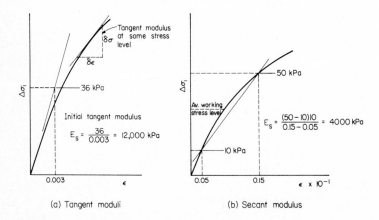

Figure 14-2 Computing stress-strain modulus E_s by several methods.

Secant modulus. Modulus based on the slope of a secant line. A secant line cuts the stress-strain curve at two points. When used, the two points are commonly equally spaced from the working stress.

The tangent and secant moduli are illustrated in Fig. 14-2. In finite element computations, a tangent modulus may be used with the tangent taken at the stress level under consideration. Unfortunately, with the finite element procedure, the stress depends on the modulus of elasticity, and thus iteration is necessary.

Soil is commonly anisotropic rather than isotropic due to formation by sedimentation. This produces a situation where E_h is generally not equal to E_v. The determination of E_s should consider the possibility of anisotropy and orient the test samples to obtain the correct stress-strain modulus.

14-3 POISSON'S RATIO

Poisson's ratio is a property of elastic materials defined (see Sec. 10-5) as

$$\mu = \frac{\varepsilon_3}{\varepsilon_1} \tag{14-2}$$

where ε_1 = strain colinear with stress of interest
ε_3 = strain orthogonal to stress of interest

Poisson's ratio is often assumed to be 0.2 to 0.4 in soil mechanics work. A value of 0.5 is usually used for saturated soils, and 0.0 is often used for dry soils and others as a computational convenience. This is because a value of Poisson's ratio is difficult to obtain for soils. The vertical strain is relatively easy; the lateral strain is particularly difficult as compared with any metal where a strain gage may be readily attached. A few soils laboratories have modified triaxial cells to allow diameter measurements to be taken as the test proceeds. We may also take volumetric measurements during a test. This is usually done by filling the cell with a liquid and during consolidation adding liquid to maintain the volume constant. It is immediately evident that a "constant volume" test can be fairly easily done by filling the cell completely with liquid which is relatively incompressible and will control any increase in volume. If a decrease in volume occurs, we have no means to control this except by having a saturated sample in a closed system.

If we measure the volumetric change ΔV we can compute the volumetric strain as

$$\varepsilon_v = \frac{\Delta V}{V_0} = \varepsilon_1 + \varepsilon_2 + \varepsilon_3$$

Also we have $\sigma_2 = \sigma_3$ = cell pressure and at failure a value of σ_1. Substituting into Eqs. (10-7) and adding the three strains, we obtain

$$\varepsilon_v = (\sigma_1 + 2\sigma_3) \frac{1 - 2\mu}{E_s} = \frac{\sigma_1 + 2\sigma_3}{3E_b} \tag{14-3}$$

where E_b = bulk modulus from Eq. (10-9a).

Plane strain strength parameters in addition to ϕ_{ps} ($> \phi_{tr}$) which were noted in Eq. (13-10) are sometimes used instead of triaxial values. Sometimes in finite-element analyses plane stress parameters are also used. We will look at the plane strain values since they represent a very common in situ stress-strain condition. A plane strain condition is where one strain is zero (but not the stress) and taking $\varepsilon_2 = 0$ and solving the corresponding equation from Eqs. (10-7) we obtain

$$\sigma_2 = \mu(\sigma_1 + \sigma_3)$$

Now substitute this value of σ_2 into either of the other two equations (and here using ε_1) and subscripting for plane strain we obtain after some manipulation

$$\varepsilon_{1_p} = \frac{1 - \mu^2}{E_s} \left(\sigma_1 - \sigma_3 \frac{\mu}{1 - \mu} \right) \tag{14-4}$$

Let us define plane strain parameters as

$$E_p = \frac{E_s}{1 - \mu^2} \qquad \mu_p = \frac{\mu}{1 - \mu}$$

and now rewrite Eq. (14-4) as

$$\varepsilon_{1_p} = \frac{1}{E_p} (\sigma_1 - \mu_p \sigma_3) \tag{14-4a}$$

This illustrates that the plane strain values of E_p and μ_p are larger than the corresponding triaxial values.

This type of analysis is particularly difficult in anisotropic soils because there are multiple values of E and μ—even in a triaxial state.

While it is conventional practice to obtain the slope of the stress-strain curve as the stress-strain modulus E_s and assume a value of Poisson's ratio, we may compute both values as follows. First, it is necessary to assume that in a sufficiently close strain interval the stress-strain curve is linear. Next, we may use the strain equation (ε_1) from Eq. (10-7), and for the triaxial test we have

$$\Delta\varepsilon_1 = \frac{1}{E_s} (\Delta\sigma_1 - 2\mu\sigma_3)$$

which can be solved for two adjacent values of $\Delta\varepsilon_1$ and $\Delta\sigma_1$ to obtain E_s and μ. This type of computation can be made up to the point where negative values or values of $\mu > 0.5$ are obtained. No real material appears to have $\mu < 0$, and a value of $\mu > 0.5$ indicates plastic behavior of an elastic material. In making this type of computation one finds that $\mu > 0.5$ occurs at relatively low strain levels, as one might intuitively suspect from this type of material. We might note also that most computations involving elastic materials are valid only for small distortions. This computation illustrates that μ is not a constant for a soil but depends on stress level. The secant modulus of elasticity is used here.

Example 14-1 Given the stress-strain curve shown (Fig. E14-1) for a soft clay soil.

REQUIRED Compute a value of E_s and μ.

SOLUTION At $\Delta\sigma_1 = 250$ kPa, obtain $\varepsilon_1 = 0.011$. At $\Delta\sigma_1 = 375$ kPa, obtain $\varepsilon_1 = 0.018$. Substituting,

$$0.011E_s = 250 - 2\mu(100)$$

$$0.018E_s = 375 - 2\mu(100)$$

By elimination, obtain

$$E_s = \frac{125}{0.007} = 17\,857 \text{ kPa}$$

Back-substitution into the first equation gives

$$\mu = \frac{250 - 196}{200} = 0.27$$

Strictly, unless the curve is nearly linear as here we should reference the 2d point to the first point, etc., rather than to the origin in computing $\Delta\varepsilon$.

////

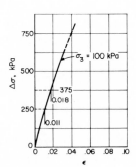

Figure E14-1

14-4 FACTORS AFFECTING THE STRESS-STRAIN MODULUS AND APPROXIMATIONS

Any factor which will modify the slope of the stress-strain curve will affect the stress-strain modulus and Poisson's ratio. These factors include:

Cell, or in situ confining, pressure (qualitatively illustrated in Fig. 14-1b)
Soil unit weight
Geologic history (producing a brittle or progressive failure)
Grain shape
Sample size
Type of test, as U, CU, or CD
Sample disturbance

For these several reasons, it may be convenient to estimate the value of E_s, at least for preliminary design. For cohesive soils,

Soft clay: $E_s \cong 100$ to $750 s_u$ Cone: $E_s = 6$ to $8 q_c$

Stiff clay: $E_s \cong 750$ to $1500 s_u$ —

where s_u = undrained shear strength
     q_c = Dutch cone point resistance
Sand is so variable that some kind of test should be performed to estimate a value. Two estimates based on the SPT are

$$E_s \cong 30\,000 \text{ to } 50\,000 \log N$$

$$E_s \cong 18\,000 + 750\,N$$

where N = standard penetration test number as outlined in Sec. 6-9. Using the cone penetration resistance (fine to medium sand) an estimate is:

$$E_s = 2 \text{ to } 4 q_c$$

14-5 RESILIENT MODULUS

The resilient modulus is defined as the initial tangent modulus of a triaxial test stress-strain curve which has been cycled several times with a deviator stress $\Delta \sigma_1$ level approximating the working stress, as shown in Fig. 14-3. Alternatively, a value of $\Delta \sigma_1 \cong \Delta \sigma_1 / 2$ (maximum) is sometimes used.

This test was initially used for pavement subgrades, but there is some body of opinion that this value may be more appropriate for many foundation settlement problems. This value is useful partially because the E value increases somewhat in the test, due to soil particle readjustment, to a more stable structure on successive load-unload cycles. This produces a larger initial tangent modulus on the last cycle and a correspondingly smaller computed settlement which is more in line with observed settlements.

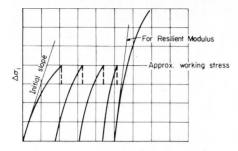

Figure 14-3 Method of obtaining the resilient modulus of deformation.

14-6 DYNAMIC SOIL STRESS-STRAIN MODULUS

The dynamic stress-strain modulus is useful in predicting vibration displacements of foundations subjected to cyclic loads. Machine, pump, radar tower, compressor, and turbine foundations are typical foundations which are loaded dynamically. Earthquakes also produce dynamic loadings which require displacement and strength predictions.

Soil dynamic problems fall in two general categories of strain magnitudes:

	Range of strain amplitude ε
Machine foundations	10^{-4} to 10^{-3}
Earthquake	10^{-2} to 10^{-1} and larger

These ranges of strain amplitudes require separate consideration to determine the dynamic properties of stress-strain modulus, damping, and soil strength. This is because the modulus and strength decrease, often markedly, with increasing strain amplitude. The values of stress-strain modulus obtained from the usual triaxial compression test tend to be quite low (perhaps on the order of $\frac{1}{2}$ to $\frac{1}{10}$) compared with dynamic stress-strain moduli determined from low amplitude strain tests. This is because the first sample strain value in the standard triaxial test is often on the order of 0.005 m/m, which may be satisfactory for an earthquake analysis and static analyses in general but may be an order of magnitude too large for machine foundation analysis. With machine foundations, dynamic soil strength is not generally a critical factor, whereas with earthquake studies this will be the principal design consideration, as shown in the next section.

It is preferable to determine the stress-strain properties in situ for machine foundations with low strain amplitudes. Where this is not practical, several laboratory methods are available, including cyclic triaxial testing (Park and Silver, 1975; Weissman and Hart, 1961), cyclic direct simple shear tests (Kovacs et al., 1971), hollow cylinder torsion tests (Ishihara and Yasuda, 1975), and resonant column testing (Hardin and Music, 1965).

The resonant column test has been used considerably for determining the dynamic modulus for machine foundations. This test utilizes the time of travel of a shear wave through a column of soil which is either hollow or solid. The frequency at resonance is obtained and the stress-strain modulus computed. General details of test details and equipment are beyond the scope of this text, and the reader is referred to Hardin and Music (1965) and Richart et al. (1970) for further details.

In situ determination of the dynamic stress-strain modulus requires measuring either shear, Rayleigh, or compression waves and back computing the modulus. Elastic, homogeneous ground stressed with a vibration produces three elastic waves traveling outward from the source at different speeds. These waves are:

Primary or compression waves, called *P waves* (have the highest velocity)
Secondary or shear waves, called *S waves*
Near ground surface waves termed *Rayleigh waves*, or *R waves*

The velocity of the *R* wave is about 10 percent less than that of the *S* wave, but the *R* wave is easier to interpret on an oscilloscope (or from an oscillograph) and is usually used.

The in situ determination of the dynamic stress-strain modulus uses seismic techniques, i.e., a shock source and an electronic pick-up unit. The pick-up unit often inputs into an oscilloscope so that the operator may either visually observe the arrival of the *R* wave (and others) to the pickup or photograph the wave trace for later data reduction. The vibration is commenced at the shock point, a known distance d_o from the pick-up unit, and the time of arrival to the pick-up unit is observed (see Fig. 6-11). The wave velocity is computed as

$$v_c = \frac{d_o}{t_c} \quad \text{or} \quad v_s = \frac{d_o}{t_s} \cong \frac{d_o}{t_R}$$

where t_i is the elapsed time for the compression, shear, or Rayleigh wave, respectively. Relationships between wave velocity and dynamic shear modulus are as follows:

$$v_c = \sqrt{\frac{E_s(1 - \mu)}{\rho(1 + \mu)(1 - 2\mu)}} \tag{14-5}$$

$$v_s \cong v_R \cong \sqrt{\frac{G}{\rho}} \tag{14-6}$$

where ρ = soil density = γ/g, and other terms are as defined earlier. The shear modulus G was given in Eq. (10-8) as

$$G = \frac{E_s}{2(1 + \mu)} \tag{10-8}$$

The shear modulus is generally required in computations of vibration amplitudes for machine foundations. Equation (14-6) is generally used to determine G directly, as the use of Eqs. (14-5) and (10-8) requires estimating Poisson's ratio, and the small error due to differences in v_s and v_R is less than that in estimating μ.

Empirical equations for G have been proposed by Hardin and Richart (1963) as:

For rounded sand grains,

$$G = \frac{6900(2.17 - e)^2}{1 + e}\,\sigma_o^{0.5} \quad \text{kPa} \tag{14-7}$$

For angular sands and normally consolidated clays of low activity,

$$G = \frac{3230(2.97 - e)^2}{1 + e}\,\sigma_o^{0.5} \quad \text{kPa} \tag{14-8}$$

For overconsolidated clays with overconsolidation ratio OCR (Hardin and Drnevich, 1972),

$$G = \frac{3230(2.97 - e)^2}{1 + e}\,\text{OCR}^M \sigma_0^{0.5} \tag{14-9}$$

In all the above equations,

$e =$ in situ (or test) void ratio

$\sigma_o = \dfrac{J_1}{3} = \dfrac{1}{3}(\sigma_1 + \sigma_2 + \sigma_3) = \textit{effective}$ confining stress, kPa

$M =$ exponent depending on the plasticity index I_P from Fig. 14-4

Example 14-2 Estimate the shear modulus for the following in situ soil conditions: $e = 0.59$; $\phi' \cong 36°$; subangular silty sand (fairly dense), $\gamma = 18.1$ kN/m^3. $w_N = 10$ percent.

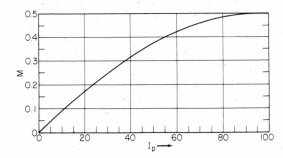

Figure 14-4 Values of exponent M for the OCR given in Eq. (14-9). (*After Hardin and Drnevich, 1972.*)

SOLUTION Estimate K_o as

$$K_o = 1 - \sin \phi = 0.41 \qquad \text{[this is Eq. (15-2)]}$$

Estimate the effective depth of vibration influence = 3 m (total range from 0 to about 10 m):

$$\sigma_v = 3(18.1) = 54.3 = \sigma_1 \qquad \text{(vertical pressure)}$$

$$\sigma_2 = \sigma_3 = K_o \sigma_1 = 0.41\sigma_1 = 22.3 \text{ kPa} \qquad \text{(lateral pressure)}$$

$$\sigma_o = \tfrac{1}{3}(54.3 + 22.3 + 22.3) = 32.9$$

$$G = \frac{3230(2.97 - 0.59)^2}{1 + 0.59} (32.9)^{1/2} = 66\,000 \text{ kPa}$$

////

14-7 CYCLIC MODULUS OF DEFORMATION AND LIQUEFACTION

The widespread interest in nuclear power has produced a need to investigate potential plant sites for geologic faults and possible soil liquefaction during an earthquake. Liquefaction was defined in Sec. 8-8 as a soil state caused by a buildup of pore pressure from a cyclic or dynamic loading until the effective pressure is zero. This is a phenomenon of fine- to medium-fine-grained sands. It has not been observed in gravels and is difficult to develop in fine silty sands or medium to coarse sands. Three types of tests have been used in the last 10 years to study the structural stability (strength) of sands. These include the cyclic triaxial test and direct simple shear test cited in the previous section and the shaking table test (Seed et al., 1977).

These studies are made to estimate the magnitude and number of cycles of stress (or strain) necessary to cause a pore pressure buildup sufficient to reduce the effective pore pressure to a low enough value that, under the soil loads, the material behaves as a viscous fluid, or liquefies. As was pointed out in Sec. 13-15, researchers have discovered that the residual pore pressure tends to increase on successive stress cycles in a clay. It has also been found that pore pressure increases with increasing strain. This phenomenon would be expected in a fully saturated soil in undrained shear. In natural soils where $S < 100$ percent even with drainage, this phenomenon may also be induced if the pore pressure buildup is larger than the drainage rate, as may occur during earthquakes, where relatively large strains are developed. Note also that "progressive" failure does not require the entire mass to be simultaneously in a state of liquefaction. A local point of liquefaction may result in sufficient load transfer to an adjacent point that the soil is overstressed, resulting in a mass failure.

The cyclic direct simple shear and cyclic triaxial tests can be used also to determine the dynamic modulus and the damping factor for a soil under the conditions of the test (void ratio, confining pressure, cementation, sample disturbance, strain or stress ratio, etc.). The data from these tests are usually dis-

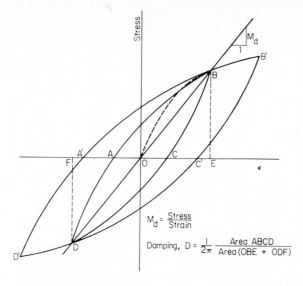

Figure 14-5 Hysteresis loop for a dynamic simple shear test. Note that this loop will locate about the origin of axis as above. A triaxial test will locate to the right of the axis as in Example 14-3. The above are actual data, with curve *ABCD* for the first cycle and *A'B'C'D'* for 10 cycles. This soil has a reduced stress-strain modulus with increasing stress cycles. This curve is obtained by reversing P_h (see Fig. 13-5*c* or 13-5*d*) on the order of 1 to 10 times per second.

played as shown in Fig. 14-5. Generally the strain and stress are electronically monitored, the output is directed to an oscillograph, and the plot shown is made. The plot can also be made by hand by obtaining the stresses and corresponding strains. Generally, since the cycling is on the order of $\frac{1}{6}$ to 10 Hz, it is only practical to use an oscillograph.

From the hysteresis loop shown, the stress-strain modulus is

$$M_d = \frac{\text{stress}}{\text{strain}} \tag{14-10}$$

where $M_d = E_s$ for axial strain and deviator stress
$\qquad = G$ for shear strain and shear stress

Damping is computed as

$$D = \frac{1}{2\pi} \left[\frac{\text{area of loop } ABCD}{\text{area of both triangles } (OBE + ODF)} \right] \tag{14-11}$$

The area of the loop and of the triangles is conveniently obtained using a planimeter.

The cyclic shear test is also used to determine the number of cycles N_c at some deviator stress, generally expressed as a stress ratio, i.e., normalized, where

$$R = \frac{\Delta\sigma_1}{2\bar{\sigma}_3}$$

where $\bar{\sigma}_3$ = effective confining pressure that will cause the measured pore pressure, starting from consolidated-drained conditions ($u = 0$), to equal the confining pressure ($u = \bar{\sigma}_3$). At this pore pressure condition, "liquefaction" is assumed. For the direct simple shear test, use Δs = cyclic shear stress and the applied normal stress σ_v for $\Delta\sigma_1$ and σ_3, respectively.

For the cyclic triaxial test,

1. Build a sample to the desired density (or void ratio). Use a damp sand of known water content and weighed into 5 to 10 separate equal parts to produce a uniformly dense sample by tamping the equal weights into equal sample volumes. Several trial samples may be necessary before a satisfactory sample is obtained. A similar procedure would be used for remolded cohesive soils. Alternatively, where or when practical, undisturbed samples may be used.

2. For liquefaction studies, saturate and consolidate the sample using a back pressure σ_b and check that the B parameter of Sec. 13-15 is 0.98^+. The effective consolidation pressure is obtained as the difference between cell pressure and back pressure,

$$\bar{\sigma}_3 = \sigma_3 - \sigma_b$$

For example, with $\sigma_3 = 650$ kPa and a back pressure on the sample pore pressure fluid reservoir of 500 kPa, the effective consolidation pressure is 150 kPa. A positive back pressure should be maintained during the test to ensure $S \rightarrow 100$ percent by keeping the air dissolved in the pore water.

   For determining the dynamic modulus, the in situ water content, or damp sand in a "drained" condition of pore lines open for drainage, should be used.

3. Place the cell into a triaxial machine modified for cyclic load application. A modification of existing equipment can be made at modest cost where competent laboratory technicians are available (Chan and Mulilis, 1976; Cullingford et al., 1972). Equipment is available from several commercial laboratory suppliers and/or can be custom built.

4. Apply a strain increment and observe the maximum deviator load (stress). Generally, a strain of 2.5, 5, and/or 10 percent is used in liquefaction studies. For machinery foundations, the strains may be on the order of 0.025 to 0.1 percent or less, and the sample is not cycled to failure; rather, one to four cycles are recorded, then the strain increment is reset and the test repeated.

   Adjust the equipment so that a portion of the stabilized stress (or strain) is cycled. It is evident that the cycling must be such that sample contact is maintained on the $(-)$ stress part of the hysterisis loop.

5. Allow several cycles to be superimposed, for a statistical averaging for computation of dynamic moduli. For liquefaction studies, the cycles are counted to liquefaction ($u = \sigma_3$), and a plot of stress ratio versus N_c is made as in Fig. 14-6 for several stress ratios with $\varepsilon = $ constant. For example, with $\sigma_3 = 600$ kPa, the initial deviator stress $\Delta\sigma_1$ to produce $\varepsilon = 0.025$ (2.5 percent strain) might be 450 kPa. We may run tests as:

$\Delta\sigma_1$	Cycles N_c to liquefaction	Stress ratio R_c
100 kPa	500	$100/(2 \times 600) = 0.08$
300	30	0.25
400	8	0.33

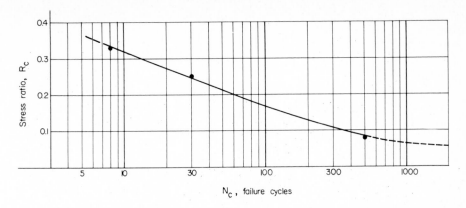

Figure 14-6 Plot of stress ratio R versus number of cycles to liquefaction N_c.

This is part of the data plotted in Fig. 14-6.

Example 14-3 Given the cyclic triaxial test data shown in Fig. E14-3 for a medium dense, coarse sand ($D_r = 0.42$) and $G_s = 2.68$. The test is stress-controlled as much as is practical. Cell pressure $\sigma_3 = 140$ kPa and the initial strain is 0.37, producing an initial $\Delta\sigma_1 = 50$ kPa and $\Delta\sigma_1/\sigma_3 = 0.36$.

REQUIRED Initial tangent modulus, dynamic stress-strain modulus, and damping factor D.

SOLUTION Draw lines $O'A'$, AOC, AB, and CD as shown.

$$\text{Static } E_s = \text{slope } O'A' = \frac{0.7(140)}{0.25}$$

$$= 390 \text{ kPa}$$

$$\text{Dynamic } E_s = \text{slope } OA$$

$$= \frac{1.50(140)}{0.68 - 0.27}$$

$$= 510 \text{ kPa}$$

To obtain the damping factor, use a planimeter

$$\text{Area of } AB'CD' = 23 \text{ units} \qquad (3 \text{ trials})$$

$$\text{Area of } OAB = 23.5 \text{ units}$$

$$\text{Area of } OCD = 10.5 \text{ units}$$

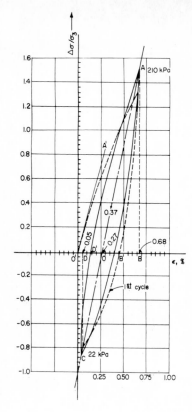

Figure E14-3

Substitition into Eq. (14-11) gives

$$D = \frac{1}{2\pi}\left(\frac{23}{23.5 + 10.5}\right) = 0.11$$

////

Laboratory testing indicates that sample preparation, grain size, void ratio (or relative density), effective confining pressure, and previous strain history all considerably influence the laboratory data. For these reasons, in using laboratory data for design, a correction factor is introduced,

$$\left(\frac{s_e}{\bar{p}_o}\right)_{field} = C_r\left(\frac{\Delta\sigma_1}{2\bar{\sigma}_3}\right)_{lab} \tag{14-12}$$

and dividing, we obtain the apparent safety factor as

$$F = \frac{C_r(\Delta\sigma_1/2\bar{\sigma}_3)_{lab}}{(s_e/p_o)_{field}} \tag{14-13}$$

where s_e = field shear stress caused by the earthquake (maximum)

p_o = in situ effective overburden stress

C_r = adjustment factor, ranging from about 0.57 for normally consolidated sands to 1 for sands with OCR → 8

σ_3 = laboratory effective consolidation stress

Note that Eq. (14-13) is to be compared at the same number of cycles.

The field shear stress can be computed (Seed and Idriss, 1971) as

$$s_e = 0.65(r_d)ma \qquad (14\text{-}14)$$

where m = mass of unit column of soil to the depth of interest h, such as location of bedrock (origin of earthquake) = $\gamma h/g$

a = estimated acceleration of earthquake, depending on magnitude and distance from epicenter

r_d = reduction factor to account for the accelerated mass m not being a rigid body; use Fig. 14-7 for values of r_d

The factor 0.65 is used to statistically average the shear stress over the cycle duration.

The number of field cycles for comparison with the laboratory curve may be estimated from Fig. 14-8. This curve is based on a scatter of at least ± 1 standard deviation.

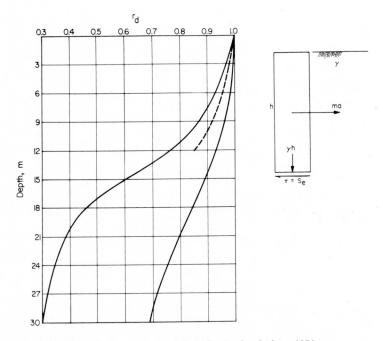

Figure 14-7 Values of r_d versus depth h. *(After Seed and Idriss, 1971.)*

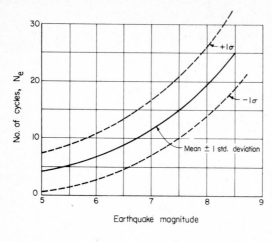

Figure **14-8** Earthquake magnitude (Richter scale) versus number of resulting stress cycles. *(After Valera and Donovan, 1977.)*

Figure 14-9 illustrates a summary of data for a clean, isotropically consolidated sand with the mean (D_{50}) grain size = 0.2 mm. This may be used to indicate the general range of stress ratio R to be expected. It also indicates variation of R with relative density and number of cycles to cause liquefaction for either stress ratio or relative density.

Figure 14-10 illustrates the effect of grain size on the strength ratio and may be used to adjust data from Fig. 14-9 for grain sizes other than 0.2 mm. These data were obtained from test samples formed by sedimentation and light, wet compaction. Moist tamping gives cyclic strengths as much as 140 percent higher. Confining pressures significantly higher than 100 to 200 kPa may give lower cyclic strengths. Naturally cemented soils may possess higher cyclic strengths.

Clay soils generally do not liquefy in cyclic loading; instead, the strain progressively increases. An exception is noted with some naturally cemented, sensitive clays, which develop sharply defined shear planes on which the soil remolds

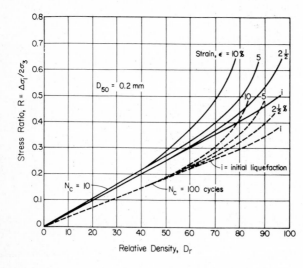

Figure **14-9** Stress ratio versus relative density for cycles to liquefaction. Initial liquefaction is the stage where the first effects of partial liquefaction are detected and is not "failure." *(After Lee and Seed, 1967.)*

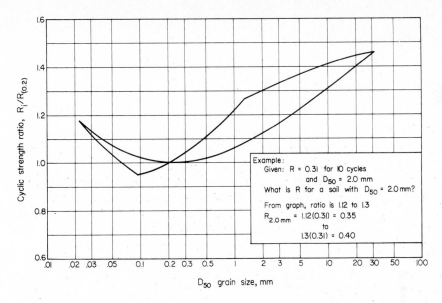

Figure 14-10 Relationship between mean (D_{50}) grain size and the cyclic strength ratio. Data from several sources.

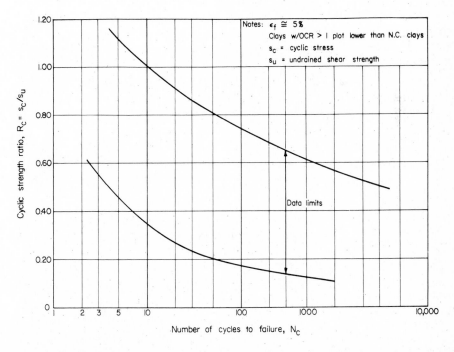

Figure 14-11 Relationship between number of cycles to failure and cyclic strength ratio for clay soils. *(After Lee and Focht, 1975.)*

to the much lower residual strength. Figure 14-11 illustrates the stress ratio of clay, defined as

$$R_c = \frac{s_c}{s_u}$$

where s_c = cyclic shear strength (also the deviator stress applied)

s_u = undrained shear strength

This figure is based on a wide range of clay soils (Lee and Focht, 1975). Equation (14-13) may be used for clay soils with $C_r = 1$.

Example 14-4 Estimate the safety factor of a nuclear power plant site for the conditions shown in Fig. E14-4 and an earthquake of magnitude 8 on the Richter scale.

SOLUTION

Step 1 Preliminary computations to obtain e and relative density of the sand:

Estimate $G_s = 2.68$. From the gravimetric-volumetric relationships in Chap. 2, derive for γ, in grams per cubic centimeter,

$$\gamma_{dry} = \frac{G_s(\gamma_{sat} - 1)}{G_s - 1} = \frac{2.68(1.937 - 1)}{2.68 - 1}$$

$$= 1.495 \text{ g/cm}^3$$

Obtain the volume of solids and voids as

$$V_s = \frac{1.495}{2.68} = 0.558 \qquad V_v = 1 - 0.558 = 0.442$$

and the void ratio is

$$e = \frac{V_v}{V_s} = \frac{0.442}{0.558} = 0.792$$

To compute the relative density D_r using Eq. (5-1), we must estimate a minimum and maximum void ratio. Assume these as 0.4 and 1.2, respectively, and

$$D_r = \frac{e_{max} - e_n}{e_{max} - e_{min}} = \frac{1.2 - 0.792}{1.2 - 0.4} = 0.51 \text{ or 51 percent}$$

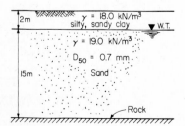

Figure E14-4

Step 2 Find the s_e/p_o ratio. Compute the average effective vertical pressure p_o to the middle of the sand stratum as

$$p_o = 2(18.0) + 7.5(19.0 - 9.807) = 105 \text{ kPa}$$

We see that this is between 100 and 200 kPa of the test data ranges used to develop Figs. 14-9 and 14-10. Now using Eq. (14-14), and with reference to the inset in Fig. 14-7, we compute the estimated earthquake shear s_e

$$m = \frac{p_o(1000)}{9.807} = \frac{105(1000)}{9.807} = 18\,201 \text{ kg/m}^2$$

We must estimate the earthquake acceleration: assume that

$$a = 0.1 \text{ g} = 0.1(9.807) = 0.981 \text{ m/s}^2$$

From Fig. 14-7 estimate $r_d = 0.92$ at $h = 6$ to 9 m. Now compute the s_e/p_o ratio:

$$s_e = 0.65 r_d ma = 0.65(0.92)(18\,201)(0.981) = 10\,700 \text{ kg·m/(s}^2 \cdot \text{m}^2)$$

For units: $10\,700 \text{ kg·m/(s}^2 \cdot \text{m}^2) = 10\,700 \text{ N/m}^2 = 10.7 \text{ kN/m}^2$ and $s_e/p_o = 10.7/105 = 0.10$.

Step 3 Find the safety factor:

(a) For earthquake magnitude $= 8$ obtain $N_e \cong 20$ cycles from Fig. 14-8.
(b) From Fig. 14-9 with $N_e = 20$ and $D_r = 0.51$, obtain $R_c = 0.25$ at $\varepsilon \cong 2.5$ percent. This strain level is arbitrary but does not greatly change R_c if 10 percent were selected.
(c) From Fig. 14-10 at $D_{50} = 0.7$ mm, interpolate between range lines to obtain the factor 1.06 relating the given grain size to the $R_{0.2}$ on which the table for R_c was based.

$$R_c = R_{c(0.2)} \times 1.06 = 0.25 \times 1.06 = 0.27$$

Note that this value of R_c is equivalent to $\Delta\sigma_1/2\sigma_3$ in Eq. (14-13).
(d) Now compute the safety factor using Eq. (14-13) with $C_r = 0.57$:

$$F = \frac{C_r R_c}{s_e/p_o} = \frac{0.57(0.27)}{0.10} = 1.54 \qquad \text{(which appears adequate)}$$

If F were on the order of 1.1 to 1.2, it would be necessary to reevaluate the several assumptions and/or perform cyclic tests on undisturbed soils. If this is not possible, consider a grout wall around the site to maintain the exterior water table to satisfy environmental concerns and dewater the interior to circumvent the problem of liquefaction.

////

14-8 SUMMARY

This chapter has introduced means of computing stress-strain moduli and Poisson's ratio for soils.

Both static, as in conventional triaxial tests, and dynamic stress-strain moduli are considered, and the reader should be aware that "static" values at low strains may be less than 50 percent of the dynamic values. At larger static or dynamic strains (as "residual" strength or liquefaction is developed) $E_s \to 0$.

The concept of liquefaction as used in earthquake studies was introduced. Cyclic shear tests as used to evaluate earthquake resistance were presented. This chapter has introduced the reader to considerable empirical data, compared with earlier chapters. This is due to the highly uncertain nature of earthquake intensity, the number of stress cycles likely to be introduced, and the mass, or quantity, of soil involved. These are coupled with the problems of relating laboratory prepared samples to field conditions with regard to methods of sample preparations, degree of saturation, confining pressure, type of soil, equipment limitations, accuracy of test simulation, and interpretation of data.

In general, however, it has been found that

1. Cyclic stresses that will induce liquefaction are much smaller than those required under static loading conditions.
2. The frequency is not particularly critical in that frequencies of $\frac{1}{6}$ to 10 Hz produce essentially the same results.
3. Cyclic stresses will produce liquefaction over a considerable range of unit weight γ.
4. When sands liquefy under cyclic stresses, the deformations immediately become very large. This causes some problems in the laboratory in determining "failure."
5. Partial liquefaction of dense sand produces a condition of near zero stress at low strains, but if the strain is increased, an appreciable resistance is recovered.
6. The higher the stress ratio, the smaller the number of cycles to liquefaction.
7. The lower the confining pressure, the lower the number of cycles required to cause liquefaction.

PROBLEMS

14-1 What is the approximate initial tangent modulus in Example 14-1?
    Ans.: $E_s \cong 25\,000$ kPa

14-2 What is the secant modulus using the two points in Example 14-1?
    Ans.: $E_s \cong 17\,800$ kPa.

14-3 Using the normalized curve (b) from Fig. 13-27, compute the stress-strain modulus at a strain of 1 percent for $\sigma_3 = 100$ and 200 kPa.
    Ans.: At 100 kPa, $E_s = 7500$ kPa

14-4 Plot the stress-strain data of Prob. 13-11, tests 1 and 3, and compare the initial tangent moduli. Compare the resulting shear modulus G for several arbitrary values of Poisson's ratio μ.

14-5 Plot the stress-strain data of Prob. 13-11, tests 4 and 6, and compare the initial tangent moduli.

14-6 Plot the assigned test data of Prob. 13-11, and find the strain at which Poisson's ratio μ is either $(-)$ or >0.50. Comment on the strain level at which this occurs. Be sure to use a large enough scale or enlarge the initial part of the graph. Also, use the smooth curve and not point-to-point of data.

14-7 Redo Ex. 14-4 for N_e at $+1$ standard deviation ($+1\alpha$), instead of the mean.

14-8 Redo Ex. 14-4 for N_e at -1 standard deviation (-1α), instead of the mean.

14-9 Redo Ex. 14-4 if the sand density is increased by vibroflotation to 20.5 kN/m^3, but all other data are the same.

    Ans.: $F \cong 3.0^+$

14-10 Redo Ex. 14-4 and plot F versus ground acceleration a. Use values of $a = 0.05$ to 1.0. If the earthquake intensity and site are in your area, what is the F value?

14-11 For the hysteresis loop shown in Fig. P14-11, compute the dynamic stress-strain modulus and indicate whether it is G or E and the damping factor D.

    Ans.: Modulus = 2300 kPa, $D = 0.17$.

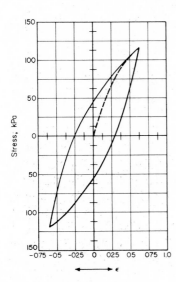

Figure P14-11

14-12 For the hysteresis loop shown in Fig. P14-12, compute the dynamic stress-strain modulus and the damping factor.

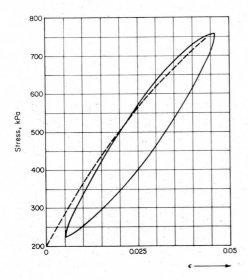

Figure P14-12

LATERAL PRESSURES, BEARING CAPACITY, AND SETTLEMENT

15-1 INTRODUCTION

Principal areas of geotechnical engineering involves estimation of

1. Lateral pressures against walls used to retain earth, coal, grain, etc. as well as providing lateral support around the perimeters of excavations.
2. Bearing capacity—or the allowable pressure beneath foundations to avoid a shear failure.
3. Settlements—including consolidation and creep described in Chaps. 11 and 12 as well as "elastic" or immediate settlements.

There are several other problem areas, but this chapter will briefly focus on these three as they have the greatest application. A more complete treatment of these and other foundation problems is found in texts and handbooks on foundation engineering such as Bowles (1982), Winterkorn and Fang (1975), and Peck et al. (1974).

15-2 SOIL STRESSES AT A POINT—K_o CONDITIONS

Soil formed into a residual or sedimentary deposit produces a column of soil over any element, as in Fig. 15-1a. The vertical pressure is $\sigma_v = p_o = \gamma h$, as shown.

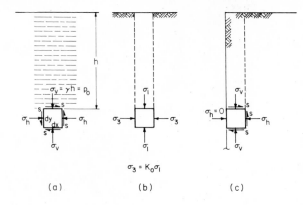

Figure 15-1 Development of in situ stresses: (a) stresses during deposition and formation of deposits; (b) stable condition stresses; (c) stresses at face of excavation.

During the formation of the deposit, the element will be consolidating under the pressure σ_v. The vertical stress produces a lateral flow into the surrounding soil due to the Poisson's ratio effect. The surrounding soil resists the lateral flow effect with a developed lateral stress σ_h. Over geological periods, consolidation, and both vertical and lateral creep strains, will become zero. At this time a stable stress state will develop in which σ_h and σ_v will become principal effective stresses, since zero displacements will produce zero shear stresses on the vertical and horizontal planes defining the soil element. The equilibrium in situ condition produced at this stress state is commonly termed the K_o condition.

The ratio of the vertical to lateral pressure has been considered in both Chaps. 10 and 13 using the symbol K which was defined as

$$K = \frac{\sigma_h}{\sigma_v} \qquad (a)$$

The K_o condition, in particular, was a principal stress (or steady) state where

$$K_o = \frac{\sigma'_h}{\gamma' h} = \frac{\sigma'_h}{p_o} \qquad (15\text{-}1)$$

Here the *effective* stresses are used as compared to Eq. (*a*) above, where K defines any lateral/vertical stress ratio. Note that the slope α of the K_f or K_o line obtained from stress path analyses is not K but is, according to Eq. (13-7),

$$\tan \alpha = \frac{1 - K_i}{1 + K_i}$$

In many cases $K_o < 1$ but is generally as follows:

$\quad K_o < 1$ for *normally* consolidated soil

$\quad K_o < 1$ for *overconsolidated* soil $\left(\text{OCR} = \dfrac{p_c}{p_o} < 3 \text{ approximately} \right)$

$\quad K_o > 1$ for *overconsolidated* soil (OCR > 3 approximately)

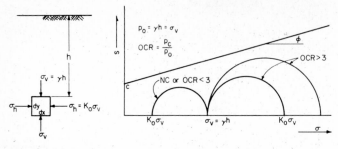

(a) Stressed point in situ.   (b) Mohr's circles for in situ stress conditions.

Figure 15-2 Qualitative representation of in situ stresses at a point.

The K_o condition can be illustrated by the several Mohr's circles, as in Fig. 15-2, for conditions of both normally and overconsolidated (OC) soils. The in situ stress conditions represent some elastic equilibrium state, since with no displacements, the soil may be considered an elastic continuum. The application of additional stresses will produce changes in the size of Mohr's circle, and when the shear stresses are of sufficient magnitude, the soil fails. Figure 15-3 illustrates the failure condition as sufficient additional stresses are imposed on the soil mass.

The determination of K_o by measuring σ_h in situ is nearly impossible, since it is irrecoverably lost when a cavity is excavated alongside the element. It is altered even when one is close, as in Fig. 15-1c, where $\sigma_h = 0$ on the left side. This is true no matter how carefully one were to place a measuring device into the cavity. Wroth (1975) reports on a device which removes the soil and simultaneously self-inserts into the cavity; however, this may not completely recover the in situ stress σ_h due to excavation disturbance. The next section gives an analytical reason for loss of σ_h and not being able to "push" the soil back into place. On a nonanalytical basis, however, the reader should be aware that due to the particulate nature of the material, any displacement produces a new structure (material) and interparticle stresses.

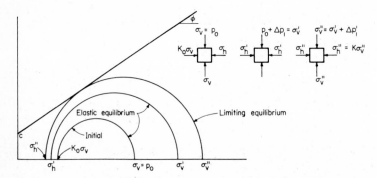

Figure 15-3 Mohr's circle representation of soil stress changes from K_o to the limiting equilibrium state as the vertical stress is increased from p_o to $p_o + \Delta p_1$. In design, Δp_1 should produce σ_v', not σ_v''. Note that there is both a decrease in lateral pressure and an increase in vertical pressure.

Based on observations of grain pressure in silos, Jaky (1948), and later Brooker and Ireland (1965), using a large series of laboratory tests on five clay soils, suggested the following equation for the in situ lateral earth pressure:

$$K_o = M - \sin \phi' \qquad (15\text{-}2)$$

where $M = 1$ for normally consolidated, cohesionless and cohesive soils; also for the lateral pressures developed in piles of grain, such as corn, barley, wheat, etc.

$\quad = 0.95$ for overconsolidated clays on the order of $OCR > 2$

$\phi' = $ *effective* angle of internal friction

More recently Sherif and Ishibashi (1981) have related K_o to the liquid limit using factors α and λ. These factors were given in chart form but the following equation computes them with sufficient precision.

$$K_o = \lambda + \alpha(OCR - 1) \qquad (15\text{-}3)$$

where $\lambda = 0.54 + 0.00444(w_L - 20)$

$\quad \lambda = 1.0$ for $w_L > 110\%$

$\quad \alpha = 0.09 + 0.00111(w_L - 20)$

$\quad \alpha = 0.19$ for $w_L > 110\%$

$OCR = $ overconsolidation ratio p_c/p_o defined earlier

15-3 ACTIVE AND PASSIVE EARTH PRESSURES

The concept of active and passive earth pressure is of particular importance in soil stability problems, bracing of excavations, design of retaining walls, and development of pullout resistances using various types of anchorage devices. Consider first the statics of a block of material on a plane at a slope angle of ρ, as in Fig. 15-4. The friction coefficient between the block and plane is v, defined in the figure. Consider the three cases shown as follows:

Case I $P_h = 0$. From summing forces parallel to the plane, obtain

$$T - vN = 0$$

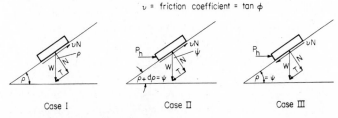

$v = $ friction coefficient $= \tan \phi$

Case I   Case II   Case III

Figure 15-4 Block sliding down a plane to develop the concept of active, at rest, and passive earth pressure.

and substituting $v = \tan \phi$ and the W components for T and N,

$$W \sin \phi - W \cos \rho \tan \phi = 0$$

from which obtain

$$\tan \rho = \tan \phi$$

or the block is on the verge of slip when the slope of the plane is $\rho = \phi$. Note the similarity of this with Fig. 5-1.

Case II. When the slope angle is $\psi = \rho + \Delta\rho$. Again summing forces parallel to the plane and with the block just restrained against sliding down the plane with the external force P_h, we have

$$P_h \cos \psi + W \cos \psi \tan \phi - W \sin \psi = 0$$

from which obtain

$$P_h = W(\tan \psi - \tan \phi) \qquad (\psi \geq \phi)$$

In this case P_h will be some minimum value, since the friction resistance is aiding P_h to restrain the block.

Case III. The slope angle is the same as in Case II, but we wish to push the block up the plane. Summing forces parallel to the plane, obtain

$$P_h \cos \psi - W \cos \psi \tan \phi - W \sin \psi = 0$$

from which

$$P_h = W(\tan \psi + \tan \phi) \qquad (\psi \geq 0)$$

In this case P_h is a maximum, since friction resistance as well as the tangential component of the weight vector must be overcome to have a movement up the plane.

These three cases in a soil represent approximately:

Case I. K_o conditions.

Case II. Active earth pressure conditions—note that to develop the limiting friction, there must be a slight movement down the plane.

Case III. Passive earth pressure conditions—note that to develop the limiting resisting friction, there must be a slight movement up the plane. In soil the angle ψ changes in value from the active to the passive case.

Now let us investigate an impossible condition of the insertion of a frictionless wall of substantial rigidity and zero volume into a cohesionless soil mass along the vertical plane AB in Fig. 15-5a, then excavate the soil from the left side of the wall as shown in Fig. 15-5b while simultaneously fixing the wall against any movement whatsoever. The stresses acting against the wall at this point would be K_o stresses, since the conditions imposed have not produced any strains

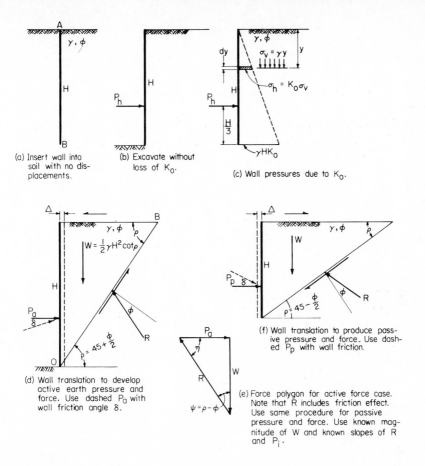

Figure 15-5 Development of active and passive earth pressures in a cohesionless soil mass.

in the soil mass behind the wall. The wall force necessary to hold the soil/wall system in plane is a K_o force, and from Fig. 15-5c, obtain

$$\sigma_h = K_o \gamma y$$

The wall force is the summation of the unit stresses over the differential of area, to obtain

$$P_h = \int_0^H \sigma_h \, dy = \int_0^H K_o \gamma y \, dy$$

$$= \tfrac{1}{2} H^2 K_o \tag{b}$$

This is the force necessary to hold the wall in place against any lateral movement. Since the pressure diagram is triangular, the force P_h acts through the centroid of the pressure area or at the $H/3$ location from the bottom of the wall, as shown.

If the wall is allowed to translate into the excavation with sufficient soil strains that the full frictional resistance is mobilized and without destroying the soil structure, the force P_h will become some minimum value (somewhat as Case II earlier). The approximate failure zone will develop behind the wall, as shown in Fig. 15-5d. Friction will be mobilized along the plane OB and in the direction shown and will be a limiting value, since some slip has occurred.

The forces on the sliding wedge of Fig. 15-5d produce the force polygon of Fig. 15-5e. Solving for $P_a = P_h$, obtain

$$P_a = \tfrac{1}{2}\gamma H^2 \cot \rho \tan (\rho - \phi) \tag{c}$$

The maximum value of P_a is obtained from $dP_a/d\rho = 0$, from which we can obtain

$$\rho = 45° + \frac{\phi}{2}$$

With this value of ρ in Eq. (c), and substitution of $\cot (45 + \phi/2) = \tan (45 - \phi/2)$, obtain

$$P_a = \frac{1}{2}\gamma H^2 \tan^2 \left(45 - \frac{\phi}{2} \right) \tag{15-4}$$

which may be written as

$$P_a = \tfrac{1}{2}\gamma H^2 K_a \tag{15-4a}$$

Since $P_a = \displaystyle\int_0^H \sigma_a \, dy$ as in Eq. (b), the *active* earth pressure is

$$\sigma_a = \gamma y \tan^2 \left(45 - \frac{\phi}{2} \right) = \gamma y K_a \tag{15-5}$$

A similar analysis for passive pressure, as when the wall is forced into the soil (Fig. 15-5f), gives

$$\rho = 45° - \frac{\phi}{2}$$

and for P_p,

$$P_p = \frac{1}{2}\gamma H^2 \tan^2 \left(45 + \frac{\phi}{2} \right) = \frac{1}{2}\gamma H^2 K_p \tag{15-6}$$

and the passive earth pressure is

$$\sigma_p = \gamma y \tan^2 \left(45 + \frac{\phi}{2} \right) = \gamma y K_p \tag{15-7}$$

Note that the active earth pressure is a condition of loosening strains where the friction resistance is mobilized to reduce the force necessary to hold the soil in position. Passive earth pressure is a condition of densifying the soil by a lateral movement into the soil mass, with the friction mobilized to increase the force necessary to cause strain. Note also that the failure slope angle decreases by ϕ over the active case, producing a much larger failure wedge. For $\phi = 30°$, the possible range of earth pressures is as follows:

Earth pressure	Symbol	Computed as	K Coefficient
Active	K_a	$\tan^2\left(45 - \dfrac{\phi}{2}\right)$	0.333
At rest	K_o	$1 - \sin\phi$	0.50
Passive	K_p	$\tan^2\left(45 + \dfrac{\phi}{2}\right)$	3.000

Thus, passive earth pressures are on the order of $K_p \rightarrow 10K_a$.

Observations and model wall tests indicate that the computations are valid for K_o and K_a pressures and that the failure slope is nearly planar, as in Fig. 15-5.

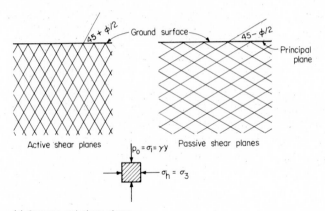

(a) Stresses and shear planes.

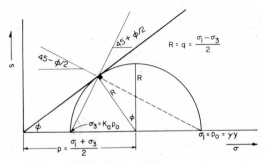

(b) Mohr's circle to orient shear planes.

Figure 15-6 Rankine stress state using Mohr's circle.

The K_p values tend to be too large, especially when wall geometry differs from that used in Fig. 15-5 when ϕ is larger than, say, about 38°. Available evidence indicates that for a sloping backfill and/or a rough wall where significant friction resistance develops, the failure surface is partly curved near the wall base. In general, the ground surface being a principal plane, the failure surface angle is $45 + \phi/2$ for the active and $45 - \phi/2$ for the passive condition (see Fig. 15-6a).

In terms of stress paths, active pressure corresponds to path $O2$ in Fig. 13-10 and passive pressure corresponds to stress path $O4$.

15-4 PRESSURES AGAINST WALLS

There are two common methods of computing lateral earth pressures against walls. The earliest analytical solution was the Coulomb method, developed by C. A. Coulomb in 1776. Later Rankine (ca. 1857) proposed a procedure for cohesionless soils which is actually a simplification of the Coulomb method. Equations (15-4) and (15-6) describe both the Rankine and Coulomb methods for horizontal ground surfaces; dry, cohesionless soils; and smooth walls. No wall friction is developed between a smooth wall and soil during wall movement; thus, the wall pressure becomes a principal stress. Only the Rankine method will be considered further in this text due to its simplicity and widespread use.

The Rankine lateral earth pressure can be studied using Mohr's circle. Consider Fig. 15-6a, where an element of soil is acted on by $p_o = \gamma y$ and is in a state of incipient failure. The "active state" shear planes are as shown (also shown are shear planes for the passive state for reference). This orientation can be obtained from Mohr's circle at an angle of $45 + \phi/2$ with the horizontal as in Fig. 15-6b.

The major effective principal stress is $p_o = \gamma y$ and the minor principal stress is $\sigma_3 = K_a p_o$, and the active earth pressure coefficient is

$$K_a = \frac{\sigma_3}{\sigma_1} = \frac{\sigma_3}{p_o}$$

Also using p and q as defined in Sec. 13-5, obtain

$$\sin \phi = \frac{q}{p} = \frac{(\sigma_1 - \sigma_3)}{(\sigma_1 + \sigma_3)}$$

With ϕ known, using γy for p_o, and with some rearranging, obtain

$$\sigma_a = \sigma_3 = \gamma y \frac{1 - \sin \phi}{1 + \sin \phi} = \gamma y \tan^2 \left(45 - \frac{\phi}{2} \right)$$

With passive pressure, the values of ϕ and $\sigma_3 = \gamma y$ are also known, and rearranging, obtain

$$\sigma_p = \sigma_1 = \gamma y \tan^2 \left(45 + \frac{\phi}{2} \right)$$

Note again that the shear planes intercept the ground surface (principal plane) at $45 + \phi/2$ for the active earth pressure condition and $45 - \phi/2$ for the passive pressure condition.

Alternative values for K_p are sometimes used, including those of Caquot and Kerisel (1948), Sokolovski (1965), and more recently Rosenfarb and Chen (1972), which are somewhat smaller than the values here when the backfill slope β of Fig. 15-7 is $> 15°$, wall friction δ is included, and $\delta > \phi/2$ and $\phi > 36$ to $38°$. Generally, unless all these conditions are met the Rankine value for K_p is satisfactory to use. The Rosenfarb and Chen values are readily available in Bowles (1982).

15-5 INCLINED COHESIONLESS GROUND

With inclined ground the Rankine method considers static equilibrium of an element at a depth y. The soil weight acts vertically, and the lateral earth pressure is conjugate to the weight as in Fig. 15-7a. Thus, the lateral earth pressure acts parallel to the ground surface. Since the Rankine method assumes a frictionless

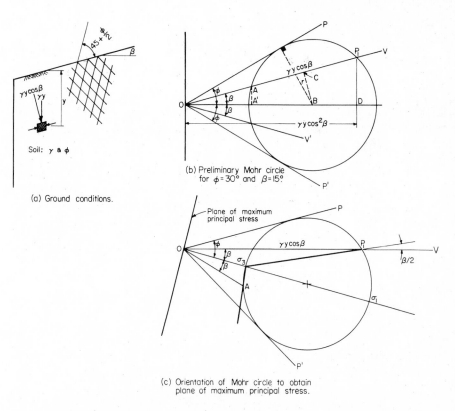

(a) Ground conditions.

(b) Preliminary Mohr circle for $\phi = 30°$ and $\beta = 15°$.

(c) Orientation of Mohr circle to obtain plane of maximum principal stress.

Figure 15-7 Using Mohr's circle to obtain Rankine earth pressure for an inclined ground surface.

wall the stresses on the vertical faces of the element are principal stresses. Rankine made an analytical solution of this case to obtain

Active pressure:

$$K_a = \cos \beta \, \frac{\cos \beta - \sqrt{\cos^2 \beta - \cos^2 \phi}}{\cos \beta + \sqrt{\cos^2 \beta - \cos^2 \phi}} \qquad (15\text{-}8)$$

Passive pressure:

$$K_p = \cos \beta \, \frac{\cos \beta + \sqrt{\cos^2 \beta - \cos^2 \phi}}{\cos \beta - \sqrt{\cos^2 \beta - \cos^2 \phi}} \qquad (15\text{-}8a)$$

Inspection of Eq. (15-8) indicates that the maximum stable slope occurs when $\beta = \phi$ for a cohesionless soil. When $\beta = 0$ the above gives K_a from Eq. (15-5) or K_p from Eq. (15-7).

The active earth pressure can be obtained directly from Mohr's circle as follows:

1. Draw a set of orthogonal axes and locate OV and OV' as shown in Fig. 15-7b.
2. Compute $\gamma y \cos \beta$ and locate point P_1 along OV as shown.
3. Lay off lines OP and OP' at $\pm \phi$ as shown. At this point we do not know enough data to directly construct a Mohr's circle. However, since we recognize that the circle must be tangent to OP and OP' and pass through point P_1 to satisfy the stress condition and shear strength simultaneously, we may continue.
4. By trial find a circle tangent to OP and OP' and through point P_1.
5. Construct a new set of axes as in Fig. 15-7c. Line OV is horizontal, with point P_1 located using $\gamma y \cos \beta$ as in step 2. Lay off lines $O\sigma_1$ at β to OV and locate OP and OP' at $\pm \phi$ from $O\sigma_1$. Also locate vector OA at β from $O\sigma_1$ as shown. Note that $\sigma_1 = \gamma y$.
6. Scale the circle center from step 4 and locate it on $O\sigma_1$. Use radius from step 4 and draw circle as shown.
7. Scale σ_a as the distance OA. The slope of $\sigma_3 P_1$ is $\beta/2$ and slope of a line through σ_3 and σ_a deviates from the vertical by $\beta/2$.

The potential shear (or slip) planes are inclined to the ground surface at $45 \pm \phi/2$, as shown in Fig. 15-7a.

An analytical solution can be obtained for the Rankine equations using Mohr's circle as follows:

Referring to Fig. 15-7b, K_a is

$$K_a = OA' = OA \cos \beta \qquad (d)$$

The conjugate stresses are

$$K = \frac{OA}{OP_1} = \frac{OC - CA}{OC + CP_1} \qquad (e)$$

But the following relations also can be obtained from Fig. 15-7b:

$$OC = OB \cos \beta$$

$$CA = CP_1 = \sqrt{r^2 - (BC)^2} \qquad \text{(since } BC \perp \text{ to } OP_1)$$

$$BC = OB \sin \beta$$

$$r = OB \sin \phi$$

Substituting these values into Eqs. (e) and (d), and using $\sin^2 \phi = 1 - \cos^2 \phi$, we obtain Eq. (15-8). Notice, however, that we have ignored the shear stress AA' and P_1D, introducing a small approximation into the solution.

Example 15-1 Given a cohesionless soil, $\phi = 32°$; $\gamma = 17.5$ kN/m^3.

REQUIRED What is the safe slope angle for an excavation?

SOLUTION From Eq. (15-8), we note that the $\sqrt{}$ term is negative for values of $\beta = \rho > 32°$. From Fig. 15-4 we note that in Case I the maximum value of $\rho = \phi$. Therefore the maximum safe slope is

$$\rho = 32°$$

////

Example 15-2 Given a wall in cohesionless soil; $\phi = 34°$, $\beta = 10°$, $\gamma = 17.9$ kN/m^3, $H = 4$ m.

REQUIRED What are the active and passive wall forces?

SOLUTION Use the Rankine equations [Eqs. (15-8), (15-8a)].

$$K_a = \cos 10 \, \frac{\cos 10 - (\cos^2 10 - \cos^2 34)^{1/2}}{\cos 10 + (\cos^2 10 - \cos^2 34)^{1/2}} = 0.2944$$

$$K_p = \cos 10 \, \frac{\cos 10 + (\cos^2 10 - \cos^2 34)^{1/2}}{\cos 10 - (\cos^2 10 - \cos^2 34)^{1/2}} = 3.2946$$

The active wall force is

$$P_a = \int_0^H \gamma h K_a \, dh = \tfrac{1}{2}\gamma H^2 K_a = 42.2 \text{ kN/m of wall width}$$

Alternatively from Fig. E15-2, $P_a = $ area of pressure diagram $= \tfrac{1}{2}H(\gamma H K_a)$. The passive pressure is

$$P_p = \int_0^H \gamma h K_p \, dh = \tfrac{1}{2}\gamma H^2 K_p = 471.8 \text{ kN/m}$$

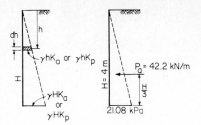

Figure E15-2

The location is at the one-third height from base, since the pressure diagram is triangular as in Fig. E15-2. The location from the top of the wall is:

$$P_a \bar{y} = \int_0^H \gamma h K_a h\ dh = \frac{\gamma H^3}{3} K_a$$

Substituting for $P_a = \frac{1}{2}\gamma H^2 K_a$, we obtain

$$\bar{y} = \frac{\gamma H^3 K_a/3}{\gamma H^2 K_a/2} = \frac{2H}{3} \qquad \text{from top of wall}$$

////

15-6 LATERAL EARTH PRESSURE FOR COHESIVE SOILS

Bell (1915), working with clay soils and the Rankine and Coulomb equations, recognized that Mohr's circle could be used to obtain Eqs. (15-5) and (15-7), which are either the Rankine or Coulomb equations for a smooth wall, cohesionless soil, and horizontal backfill.

Using Mohr's circle in a similar manner for a cohesive soil, one can, through some trigonometric manipulations, obtain:

Active case:

$$\sigma_a = \gamma h \tan^2\left(45 - \frac{\phi}{2}\right) - 2c \tan\left(45 - \frac{\phi}{2}\right) \tag{15-9}$$

Passive case:

$$\sigma_p = \gamma h \tan^2\left(45 + \frac{\phi}{2}\right) + 2c \tan\left(45 + \frac{\phi}{2}\right) \tag{15-9a}$$

Inspection of Eq. (15-9) indicates that a vertical excavation can be made in a cohesive soil (it is impossible in cohesionless soil) as in Fig. 15-8 according to the following:

At the ground surface, $\sigma_v = \gamma h = 0$, and

$$\sigma_a = -2c \tan\left(45 - \frac{\phi}{2}\right) = -2c\sqrt{K_a}$$

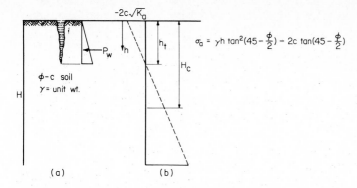

Figure 15-8 Excavation in a cohesive soil: (*a*) effect of a tension crack filling with surface water; (*b*) theoretical pressure diagram for active pressure conditions.

At the point h_t, we have $\sigma_a = 0$, and

$$h_t = \frac{2c}{\gamma\sqrt{K_a}}$$

The theoretical maximum excavation depth H_c is obtained where $P_a = 0$ as

$$P_a = 0 = \int_0^{H_c} \sigma_a \, dh = \frac{4c}{\gamma\sqrt{K_a}} = 2h_t$$

Since the soil will be in tension through the depth h_t, this depth may, and often does, form tension cracks which can be readily observed along and near the edges of vertical cuts (see Fig. 8-13*a*).

The theoretical depth H_c is not used for design due to the possibility of tension cracks into which rainwater can enter, reducing the cohesion as well as causing an additional lateral force from hydrostatic pressure. There is usually a reduction in cohesion when clay is exposed due to various environmental factors, which also reduces the theoretical depth H_c.

The lateral force/unit width against a wall in cohesive soil is computed as

$$P_i = \int_0^H \sigma_i \, dh$$

For the active earth pressure case, this produces

$$P_a = \tfrac{1}{2}\gamma H^2 K_a - 2cH\sqrt{K_a}$$

The lateral force may be computed directly from the pressure diagram using the appropriate equations for pressure areas. A typical active earth pressure diagram is shown in Fig. 15-8*b*.

Example 15-3 Given cohesive soil, $\gamma = 18.0$ kN/m³; undrained shear strength $s_u = c = 30$ kPa; wall height = 4 m.

REQUIRED Wall pressure profile and the location and magnitude of active force.

SOLUTION For undrained shear conditions, $\phi = 0$ and

$$K_a = \tan^2 (45 - 0) = 1.00 \qquad \text{(same as water)}$$

At top of wall,

$$\sigma_a = -2c = -60 \text{ kPa}$$

At base of wall,

$$\sigma_a = \gamma H - 2c = (18)(4) - 2(30) = 12 \text{ kPa}$$

$$h_t = \frac{2c}{\gamma K_a} = \frac{60}{18} = 3.333 \text{ m} \qquad y' = \frac{12}{18} = 0.667 \text{ m}$$

These data are shown on Fig. E15-3.

$$P_a = \text{area of pressure diagram}$$

$$= \frac{-60(3.333)}{2} + 12\left(\frac{0.667}{2}\right)$$

$$= -96 \text{ kN/m}$$

Alternatively, integrate Eq. (15-9) to obtain

$$P_a = \tfrac{1}{2}\gamma H^2 K_a - 2cH$$

$$= \tfrac{1}{2}(18)(4)^2 - 2(30)(4)$$

$$= -96 \text{ kN/m}$$

The location of P_a is found by summing moments about point O of the figure.

$$-96\bar{y} + 4\left(\frac{0.667}{3}\right) - 100(2.889) = 0$$

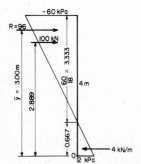

Figure E15-3

Clearing up and solving for $\bar{y}$ obtain

$$\bar{y} = -\frac{288}{96} = -3.0 \text{ m} \qquad \text{(below point } O)$$

Since this type of resultant is meaningless it is common to ignore the "tension" zone so that $P_a = 4$ kN located at $\bar{y} = 0.667/3$ above O.

////

Example 15-4 Given soil in Example 15-3.

REQUIRED What is the depth of tension crack? What is the theoretical depth of unbraced excavation?

SOLUTION As shown on Fig. E15-3,

$$h_t = \frac{2c}{\gamma} = 3.333 \text{ m}$$

Note that $\sigma_a = 0$ at this point.

$$H_c = 2h_t = 2(3.333) = 6.67 \text{ m}$$

It may be possible (but OSHA regulations may disallow this) to excavate, say,

$$H_e = \frac{H_c}{F} = \frac{H_c}{2} = \frac{6.7}{2} = 3 \text{ m}$$

without bracing if the cut is very temporary and if no heavy material is stored along the bank.

////

15-7 THE TRIAL WEDGE SOLUTION

The force polygon in Fig. 15-5e is a graphical solution for the *trial wedge* of Fig. 15-5d if the weight vector is drawn to scale and the angles η and ψ are measured. The maximum value of P_a or P_p can be obtained from several trial failure wedges defined by the wall, backfill slope, and trial failure slope angle ρ.

The more general trial wedge for either cohesionless or cohesive soil, including backfill slope, tension crack, wall friction, and cohesion effect on the wall (termed *adhesion*), is shown in Fig. 15-9. It is necessary to make several trials to obtain the maximum value of P_i, as in Fig. 15-9c for P_a. This general case may be used to obtain a solution for irregular backfill surfaces.

The wall adhesion c_a is some value intermediate between c and, say, $0.6c$, depending on an engineering assessment of how well the soil adheres to the wall. The soil-to-soil interface uses the values of ϕ and c. The wall angle δ may be obtained from tables found in Bowles (1982) or elsewhere. For rough concrete or

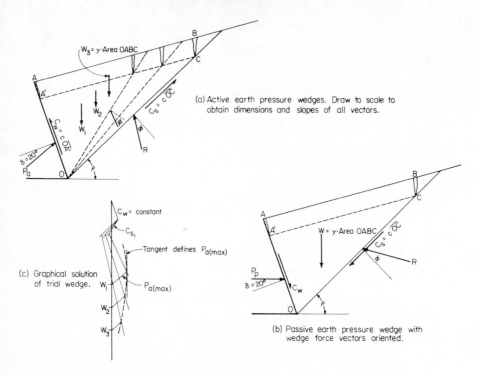

(a) Active earth pressure wedges. Draw to scale to obtain dimensions and slopes of all vectors.

(c) Graphical solution of trial wedge.

(b) Passive earth pressure wedge with wedge force vectors oriented.

Figure 15-9 General trial wedge solution for any soil.

masonry walls, $\delta \to \phi$; for wood or steel sheet pile walls, δ ranges from about 15 to 26°. It may be measured in the laboratory by a direct shear type test, i.e., pulling a piece of the material across the soil and simultaneously applying a normal pressure.

The trial wedge may be programmed for the digital computer to compute the weight vector and values of C_s, C_w, and h_t. The R and P_i vectors can be solved by a pair of simultaneous equations, since $\sum F_h$ and $\sum F_v$ must be satisfied for closure of the force polygon. By incrementing ρ in intervals of, say, 1°, a satisfactory maximum value of P_i can be readily found.

The trial wedge for cohesionless soils is considerably simpler than that for cohesive materials. In these soils both C_s and C_w are zero and there is no tension crack.

15-8 LOGARITHMIC SPIRAL AND ϕ-CIRCLE METHODS FOR PASSIVE PRESSURE IN COHESIONLESS SOIL

It has been shown by several investigators that the slope of at least a part of the slip surface for *passive pressure* is part of a logarithmic spiral for an ideal plastic material. This may be true also for active pressure, but the difference—if any—is so small that no significant error is introduced by considering a plane slip surface.

The equation of the logarithmic spiral as used in soil mechanics work is

$$r = r_0\, e^{\theta \tan \phi}$$

where terms are identified in Fig. 15-10a and ϕ = angle of internal friction of the soil. For $\phi = 0$ soils, the spiral becomes a circle. A circle is sometimes assumed as a computational convenience for other soils. Regardless of the assumed shape of the lower part of the failure surface, the upper part is an approximate Rankine solution with a ground surface (principal plane) intersection of $45 - \phi/2$ and with the maximum principal stress as shown in Fig. 15-10b.

The logarithmic spiral method is illustrated in Fig. 15-10b. The steps to obtain the passive pressure include:

1. Select a trial origin O with $OB = r_0$. At A lay off a line AD at an angle with the ground surface of $45 - \phi/2$ as shown.

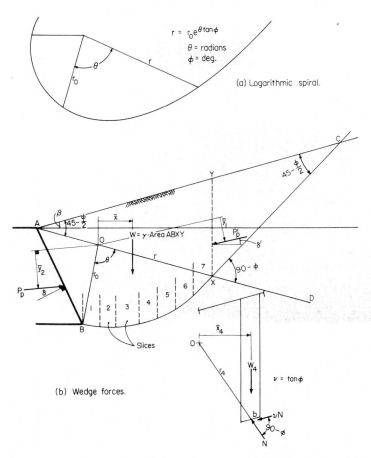

(a) Logarithmic spiral.

(b) Wedge forces.

Figure 15-10 Passive earth pressure using the logarithmic spiral as a part of the assumed failure surface.

2. Extend a trial spiral from B to intersect line AD at some point X.
3. At X and tangent to the spiral, lay off XC such that the ground surface is intersected with an angle of $45 - \phi/2$, as shown. Readjust point O and the spiral orientation until these conditions are met. Draw vertical XY to ground surface.
4. Compute the passive resistance of wedge XYC on the curved block using the Rankine method and apply at the one-third point as shown.
5. Subdivide $ABXY$ into four to six strips which can be treated as trapezoids, as shown for element 4 in Fig. 15-10b.
6. For each trapezoidal element, compute W_i and scale the moment arm $\bar{x}_i$ and r_i. Also compute the friction resistance as $N \tan \phi$. Note that the normal force N acts at $90 - \phi$ to the tangent of the spiral at any point (including X). Since this direction passes through the spiral origin, no rotational moment is caused.
7. Scale the moment arms of P_p and P'_p as $\bar{y}_2$ and $\bar{y}_1$ with respect to point O.
8. Find $\sum M_o = O$ and obtain

$$P_p \bar{y}_2 - \sum [W_i \bar{x}_i + (vN)r_i] - P'_p \bar{y}_1 = 0$$

from which P_p may be readily obtained.

Alternatively, one may use a circular arc for the part BX with little loss of accuracy, as shown in Fig. 15-11a. The use of a circular arc greatly simplifies the work as follows:

1. Draw wall system AB and ground surface AC. Note that AC may slope as in Fig. 15-10 but is horizontal here for simplicity.
2. Lay off line AD at $45 - \phi/2$ to ground surface. Also select a trial failure surface XC which intersects the ground surface at $45 - \phi/2$ as shown.
3. At point X, which is tangent to the circular part BX, erect a perpendicular to XC and, by trial, locate a point O which produces a radius through X and B and draw arc BX.
4. Erect vertical XY as shown and compute the passive earth force acting on this vertical plane as P'_p using the Rankine method. Locate P'_p at $XY/3$ as shown.
5. By some means find the center of area (c.g.a) of area $ABXY$. A cardboard cutout suspended by a thread at several points is recommended. Use a planimeter to measure the area of $ABXY$ to compute the weight W.
6. Compute the radius of the ϕ circle as $r \sin \phi$ where $r = OX$, and draw the circle about point O as shown.
7. Extend the line of action of P'_p to intersect the W vector at point E. Start the force polygon for the system of forces as shown in Fig. 15-11b by drawing W from point 1 to 2 and P'_p from 2 to 3, and obtain the vector R'.
8. Through E and using the slope of R', extend the vector, and from P_p extend the vector to intersect at point Z.
9. Through Z and tangent to the ϕ circle, draw the R vector.

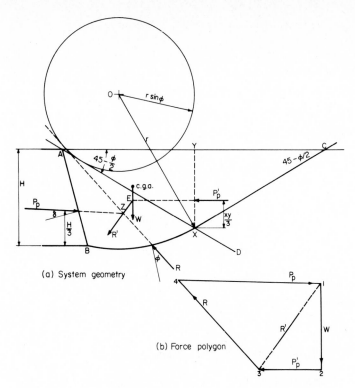

Figure 15-11 ϕ-Circle method for passive pressure in cohesionless soil.

10. Now complete the force polygon by transferring the slope of R and P_p. Note that R continues from point 3 and P_p must terminate on point 1 to close the polygon. The intersection of R and P_p at point 4 allows scaling of P_p for magnitude.

It will be necessary to make several trials to obtain the maximum value of P_p for either the log spiral or the circular arc method. In all trials, the arc will intersect the line AD at some point X, and the slopes of both AD and XC are constant for the given soil conditions and ground geometry.

In cohesive soils there will be an additional resistance due to wall adhesion and due to cohesion along the slip surface BX. The effect of cohesion along XC will be included in P'_p. It will be easier and sufficiently accurate to always use BX = circular arc in cohesive soil.

15-9 BEARING CAPACITY—THEORETICAL

The ultimate bearing capacity of soil beneath a foundation load depends primarily on the shear strength. The working, or allowable, value for design will take into consideration both strength and deformation characteristics.

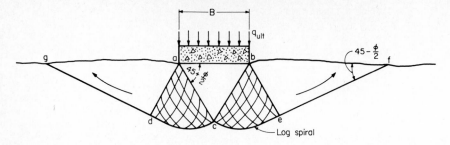

Figure 15-12 Assumed failure surface for bearing capacity failure for a footing on the ground surface. Note the upward assumed soil movement adjacent to the footing.

Most of the currently used bearing capacity theories are based on plasticity theory. Prandtl (ca. 1920) developed expressions from analysis of the assumed flow conditions of Fig. 15-12. The curved part of the arc *ed* or *ce* is assumed to be a part of a log spiral. For foundations on saturated clay, it is usual to assume undrained ($\phi = 0$) conditions; from the preceding section, we have a circular arc, and the ultimate bearing capacity by Prandtl's method is

$$q_{\text{ult}} = (\pi + 2)c = 5.14c$$

Others have found values of 5.64 to 5.74c compared to this value of 5.14 for surface footings.

Terzaghi (1943) modified Prandtl's problem and obtained for strip footings

$$q_{\text{ult}} = cN_c + \gamma D N_q + \tfrac{1}{2}\gamma B N_\gamma \tag{15-10}$$

for square footings

$$q_{\text{ult}} = 1.3cN_c + \gamma D N_q + 0.4\gamma B N_\gamma \tag{15-10a}$$

for round footings

$$q_{\text{ult}} = 1.3cN_c + \gamma D N_q + 0.6\gamma R N_\gamma \tag{15-10b}$$

where D = footing depth
B = footing width (least dimension)
R = footing radius
γ = *effective* unit weight of soil and may be different for the N_q and N_γ parts of Eq. (15-10) depending on the water table location
N_i = bearing capacity factors, shown in Fig. 15-13

In general, the Terzaghi equations are applicable for shallow foundations where $D \le B$. Using Terzaghi's equations and $\phi = 0$, we obtain for the N_i terms

$$N_c = 5.74 \qquad N_q = 1.00 \qquad N_\gamma = 0.0$$

The Terzaghi value of 5.74 instead of 5.14 for N_c is an increase due to friction between soil and footing base. The N_q term is for the surcharge effect due to overburden when the footing is below the ground surface as with the inset to

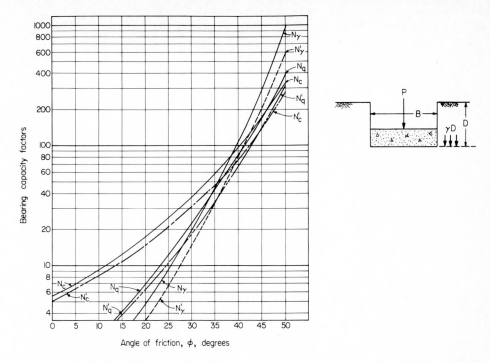

Figure 15-13 Bearing capacity factors for Terzaghi's equation N_i and the Hansen values N_i'.

Fig. 15-13. This term is often a major contribution to bearing capacity—particularly for soils with little cohesion.

Bearing capacity does not increase without bound (with the $\gamma D N_q$ term), so either the footing depth is limited to about $2B$ or reduced values for N_q are used for greater depths.

The N_γ term reflects the contribution of the punch zone abc in Fig. 15-12 and is seldom a large contribution to bearing capacity unless the footing is very wide.

Since the effective unit weight of the soil is used in both the N_q and N_γ parts of the bearing capacity equations, the bearing capacity will be reduced if the water table is between point c in Fig. 15-12 and the ground surface. If the water table is between b and c, the unit weight used in the N_γ term is adjusted. If the water table is between b and ground surface, the submerged unit weight γ' is used with the N_γ term and the effective unit weight with N_q.

Numerous bearing capacity equations have been proposed as better solutions to the bearing capacity problem than the Terzaghi equations, which are generally taken as overly conservative. This is not serious, however, as bearing capacity seldom controls—rather, it is settlement that usually determines the allowable soil pressure beneath a foundation element.

One set of bearing capacity equations which is widely used are those of Hansen (1970). These include shape and depth factors which give them some additional generality over the earlier Terzaghi equations. The generalized Hansen bearing capacity equation is

$$q_{ult} = cN_c s_c d_c + \gamma DN_q s_q d_q + \tfrac{1}{2}\gamma BN_\gamma s_\gamma d_\gamma \qquad (15\text{-}11)$$

where $N_q = \tan^2(45 + \phi/2)\exp(\pi\tan\phi)$
$N_c = (N_q - 1)\cot\phi$
$N_\gamma = 1.5(N_q - 1)\tan\phi$

and the shape and depth factors are approximately

$$s_c = 1 + \frac{N_q B}{N_c L} \qquad\qquad d_c = 1 + \frac{0.4D}{B}$$

$$s_q = 1 + \frac{B}{L}\tan\phi \qquad d_q = 1 + 2\tan\phi(1 - \sin\phi)^2\frac{D}{B}$$

$$s_\gamma = 1 - \frac{0.4B}{L} \qquad\qquad d_\gamma = 1.00$$

Other terms are the same as in the Terzaghi equations, and L = footing length. For a round footing, use B = diameter. The N_i factors are shown on Fig. 15-13 with the Terzaghi values for comparison.

Skempton (1951) made a comprehensive study of foundations in London clay and proposed the N_c bearing capacity factor as shown in Fig. 15-14. Using Fig. 15-14, the bearing capacity is

$$q_{ult} = cN_c + \gamma DN_q$$

Later work in the United States and elsewhere indicates that $N_c = 9$ is satisfactory for deep foundations ($D/B \geq 5$) in clay for such round footings as bases of

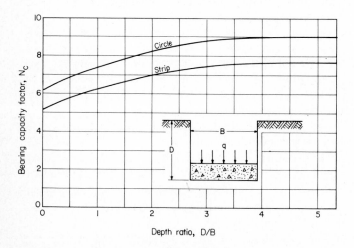

Figure 15-14 Bearing capacity factor N_c for a deep foundation in clay. (*After Skempton, 1951.*)

caissons or belled piers. Since this is a $\phi = 0$ (undrained) condition, the above equation for large D/B ratios becomes

$$q_{ult} = 9c + \gamma D$$

In all cases the allowable bearing capacity is reduced from the ultimate value by a suitable factor of safety F to obtain

$$q_a = \frac{q_{ult}}{F} \qquad (15\text{-}12)$$

where $F = 2.0$ for cohesionless soils (usually)
$= 3.0$ for cohesive soils

Strictly the angle of internal friction should be the triaxial value for square footings, the plane strain value [see Eq. (13-10)] for strip footings and some value intermediate between triaxial and plane strain for rectangular footings. In practice, this refinement is not often made both because ϕ may be estimated and the bearing capacity equations are not considered to be "exact" solutions. For these reasons bearing capacity charts such as Fig. 15-13 are often used rather than obtaining the N factors from computed tables.

In terms of stress paths, the footing load produces stress path $O1$ in Fig. 13-10. Excavation of an area is a negative foundation load producing stress path $O3$ in Fig. 13-10.

Example 15-5 Given square $(B/L = 1)$ footing with $B = 1$ m, $D = 0.5$ m $(D/B = 0.5)$, $\phi = 20°$, $c = 30$ kPa, $\gamma = 17.6$ kN/m³.

REQUIRED Estimate the allowable bearing capacity using Hansen's equation [Eq. (15-11)].

SOLUTION From Fig. 15-13, obtain $N_c = 15$, $N_q = 6.2$, and $N_\gamma = 3.5$. Compute

$$s_c = 1 + \frac{6.2}{15} = 1.4 \qquad s_q = 1 + \tan\phi = 1.4$$

$$s_\gamma = 1 - \frac{0.4B}{L} = 0.6 \qquad d_c = 1 + 0.4(0.5) = 1.2$$

$$d_q = 1 + 2\tan 20(1 - \sin 20)^2 \frac{D}{B} = 1 + 0.315(0.5) = 1.2$$

$$d_\gamma = 1.0$$

Substituting into Eq. (15-11),

$$q_{ult} = 30(15)(1.4)(1.2) + 0.5(17.6)(6.2)(1.4)(1.2)$$
$$+ \tfrac{1}{2}(17.6)(1)(3.5)(0.6)(1)$$
$$= 756 + 92 + 18 = 866 \text{ kPa}$$

Obtain the allowable bearing capacity using $F = 3$ for a cohesive soil as

$$q_a = \frac{q_{ult}}{3} = \frac{866}{3} = 289 \text{ kPa} \qquad (\text{use 275 or 300 kPa})$$

////

15-10 BEARING CAPACITY BY EMPIRICAL METHODS

Several empirical methods are used to either directly or indirectly obtain the foundation bearing capacity. In cohesive soil we may use the unconfined compression test strength q_u, the Terzaghi value of N_c and a safety factor $F = 3$ (cohesive soil) to obtain the allowable bearing capacity as

$$q_a = \frac{q_{ult}}{3} = \frac{1.3cN_c}{3} + \frac{\gamma DN_q}{3}$$

With $N_q = 1$ and $c = q_u/2$ we obtain

$$q_a = \frac{1.3q_u(5.7)}{(2)(3)} + \frac{\gamma D}{3} \cong q_u + \frac{\gamma D}{3} \qquad (15\text{-}13)$$

Taking the unconfined compression strength q_u as the allowable bearing capacity (dropping $\gamma D/3$ as shown here) is a very common practice.

In cohesionless soils a q_a for footings to limit settlement to (hopefully) not more than 25 mm (or 1 in) has been given by Meyerhof (1956, 1974) as

$$q_a = \frac{N}{F_1} K_d \qquad\qquad B \leq F_4$$

$$q_a = \frac{N}{F_2} \left(\frac{B + F_3}{B}\right)^2 K_d \qquad B > F_4 \qquad (15\text{-}14)$$

where B = least lateral dimension of the footing, ft or m
D = footing depth in units of B
$K_d = 1 + 0.33D/B \leq 1.33$
N = standard penetration test (SPT) number (Use average value in the depth to about 0.75B beneath the footing.)
F_i = constant which depends on units used

F	SI, m	FPS, ft
1	0.05	2.5
2	0.08	4.0
3	0.30	1.0
4	1.20	4.0

For mat foundations which have a very large B,

$$q_a = \frac{N}{F_2} K_d \qquad (15\text{-}15)$$

where terms are as defined in Eq. (15-14). For settlements other than 25 mm, assume a linear variation in q_a for both Eqs. (15-14) and (15-15).

Parry (1977) suggested that the ultimate bearing pressure on cohesionless soils using the SPT number N could be computed as

$$q_{\text{ult}} = 30N \qquad (\text{kPa}) \qquad (15\text{-}16)$$

The cone penetration test (CPT) can be used in cohesionless soils according to Meyerhof (1956) to obtain

$$q_a = \frac{q_c}{30} \qquad B \leq F_4$$

$$q_a = \frac{q_c}{50} \left(\frac{B + F_3}{B} \right)^2 \qquad B > F_4 \qquad (15\text{-}17)$$

where q_c = cone resistance, kPa or k/ft^2, and other terms as defined in Eq. (15-14).

In cohesive soils, the undrained shear strength s_u can be estimated according to Begemann (1974) as

$$s_u = \frac{q_c - p_o}{N_c'} \qquad (15\text{-}18)$$

where p_o = effective overburden pressure at CPT depth

N_c' = constant ranging from 5 to 70 depending on deposit and OCR (Values ranging from 9 to 15 are most common.)

Since $s_u = c$, we can use Eqs. (15-10) to obtain q_{ult}. This rough approximation would not justify the use of "refined" bearing capacity equations such as Eq. (15-11). The cone resistance q_c should be an average obtained in a zone similar to N of about 0 to 0.75B below the estimated footing depth.

Example 15-6 The average SPT blow count N in the effective zone beneath a footing is 15. Estimate the allowable bearing capacity q_a, assuming that the footing depth D is 3.3 m and the soil in the zone of influence is a medium dense, coarse sand.

SOLUTION Since q_a depends on B we will make a graph as in Fig. E15-6 of q_a versus B. We should also be aware that while the value of q_a should limit sand settlement to 25 mm, if there is a soft clay layer in close proximity which may consolidate, the total settlement will be more than 25 mm unless q_a is further reduced.

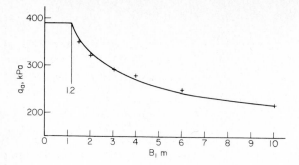

Figure E15-6

Using Eqs. (15-14) we have $F_1 = 0.05$, $F_2 = 0.08$, $F_3 = 0.30$, and $F_4 = 1.2$. We will program the second of Eqs. (15-14) on a programmable calculator. Set up the following table:

$B = \leq 1.2$	1.5	2.0	3.0	4.0	6.0	10.0	
K_d 1.3	1.3	1.3	1.3	1.4	1.2	1.1	(computed)
q_a 390	351	322	295	282	248	219	kPa

Plotting q_a versus B and "smoothing" the curve, we obtain Fig. E15-6. This type of curve is suitable for a client where footing widths B are not known in advance.

$////$

15-11 OTHER BEARING CAPACITY PROBLEMS

Footings may be located on or adjacent to slopes which will reduce the bearing capacity on the slope side (unless the footing is sufficiently far from the edge of the slope—usually 3 to 4B). Both the N_c and N_q values will be affected by the slope. Tables of reduced bearing capacity factors which tend to allow for this reduction are available in Bowles (1982).

Footings may be loaded with inclined or eccentric loads. An inclined load will tend to cause the footing to translate laterally unless there is sufficient base resistance. An eccentric load tends to reduce the effective base contact area. Both types of loads tend to reduce the "ultimate" bearing pressure. Methods to evaluate the resulting reduced bearing pressure are also available in Bowles (1982).

One obvious solution to these several problems is to arbitrarily reduce the bearing capacity computed from Eq. (15-11) by 30 to 50 percent. In a few cases this reduction would be overly conservative, but in most cases the result is about as reliable as the soil data or "theoretical" methods can justify.

15-12 DEEP FOUNDATIONS

A deep foundation refers to the use of a pile or caisson to distribute the load vertically (via skin resistance) in the strata or to transfer the load through poor materials to a bearing on competent soil (end bearing). Piles are usually driven but, similar to caissons, they can be of concrete which is poured into a cavity to produce a vertical structural member with considerable load capacity. A caisson is often considered to be a member with a diameter of 75 cm or larger.

Piles and caissons are analyzed similarly for static load capacity. Piles may also be analyzed for dynamic load capacity. Piles (and sometimes caissons) may be load tested to verify the load capacity obtained by computational methods.

Load tests are considered to be the most reliable method of determining pile (or caisson) capacity, and most projects require one or more load tests prior to full scale installation so any design deficiencies can be corrected. Usually if the load test piles are satisfactory they are used in the foundation.

Static Pile-Caisson Capacity

In nearly all cases, regardless of the assumption of a skin resistance (also termed *floating*) or an end bearing pile-caisson, the static load Q_{ult} is a combination of skin resistance Q_s and end bearing Q_p so that

$$Q_{ult} = Q_s + Q_p \qquad (15\text{-}19)$$

Using the safety factor F, the allowable pile capacity is

$$Q_a = \frac{Q_{ult}}{F}$$

The allowable capacity is the sum of the foundation loads and the pile self-weight which should be considered—particularly for concrete piles—since this may be substantial. In general,

$$Q_a \geq \text{Working loads} + \text{pile weight}$$

The problem with the simple-appearing equation for ultimate capacity [Eq. (15-19)] is that neither Q_s nor Q_p can be reliably evaluated. It is common to evaluate Q_p using the Terzaghi bearing capacity equation for a round or square footing to obtain

$$Q_p = A_p[1.3cN_c + \eta\gamma L(N_q - 1) + 0.5\gamma BN_\gamma] \qquad (15\text{-}20)$$

where
A_p = area (or projected area) of the pile-caisson point
B = pile or caisson point width or diameter
c, N_c, γ, N_γ = values previously defined in Eqs. (15-10)
L = pile length [equivalent to D in Eq. (15-10)] (For piles some are of the opinion that L should be limited to $20B$.)

N_q = factor in Eq. (15-10) but reduced by (-1) to allow for displacement of the column of earth occupied by the pile; justifies inclusion of the pile weight with the working loads to obtain a comparison with Q_a

η = factor to adjust N_q for $D \gg B$ (being a deep foundation)

Because of the uncertainty in Q_p and since Q_p is seldom a major contribution (unless end bearing), the factor η is often taken as 1.0, N_c is usually taken as 9.0, and the γN_γ contribution (except for caissons) is dropped. The values of N_q and N_γ are often taken as those shown in Fig. 15-13. Values of N_q and N_γ are available which purport to account for $D \gg B$ but generally do not increase the reliability of the point bearing contribution unless we have load tests to back compute to (Coyle and Castello, 1981).

The skin resistance contribution is always of the form

$$Q_s = \sum_{i}^{n} (c_{ai} + v\sigma_{hi})p_i(\Delta L_i) \qquad (15\text{-}21)$$

where i = ith element and taking a summation over n elements for Q_s

c_{ai} = adhesion as a fraction α of the cohesion $(=\alpha c)$. The α term is commonly 1 for normally consolidated (soft) clays and ranging downward to as little as 0.25 for very stiff clay.

v = coefficient of friction between soil and pile (We commonly use δ as the angle of friction between pile and soil so that $v = \tan \delta$. Usually $\delta < \phi$ and for steel piles δ is on the order of 20 to 25°.)

σ_{hi} = average effective lateral pressure in element length ΔL_i obtained as $\gamma' y K$

γ' = average effective unit weight of soil in depth y

y = depth from ground surface to $\Delta L_i/2$

K = lateral earth pressure coefficient; may be K_a, K_o, or K_p (A value of $K_o \leq K \ll K_p$ is usually used.)

p_i = average effective perimeter of ith element

ΔL_i = length of ith element (In practice ΔL_i is often L, i.e., $n = 1$ element.)

In a ϕ–c soil we have all of the above contributions to skin resistance. In a cohesionless soil the c_{ai} term is zero and in a $\phi = 0$ soil the friction component is zero. It is common to obtain soil properties from a combination of the SPT and field (sometimes laboratory) unconfined compression tests before the piles are driven. What the final (after driving) soil properties are is highly speculative.

Example 15-7 Given: a pile is to be driven into a site with the profile shown in Fig. E15-7. Estimate the allowable static capacity using $F = 4$. Data shown are estimated from SPT and unconfined compression tests.

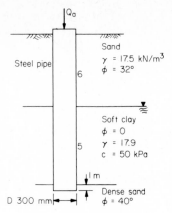

Figure E15-7

SOLUTION For the first 6 m we have $c = 0$ and perimeter $= \pi D = 3.142(0.3) = 0.94$ m. We will arbitrarily take

$$K = 1.2K_o = 1.2(1 - \sin \phi)$$

$$K = 1.2(1 - \sin 32°) = 0.56$$

$$\sigma'_{v(av)} = 17.5(6/2) = 52.5 \text{ kPa} \qquad \text{(at midheight of layer)}$$

We will assume $\delta = 20°$ for steel to soil.
From Eq. (15-20).

$$Q_{s1} = [\sigma'_v(K) \tan \delta]p \cdot L = 52.5(0.56)(\tan 20)(0.94 \times 6) = 60 \text{ kN}$$

For the next 5 m (we will neglect the 1 m of penetration into dense sand) we have only $c = 50$ kPa and taking $\alpha = 1$,

$$Q_{s2} = (\alpha c)(p \cdot L) = (1 \times 50)(0.94 \times 5) = 235 \text{ kN}$$

For the point we use Eq. (15-19) and neglect the γN_γ term. Obtain at $\phi = 40°$ from Fig. 15-13 a value $N_q = 65$ [the value could be computed by using Eq. (15-11)]; use 6 m of "clay" to allow somewhat for 1 m of penetration into dense sand:

$$Q_p = 0.7854D^2[17.5 \times 6 + (17.9 - 9.81)6](N_q - 1)$$

$$\text{(using net pressure)}$$

$$= 0.071(153.6)(64) = 698 \text{ kN}$$

$$Q_{ult} = Q_{s1} + Q_{s2} + Q_p = 60 + 235 + 698 = 993 \text{ kN}$$

$$Q_a = \frac{993}{4} = 248, \qquad \text{say, 250 kN}$$

Since most of the pile capacity is from the point, it might be called a "bearing pile" and the side resistance might be neglected. In this case

$$Q_a = \frac{698}{4} = 175 \text{ kN} \qquad \text{(call it anything from 150 to 200 kN)}$$

If Q_a is too small, we may by trial and error change the diameter using available pile sizes until a satisfactory diameter is found. We could go deeper but the driving into the dense sand would be difficult and might even be impossible so that a larger or smaller diameter, as the case might be, would be preferred.

////

Dynamic Pile Capacity

Pile capacity may be estimated during driving by using some kind of dynamic equation. Although dynamic equations are little better than guesses most of the time they are still widely used. One of the most widely used dynamic equations is the Engineering News formula (often called the Engineering News–Record or ENR formula). This formula has been around since the early 1900s and is based on the impulse-momentum principle with substantial simplification and lumping of losses to give a simple equation as follows

$$Q_{\text{ult}} = \frac{W_r h}{s + C} = \frac{e_h E_h}{s + C} \qquad (15\text{-}22)$$

where Q_{ult} = ultimate estimated capacity in units of W_r

$W_r h$ = input energy as product of ram weight and height of ram fall [Hammer manufacturers usually rate the hammer with this product directly given as E_h (kN $\cdot$ m or k/ft^2).]

e_h = hammer efficiency when E_h is given; ranges from 1.0 to about 0.6 depending on condition of hammer

s = set or penetration per blow of hammer (usually for last 5 or 10 blows as an average) (Use units consistent with E_h or h.)

C = constant depending on type of hammer and units of E_h or h.

Hammer	SI, h and s in meters	FPS, h and s in inches
Drop	0.0254	1.00
All other impact types	0.00254	0.10

The ENR formula has considerable value when used with a load test where the driving record (the set) is available. If the load test is satisfactory, then piles with similar driving resistance in terms of set should be satisfactory.

Example 15-8 A DE-30 diesel hammer (McKiernan-Terry Corporation) is used to drive a pile. The set (average of the last five blows) is 5 mm. Estimate the ultimate and allowable pile capacity using the Engineering News formula.

SOLUTION The safety factor for the ENR formula is commonly taken as $F = 6$. From manufacturer's pile hammer data tables, obtain $E_h = 30.4$ kN·m. The DE-30 is a diesel hammer for which e_h is usually taken as 1.0. Substituting into Eq. (15-22) with $C = 0.00254$, we have

$$Q_{ult} = \frac{e_h E_h}{s + C} = \frac{1(30.4)}{0.005 + 0.00254} = 4030 \text{ kN}$$

$$Q_a = \frac{4032}{6} = 670 \text{ kN}$$

One would also need to check that the driving stresses caused by (possibly) 4030 kN does not damage the pile material.

////

15-13 IMMEDIATE SETTLEMENT COMPUTATIONS

Settlements beneath loaded areas are of the three types given in Sec. 11-10 as

$$\Delta H = \text{Immediate } (\Delta H_i) + \text{consolidation } (\Delta H_c) + \text{creep } (\Delta H_s)$$

We have already considered ΔH_c and ΔH_s in Chaps. 11 and 12. We will now look briefly at the immediate settlement contribution ΔH_i. It should be evident that any one or more of the three settlement components may be negligible. Also, much of the "immediate" settlement occurs as the dead and construction loads are added during the construction sequence and may be largely built out or compensated for.

All elastic settlements are of the general form

$$\Delta H_i = \int_0^L \varepsilon \, dL \tag{15-23}$$

where L = effective stressed depth—usually on the order of 2 to $4B$
  ε = strain from Eq. (10-7) modified to $\varepsilon = (1/E_s)(q_y - \mu K_x q_y - \mu K_z q_y)$
    with $q_y = K'\sigma_o$ in Fig. 15-15c; *net* increase in pressure
  K_x, K_z = lateral earth pressure coefficients
    μ = Poisson's ratio taken as a constant

The Boussinesq pressure profiles in Figs. 10-12 and Fig. 15-15c display that most of the strain ε occurs in a zone directly beneath the footing of about 0.3 to 0.5B where the increase in pressure from the footing is greatest. These profiles also show that the depth L producing a significant strain contribution is between 2 and 4B. The empirical procedure of Schmertmann (1970) is based on this

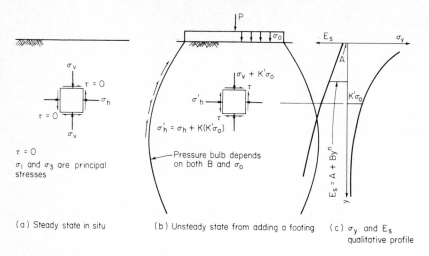

(a) Steady state in situ   (b) Unsteady state from adding a footing   (c) σ_y and E_s qualitative profile

Figure 15-15 Stress changes producing settlements beneath a footing. Also qualitative pressure and stress-strain modulus variation with depth.

concept; however, we might note that Eq. (15-12) is exact. Any deviations between computed and measured values of H_i using Eq. (15-12) are from using an incorrect L or strain value. The strain is particularly difficult since E_s usually increases with depth, q_y decreases, and K_x and K_z are unknown values which are probably larger than K_o and μ must usually be estimated at 0.3 to 0.4.

One can use the trapezoidal formula given by Eq. (10-21) to obtain average values of E_s and q_y if a vertical profile of values is available. Alternatively, one may subdivide the stratum into several increments and compute the settlement of the increment and sum settlements to obtain the total. This latter usually represents substantial computations and is often not done.

The most common equation for the elastic settlement is based on integrating Eq. (10-12) [as was done to Eq. (10-13) for stress] over a flexible loaded area of dimensions $B \cdot L$ and *net contact pressure* σ_o on the surface of a semi-infinite half space that is elastic and weightless. After suitable manipulations one obtains:

$$\Delta H_i = \sigma_o B' \frac{1 - \mu^2}{E_s} F_1 \tag{15-24}$$

where F_1 from Bowles (1982) is given as

$$F_1 = \frac{1}{\pi} \left[M \ln \frac{(1 + \sqrt{M^2 + 1})\sqrt{M^2 + N^2}}{M(1 + \sqrt{M^2 + N^2 + 1})} + \ln \frac{(M + \sqrt{M^2 + 1})\sqrt{1 + N^2}}{M + \sqrt{M^2 + N^2 + 1}} \right]$$

$$M = \frac{L'}{B'} \qquad N = \frac{H}{B'}$$

H = *effective* thickness of stratum, say, 2 to 4B below
  footing or actual H if thickness is less

and

For center F_i: $L' = \dfrac{L}{2}$; $B' = \dfrac{B}{2}$;   use $4 \cdot F_1$

For corner F_1: $L' = L$; $B' = B$;   use $1 \cdot F_1$

Generally one may take Poisson's ratio in Eq. (15-24) as:

Cohesionless materials, partially saturated clay $\mu = 0.3$ to 0.4

Saturated (or nearly saturated) clay $\mu = 0.4$ to 0.5

Poisson's ratio of 0.5 is commonly used but is valid only for the volume distortion which occurs at saturation.

Sometimes a reduction for the footing being at a depth in the soil is applied to Eq. (15-24). Christian and Carrier (1978), using finite-element studies, concluded that the reduction for depth should be used cautiously if at all.

Inspection of Eqs. (15-23) and (15-24) shows that E_s has a major influence on ΔH. This is because the Boussinesq pressure profile is fairly reliable, μ has a maximum effect of 25 percent, footing width B is known, and F_1 is not highly sensitive to depth. Unfortunately, as noted in Chap. 14, E_s is very uncertain. It usually increases with depth and is often empirically obtained so that it can easily vary by a factor of 2 or more. It is easy to see that one can obtain any desired settlement by manipulation of E_s.

Since settlements are so heavily dependent on E_s, Eqs. (15-23) or (15-24) are generally adequate for estimating ΔH_i. Finite-element analyses may also be used, but the additional refinement (and computational expense) is not warranted with uncertainty inherent in E_s (needed in FEM as well) as well as difficulties in producing a suitable FEM model of the elastic continuum.

About the only reliable way to obtain E_s is to back compute a full-scale footing load test. Small plate tests are not reliable since the scale effect (shallow $2B$) will not extract much variation of E_s with depth, which is necessary to produce the settlement of the full-size footing.

In situ measurement of E_s using a pressuremeter involves estimating Poisson's ratio and the value is for a horizontal orientation so unless the soil is isotropic the value obtained is different from the vertical E_s. Seismic methods give a dynamic value which is usually several times larger that the static E_s value. Because of these several factors E_s is often estimated empirically with equations similar to those given in Sec. 14-4.

Example 15-9 A square footing of side 3 m is located on a sand stratum which is believed to be $100 + $ m thick. The contact pressure $\sigma_o = 240$. Assume that N in the critical zone to about 3 m below footing is 30. Also assume Poisson's ratio $\mu = 0.3$.

REQUIRED:

(a) Estimate the center and corner settlements based on an effective depth of 3B.

(b) Estimate the settlement if the stratum is "infinitely" thick.

SOLUTION For part a:

Step 1 From Sec. 14-4 estimate E_s:

$$E_s = 18\,000 = 750N = 40\,500 \text{ kPa}$$

$$E_s = 40\,000 \log N = 59\,000 \text{ kPa}$$

(average value of constant; log 30 = 1.47)

Use E_s = average = 50 000 kPa

Step 2 Find center H_i:

$$B' = \frac{3}{2} = 1.5 \qquad L' = \frac{B}{2} = 1.5$$

$$M = \frac{1.5}{1.5} = 1 \qquad N = \frac{H}{B'} = \frac{3(3)}{1.5} = 6$$

Substituting into equation for F_1

$$F_1 = \frac{1}{\pi}\left[1 \ln \frac{(1 + \sqrt{1^2 + 1})\sqrt{1^2 + 6^2}}{1(1 + \sqrt{1^2 + 6^2 + 1}} + \ln \frac{(1 + \sqrt{1^2 + 1})\sqrt{1 + 6^2}}{1 + \sqrt{1^2 + 6^2 + 1}} \right]$$

$$F_1 = \frac{1}{\pi}\left(\ln \frac{14.7}{7.16} + \ln \frac{14.7}{7.16} \right) = \frac{1.44}{\pi} = 0.458$$

The center settlement (using $4 \cdot F_1$) is

$$\Delta H_i = \sigma_o B' \frac{1 - \mu^2}{E_s}(4F_1) = 240(1.5)\frac{1 - 0.3^2}{50\,000}(4 \times 0.458) = 0.012 \text{ m}$$

$$= 12.0 \text{ mm}$$

Step 3 Find corner settlement for depth of influence = 3B:

$$B' = 3 \qquad L' = B = 3 \qquad M = \frac{3}{3} = 1 \qquad N = \frac{9}{3} = 3$$

Substitution of values into F_1 obtains

$$F_1 = \frac{1}{\pi}\left(2 \ln \frac{7.63}{4.32} \right) = 0.362$$

$$\Delta H_i = 240(3)\left(\frac{0.91}{50\,000} \right)0.362 = 0.0047 \text{ m}$$

$$= 4.7 \text{ mm} \qquad \text{(corner)}$$

Note we use only $1F_1$ for the corner. Also, one would expect the corner of a 3-m square footing to be less than the center unless the footing were very thick (or rigid). Note the similarity of obtaining values for F_1 as used with Eq. (10-19).

For part b:

Find the settlement at the center of the footing if the sand depth is "infinite." Take $B' = 1.5 = L'$, $M = 1$, and $H/B' = 1000 = N$; $\sqrt{2^2 + 1000^2} = 1000$, etc.

Substituting into F_1 with the above assumption, we obtain

$$F_1 = \frac{1}{\pi}\left(\ln\frac{2414}{1001} + \ln\frac{2414}{1001}\right) = \frac{1.76}{\pi} = 0.56$$

$$\Delta H_i = 240(1.5)\left(\frac{0.91}{50\,000}\right)(4 \times 0.56) = 0.015 \text{ m}$$

$$= 15 \text{ mm} \qquad \text{(center, infinite depth)}$$

This example shows (and independent of E_s) that the settlement of a square footing that is adjusted for an effective depth of $3B$ will always compute a settlement that is 80 percent of that using an infinite stratum depth. We should further reduce the "adjusted" settlement 10 to 15 percent for effect of embedment since no footing is ever placed on the surface of the ground.

////

15-14 SUMMARY

This chapter has introduced the concept of earth pressure as

Pressure	Type
Active	Minimum earth pressure due to expansive relative movements away from soil mass
K_o	In situ equilibrium earth pressure
Passive	Maximum earth pressure due to compressive relative movements into soil mass

We note that the soil resistance aids in active earth pressure and that the volume of soil is a minimum. The soil resistance must be overcome for a larger volume of soil in developing passive earth pressure. This change in the volume of soil in the failure wedge plus the particulate nature of soil and the associated soil flow with loss of lateral pressure creates a situation where K_o conditions are difficult to impossible to measure.

The trial wedge method of determining passive earth pressure was introduced, along with use of a log spiral and the ϕ-circle methods. It was pointed out that the Rankine and Coulomb earth pressure methods are commonly used to determine wall pressures. The Rankine method was extensively covered both by use of Mohr's circle and by use of the trial wedge method. The use of a plane failure surface gives little loss of accuracy for the active earth pressure case, and for small slope angles, small angles of wall friction, and relatively small angles of internal friction gives reasonably good values for passive earth pressure.

A brief introduction to bearing capacity has been made with two of the more widely used sets of equations presented. The Terzaghi equations tend to be more conservative than those of others, but being more simple, have considerable appeal.

Pile capacity has also been briefly examined. It was suggested that side resistance can be obtained as a summation process, but it was also noted that it was common to use a single computation with "average" soil parameters as a computational convenience.

The Engineering News formula for estimating dynamic pile capacity was also presented. While it is not very reliable its simplicity gives it great appeal.

Immediate settlements of foundations was also introduced. It was suggested that the exact method would not likely give correct settlements because of the problems of obtaining the correct values of E_s and μ. This implies that empirical methods should be treated cautiously since, of necessity, some form of E_s is always used. Any computations directly dependent on E_s can be "adjusted" to give correlative results where answers are known from measurements.

Inspection of the settlement equation shows settlement is directly dependent on the contact pressure σ_o and footing width. The footing width determines the effective depth of stress influence (length over which strains occur), and the contact pressure determines the strain. It should be evident that with the contact pressure being the net increase no settlements will occur if the net increase is zero (or less). This is readily accomplished by excavating a quantity of soil equal to the weight of the superstructure and is called "floating" the foundation.

PROBLEMS

Lateral pressure

15-1 What is the lateral pressure at the base of a 5-m bin of corn which has a unit weight of 7.1 kN/m³ and $\phi = 31°$? What is the wall force per meter?
Partial ans.: Pressure at base = 11.4 kPa

15-2 For the conditions given in Figs. 15-5d and 15-5e, make several trials to obtain $P_{a(max)}$ by solving the force triangles. Take $\phi = 30°$, $\gamma = 19.00$ kN/m³, and $H = 5$ m. Make trials for $\rho = 50, 55, 60$ and $65°$. Plot ρ versus P_a and obtain a graphical maximum for P_a and the corresponding ρ.
Ans.: $P_a \cong 80$ kN

15-3 Do Prob. 15-2 for P_p. Try values of $\rho = 20, 25, 30, 35$ and $40°$.
Ans.: $P_p \cong 712$ kN

15-4 Compute the active earth force for Prob. 15-2, using the Rankine equation. Draw a neat sketch showing the pressure profile P_a and its location.

15-5 Compute the pressure profile and the active and passive wall forces for the following conditions: $H = 6$ m; $\gamma = 18.5$ kN/m³; $\phi = 34°$; and $\beta = 10°$.

15-6 What is the active earth force for a soil with $s_u = c = 30$ kPa and $\gamma = 18.5$ kN/m^3? The wall height is 6 m. Also compute the theoretical depth of tension crack and the theoretical height a vertical cut would stand without requiring a wall.

Partial ans.: $h_t = 3.24$ m; $P_a = 70.4$ kPa (neglecting tension)

Bearing capacity

15-7 Compare the Terzaghi and Hansen bearing capacities for a footing 2.0 m square on a cohesionless soil with $\phi = 34°$, $\gamma = 18.00$ kN/m^3 and D = 2 m.

Partial ans.: $q_{ult(Hansen)} = 1273$ kPa (N_i computed)

15-8 Do Prob. 15-7, assuming that the soil also has cohesion $c = 20$ kPa.

Partial ans.: $q_a = 1050$ kPa ($F = 3$)

15-9 What is the allowable and ultimate bearing capacity for a cohesionless soil where the SPT value N in the zone of influence averages 12?

Partial ans.: $q_{ult} = 360$ kPa

15-10 What is the allowable bearing capacity for a soil where the unconfined compression strength $q_u = 125$ kPa? Estimated footing widths B are 2 to 4 m, and the footing depth will average 1.5 m.

15-11 Compare the bearing capacity using the Hansen equation for a cohesionless soil with $\phi = 36°$ from a triaxial test for a square footing $B = 2$ m versus a strip footing with an L/B ratio of 4 and $B = 1$ m. The footings will be 1 m in the ground and the average unit weight $\gamma = 18.5$ kN/m^3.

Deep foundations

15-12 A DE-40 hammer with $E_h = 43.4$ kN · m is used to drive a pile. What is Q_a if set $s = 8$ mm?

15-13 Make a plot of set (mm) versus Q_{ult} using the data in Example 15-8 for a range of s from 1 to 50 mm.

15-14 What set s would produce "yield" in a steel pipe pile with cross-sectional area $A = 0.010$ m^2 if the yield stress is 250 MPa using data from Example 15-8?

Ans.: $s \leq 9.6$ mm

15-15 Do Example 15-7, assuming that there is 15 m of loose sand and no clay with the water table at -3 m. Loose and dense sand parameters are as given in the example.

Ans.: $Q_a = 180$ kN (using no side resistance; why?)

Immediate settlements

15-16 Verify $F_1 = 0.362$ for the corner computation given in Example 15-9.

15-17 What is F_1 for an "infinite" stratum if $N = H/B' = 2000$ is taken in part b of Example 15-9?

Ans.: $F_1 = 0.56$

15-18 Do Example 15-9 for the center of the 3-m square footing based on using the strain in the zone $3B$. Use the strain expression with $K_x = K_y = 0.6$ and obtain $q_y = K'\sigma_o$ using the Boussinesq method. Use $E_s = 50\,000$ kPa and $\mu = 0.3$ both constant within the depth $3B$.

Ans.: $\Delta H_i \cong 8$ mm

Computer programming

15-19 Program Eqs. (15-8) and (15-8a) and make a table of active and passive earth pressure coefficients for $\phi = 25–45°$. Use ϕ in increments of 2°. Increment β from 0–20° in 5° increments.

15-20 Program the Hansen bearing capacity equations for the N_i factors. Include d_q (excluding the D/B term) and the ratio N_q/N_c in the output. Develop table for ϕ from 0–44° in 2° increments.

SIXTEEN

STABILITY OF SLOPES

16-1 GENERAL CONSIDERATIONS IN STABILITY OF SLOPES

Slopes may be man-made, as in:

Cuts and fills for highways and railroads
Earth dams
River levees
Dikes for containment of water, including leachates, industrial wastes, and
    sewerage
Landscaping operations for industrial or other development
Banks of canals and other water conduits
Temporary excavations

Slopes may also be naturally formed as hillsides or stream banks.

In any case, the ground not being level results in gravity components of the weight tending to move the soil mass from a higher to a lower elevation. Seepage may be a very important consideration in moving the soil where water is present. Earthquake forces may also be important in stability analysis on occasion.

These several forces produce shear stresses throughout the soil mass, and a movement will occur unless the shearing resistance on every possible failure surface throughout the mass is sufficiently larger than the shearing stress. The shearing resistance depends on the shear strength of the soil and other natural factors, such as instant presence of water from seepage and/or rainfall infiltration as well as roots, ice lenses, frozen ground, or rocks which must be severed along the slip surface. Animal, worm, and reptile burrows or decayed roots may produce a progressive failure mechanism which initiates a slope failure.

The major factor, however, is the shear strength of the soil, which may:

1. Be undrained (s_u) for some cases of loading
2. Be effective (ϕ', c') for some cases of loading
3. Increase with time (as consolidation) or with depth
4. Decrease with time due to later saturation, development of excess pore pressure, as when a downward sloping pervious stratum has the exit blocked and fills with water, or loss of negative pore pressure (see Sec. 13-7)

A stability analysis involves making an estimate of both the failure model and the shear strength. The failure model will require prediction of the weights (or loads) to be resisted and the effect of water. The water estimate requires consideration of seepage forces and saturated and effective unit weights. The shape of the failure model can usually be reasonably well defined; however, for the center of rotation, it may require numerous trials to find the worst case.

The solution is highly sensitive to the shear strength, and the shear strength is generally the most difficult parameter in the analysis to predict. This is because of its variation with depth and the difficulty of deciding whether to use the effective or undrained strength. It has already been pointed out in earlier chapers that the shear strength is sensitive to disturbance and testing procedures (refer to Figs. 13-1 and 13-10). It is also difficult to predict changed soil water conditions.

Care must be exercised to produce an economical solution, since one could always insert overly conservative values of shear strength into the problem.

Some solutions entail careful monitoring of the construction (fill) using piezometers to ensure that pore pressures do not build sufficiently to lower the shear stress to a failure. This procedure is routinely done at present on large earth dam fills as a result of at least two major construction failures involving several million cubic meters of soil (Casagrande, 1965). It is important to have the piezometers located where they can measure critical pore pressures, as in the Fort Peck slide (in 1938) piezometers were installed, but not into the zone where the pore pressures caused failure.

In general, slope stability is a plane strain problem, i.e., the length compared to cross section is very large. It is usual to investigate a typical cross section which is 1 unit thick with plane strain, ignoring the perpendicular strains (and stresses). Many small slope failures, as can be readily observed along road cuts in mountainous areas, are nearly equal in lateral and vertical dimensions; nevertheless, plane strain conditions are commonly assumed in their analyses.

16-2 INFINITE SLOPES

Figure 16-1a illustrates the cross section of an infinite slope in a cohesionless soil without seepage. The thickness perpendicular to the plane of the paper is 1 unit

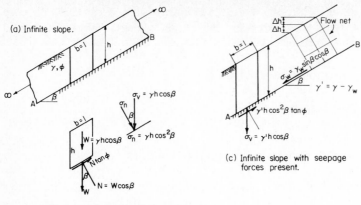

(a) Infinite slope.

(b) General free-body of slope element.

(c) Infinite slope with seepage forces present.

Figure 16-1 Infinite slope with cohesionless soil.

(1 m). The soil is often assumed to be homogeneous; however, in real situations, the soil may be highly stratified with widely varying shear strengths. If we isolate an element, as in Fig. 16-1b, and examine the forces for stability, $\sum F$ parallel to the slope gives

$$W \tan \beta - W \cos \beta \tan \phi = 0$$

and solving we obtain

$$\beta = \phi$$

for stability. This is the same as the Case 1 solution of Sec. 15-4.

With a constant seepage and with the water table at the ground surface, we have from Fig. 16-1c

$$\sigma_n = \gamma'h \cos \beta \cos \beta = \gamma'h \cos^2 \beta$$
$$\sigma_t = \gamma'h \sin \beta \cos \beta$$

The pore pressure stress is

$$\sigma_w = \gamma_w h \sin \beta \cos \beta$$

and summing stresses parallel to the plane AB, obtain

$$\gamma_w h \sin \beta \cos \beta + \gamma'h \sin \beta \cos \beta - \gamma'h \cos^2 \beta \tan \phi = 0$$

Solving for the critical slope angle β, obtain

$$\tan \beta = \frac{\gamma'}{\gamma' + \gamma_w} \tan \phi$$

or

$$\beta = \tan^{-1} \left(\frac{\gamma'}{\gamma' + \gamma_w} \tan \phi \right)$$

For example, with no water, $\gamma' = \gamma_t$, $\gamma_w = 0$, and $\beta = \phi$; with water and $\gamma' = 17.8 - 9.8 = 8.0 \text{ kN/m}^3$ and $\phi = 32°$, the critical slope angle is

$$\beta = \tan^{-1}\left(\frac{8}{17.8} \tan 32\right) = 15.7°$$

Thus with water the theoretical safe slope angle β for a cohesionless soil is only about half that without water.

16-3 STABILITY OF INFINITE COHESIVE SLOPES

Referring to Fig. 16-2, and with no water, we have

$$\sigma_t = \gamma h \sin \beta \cos \beta \qquad \text{(stress)} \qquad (a)$$

$$\sigma_n = \gamma h \cos^2 \beta \qquad \text{(stress)} \qquad (b)$$

The resisting stress is

$$s = c_d + \sigma \tan \phi_d \qquad (c)$$

where c_d and ϕ_d are design shear strength parameters and not the actual soil values unless the safety factor $F = 1$. At $F = 1$, the shear strength $s = \sigma_t$, and by substitution of Eqs. (a) and (b) into (c), obtain

$$\gamma h \sin \beta \cos \beta = c_d + \gamma h \cos^2 \beta \tan \phi_d$$

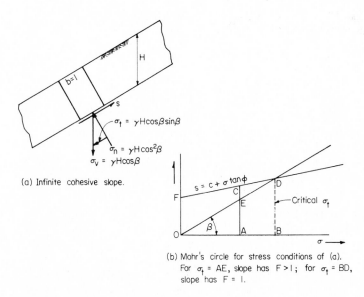

(a) Infinite cohesive slope.

(b) Mohr's circle for stress conditions of (a). For $\sigma_t = AE$, slope has $F > 1$; for $\sigma_t = BD$, slope has $F = 1$.

Figure 16-2 Infinite slope in cohesive soil.

or the design cohesion is

$$c_d = \gamma h \cos^2 \beta (\tan \beta - \tan \phi_d)$$

The critical value of clay thickness is

$$H = \frac{c_d}{\gamma} \frac{\sec^2 \beta}{\tan \beta - \tan \phi_d} \qquad (16\text{-}1)$$

This can be illustrated by Mohr's circle as in Fig. 16-2b, where OA represents the normal stress at some height less than critical and OB represents the normal stress at the critical height, at which time $\sigma_t = BD$. We note that if $\beta < \phi$ the rupture line is never intersected (theoretically safe). Alternatively, we may write

$$N_s = \frac{c_d}{\gamma H} = \cos^2 \beta (\tan \beta - \tan \phi_d) \qquad (16\text{-}2)$$

where N_s = stability number as commonly used in the literature (some writers have used N_s as $\gamma H/c$, so that the reader should be careful to see how the term is used or strange results may be obtained). The stability number, being dimensionless, allows the combining of three problem parameters into a single value and allows the use of simple charts for representing stability relationships.

With seepage the full depth of interest, the stability number becomes

$$N_s = \cos^2 \beta \left(\tan \beta - \frac{\gamma'}{\gamma_{\text{sat}}} \tan \phi_d \right) \qquad (16\text{-}2a)$$

If the top flow line is a distance h_1 below and parallel to the ground surface and the soil weight in zone h_1 is γ_1, we obtain

$$N_s = \cos^2 \beta \left[\left(1 - \frac{h_1}{H} \frac{\gamma_{\text{sat}} - \gamma_1}{\gamma_{\text{sat}}} \right) \tan \beta - \left(\frac{\gamma'}{\gamma_{\text{sat}}} + \frac{h_1}{H} \frac{\gamma_1 - \gamma'}{\gamma_{\text{sat}}} \right) \tan \phi_d \right] \quad (16\text{-}3)$$

16-4 CIRCULAR ARC ANALYSIS

Where slopes are finite in extent, as most artificial embankments and roadway cuts are, the failure surface is curved. Various workers have suggested that the curved surface is a part of a circular arc or a log spiral. Observed slip surfaces tend to be a combination of a circular arc and a log spiral, somewhat oval with relative flat arcs on each end and a sharper arc on the interior. There may be plane discontinuities if the surface intersects a hard zone such as very stiff clay, dense sand, or a rock surface.

The circular arc is the simplest solution and the only one considered here. In the author's opinion, the errors in a slope analysis are not so much in the shape of the assumed failure surface but in the soil properties and the search for the critical failure location. If one can find the critical location (which depends on both geometry and soil properties) from among the infinite number possible, it will be a happy coincidence.

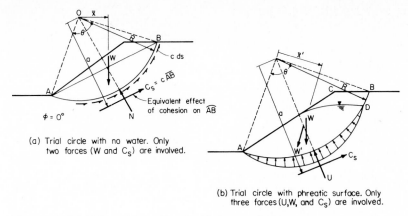

(a) Trial circle with no water. Only two forces (W and C_s) are involved.

(b) Trial circle with phreatic surface. Only three forces (U, W, and C_s) are involved.

Figure 16-3 Trial failure circles with and without water forces.

When the slope is a homogeneous clay and an undrained strength analysis is used, the stability analysis is relatively straightforward. Figure 16-3 illustrates trial failure circles and the forces involved. Figure 16-3a is for a case with no water, and Fig. 16-3b includes seepage forces.

When seepage forces are present, it is necessary to locate the phreatic line and sketch in a flow net. The equipotential lines intersect the trial arc, and with the head known, the pressure at these points can be computed to give the pressure profile as shown in Fig. 16-3b. A numerical integration of this area can be made to obtain the total water force U, which has a line of action through the circle center O. This value of U can be added vectorially to the weight vector W to obtain the new force vector W' with a new line of action and scaled moment arm $\bar{x}'$. The resulting factor of safety is

$$F = \frac{R(cAB)}{W'\bar{x}'} \tag{16-4}$$

A circular arc analysis for a homogeneous slope (Fig. 16-3a) can be made as follows:

1. Obtain the weight of the failure mass W and its moment arm with respect to point O. This may be done by using a planimeter for the area (and weight) and making a cardboard cutout which is suspended by a thread at two or more points to find the center of area.
2. Measure angle θ, convert to radians and compute the arc length as $AB = R\theta$.
3. Compute the factor of safety as

$$R = \frac{\sum \text{ resisting moments}}{\sum \text{ overturning moments}} = \frac{R(cAB)}{W\bar{x}}$$

It is necessary to make several trial circle analyses with the safety factor plotted on the center point so that contours of F can be made which will, hopefully, give the minimum value.

16-5 THE ϕ-CIRCLE METHOD

The friction circle concept may be used for the particular slope condition of a homogeneous soil with a shear strength of

$$s = c_d + \sigma \tan \phi_d$$

where c_d, ϕ_d = design shear strength parameters and σ is the normal stress on the slip surface.

Figure 16-4 illustrates the general concept. Figure 16-4a illustrates the trial circle and all the forces involved in a soil where no water is present. The shear resistance is obtained from summing the cohesion and friction forces along the arc to obtain (it being understood we are using the design values of c and ϕ)

$$F_R = \int c \, ds + \int dN \tan \phi$$

or, alternatively,

$$F_R = C_s + F_f$$
$$= cAB + N \tan \phi$$
$$= cAB + W \cos \phi \tan \phi$$

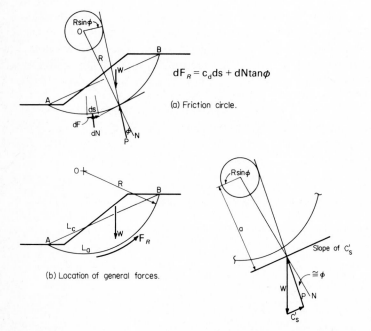

$$dF_R = c_d ds + dN \tan \phi$$

(a) Friction circle.

(b) Location of general forces.

(c) Force polygon-- if forces correctly computed and located, the resultant P is nearly tangent to the ϕ-circle.

Figure 16-4 The ϕ-circle force system.

as shown in Fig. 16-4c. Note, as in Chap. 15, that the friction resistance combines with N to give a vector P which acts at the angle ϕ to the tangent to the arc. The line of action of P extends and gives a moment arm with respect to point O of $\bar{x}$, computed as

$$\bar{x} = R \sin \phi$$

For all the vectors of dP, the moment arms $\bar{x}$ trace a portion of a small circle with radius $R \sin \phi$. This small circle about point O is called the ϕ *circle*.

The cohesion term is independent of ϕ, and noting that a general tangent can be obtained for arc AB which is parallel to the chord, and further noting that the normal component of C_s cancels, by equating moments along the arc L_a and with respect to the chord length L_c, we obtain (refer to Figs. 16-4b and 16-3a)

$$cL_c a = cL_a R$$

which produces an equivalent moment arm of

$$a = \frac{L_a}{L_c} R$$

With this value of moment arm, the equivalent cohesion force is computed as

$$C'_s = cL_c$$

For a force system in moment and static equilibrium, the system must be concurrent, with both $\sum F_H$ and $\sum F_v = 0$. A force polygon can be used for the forces as shown in Fig. 16-4c, with the slope of P obtained from the intersections of the line of action of W and C'_s and approximately tangent to the ϕ circle. Usually we assume that P is exactly tangent to the ϕ circle, as the maximum difference is generally less than 7 percent. The body force W is obtained by planimetering the area or by analysis for simple geometric sections. The line of action is obtained by cutting a cardboard model and suspending by a thread at two or more points. Several trial circles are necessary to obtain the minimum F, which is computed as before, i.e.,

$$F = \frac{\sum \text{ resisting moments}}{\sum \text{ overturning moments}}$$

Inspection of the force polygon in Fig. 16-4c indicates that

$$W = f(\beta, H, \phi)$$

$$C'_s = f(c)$$

$$P = f(\phi)$$

Thus there are five variables. We may combine H, γ, and c into the single variable N_s as earlier and reduce the variables to three. In practice, four of the variables must be given; however, by use of the dimensionless variable N_s, a parametric study for a particular slope can be made rather easily.

16-6 SLOPE STABILITY CHARTS

A number of workers starting with Taylor (1937) have solved some form of Eq. (16-4) to produce design aids in the form of stability charts to reduce the search effort to find the critical circle and the minimum factor of safety. This work has shown that it is possible to combine several of the problem parameters into a single term (such as N_s) to simplify the work.

Cousins (1978) solved the simple slope including pore pressure effects but not including any zoned strengths (given in Fig. 13-1). He used a modified form of Eq. (16-4) and programmed the search on the computer. The results were summarized in charts in terms of:

1. Slope angle α (this and other terms are defined in Fig. 16-5).
2. Pore pressure ratio r_u, defined as

$$r_u = \frac{u}{\gamma H}$$

which produces the effective stress as

$$\sigma'_y = (1 - r_u)\gamma H$$

The pore pressure ratio r_u is an average value since u varies along the critical circle in some manner (qualitatively illustrated in Fig. 16-3). Its principal use is to obtain a quantitative insight into the effect of some pore pressure on the safety factor F.

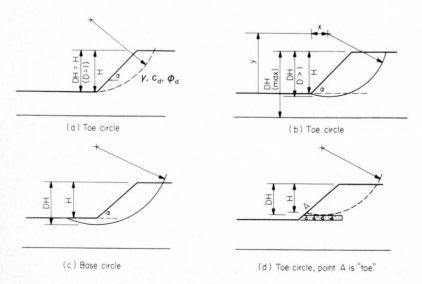

Figure 16-5 Toe and base circle configurations for use with Figs. 16-6 and 16-7. Note that point A in part d requires estimating the point at some distance above the drain. One or more trials may be required for this situation. Soil properties represent "average" values in the zone of interest.

3. Factor $\lambda_{c\phi}$ defined as

$$\lambda_{c\phi} = \frac{1 - r_u}{c} \tan \phi_d = \frac{1 - r_u}{N_s} \tan \phi_d$$

where N_s = stability number defined by Eq. (16-2)
ϕ_d = design value of angle of internal friction
4. Factor N_F defined as

$$N_F = \frac{F}{N_s}$$

where F = factor of safety

In this analysis the same factor F is applied to both c and ϕ, which is commonly done in slope analysis.

The Cousin's charts (Figs. 16-6 and 16-7) have the advantage of not being overly congested and at the same time allow easy determination of the safety factor F. Also, the charts allow easy determination of the slope height or angle for a required safety factor.

Where the soil is stratified, it is probably best to analyze the slope using a computer program; however, for preliminary studies one can "average" the soil properties and obtain a reasonable estimate. Where the soil is homogeneous, the charts are preferable since their results are as accurate as a computer analysis and considerably cheaper.

Figure 16-6 gives charts for three pore pressure ratios r_u for a "toe circle," where a toe circle is as defined in Fig. 16-5. Figure 16-6d is an alternative presentation of $r_u = 0$ data. Note that a toe circle is possible only for $D = 1$ (or slightly greater). All other values of $D > 1$ are for base circles. Observe that the "toe circles" line in Fig. 16-6a is coincident with the $D = 1$ line in Fig. 16-6d. Actually, either Fig. 16-6a or 16-6d could be used for the same problem; however, Fig. 16-6a gives a closer estimation of D for many cases.

The chart range to $\alpha = 45°$ is adequate for most cases since few slopes are built much steeper than this (which is 1 on 1).

Figure 16-7 is used to approximately locate the center coordinates of the critical circle referenced to the toe intersection as in Fig. 16-5b. It should be evident that a scaled drawing with these coordinates and using the DH value should approximately locate the critical circle and display whether it is a toe or base type.

Several additional cases extending Figs. 16-6d and 16-7 for $r_u = 0.25$ and 0.5 are available in Cousins (1978). More recently (Cousins, 1980) the charts have been extended to include tension cracks and water in the tension cracks. These charts show that a tension crack tends to reduce F by 8 to 10 percent and water in the tension crack makes an additional reduction of another 8 to 10 percent— depending on soil and slope geometry.

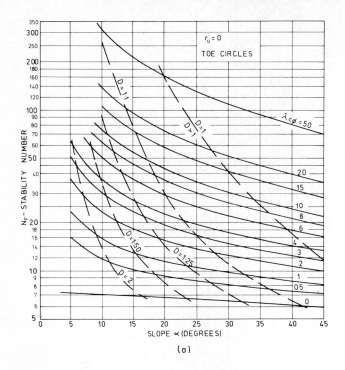

(a)

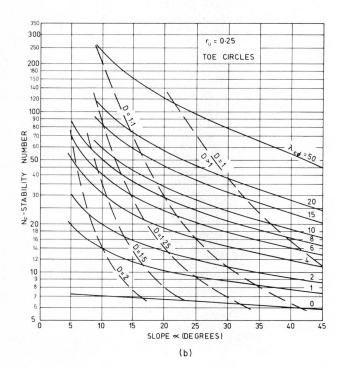

(b)

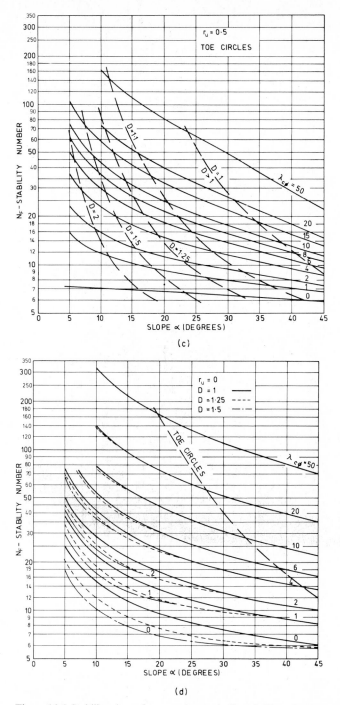

Figure 16-6 Stability charts for several cases indicated. Note that a toe circle is only obtained when $D = 1$ or is very close regardless of chart label. *(From Cousins, 1978.)*

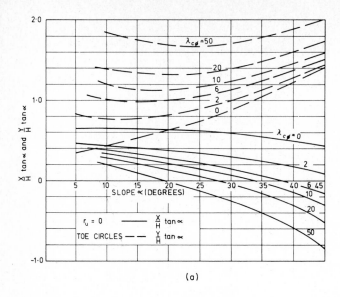

(a)

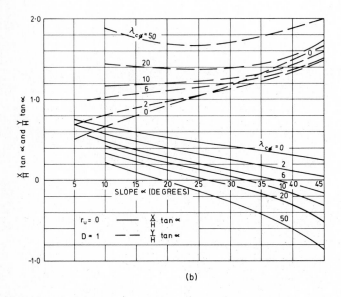

(b)

Figure 16-7 Charts used for location of critical circle center for cases indicated. More complete charts for other r_u are available in the reference. *(From Cousins, 1978.)*

Example 16-1 Given: a homogeneous slope with a slope of 1V on 1.5H has $H = 12$ m, $\gamma = 18.0$ kN/m^3 and $s_u = c = 30$ kPa, r_u is assumed $= 0.0$.

REQUIRED What is F if (a) $\phi = 0°$ and (b) $\phi = 15°$ and show location of critical circle?

SOLUTION For part a with $\phi = 0°$:

$$\alpha = \tan^{-1} \frac{1}{1.5} = 34°$$

$$N_s = \frac{c}{\gamma H} = \frac{30}{18(12)} = 0.139$$

$$\lambda_{c\phi} = \frac{1 - r_u}{N_s} \tan \phi = \frac{1 - 0}{N_s} \tan 0° = 0$$

From Fig. 16-6a entering at $\alpha = 34°$ and projecting to $\lambda_{c\phi} = 0$ and horizontal to the N_F axis, we obtain

$$N_F = 6.3 = \frac{F}{N_s}$$

$$F = N_F N_s = 6.3(0.139) = 0.9 < 1 \qquad \text{(unsafe)}$$

For part b: $\phi = 15°$, but other data are the same; and

$$\lambda_{c\phi} = \frac{1 - 0}{N_s} \tan 15° = \frac{1}{0.139} \tan 15° = 1.93$$

From Fig. 16-6a, or 16-6d at $\alpha = 34°$ and projecting to the estimated location of $\lambda_{c\phi} = 1.93$ and horizontally to the N_F axis, we obtain

$N_F = 11.2$

$F = 11.2(0.139) = 1.56$ \qquad (would generally be considered very safe)

Interpolating at $\lambda_{c\phi} = 1.93$ between D lines, we obtain $D = 1.08$ so that

$$DH = 1.08(12) = 13 \text{ m}$$

The center coordinates for the failure circle (using Fig. 16-7a) are approximately:

$$\frac{x}{H} = 0.25 \qquad x = 0.25(12) = 3 \text{ m}$$

$$\frac{y}{H} = 1.09 \qquad y = 1.09(12) = 13.1 \text{ m}$$

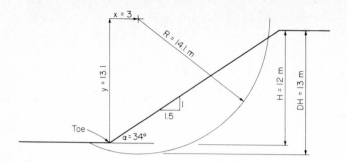

Figure E16-1

Using these data, $H = 12$ m and a slope of 1 on 1.5, the slope of Fig. E16-1 can be drawn with a failure circle tangent to the DH line as shown.

////

Example 16-2 What H will give a safety factor $F = 1.5$ for Example 16-1 for the $\phi = 0°$ case?

SOLUTION From Example 16-1 we have $N_s = 0.139$, $F = 0.9$ and $N_F = 6.3$. Since $\lambda_{c\phi} = 0$ does not change, N_F will not change for any N_s, so the only way to increase F is to change N_s:

$$F = 6.3N_s = 1.5 \qquad N_s = \frac{1.5}{6.3} = 0.238$$

$$N_s = \frac{c}{\gamma H} = 0.238 = \frac{30}{(18)(H)} \text{ (note } c \text{ and } \gamma \text{ are constant)}$$

$$H = \frac{30}{18(0.238)} = 7.0 \text{ m} \qquad \text{(which is } < 12 \text{, as it should be)}$$

////

Example 16-3 What slope angle will give an $F = 1.5$ for Example 16-1 for the $\phi = 15°$ case?

SOLUTION From Example 16-1:

$$N_s = 0.139$$

$$\lambda_{c\phi} = 1.93$$

Since neither of these depends on the slope angle α, they are unchanged

$$F = N_F N_s = 1.5 \text{ (new condition)}$$

$$N_F = \frac{1.5}{0.139} = 10.8$$

Enter Fig. 16-6a or 16-6d ($r_u = 0$) at $N_F = 10.8$ and read horizontally to the interpolated location of $\lambda_{c\phi} = 1.93$ and vertically to obtain $\alpha = 40°$ (which is larger as it should be since F is less).

////

16-7 SLOPE ANALYSIS BY METHOD OF SLICES

Most natural slopes and many artificially created slopes consist of more than one soil, or the soil properties vary so much that some type of finite element solution is mandated. The finite element method generally used is to divide the failure section $ADCB$ into a series of vertical slices as illustrated in Fig. 16-8a.

The slice width is sufficiently small that the actual shape can be replaced with a trapezoid, as shown in Fig. 16-8b. It is assumed that the slice weight W_i acts through the midpoint of the area as shown. With this assumption the following relationships are developed:

$$N_i = (W_i + V_i) \cos \alpha$$

$$T_i = (W_i + V_i) \sin \alpha$$

$$F_s = N_i \tan \phi + cb$$

$$= (W_i + V_i) \cos \alpha \tan \phi + c \frac{\Delta x}{\cos \alpha}$$

and

$$\alpha = \tan^{-1} \frac{\Delta y}{\Delta x}$$

It is usual practice to neglect the interelement forces of X_i and P_i. Some persons have used these forces, but the point of application and line of action of the P forces are indeterminate in stratified soils or where the soil properties (ϕ, c, γ) vary with depth. In these cases, about all that is known for certain is that the line of action of P is inside the failure surface. The vertical force depends on both P and the soil properties. Several researchers have shown that little error is introduced by neglecting the X and P forces. Note, too, that at slip the soil properties at the trial circle boundary are all that are valid—those inside the failure zone are those of a highly remolded soil or are in transition.

Moment equilibrium about point O, using a summation for all the slices in the failure circle with attention to signs, gives

$$\sum RF_s - \sum R(W_i + V_i) \sin \alpha = 0$$

The resisting moment is $\sum RF_s$, and the safety factor F is

$$F = \frac{\sum \text{resisting moments}}{\sum \text{overturning moments}} = \frac{\sum RF_s}{\sum R \sin (W_i + V_i)}$$

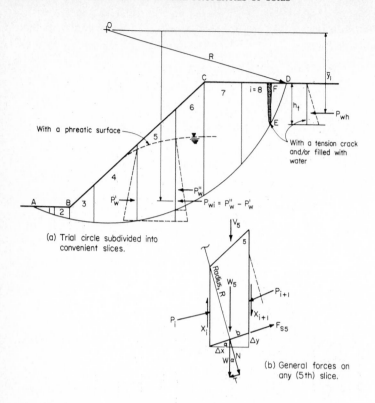

(a) Trial circle subdivided into convenient slices.

(b) General forces on any (5th) slice.

Figure 16-8 Geometry of the method of slices.

Cancelling R and inserting for the shear strength s, obtain

$$F = \frac{\sum (cb + (W_i + V_i) \cos \alpha \tan \phi}{\sum (W_i + V_i) \sin \alpha} \qquad (16\text{-}5)$$

We may use either effective or total stresses and with the appropriate c and ϕ parameters in evaluating Eq. (16-5). The effective stress is most conveniently developed by using γ and γ' as appropriate in computing the weight vector W.

It is considered more correct to apply F to the soil parameters in Eq. (16-5) as was done earlier. If we do this, we obtain

$$F = \frac{\sum [cb/F + (W_i + V_i) \cos \alpha(\tan \phi)/F]}{\sum (W_i + V_i) \sin \alpha} \qquad (16\text{-}6)$$

Since $b = \Delta x/\cos \alpha$, we have the α angle producing a significant role in the problem. Bishop (1955) suggested that the effect of α could be reduced by an alternative method of determination of the normal force. Referring to Fig. 16-8b, $\sum F_v$ on the element (and neglecting the X's) is

$$N_i \cos \alpha - (W_i + V_i) + F_{si} \sin \alpha = 0 \qquad (a)$$

The friction resistance due to ϕ is

$$F_{si} = \frac{N_i \tan \phi}{F} \qquad (b)$$

Substituting Eq. (b) into (a) and dividing by cos α, obtain

$$N_i + \frac{N_i \tan \phi \tan \alpha}{F} = \frac{W_i + V_i}{\cos \alpha}$$

or

$$N = \frac{W_i + V_i}{\cos \alpha} \frac{1}{1 + (\tan \phi \tan \alpha)/F} \qquad (c)$$

Since $(W_i + V_i) \cos \alpha$ in Eq. (16-5) is N of Eq. (c), substitution for N gives:

$$F = \frac{\sum [c \, \Delta x + (W_i + V_i) \tan \phi] \dfrac{\sec \alpha}{[1 + (\tan \phi \tan \alpha)/F]}}{\sum (W_i + V_i) \sin \alpha} \qquad (16\text{-}7)$$

This form of the equation has been programmed by the author (Bowles, 1974). Note again that when effective stresses are used, the weight vector W_i uses the appropriate unit weight (γ or γ') in each slice.

It may be necessary to consider tension cracks, water in the tension crack, and/or unbalanced water pressure (as a seepage force) on each side of the slices. Noting that the radius R effect has been factored both top and bottom and that the numerator represents the resistance in Eq. (16-7), we may readily adjust the denominator with reference to Fig. 16-8a as follows. Let

$$D = \sum (W_i + V_i) \sin \alpha$$

Also, let the unbalanced water pressure on any slice be P_{wi} and the water force in the tension crack be P_{wh}. Moment arms are $\bar{y}$ and $\bar{y}'$, respectively, and a new denominator D' can be written as

$$D' = \sum \left[(W_i + V_i) \sin \alpha + \frac{P_w \bar{y}}{R} \right] + \frac{P_{wh} \bar{y}_1}{R}$$

The value of D' is now used instead of D in Eq. (16-7). The failure section is now defined by sector $AEFCB$ of Fig. 16-8a.

An iterative analysis is necessary to obtain F in Eq. (16-7), since it is on both sides of the equality sign. Programming on the computer allows a rapid solution after only a few cycles (usually 2 to 3) by assuming $F = 1$ initially for the right-hand side and computing F on the left. This value of F is compared with the assumed value; if it is not sufficiently close, the just computed F is used in the next iteration and the cycle repeated. Inspection of Eq. (16-7) shows that for $\phi = 0$ soils an iteration is not necessary, as the equation reduces to Eq. (16-5).

A computer program must develop the arc based on initializing point O and the area entrance coordinates. The arc is divided into i slices and α, b, and $\bar{y}$ (if necessary) are computed for each slice. The weight is computed based on the soil(s) in the slice and/or making allowance for the phreatic surface. The quantities cb and $(W_i + V_i) \tan \phi$ are computed using soil parameters for the soil on the base of the slice. Once F is computed, either the arc entrance or circle center coordinates are incremented and additional analyses made for obtaining a minimum F.

An approximate analysis can be made by hand. The work is generally too prohibitive to iterate or to make an extensive critical circle search. By hand, the trial circle is drawn and subdivided into convenient slices, as shown in Example 16-4. The weight W_i (and V_i) is computed and plotted to scale, assuming that the vector acts through the midpoint of each slice as shown. The normal and tangent components can be directly obtained graphically from a radius through the projection of W_i onto the failure arc. The normal acts through point O and can be neglected. Compute the overturning moment, using the scaled values of T_i and the correct sign depending on which side of O the slice is located, as

$$M_{\text{overturning}} = R \sum T$$

Compute the resisting moment as

$$M_{\text{resisting}} = R \sum (cb + N \tan \phi)$$

The factor of safety is computed as

$$F = \frac{M_{\text{resisting}}}{M_{\text{overturning}}} = \frac{\sum (cb + N \tan \phi)}{\sum T} \tag{16-8}$$

Example 16-4 Hand solution of method of slices is given in this example. The soil properties are: $\gamma = 17.6 \text{ kN/m}^3$, $c = 40 \text{ kPa}$, and $\phi = 24°$.

			Scaled				
Slice	Area	W	b	T	N	$N \tan \phi$	cb
1	$(0 + 2.8)(4.5/2) = 6.3$	110.9	5.3	-60	95	42	212
2	$(2.8 + 16.1)(15.3/2) = 144.6$	2544.7	16.0	-800	2440	1086	640
3	$(16.1 + 23.3)(12.6/2) = 248.2$	4368.7	12.8	0	4369	1945	512
4	$(23.3 + 27.0)(12.6/2) = 316.9$	5577.3	13.3	1600	5250	2337	532
5	$(27.0 + 25.5)(12.6/2) = 330.8$	5821.2	15.5	3260	4850	2159	620
6	$(25.5 + 0)(13.6/2) = 173.4$		28.9	2710	1800	801	1156
	$173.4(17.6 = 3051.8 + 200 = 3251.8$			6710		8370	3762

$$F = \frac{8370 + 3762}{6710} = 1.81$$

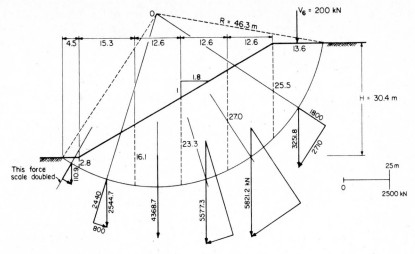

Figure E16-4 Hand solution of method of slices.   ////

16-8 WEDGE BLOCK ANALYSIS

A sliding block or wedge may be a more appropriate cross section for many stability problems (Fig. 16-9) where the failure surfaces may be defined by a series of broken lines. This method may be extended to analysis of rock slopes, particularly in layered strata.

Figure 16-10 illustrates how to apply the wedge analysis method. The reader should make the necessary adjustments in active and passive pressures and the weight of the block for phreatic surfaces, stratification, etc.

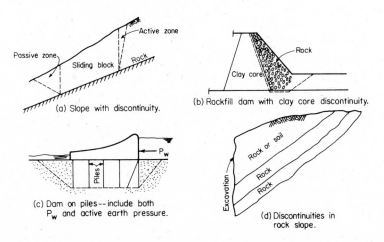

Figure 16-9 Typical situations where a wedge analysis may be appropriate.

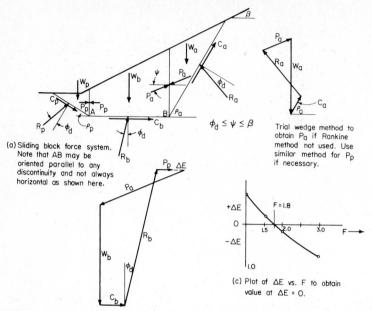

(a) Sliding block force system.
Note that AB may be
oriented parallel to any
discontinuity and not always
horizontal as shown here.

$\phi_d \leq \psi \leq \beta$

Trial wedge method to
obtain P_a if Rankine
method not used. Use
similar method for P_p
if necessary.

(b) Force polygon for central block to obtain ΔE.

(c) Plot of ΔE vs. F to obtain
value at $\Delta E = 0$.

Figure 16-10 The wedge block analysis. Note that ΔE may be oriented other than horizontally, with either the value or the horizontal component used in part c to obtain F.

The active and passive earth forces are computed using either the Rankine earth pressure or the trial wedge method. The Rankine method, with $\rho = 45 \pm \phi/2$, can be used where the soil is homogeneous in these zones, but the trial wedge method is necessary where a fault or discontinuity locates the failure surface and may be used for stratified soils. An assumed F is used to compute the values of

$$c_d = \frac{c}{F} \qquad \phi_d = \frac{\phi}{F}$$

for computing the active and passive Rankine earth pressure forces, and the base resistance of the block of C_s and $F_f = N \tan \phi_d$. The problem is conveniently solved using a force polygon for each assumed value of F, as in Fig. 16-10b for the sliding block, with the closure error scaled as ΔE. A plot of ΔE versus F can be made as in Fig. 16-10c; the value of F at $\Delta E = 0$ is the desired value. This method of analysis is illustrated in Example 16-5.

A trial wedge type solution can also be obtained using the "method of slices" where adjustments are made to identify the slip plane which is intersected by the trial circle. This produces arc portions at the entrance and exit ends of the failure zone and a sliding block defined by the slip plane for the interior. The line of action of the normal and tangential forces and the moment arms along the wedge requires additional computations but when these are made a computer program for the "method of slices" can also solve the sliding wedge problem.

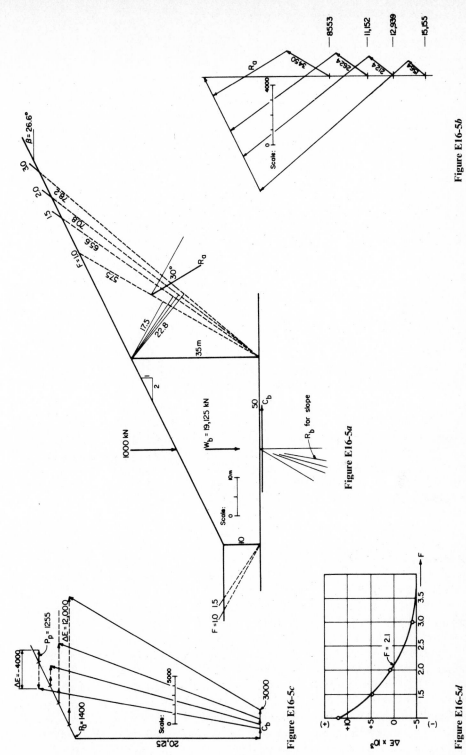

Figure E16-5a

Figure E16-5b

Figure E16-5c

Figure E16-5d

Example 16-5 Given the sliding block geometry of Fig. E16-5a and the following soil data: $\gamma = 17.0 \text{ kN/m}^3$; $c = 60 \text{ kPa}$; $\phi = 30°$.

REQUIRED Make a sliding block analysis and find the factor of safety.

SOLUTION Make the following assumptions: (1) no tension crack; (2) for slope of P_a, take $\psi = \beta$; and (3) use Rankine K_p for passive earth force P_p.
Compute

$$\beta = \tan^{-1} \tfrac{1}{2} = 26.6°$$

$$K_p = \tan^2\left(45 + \frac{\phi}{2}\right) \qquad P_p = 0.5\gamma H^2 K_p + 2cH\sqrt{K_p}$$

Scale selected dimensions from Fig. E16-5a, including the dimensions of base and height of the active pressure wedges. It will be necessary to use the trial wedge method to obtain P_a, since $\beta > \phi$ when F is applied to ϕ. Set up the following table of values:

F	ϕ_d	K_p	c_d	P_p, kN	b, m	h, m	$W_a = \tfrac{1}{2}\gamma bh$	C_a	P_a	C_b
					\multicolumn{2}{c	}{Scaled}				
1	30°	3.00	60	4628	57.5	17.5	8 553	3450	1400	3000
1.5	20	2.04	40	2876	65.6	20.0	11 152	2624	3950	2000
2.0	15	1.70	30	1835	70.8	21.5	12 939	2124	5900	1500
3.0	10	1.42	20	1684	78.2	22.8	15 155	1564	8450	1000

In Table: $C_a = c_d b$, $C_b = 50c_d$.

P_a is obtained graphically from Fig. E16-5b using W_a, C_a, and the slopes of P_a and R_a obtained from Fig. E16-5a.
The values of P_p, P_a, C_b, and W_b are used, together with the slope of R_b from Fig. E16-5a, to plot the force polygon of Fig. E16-5c to obtain values of ΔE for the corresponding F used.
A plot of ΔE versus F is made in Fig. E16-5d to obtain $F = 2.1$.

////

16-9 VALIDITY OF SLOPE STABILITY ANALYSES

There are a number of computer program variations of the Bishop simplified slope stability method. All these methods are based on "limiting equilibrium" (also called "limit states" or "critical states") using either forces or moments. Strictly, the method should satisfy both moment and force equilibrium, but this is not possible except by making some assumptions to reduce the number of unknowns so certain arbitrary assumptions of force distribution or the point of application of element forces is made. Since the forces on the interior of a mass on the verge of failure (and if progressive failure is a mechanism some zones are already failed and indeterminate) are highly speculative, it hardly seems necessary to greatly complicate the problem to gain 5 to 10 percent differences (not necessarily improvements) in computed safety factors. It appears that slope stability analyses are not overly sensitive to including the assumptions.

Some computer program owners claim that the critical surface can be statistically located with great precision, others claim that their program produces a better slip line by generating other than a circular arc for the surface of sliding. Once several slope failures have been inspected, it becomes readily apparent that at best a slope analysis is speculative, and thus the simplest analysis is generally preferable.

In passing, we might note three important factors:

1. Extensive comparative studies have shown little differences in F from the Bishop method and other "sophisticated" methods.
2. Extensive studies of slope failures show that

F	Event
$F < \sim 1.07$	Failures are common
$1.07 < F \leq 1.25$	Failures do occur
$F > 1.25$	Failures almost never occur

In these studies the Bishop method was generally used.
3. More improvement in slope stability analysis will come from use of better values for the soil strength parameters. We see from "2" that failures occur with high safety factors, and from "1" we find that making sophisticated analyses changes F only a small amount; thus the "failures" at high values of F must be from using incorrect soil parameters.

Certainly the correct soil parameters have been long recognized as critical to making a reliable slope analysis. Unfortunately, this is another area where the computational method is far ahead of the soil data. We might note that some improvement might be obtained by zoning the strength parameters as illustrated in Fig. 13-1; however, this is not simple since how much of the slip line goes with the $CK_o U$ (compression) test and how much with the DSS test and lastly, how much with the $CK_o E$ (extension) test are not known. We see in Fig. E16-1 that all three zones are applicable, but in many other cases only the $CK_o U$ with perhaps a small zone of DSS would apply.

16-10 SUMMARY

The mechanics of slope stability have been considered in some detail. We note that in any stability analysis two factors are of paramount importance:

1. Soil properties
2. Shape, and the instant center, of the potential failure mass

Where the slope is isotropic and homogeneous, solutions may be tabulated as in Fig. 16-6 but are relatively easy to obtain analytically.

Where the soil is stratified, contains a phreatic surface, is nonisotropic, or has a discontinuity, some type of finite element solution is necessary.

The method of slices is commonly used where no clearly defined discontinuities are present. The sliding wedge solution is used where a discontinuity forces the location of a part of the failure surface.

If the safety factor is $F > 1.25$, we may have considerable confidence that the slope is safe. If F is less than about 1.07, we may expect a slope failure.

PROBLEMS

16-1 What is F for $c = 0°$ if $\phi = 25°$ in Example 16-1a?
   Ans.: $F = 0.9$

16-2 What value of c in Prob. 16-1 will give $F = 1.25$?
   Ans.: $c = 40$ kPa

16-3 Make a plot of F versus r_u for Example 16-1b, using $F = 1.56$ for $r_u = 0$ and computing the other values using Fig. 16-6b and 16-6c.
   Partial ans.: $F = 1.08$ for $r_u = 0.5$

16-4 For Fig. P16-4, what is the maximum value of H for $F = 1.25$ for $r_u = 0$? (*Hint:* You must find H by trial.)

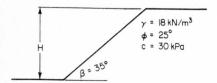

$\gamma = 18$ kN/m³
$\phi = 25°$
$c = 30$ kPa

$\beta = 35°$

Figure P16-4

16-5 For Fig. P16-4, what is the maximum value of α for $F = 1.0$, $H = 30$ m, and $r_u = 0.5$?
   Ans.: $\alpha \cong 16°$

16-6 What is F for the slope conditions in Fig. P16-6, using the circle center coordinates assigned from the table and all entrance points at B?

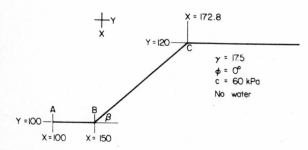

$X = 172.8$

$Y = 120$ — C

$\gamma = 17.5$
$\phi = 0°$
$c = 60$ kPa
No water

$Y = 100$ — A B β

$X = 100$ $X = 150$

Figure P16-6

No.	(a)	(b)	(c)	(d)	(e)	(f)	(g)	(h)	
X	150	150	150	150	155	155	155	155	
Y	140	145	150	155	140	145	150	155	
	—	—	1.465	1.397	—	—	1.331	1.304	Ans. for 16-6
	2.545	2.384	—	2.342	—	—	2.400	2.444	Ans. for 16-8

16-7 Compute minimum F using the assigned problem number of Prob. 16-6.

16-8 Compute F for Prob. 16-6 as assigned if the soil properties include $\phi = 25°$.

16-9 Compute F for the assigned problem of 16-6 with $\phi = 25°$ and including a tension crack to the right of point C such that the largest part of the crack intersects the failure circle. Do not include water in the crack unless specifically assigned.

16-10 Verify Example 16-5 using a suitably large scale but fit all of the drawing on a 28×45 cm (11×17 in) drawing sheet.

16-11 Do Example 16-5 using a suitably large drawing scale, assuming that the concentrated force is increased from 1000 to 5000 kN.

16-12 Do Example 16-3, assuming that the base of the sliding block is at an angle α with the horizontal of $10°$ (it is now $0°$). Also assume that there is no surcharge force and that the soil properties are: $\gamma = 18$ kN/m^3, $\phi = 32°$, and $c = 80$ kPa.

BIBLIOGRAPHY

The following abbreviations are used to simplify the reference list:

HRB	Highway Research Board [presently Transportation Research Board (TRB), Washington, D.C.]. Publications available from TRB if in print
HRR	Highway Research Record. Publications available from TRB if in print
ASCE	American Society of Civil Engineers
JGED	Journal of Geotechnical Engineering Division, ASCE (1974–)
JSMFD	Journal of Soil Mechanics and Foundations Division ASCE (1956–1973)
PSC	Proceedings ASCE Specialty Conferences
	1st PSC: Shear Strength of Cohesive Soils (1960)
	3d PSC: Placement and Improvement of Soil to Support Structures (1968)
	5th PSC: Performance of Earth and Earth Supported Structures (1972)
	6th PSC: In Situ Measurement of Soil Properties (1975)
ASTM	American Society for Testing and Materials, Philadelphia, Pa.
STP	Special Technical Publication published by ASTM.
CGJ	Canadian Geotechnical Journal, Ottawa, Canada.

ICSMFE   Proceedings of International Conference on Soil Mechanics and Foundation Engineering

ISSS   Proceedings of International Symposium on Soil Structure, Gothenberg, Sweden

Geotechnique is published by the Institution of Civil Engineers, London.

Abbot, M. B. (1960): One-Dimensional Consolidation of Multilayered Soils, *Geotechnique*, vol. 10, no. 4, pp. 151–165.

Aboshi, H., and H. Monden (1961): Three-Dimensional Consolidation of Clay, *5th ICSMFE*, vol. 1, pp. 550–562.

Arman, A., J. K. Poplin, and N. Ahmad (1975): Study of the Vane Shear, 6th PSC, vol. 1, pp. 93–120.

ASCE (1982): Grouting in Geotechnical Engineering, "*Proceedings of a Specialty Conference*," published by ASCE.

——— (1980): Expansive Soils, *4th International Conference*, proceedings published by ASCE, 2 vols., 917 pages.

——— (1978): *Symposium on Earth Reinforcement*, symposium published by ASCE.

——— (1968): Placement and Improvement of Soil to Support Structures, *3d PSC, ASCE*, 440 pages.

ASTM (1983): "Annual Book of Standards: Soil and Rock; Building Stones; Peats," vol. 4.08, ASTM, Philadelphia, Pa. (prior to 1983 Pt. 19).

——— (1973): "Evaluation of Relative Density and its Role in Geotechnical Projects Involving Cohesionless Soils," *STP* no. 523, 510 pages.

Azzouz, A. S., R. J. Krizek, and R. B. Corotis (1976): Regression Analysis of Soil Compressibility, *Soils and Foundations*, Tokyo, vol. 16, no. 2, pp. 19–29.

Banks, D. C., and B. N. MacIver (1969): "Variation in Angle of Internal Friction with Confining Pressure," Misc. Paper S-69-12, U.S. Army Waterways Experiment Station, Vicksburg, Miss., 30 pages.

Barden, L., and R. J. McDermott (1965): Use of Free Ends in Triaxial Testing of Clays, *JSMFD, ASCE*, vol. 91, SM 6, pp. 1–23.

Begemann, H. (1974): General Report: Central and Western Europe, *Proc., Conference on Penetration Testing*, Stockholm, Vol. 2.1, pp. 29–39.

Bell, A. L. (1915): The Lateral Pressure and Resistance of Clay and the Supporting Power of Clay Foundations, in "A Century of Soil Mechanics," Institution of Civil Engineers, London, pp. 93–134.

Berry, P. L., and W. B. Wilkinson (1969): The Radial Consolidation of Clay Soils, *Geotechnique*, vol. 19, no. 2, pp. 253–284.

Bertram, G. E. (1940): "An Experimental Investigation of Protective Filters," publication of the Graduate School of Engineering, Harvard University, no. 267.

Bishop, A. W. (1955): The Use of the Slip Circle in the Stability Analysis of Slopes, *Geotechnique*, vol. 5, no. 1, pp. 7–17.

——— and L. D. Wesley (1975): A Hydraulic Triaxial Apparatus for Controlled Stress Path Testing, *Geotechnique*, vol. 25, no. 4, December, pp. 657–670.

Bjerrum, L. (1972): Embankments on Soft Ground, *Proc. 5th PSC*, vol. 2, pp. 1–54.

——— and N. E. Simons (1960): Comparison of Shear Strength Characteristics of Normally Consolidated Clays, *Proc. 1st PSC*, pp. 711–726.

——— A. Casagrande, R. B. Peck, and A. W. Skempton (1960): "From Theory to Practice in Soil Mechanics: Selections from the writings of K. Terzaghi," John Wiley and Sons, Inc., New York, 425 pages.

Blessey, W. E. (1970): Pile Foundations Loads and Deformations—Mississippi River Deltaic Plain, *Proc. Conf. Design and Installation of Pile Foundations and Cellular Structures, Lehigh University*, pp. 1–26.

Bowles, J. E. (1982): "Foundation Analysis and Design," 3d ed., McGraw-Hill Book Company, New York, 816 pages.

—— (1978): "Engineering Properties of Soils and Their Measurement," 2d ed., McGraw-Hill Book Company, New York, 213 pages.

—— (1974): "Analytical and Computer Methods in Foundation Engineering," McGraw-Hill Book Company, New York, 519 pages.

Broms, B. B. (1980): Soil Sampling in Europe: State-of-the-Art, *JGED, ASCE*, vol. 106, GT 1, pp. 65–98.

Brooker, E. W., and H. O. Ireland (1965): Earth Pressures at Rest Related to Stress History, *CGJ*, vol. 2, no. 1, pp. 1–15.

Campanella, R. G., and Y. P. Vaid (1972): A Simple K_o Triaxial Cell, *CGJ*, vol. 9, no. 3, August, pp. 249–260.

Caquot, A., and J. Kerisel (1948): "Tables for the Calculation of Passive Pressure, Active Pressure, and Bearing Capacity of Foundations," trans. M. A. Bec, Gauthier-Villars, Paris.

Caroll, D. (1970): "Rock Weathering," Plenum Publishing Corporation, New York, 203 pages.

Casagrande, A. (1965): Role of the "Calculated Risk" in Earthwork and Foundation Engineering, *JSMFD, ASCE*, vol. 91, SM 4, pp. 1–40.

—— (1958): Notes on the Design of the Liquid Limit Device, *Geotechnique*, vol. 8, no. 2, pp. 84–91.

—— (1948): Classification and Identification of Soils, *Trans. ASCE*, vol. 113, pp. 901–930.

—— (1937): Seepage through Dams, in "Contributions to Soil Mechanics 1925–1940," Boston Society of Civil Engineers, pp. 295–336.

—— (1936): The Determination of the Preconsolidation Load and Its Practical Significance, *1st ICSMFE*, vol. 3, pp. 60–64.

—— (1932): Research on the Atterberg Limits of Soils, *Public Roads* vol. 13, no. 8, pp. 121–136.

—— and R. E. Fadum (1944): Application of Soil Mechanics in Designing Building Foundations, *Trans. ASCE*, vol. 109, pp. 383–490.

Cedergren, H. (1977): "Seepage, Drainage and Flow Nets," 2d ed., John Wiley and Sons, Inc., New York, 534 pages.

Chan, C. K., and J. P. Mulilis (1976): Pneumatic Sinusoidal Loading System, *JGED, ASCE*, vol. 102, GT 3, pp. 277–282.

Chassie, R. G., and R. D. Goughnour (1976): States Intensifying Efforts to Reduce Highway Landslides, *Civil Engineering*, April, pp. 65–66.

Christian, J. T., and W. D. Carrier (1978): Janbu, Bjerrum and Kjaernsli's Chart Reinterpreted, *CGJ*, vol. 15, no. 1, February, pp. 123–128.

Clevenger, W. A. (1958): Experiences with Loess as a Foundation Material, *Trans. ASCE*, vol. 123, pp. 151–194.

Cluff, L. S. (1971): Peru Earthquake of May 31, 1970; Engineering Geology Observations, *Seismol. Soc. Am. Bull.*, vol. 61, no. 3, June, pp. 511–521.

Collins, K., and A. McGown (1974): The Form and Function of Microfabric Features in a Variety of Natural Soils, *Geotechnique*, vol. 24, no. 2, pp. 223–254.

Cousins, B. F. (1980): Stability Charts for Simple Earth Slopes Allowing for Tension Cracks, *Proc. 3rd Australia–New Zealand Conference on Geomechanics*, pp. 1–5.

—— (1978): Stability Charts for Simple Earth Slopes, *JGED, ASCE*, vol. 104, GT 2, February, pp. 267–279.

Coyle, H. M., and R. R. Castello (1981): New Design Correlations for Piles in Sand, *JGED, ASCE*, vol. 107, GT 7, July, pp. 965–986.

Cullingford, G., A. K. Lashine, and G. B. Parr (1972): Servo-Controlled Equipment for Dynamic Testing of Soils, *Geotechnique*, vol. 22, no. 3, pp. 526–529.

Cummings, A. E. (1936): Distribution of Stresses Under a Foundation, *Trans. ASCE*, vol. 101, pp. 1072–1134.

Deere, D. U. (1968): Chap. 1: Geological Considerations, in "Rock Mechanics in Engineering Practice," K. G. Stagg and O. C. Zienkiewicz, eds., John Wiley and Sons, Inc., New York, pp. 1–20.

De Leeuw, E. H. (1965): The Theory of Three-Dimensional Consolidation Applied to Cylindrical Bodies, *6th ICSMFE*, vol. 1, pp. 287–290.

De Mello, V. F. B. (1971): The Standard Penetration Test, State-of-Art Report, *Proc. 4th Panamerican Conference on Soil Mech. and Found. Engrg.*, vol. 1, pp. 1–86.

Desai, C. S. (1979): "Elementary Finite Element Method," Prentice-Hall Book Co., Englewood Cliffs, N.J., 434 pages.

DOT (1976): "Flyash: A Construction Material," U.S. Department of Transportation, Washington, D.C., 198 pages.

Dunn, I. S., L. R. Anderson, and F. W. Kiefer (1980): "Fundamentals of Geotechnical Analysis," John Wiley and Sons, Inc., New York, 414 pages.

Eden, W. J. (1965): An Evaluation of the Field Vane Test in Sensitive Clay, *ASTM STP* 399, pp. 8–17.

Fadum, R. E. (1948): Influence Values for Estimating Stress in Elastic Foundations, *2d ICSMFE*, vol. 3, pp. 77–84.

Fardis, M. N., and D. Veneziano (1981): Estimation of SPT-N and Relative Density, *JGED, ASCE*, vol. 107, GT 10, October, pp. 1345–1359.

Flint, R. F. (1970): "Glacial and Quaternary Geology," John Wiley and Sons, Inc., New York, 892 pages.

Gibson, R. E., and P. Lumb (1952): Numerical Solution of Some Problems in Consolidation of Clay, *Proc. Inst. Civil Engineers*, London, pt. 1, no. 2, March, pp. 182–198.

Griffiths, D. H., and R. F. King (1965): "Applied Geophysics for Engineers and Geologists," Pergamon Press, Ltd., Oxford, 223 pages.

Grimm, R. E. (1968): "Clay Mineralogy," 2d ed., McGraw-Hill Book Co., New York, 596 pages.

Gromko, G. J. (1974): Review of Expansive Soils, *JGED, ASCE*, vol. 100, GT 6, pp. 667–687.

Hammitt, G. M., (1966): "Statistical Analysis of Data From a Comparative Laboratory Test Program Sponsored by ACIL," Misc. Paper No. 4-785, Waterways Experiment Station, Vicksburg, Miss., 62 pages.

Hansen, J. B. (1970): A Revised and Extended Formula for Bearing Capacity, *Danish Geotechnical Inst. Bull.* 28, Copenhagen, 21 pages.

Hardin, B. O., and V. P. Drnevich (1972): Shear Modulus and Damping in Soils: Design Equations and Curves, *JSMFD, ASCE*, vol. 98, SM 7, pp. 667–692.

—— and J. Music (1965): Apparatus for Vibration of Soil Specimens During a Triaxial Test, *ASTM STP* no. 392, pp. 55–73.

—— and F. E. Richart Jr. (1963): Elastic Wave Velocities in Granular Soils, *JSMFD, ASCE*, vol. 89, SM 1, pp. 33–65.

Harr, M. E. (1962): "Groundwater and Seepage," McGraw-Hill Book Company, New York, 315 pages.

Hazen, A. (1911): Discussion: Dams on Sand Foundations, *Trans. ASCE*, vol. 73, p. 99.

Henkel, D. H. (1960): The Shear Strength of Saturated Remolded Clays, *1st PSC, ASCE* pp. 533–553.

Herrero, O. (1983): Universal Compression Index Equation, (errata) *JGED, ASCE*, vol. 109, GT 10, p. 1349.

Hiriart, F., and R. J. Marsal (1969): The Subsidence of Mexico City, in "Nabor Carillo Volume: The Subsidence of Mexico City and Texcoco Project," Secretaria de Hacienda Credito Publico Fiduciacia: Nacional Financiera, S.A., Mexico, pp. 109–147.

Holtz, W. G. (1980): Public Awareness of Homes Built on Shrink-well Soils, *Proc. 4th Int. Conf. on Expansive Soils, ASCE*, vol. 1, pp. 617–638.

—— and H. J. Gibbs (1956): Engineering Properties of Expansive Clays, *Trans. ASCE*, vol. 121, pp. 641–677.

Hvorslev, M. J. (1949): "Subsurface Exploration and Sampling of Soils for Civil Engineering Purposes," Waterways Experiment Station, Vicksburg, Miss. (available from Engineering Foundation, New York), 521 pages.

—— (1937): "Physical Properties of Remolded Cohesive Soils," Ph.D. thesis translated and published (1969) as Translation No. 69-5 by Waterways Experiment Station, Vicksburg, Miss., 165 pages.

Ishihara, K., and S. Yasuda (1975): Sand Liquefaction in Hollow Cylinder Torsion under Irregular Excitation, *Soils and Foundations*, Tokyo, vol. 15, no. 1, pp. 45–59.

Jaky, J. (1948): Pressure in Silos, *2d ICSMFE*, vol. 1, pp. 103–107.

Janbu, N. (1965): Consolidation of Clay Layers Based on Nonlinear Stress-Strain, *6th ICSMFE*, vol. 2, pp. 83–87.

Johnson, A. W., and J. R. Sallberg (1960): Factors That Influence Field Compaction of Soils, *HRB Bull.* no. 272, 206 pages.

Jones, D. E., and W. G. Holtz (1973): Expansive Soils—The Hidden Disaster, *Civil Engineering*, ASCE, pp. 49–51.

Karlsson, R., and L. Viberg (1967): Ratio *c/p'* in Relation to Liquid Limit and Plasticity Index, with Special Reference to Swedish Clays, *Proc. Geotechnical Conf., Oslo, Norway*, vol. 1, pp. 43–47.

Kaufman, R. I., and W. C. Sherman, Jr. (1964): Engineering Measurements for Port Allen Lock, *JSMFD, ASCE*, vol. 90, SM 5, September, pp. 221–247.

Kiersch, G. A. (1964): Vaiont Reservoir Disaster, *Civil Engineering, ASCE*, vol. 34, no. 3, March, pp. 32–39.

Kovacs, W. D., H. B. Seed, and C. K. Chan (1971): Dynamic Moduli and Damping Ratios for Soft Clay, *JSMFD, ASCE*, vol. 97, SM 1, pp. 59–75.

Ladd, C. C., et al. (1977): Stress-Deformation and Strength Characteristics, *State-of-Art Report, 9th ICSMFE*, vol. 2, pp. 421–494.

Lade, P. J., and K. L. Lee (1976): "Engineering Properties of Soils," Soil Mech. Laboratory, UCLA-Eng 7652, May, 145 pages.

Lambe, T. W. (1958): The Engineering Behavior of Compacted Clay, *JSMFD, ASCE*, vol. 84, SM 2, May, pp. 1655–1–35.

———— and R. V. Whitman (1969): "Soil Mechanics," John Wiley and Sons, Inc., New York, 553 pages.

Lee, K. L. (1970): Comparison of Plane Strain and Triaxial Tests on Sand, *JSMFD, ASCE*, vol. 96, SM 3, May, pp. 901–923.

———— and J. A. Focht, Jr. (1975): Strength of Clay Subjected to Cyclic Loading, *Marine Geotechnology*, New York, vol. 1, no. 3.

———— and H. B. Seed (1967): Cyclic Stresses Causing Liquefaction of Sand, *JSMFD, ASCE*, vol. 93, SM 1, pp. 47–70.

———— and A. Singh (1971): Relative Density and Relative Compaction, *JSMFD, ASCE*, vol. 97, SM 7, pp. 1049–1052.

Leggett, R. F. (1962): "Geology and Engineering," 2d ed., McGraw-Hill Book Company, New York, 884 pages.

———— (1979): Geology and Geotechnical Engineering, *JGED, ASCE*, vol. 105, GT 3, March, pp. 342–391.

Leonards, G. A. (1962): Engineering Properties of Soils, in "Foundation Engineering," Chap. 2, McGraw-Hill Book Co., New York.

———— W. A. Cutter, and R. D. Holtz (1980): Dynamic Compaction of Granular Soils, *JGED, ASCE*, vol. 106, GT 1, January, pp. 35–44.

Lo, K. Y. (1969): The Pore Pressure-Strain Relationship of Normally Consolidated Undisturbed Clays, Parts I and II, *CGJ*, vol. 6, no. 4, pp. 383–412.

McDonald, A. B., and E. Sauer (1970): The Engineering Significance of Pleistocene Stratigraphy in Saskatoon Area, Saskatchewan, Canada, *CGJ*, vol. 7, no. 2, May, pp. 116–126.

McNown, J. S., E. Hsu, and C. Yih (1955): Applications of the Relaxation Technique in Fluid Mechanics, *Trans. ASCE*, vol. 120, pp. 650–686.

Mansur, C. I., and R. I. Kaufman (1962): Dewatering, in "*Foundation Engineering*," Chap. 3, McGraw-Hill Book Co., New York.

Marchetti, S. (1980): In Situ Tests by Flat Dilatometer, *JGED, ASCE*, vol. 106, GT 3, pp. 299–321.

Marcusson, W. F., and W. A. Bieganousky (1977): SPT and Relative Density in Coarse Sands, *JGED, ASCE*, vol. 103, GT 11, pp. 1295–1309.

Martins, J. B. (1965): Consolidation of a Clay Layer of Non-Uniform Coefficient of Permeability, *6th ICSMFE*, vol. 1, pp. 308–312.

Massarsch, K. R., and B. Broms (1976): Lateral Earth Pressure at Rest In Soft Clay, *JGED, ASCE*, vol. 102, GT 10, pp. 1041–1047.

Matsuo, S., and M. Kamon (1973): Microscopic Research on the Consolidated Samples of Clayey Soils, *Proc. ISSS*, pp. 194–203.

Meehan, R. L. (1967): The Uselessness of Elephants in Compacting Till, *CGJ* vol. 4, no. 3, pp. 358–360.

Menard, L., and Y. Broise (1975): Theoretical and Practical Aspects of Dynamic Consolidation, *Geotechnique*, vol. 25, no. 1, pp. 3–18.

Mesri, G., and P. M. Godlewski (1977): Time and Stress-Compressibility Relationship, *JGED*, *ASCE*, vol. 103, GT 5, pp. 417–430.

Meyerhof, G. G. (1974): General Report: Outside Europe, *Proc. Conf. on Penetration Testing*, Stockholm, vol. 2.1, pp. 40–48.

——— (1956): Penetration Tests and Bearing Capacity of Cohesionless Soils, *JSMFD*, *ASCE*, vol. 82, SM 1, pp. 1–19.

Middlebrooks, T. A. (1942): Fort Peck Slide, *Trans. ASCE*, vol. 107, pp. 723–742.

Miller, R. M. (1938): Soil Reactions in Relation to Foundations on Piles, Trans. ASCE, vol. 103, pp. 1193–1216.

Mills, John S. (1913): "The Panama Canal," Thomas Nelson and Sons, London, 344 pages.

Mirza, C. (1982): A Case for the Extension of the Unified Soil Classification System, *CGJ*, vol. 19, no. 3, August, pp. 388–391.

Mitchell, J. K. (1976): "Fundamentals of Soil Behavior," John Wiley and Sons, Inc., New York, 422 pages.

——— V. Vivatrat, and T. W. Lambe (1977): Foundation Performance of Tower of Pisa, *JGED*, *ASCE*, vol. 103, GT 3, pp. 227–249.

Morgenstern, N. R. (1967): Shear Strength of Stiff Clay, *Proc. Geotechnical Conf.*, Oslo, vol. 2, pp. 59–69.

NAFAC (1971): "Design Manual: Soil Mechanics, Foundations and Earth Structures," NAFAC DM-7, Department of the Navy, Washington, D.C.

Newmark, N. M. (1942): Influence Charts for Computation of Stresses in Elastic Foundations, University of Illinois Engineering Experiment Station Bull. no. 338 (reprinted as vol. 61, no. 92, June, 1964; also in "Selected Papers by Nathan M. Newmark," *ASCE*, 1976).

——— (1935): Simplified Computation of Vertical Pressures in Elastic Foundations, University of Illinois Engineering Experiment Station Circular no. 24, pp. 5–19 (also in ASCE 1976 publication above, pp. 43–57).

Olsen, R. E., and C. C. Ladd (1979): One-Dimensional Consolidation Problems, *JGED*, *ASCE*, vol. 105, GT 1, pp. 11–30.

Park, T. K., and M. L. Silver (1975): Dynamic Triaxial and Simple Shear Behavior of Sand, *JGED*, *ASCE*, vol. 101, GT 6, pp. 513–529.

Parry, R. H. G. (1977): Estimating Bearing Capacity of Sand From SPT Values, *JGED*, *ASCE*, vol. 103, GT 9, pp. 1014–1019.

PCA (1959): "Soil-Cement Laboratory Handbook," Portland Cement Association, Skokie, Illinois, 62 pages.

Peck, R. B., and F. G. Bryant (1952–1953): The Bearing Capacity Failure of the Transcona Elevator, *Geotechnique*, vol. 3, pp. 201–208.

——— W. E. Hanson, and T. H. Thornburn (1974): "Foundation Engineering," 2d ed., John Wiley and Sons, Inc., New York, 514 pages.

Poskitt, T. J. (1969): The Consolidation of Clay with Variable Permeability and Compressibility, *Geotechnique*, vol. 19, no. 2, pp. 201–208.

Poulos, H. G., and E. H. Davis (1974): "Elastic Solutions for Soil and Rock Mechanics," John Wiley and Sons, Inc., New York, 411 pages.

Proctor, R. R. (1933): Fundamental Principles of Soil Compaction, *Engineering News-Record*, Aug. 31, Sept. 7, Sept. 21, and Sept. 28.

Richart, F. E., Jr., J. R. Hall, and R. D. Woods (1970): "Vibrations of Soils and Foundations," Prentice-Hall, Inc., Englewood Cliffs, N.J., 414 pages.

Rosenfarb, J. L., and W. F. Chen (1972): Limit Analysis Solution of Earth Pressure Problems, *Fritz Engineering Laboratory Report* 355.14, Lehigh University, 53 pages.

Rowe, P. W. (1964): The Calculation of the Consolidation Rates of Laminated, Varved or Layered Clays with Particular Reference to Sand Drains, *Geotechnique*, vol. 14, no. 4, pp. 321–339.

Rumer, R. R. (1964): Discussion: Laminar and Turbulent Flow through Sand, *JSMFD, ASCE* vol. 90, SM 2, pp. 205–207.

Rutledge, P. C. (1944): Relation of Undisturbed Sampling to Laboratory Testing, *Trans. ASCE*, vol. 109, pp. 1155–1216.

Saada, A. S., and F. C. Townsend (1981): State of the Art: Laboratory Strength Testing of Soils, *ASTM STP 740*, pp. 7–77.

Schmertmann, J. H. (1975): Measurement of In-Situ Shear Strength, State-of-Art Paper, *6th PSC*, vol. 2, pp. 57–138.

———— (1970): Static Cone to Compute Static Settlement Over Sand, *JSMFD, ASCE*, vol. 96, SM 3, pp. 1011–1043.

———— (1955): The Undisturbed Consolidation Behavior of Clay, *Trans. ASCE*, vol. 120, pp. 1201–1233.

Schnitter, N. J. (1982): Discussion: There Were Giants in Those Days, *JGED, ASCE*, vol. 108, GT 6, pp. 902–904.

Seed, H. B., and C. K. Chan (1959): Structure and Strength Characteristics of Compacted Clays, *JSMFD, ASCE*, vol. 85, SM 5, pp. 87–128.

———— (1958): Undrained Strength of Compacted Clays After Soaking, *JSMFD, ASCE*, vol. 85, SM 6, pp. 31–47.

———— K. Mori, and C. K. Chan (1977): Influence of Seismic History on Liquefaction of Sands, *JGED, ASCE*, vol. 103, GT 4, pp. 257–270.

———— and I. M. Idriss (1971): Simplified Procedure for Evaluating Soil Liquefaction Potential, *JSMFD, ASCE*, vol. 97, SM 9, pp. 1249–1273.

Shepard, E. R., and R. M. Haines (1944): Seismic Subsurface Exploration on the St. Lawrence River Project, *Trans. ASCE*, vol. 109, pp. 194–222.

Sherard, J. L., L. S. Cluff, and C. R. Allen (1974): Potentially Active Faults in Dam Foundations, *Geotechnique*, vol. 24, no. 3, pp. 367–428.

Sherif, M. A., and I. Ishibashi (1981): Overconsolidation Effects on K_o Values, *10th ICSMFE*, Stockholm, vol. 2, pp. 785–788.

Sherman, G. G., R. O. Watkins, and R. H. Prysock (1967): A Statistical Analysis of Embankment Compaction, HRR no. 177, pp. 157–185.

Simons, N. E. (1960): Comprehensive Investigation of the Shear Strength of an Undisturbed Drammen Clay, *1st PSC, ASCE*, pp. 727–745.

Skempton, A. W. (1964): Long Term Stability of Clay Slopes, *Geotechnique*, vol. 14, no. 2, pp. 77–101.

———— (1961): Effective Stress in Soils, Concrete and Rocks, *Proc. Conf. Pore Pressure and Suction in Soils*, Butterworths, London, pp. 4–16.

———— (1954): The Pore Pressure Parameters *A* and *B*, *Geotechnique*, vol. 4, no. 4, pp. 143–147.

———— (1951): "The Bearing Capacity of Clays," The Building Research Congress, London.

———— and A. W. Bishop (1954): Soils, in "Building Materials; Their Elasticity and Inelasticity," M. Reiner, ed., Chap. 10, North Holland Publishing Co., Amsterdam, pp. 417–482.

Sokolovski, V. V. (1965): "Statics of Granular Media," Pergamon Press, London, 270 pages.

Sowers, G. F. (1981): There Were Giants in Those Days, *JGED, ASCE*, vol. 107, GT 4, pp. 385–419.

Stokoe, K. H., and R. D. Woods (1972): In Situ Shear Wave Velocity by Cross-Hole Method, *JSMFD, ASCE*, vol. 98, SM 5, pp. 443–460.

Taylor, D. W. (1948): "Fundamentals of Soil Mechanics," John Wiley and Sons, Inc., New York, pp. 229–239.

———— (1937): Stability of Earth Slopes, *Contributions to Soil Mechanics 1925–1940*, Boston Society of Civil Engineers, pp. 337–386.

Terzaghi, K. (1943): "Theoretical Soil Mechanics," John Wiley and Sons, Inc., New York, 510 pages.

———— (1925): Principles of Soil Mechanics II—Compressive Strength of Clay, *Engineering News-Record*, vol. 95, no. 20, p. 799.

———— and R. B. Peck (1967): "Soil Mechanics in Engineering Practice," 2d ed., John Wiley and Sons, Inc., New York, p. 72.

Tovey, N. K., and W. K. Yan (1973): The Preparation of Soils and Other Geological Materials for the SEM, *ISSS*, pp. 59–69.

Tschebotarioff, G. P. (1936): Comparison between Consolidation, Elastic and Other Properties …, *1st ICSMFE*, vol. 1, pp. 33–36.

Turnbull, W. J. (1950): Compaction and Strength Tests on Compacted Soil, paper presented at annual ASCE meeting.

—— J. R. Compton, and R. G. Ahlvin (1966): Quality Control of Compacted Earth Work, *JSMFD, ASCE*, vol. 92, SM 1, pp. 93–103.

USBR (1968), "Earth Manual," U.S. Bureau of Reclamation, Denver, Colo., 783 pages.

Valera, J. E., and N. C. Donovan (1977): Soil Liquefaction Procedures—A Review, *JGED, ASCE*, vol. 103, GT 6, pp. 607–625.

Weber, W. G., Jr. (1969): Performance of Embankments Constructed over Peat, *JSMFD, ASCE*, vol. 95, SM 1, pp. 53–76.

Weissmann, G. F., and R. R. Hart (1961): The Damping Capacity of Some Granular Soils, *ASTM STP* no. 305, pp. 45–54.

Westergaard, H. M. (1938): A Problem of Elasticity Suggested by a Problem in Soil Mechanics, in *Contributions to the Mechanics of Solids*, Timoshenko 60th Anniversary volume, The Macmillan Co., New York.

White, L. S. (1952–1953): Transcona Elevator Failure: Eye Witness Account, *Geotechnique*, vol. 3, pp. 209–214 (see also Peck and Bryant, 1952–1953, pp. 201–208).

Wineland, J. D. (1975): Borehole Shear Device, *6th PSC, ASCE*, vol. 1, pp. 511–522.

Winterkorn, H. F., and H. Y. Fang, eds. (1975): "Foundation Engineering Handbook," Van Nostrand Reinhold Co., New York, 751 pages.

Wroth, C. P. (1975): In Situ Measurement of Initial Stresses and Deformation Characteristics, *6th PSC, ASCE*, vol. 2, pp. 181–230.

—— and D. M. Wood (1978): The Correlation of Index Properties With Some Basic Engineering Properties of Soils, *CGJ*, vol. 15, no. 2, pp. 137–145.

Wu, T. H., N. Y. Chang, and E. M. Ali (1978): Consolidation and Strength Properties of Clay, *JGED, ASCE*, vol. 104, GT 7, pp. 889–905.

Yong, R. N., and D. E. Sheeran (1973): Fabric Unit Interaction and Soil Behavior, *Proc. ISSS*, pp. 176–183.

Zienkiewicz, O. C. (1977): "The Finite Element Method," 3d ed., McGraw-Hill Book Co., New York, 787 pages.

NAME INDEX

Abbot, M. B., 411
Aboshi, H., 411
Ahlvin, R. G., 230
Ahmad, N., 460
Ali, E. M., 383
Allen, C. R., 173
Anderson, L. R., 304
Arman, A., 460
ASCE (American Society of Civil Engineers), 206, 207, 216
ASTM (American Society for Testing and Materials), 152, 185, 208, 423
Azzouz, A. S., 370, 371

Banks, D. C., 431
Barden, L., 426
Begemann, H., 523
Bell, A. L., 510
Berry, P. L., 411
Bertram, G. E., 303
Bieganousky, W. A., 187
Bishop, A. W., 429, 468, 552
Bjerrum, L., 8, 449, 456, 460, 461, 465
Blessey, W. E., 12
Bowles, J. E., 32, 121, 128, 151, 212, 227, 304, 314, 326, 359, 498, 507, 513, 524, 530, 553
Broise, Y., 233
Broms, B. B., 171, 201
Brooker, E. W., 501
Bryant, F. G., 9

Campanella, R. G., 429
Caquot, A., 507

Caroll, D., 94
Carrier, W. D., 531
Casagrande, A., 8, 11, 40, 121, 128, 134, 291, 292, 304, 367, 376, 539
Castello, R. R., 526
Cedergren, H., 254, 304
Chan, C. K., 213, 483, 486, 488
Chang, N. Y., 383
Chassie, R. G., 11
Chen, W. F., 507
Christian, J. T., 531
Clevenger, W. A., 376
Cluff, L. S., 11, 173
Collins, K., 155
Compton, J. R., 230
Corotis, R. B., 370, 371
Cousins, B. F., 544, 545, 547, 548
Coyle, H. M., 526
Cullingford, G., 488
Cummings, A. E., 341, 342
Cutter, W. A., 233

Davis, E. H., 344
Deere, D. U., 192, 193
De Leeuw, E. H., 411
De Mello, V. F. B., 185
Desai, C. S., 304
Donovan, N. C., 492
DOT (U.S. Department of Transportation), 236
Drnevich, V. P., 485
Dunn, I. S., 304

Eden, W. J., 460

Fadum, R. E., 337, 345, 376
Fang, H. Y., 498
Fardis, M. N., 187
Flint, R. F., 100
Focht, J. A., Jr., 493, 494
Foott, R., 461

Gibbs, H. J., 266
Gibson, R. E., 411
Godlewski, P. M., 375
Goughnour, R. D., 11
Griffiths, D. H., 199
Grimm, R. E., 158, 159
Gromko, G. J., 12

Haines, R. M., 196
Hall, J. R., 484
Hammitt, G. M., 121
Hansen, J. B., 519
Hansen, W. E., 498
Hardin, B. O., 483–485
Harr, M. E., 304
Hart, R. R., 483
Hazen, A., 251
Henkel, D. J., 437
Herrero, O., 371
Hiriart, F., 9
Holtz, R. D., 233
Holtz, W. G., 265, 266
Hsu, E., 304
Hvorslev, M. J., 171, 182, 184, 464

Idriss, I. M., 491
Ireland, H. O., 501
Ishibashi, I., 501
Ishihara, K., 461, 483

Jaky, J., 501
Janbu, N., 397
Johnson, A. W., 210
Jones, D. E., 265

Kamon, M., 163

Karlsson, R., 465
Kaufman, R. I., 304, 306, 377
Kerisel, J., 507
Kiefer, F. W., 304
Kiersch, G. A., 11
King, R. F., 199
Kovacs, W. D., 483
Krizek, R. J., 161, 370, 371

Ladd, C. C., 383, 461
Lade, P. J., 401, 441, 465
Lambe, T. W., 8, 213, 214, 317
Lashine, A. K., 488
Lee, K. L., 224, 400, 401, 441, 465, 492–494
Leggett, R. F., 9, 112, 172
Leonards, G. A., 233, 359
Lo, K. Y., 466–468
Lumb, P., 411

McDermott, R. J., 426
McDonald, A. B., 377
McGown, A., 155
MacIver, B. N., 431
McNown, J. S., 304
Mansur, C. I., 304, 306
Marchetti, S., 201
Marcusson, W. F., 187
Marsal, R. J., 9
Martins, J. B., 411
Massarsch, K. R., 201
Matsuo, S., 163
Meehan, R. L., 205
Menard, L., 233
Mesri, G., 375
Meyerhof, G. G., 522, 523
Middlebrooks, T. A., 11
Miller, R. M., 12
Mills, J. S., 10
Mirza, C., 119
Mitchell, J. K., 8, 158
Monden, H., 411
Morgenstern, N. R., 454
Mori, K., 486
Mulilis, J. P., 488
Music, J., 483, 484

NAFAC (U.S. Department of the Navy), 254
Newmark, N. M., 337, 345

Olsen, R. E., 383

Park, T. K., 483
Parr, G. B., 488
Parry, R. H. G., 523
PCA (Portland Cement Association), 236
Peck, R. B., 9, 364, 372, 498
Poplin, J. K., 460
Poskitt, T. J., 411
Poulos, H. G., 344, 461
Proctor, R. R., 208
Prysock, R. H., 226, 231

Richart, F. E., Jr., 484, 485
Rosenfarb, J. L., 507
Rowe, P. W., 411
Rumer, R. R., 244
Rutledge, P. C., 376

Saada, A. S., 453
Sallberg, J. R., 201
Sauer, E., 377
Schlosser, F., 461
Schmertmann, J. H., 185, 364, 376, 529
Schnitter, N. J., 8
Seed, H. B., 213, 483, 486, 491, 492
Sheeran, D. E., 155
Shepard, E. R., 196
Sherard, J. L., 173
Sherif, M. A., 501
Sherman, G. B., 226, 231
Sherman, W. C., Jr., 377
Silver, M. L., 483
Simons, N. E., 436, 456, 461, 465
Singh, A., 224
Skempton, A. W., 8, 456, 457, 466–468, 520
Sokolovski, V. V., 507

Sowers, G. F., 8
Stokoe, K. H., 197

Taylor, D. W., 304, 377, 390, 395, 544
Terzaghi, K., 121, 364, 372, 518
Thornburn, T. H., 498
Tovey, N. K., 161
Townsend, F. C., 453
Tschebotarioff, G. P., 264
Turnbull, W. J., 217, 230

USBR (U.S. Bureau of Reclamation), 254

Vaid, Y. P., 429
Valera, J. E., 492
Veneziano, D., 187
Viberg, L., 465
Vivatrat, V., 8

Wacker Corp., 222
Watkins, R. O., 226, 231
Weber, W. G., Jr., 374
Weissmann, G. F., 483
Wesley, L. D., 429
Westergaard, H. M., 348
White, L. S., 9
Whitman, R. V., 217
Wilkinson, W. B., 411
Wineland, J. D., 459
Winterkorn, H. F., 498
Wood, D. M., 371
Woods, R. D., 197, 484
Wroth, C. P., 371, 500
Wu, T. H., 383

Yan, W. K., 161
Yasuda, S., 483
Yih, C., 304
Yong, R. N., 155

Zienkiewicz, O. C., 304

SUBJECT INDEX

AASHTO (American Association of
State Highway and Transportation
Officials), soil classifications, 119,
120, 135
  group index for, 138
Active earth pressure, 501
Activity, 164
Alluvial deposits, 96
Angle of internal friction, 147, 441, 456
  residual value, 457
  triaxial vs. plane, 441
    table of, 441
Anisotropic consolidation, 429, 445, 468
  defined for soil, 325
  pore pressure in, 356, 390
  strength, 461
Anticline, 81
Aquifer, 111
Area ratio, 182
Atterberg limits, 40, 121, 266
  fall cone test for, 123
  in strength correlations, 457
Auger boring, 175

Bearing capacity:
  equations for, 518, 522
  factors for, 519
Borehole shear test, 459
Boussinesq equations, 334
  in consolidation settlement, 381
  in settlement analysis, 531
  stress factors for, 339
  using vertical pressures, 339, 341
Brucite, 159
Bulking of sand, 150
Buoyant unit weight, 52

Capillarity:
  producing shrinkage, 263
  suction from, 452
Capillary rise, 259
Capillary tubes, 261
  approximation in soil, 263
Cation exchange, 165
Classification of soil, 119
  AASHTO system, 135
  FAA system, 120
  Unified Soil Classification System,
    128
Clay:
  activity of, 164
  expansive, 216, 265
  normally consolidated, 447
  overconsolidated, 449
    fissured, 453
  pH of, 166
  structure of, 158
Clay minerals, 158
  properties of, 164
  size of, 120, 159
Cohesion, strength parameter of, 146,
    437, 448
  undrained, 444
Cohensionless soil, 39, 149, 440
  relative density used for, 151
Cohesive soils, 39, 446
  Atterberg limits for, 41
  compacted, 213
  microstructure of, 156
Compaction, 207
  benefits of, 207
  deep, 233
  equipment for, 219, 221
  of frozen soil, 239

Compaction (*Cont.*):
  relative, 223
  standard tests for, 208
  of trench backfill, 238
Compression, secondary, 374
  coefficient of, 369, 375
Compression index, 358, 368, 371
  empirical values for, 371
Concavity, coefficient of, 126
Cone penetration test (CPT), 191
Consolidated-drained test parameters,
    455
Consolidated-undrained test parameters,
    445, 451
  vane test for, 460
Consolidation, 356, 387
  anisotropic (*see* Anisotropic
      consolidation)
  coefficient of, 358, 387, 389
  dynamic, 233
  isotropic, 445
    pore pressure in, 392
  percent of, 391, 400
  as a strain rate process, 397
  time factor for, 390, 393, 396, 400
Creep (*see* Compression, secondary)
Critical states, 416
Critical void ratio, 419

Dams, earth, 287
  flow nets for, 292
  piping control in, 11, 303
Darcy's law, 243
Deformation (*see* Settlements)
Dilatancy, 140
Direct shear test, 426, 486
Dynamic stress-strain modulus, 486
  Rayleigh waves used for, 484

Earth dams (*see* Dams, earth)
Earth pressure, 52, 449
  lateral, 499, 506
    active, 501
    passive, 505
  preconsolidation, 366
  vertical, 365

Earthquake stresses, 173, 491
Effective pressure, 56, 360, 366, 499
Elasticity, modulus of, 322, 326, 482
Equipotential line, 279, 293
Eskers, 101
Exploration, soil, 170
  cone penetration test (CPT), 191
  dilatometer used for, 201
    total stress cell, 201
  drilling in, 175
  sampling, 179
    electrical resistivity, 199
    seismic, 194
  standard penetration test (SPT),
      185

Fabric, soil, 146
Failure, criteria for soil, 416
  brittle, 421
  dynamic, 482
  progressive, 461
  surface (shape), 559
Faults, 83
Federal Aviation Administration (FAA)
    soil classification system, 120
Field density tests, 227
Filters, graded, 303
Fissures, 453
Flow in wells, 305
Flow net, 280, 282, 293, 297
Flow path, 280

Glacial deposits, 100
  depth of, 106
  drift, 104
  moraines, 101
Grain size, 48
  in cyclic strength, 493
  sieve sizes used for, 125
  for soil classification, 120, 124
Groundwater, 109
  aquifers, 111
  artesian, 113
  springs from, 114
Group index, AASHTO, 138
Grouting, 206

Halloysite, 160
Hooke's stress-strain law, 320
Hydration, 164
Hydraulic gradient, 111, 247, 253
    critical, 271
Hydrometer analysis, 128

Igneous rock, 70
Illite, 160
Index:
    compression (*see* Compression index)
    group, AASHTO, 138
    liquidity, 42
    plasticity (*see* Plasticity index)
    recompression, 368
Intergranular pressure (*see* Effective
        pressure)
Invariants, stress, 317

K_0 stresses, 429, 499
    values for, 501
Kaolinite, 159
Karst areas, 116

Laminar flow, 244
Laplace equation, 279
Lateral pressure:
    active, 505
    for cohesive soil, 510
    passive, 504
    ϕ-circle for, 514
    Rankine equations for, 506
    trial wedge solution, 513
Limestone, 75
Liquefaction, 270, 486
    of clay, 486
Liquidity index, 42
Loess, 107

Median value, 18
Metamorphic rock, 68, 79
Method of slices, slope analysis by,
        551
Minerals, properties of, 66

Mohr's circle, 328, 332, 419, 500
    for strength parameters, 430
Montmorillonite (bentonite), 162
Moraines, 101

Newmark chart, 345
Normalizing parameters, 463

Octahedral stress, 314
Optimum moisture content, 210
Overconsolidation ratio (OCR), 366, 450
    for shear modulus, 485

Packing, soil particle, 147, 207
Peat, classification of, 130, 138
Peds, 155
Penetrometer, pocket, 186, 238, 458
Permeability, 242
    range of values, 252
    in stratified soils, 254
    tests for, 249
Phreatic surface, 110, 286, 289
    for $\beta > 30^0$, 290
Piezometer, 58, 246
Piles, 525
    dynamic pile capacity, 528
Piping, 14, 302
Plasticity index, 40
    for clay activity, 164
    for soil classification, 120, 130, 137
    for strength correlations, 460, 465
    in strength parameters, 457
    for volume change, 216, 266
Poisson's ratio, 321, 323, 342, 479
    in elastic settlements, 334, 530
Pore pressure, 356, 426, 448
    in liquefaction, 270
    negative, 452
Pore pressure parameters, 466
Porosity, 30, 110
    of rock, 112
Pressure:
    earth (*see* Earth pressure)
    effective, 56, 360, 366, 499

Pressure (*Cont.*):
  effective overburden, 447
    parameters for, 467
  lateral (*see* Lateral pressure)
  pore (*see* Pore pressure)
  preconsolidation, 364
    due to drying (shrinkage), 264
Pressure profiles, 341, 343
Principal strains, 313
Principal stresses, defined, 314

Quick condition (*see* Liquefaction)

Random numbers, table of, 226
Rankine equations, 504, 508
Ratio, strength orientation, 449, 461
Rayleigh waves, 484
Recompression index, 368
Relative compaction, 223
Relative density, 151, 152, 224
  correlations using, 187
  field tests for, 152
  linear plot of, 154
  in liquefaction studies, 492
Residual soil, 93
Residual strength, 311, 448
  parameters for, 419
Resilient modulus, 482
Reynolds number, 244
Rheology, 311
  models for, 351
Rock drilling, 192
Rock quality designation (RQD), 192

Scanning electron microscope (SEM),
    155, 161, 163
Sedimentary rock, 74
  distribution of, 68
Seepage and flow net theory, 277
Seepage forces, 269, 272
  sudden drawdown and, 301
Seepage quantity, 281, 289, 296, 304
  control of, 299
  when $k_x \neq k_y$, 254, 298
Seepage velocity, 248

Sensitivity, 455
Settlements:
  consolidation, 350, 358
  elastic, 349, 529
    Poisson's ratio in, 334, 530
    shape factors for, 530
  secondary, 350
Shale, 74
Shear modulus, 323, 484
Shear strength, 311, 414, 440
  affected by OCR, 419, 451
    anisotropic effects, 461
  correlations for, 465
  empirical values of, 187
  factors affecting, 462
  ratio s_u / p_o, 464, 468
  residual, 311
  soil tests for, 423
  unconfined compression, 424
  undrained (*see* Undrained shear
    strength)
  vane test for, 460
  zones, 415, 462
Shrinkage, 207
  cracks caused by, 265
Shrinkage limit, 40, 266
SI (International System of Units), 23
Sieve analysis, 123
  tables of sizes, 125
Slaking, 264
Slopes, stability of, 536
  charts for, 546
  circular arc analysis, 540, 551
  factor of safety for, 541, 543, 553,
    559
  infinite, 537
  method of slices, analysis by, 551
  wedge block analysis, 555
Soil, defined, 64, 147
Soil classification (*see* Classification of
    soil)
Soil formations, 64, 92, 145
  glacial, 100
  residual, 93
  talus, 109
Soil horizons, 93
Specific gravity, 31, 45

Specific surface, 43
Stabilization of soil, 204
  chemical, 206, 236
  by compaction, 207
  with geotextiles, 235
  static, 234
Standard deviation, 19, 121, 231
  relative density determination,
    152
Standard penetration test (SPT), 185,
    458, 482, 522
  table of empirical values, 187
Statistics, 16, 180
  for compaction control, 231
Strength, shear (see Shear strength)
Stress path, 319, 435
  plotting of, 436
Stress-strain modulus, 320, 476
  dynamic values of, 483
  empirical values for, 482
  factors affecting, 462
  secant modulus, 321, 479
  shear, 484
Stresses, 311
  earthquake, 173, 491
  effective, 54, 267, 360, 506,
    518
  invariants, 317
  isotropic, 319
  Mohr's circle for, 328
  octahedral, 314
  principal, 314
Structure, soil, 147, 155
  compacted, 213
Student's t numbers, 21
Surface tension, 44, 150, 167
Syncline, 81

Tension crack, depth of, 511
  on slopes, 547
Time factor, 390, 393
Time scale, geological, 69
Topsoil, 93
Torvane shear test, 457
Triaxial shear test, 429,
  449

Undrained shear strength, 423, 426, 445,
  458
  pore pressure effect on, 444
Unified Soil Classification System, 119,
    120, 128
  table for, 131–133
Uniform soil, 129
Uniformity coefficient, 126
Unit weight, 29, 50
  due to compaction, 208
  due to packing, 148
  submerged (effective), 52, 55
  table of typical values, 187
Units, SI, 23

Van der Waal's force, 155
Vane shear test, 459
  correction for, 460
Variance, coefficient of, 18, 231
Varved clay, 99
Velocity of flow, 244, 248,
  302
Viscosity:
  effect on consolidation, 358
  effect on permeability, 251
  tables for water, 251
Void ratio, 29, 37, 364
  critical, 419
  maximum, 148
  in quick conditions, 271
  for relative density, 151
  table of values, 147

Water content, 30, 208
  natural, 41
  optimum, 210, 224
Water table, 110, 184
Weathering of rocks, 85
  chemical, 88
Wells, hydraulics of, 305
Westergaard stresses, 348

Zero air voids (ZAV), curve of,
  210